Information Processing

VOLUME III

Introduction to Information Geometry

David J. Blower

© David J. Blower, Third Millennium Inferencing,
Pensacola, Florida, April 2016.

DEDICATED TO MY WIFE

ELIZABETH

Preface

My overall goal here is to lay down some minimal set of *coherent* and *consistent* requirements for Advanced Information Processors (AIPs). Can we support the claim that these AIPs reason in an optimal manner by generalizing logic?

The first two volumes belabored the important conceptual distinctions involved when an Information Processor uses probability theory to make an inference. In Volume I, the formal manipulation rules for probabilities were treated as the first important concept. The concern at that point was not with how numerical values got attached to abstract probabilities.

I thought it instructive to draw a close analogy between the formal manipulation rules as required by probability theory, and the traditional axiomatic treatment of Boolean Algebra. With this as background, I emphasized that probability theory generalized Classical Logic.

It was not too surprising that these formal rules did not care much about any assigned numerical values, other than requiring that they be legitimate probabilities. Given their origin from within Boolean Algebra, which does not even possess the concept of a number, the formal rules disdained any deep discussion of any rationale for numerical assignments.

Bayes's Theorem, viewed as an example of a formal manipulation rule, always worked no matter what particular numerical values might have been assigned to the probabilities for the joint statements through some particular model.

We waited until Volume II to devote considerable attention to the latter half of the conceptual divide. There, some pleasant features of the Maximum Entropy Principle (MEP) were highlighted with regard to numerical assignments.

The MEP was seen to provide the Information Processor with an algorithm for making rational and defensible numerical assignments to probabilities. Words like "rational" and "defensible" are dredged up mainly to convey some sense of a connection between notions of "information" and a "degree of belief."

By invoking the MEP, any desired "active information" could be *included* within a probability distribution. At the same time, all "missing information" could be *excluded* from that same probability distribution. And, all of this transpires as a direct consequence of some tentatively entertained model.

In this third book, I continue to dwell on the issue of how to assign legitimate numerical values to abstract probabilities. Volume II's task was to introduce an MEP algorithm, and show how it could be used for its stated purpose. There was little attempt at any deeper meaning, or at trying to fit the MEP into a broader mathematical framework. The explanations tended to rely upon some intuitive sense of words such as "information" and "entropy," together with operations such as minimizing the "distance between distributions."

But the MEP is more than just a convenient heuristic. When approached from the viewpoint of Linear Algebra, Differential Geometry, and Tensor Calculus, it acquires a deeper meaning. Over the years, various authors have leveraged relevant elements from these classical mathematical disciplines to breathe life into a new creature called *Information Geometry*.

Unfortunately, but inevitably, these originators chose to cloak their work in tightly bound arguments. To the uninitiated, the mathematical framework as the supporting rationale for Information Geometry is obscure, abstruse, and, on the whole, generally impenetrable.

Therefore, I shall attempt to unwind some of these arguments as part of the effort at an introductory tutorial to Information Geometry. The hallmark of Volume III, as I hope it was for the previous two Volumes, is a more readily accessible treatment than currently available. The tradeoff for the reader is to persevere through my lengthier discursive style, albeit interleaved with many numerical examples and solved problems. I do not know how else an introductory tutorial can achieve its objective.

This tutorial takes advantage of one feature not yet seen in the Information Geometry literature. To wit, the relatively recent appearance of *Mathematica* is a convenient tool for providing symbolic, numerical, and pictorial assistance for some of the more onerous technical mathematical aspects of Information Geometry. Who knows where we would be if the originators of Information Geometry could have tapped into this resource?

Mathematica saves us from an enormous amount of ancillary effort that most likely would have derailed the project from the start. And, as a bonus, it serves the valuable role of ultimately defining what any of the mathematical symbolism must mean operationally when incorporated into a computer program. As the creator of *Mathematica*, Stephen Wolfram has exhorted its users to explore the *computational universe*. I do try to follow his advice here with a computational approach to an initial exploration of Information Geometry.

David John Blower
Pensacola, Florida, USA
April 2016

An Author's Apologia

After our adventures in Volume II, we have a working knowledge of the Maximum Entropy Principle (MEP) and how it goes about assigning numerical values to probabilities through the auspices of some model. The effort we spent there will make the task of learning about Information Geometry somewhat easier.

Not having to begin in a disconnected vacuum is a definite advantage. Instead, we can start off in a more benign environment given our understanding of the formal aspects of probability theory augmented by the MEP algorithm. Thus, the rather difficult and purely abstract mathematical concepts inherent in Information Geometry can be fitted into this framework. I trust that this approach represents a more accessible introduction.

We start off with a brief historical survey to provide some context for how Information Geometry fits into the grand scheme of things. Next, I introduce some of the special language that Information Geometry employs.

As is true for any novice speaker of a language, we utter the sounds so that they become familiar to us. We can't claim with a straight face that we really understand what the sounds mean. Over the course of time however, through some mysterious process, we might claim some understanding of the rudiments of Information Geometry. That understanding, however, must be couched within its own specialized language.

Brief historical context

Pinpointing origins is always a tricky affair, but Information Geometry might legitimately be said to have begun with the generalization of Euclidean geometry by Gauss and Riemann in the 19th Century.

The first mention of information and its role in statistical inference seems to belong to Fisher who, in 1925, presented some results on the theoretical difficulty of estimating parameters. These ideas later became enshrined in what was labeled as the Fisher information matrix, and in theorems about the limits of the variability of estimates as exemplified by the Cramér–Rao inequality. During the late 1940s, the Indian statistician C. R. Rao was one of the first people to think and talk about statistical problems in terms of "information."

A landmark event occurred in 1948 when Claude Shannon defined information entropy in terms of probabilities, and then applied it to communication theory. During the 1950s, there followed an explosion of interest in this newly emerging field of Information Theory. All commentators remarked on Information Theory's fundamental reliance upon probabilities.

Two particular individuals stand out during this period of the 1950s for their seminal contributions and advancement of ideas later subsumed under Information Geometry. The first is Solomon Kullback who wrote a revolutionary book showing how information theory could address all the statistical techniques in common use at that time. Even today, these data analytical procedures would be ranked as among the most popular for scientists studying behavioral, medical, and econometric issues.

He gave examples of his "information–based approach" to statistical inference that involved contingency tables, regression, analysis of variance, and discriminant analysis. We can do no better half a century later except to take advantage of the sweeping generalizations afforded by Information Geometry to expand the frontier first explored by Kullback. We are all products of our time and, since the assumption of Multivariate Normality was the only game in town, Kullback restricted most of his examples to those models.

The other towering individual to whom we owe a great debt is Edwin T. Jaynes who, also writing in the 1950s, connected Shannon's definition of information with the idea of *maximum entropy*. Entropy had been introduced by physicists about a century earlier and used to great effect by Ludwig Boltzmann and J. Willard Gibbs in advancing statistical mechanics.

Again, with the hindsight of over 50 years and the diligent efforts of numerous mathematicians, we see that Jaynes's Principle of Maximum Entropy adumbrated certain notions evolving later, and independently, from within the discipline of Information Geometry. Perhaps from a conceptual point of view, Jaynes's greatest insight was the separation of the rules for the *manipulation* of probability symbols from the rules for *assigning* numerical values to probabilities. He clearly saw that the information contained in some specified model was responsible for any particular assignment. Indeed, this powerful global conceptual reorientation is exactly what I have tried, repeatedly, to emphasize in my presentation.

In my personal view, it is a shame that Jaynes did not pursue more vigorously the geometrical aspects underlying his understanding and exposition of information entropy. He was in a perfect position to do so some twenty years before anyone else, and certainly had the mathematical wherewithal to carry out this task. But Fate dictated otherwise and, except for a few hints here and there, he did not really go down the path that we now see leads to Information Geometry.

About 30 years ago, Information Geometry officially split off from its roots in probability, information theory, physics, and differential geometry to become a recognizably independent discipline. Shun-ichi Amari, along with his colleagues and students in Japan, has written extensively to translate purely mathematical ideas from Differential Geometry to applications in statistical inference. Prominent names who have treated statistical problems from the viewpoint of Information Geometry and have written books in this new language are the Danish mathematician Ole Barndorff–Nielsen collaborating with British colleagues, Robert Kass in the US, and Murray and Rice in Australia.

At first, it seems that the only lacunae are the specific translations from the theory to the inferential tools that practicing scientists employ. However, after mucking about trying to decipher the concepts various authors have relied on in their expositions, the fundamental divide between Bayesian and the Orthodox approach raised its ugly head once more. In some small measure, these are the very gaps we hope to fill.

Initial musings

The "information" part of *Information Geometry* provides a mathematically generalized and abstract framework for dealing more rigorously with the notion of missing information. Studying entropy and its maximization was a start in this direction. As such, it also refines and sheds new light on the MEP methodology for assigning numerical values to probability symbols.

The "geometry" part of *Information Geometry* also involves a mathematically generalized and abstract treatment of familiar geometrical concepts like points, curves, distances, and volumes. For example, we will be interested in the geometrical notion of a generalized *distance* between points that live in a manifold. But these "points" are considered to be entire probability distributions, so actually what we are up to is an understanding of what it means for two probability distributions to be similar or dissimilar.

There is a distinction made in these advanced geometries not only between Euclidean and non–Euclidean geometry, but also between what is termed "intrinsic" and "extrinsic" geometries. The essential idea is that for an extrinsic geometry any manifold under discussion is visualized as embedded within a higher dimensional Euclidean space. Thus, the geometry of the surface of a sphere, for example, is always visualized as being embedded in a three dimensional space with the typical Cartesian coordinates x, y, and z.

On the other hand, the intrinsic approach would prefer to completely abandon any embedding of the manifold in some larger dimensional Euclidean space. The manifold exists, so to speak, on its own with its own separately defined coordinate system. The geometry of the sphere's surface now takes place, not as previously in a three dimensional space with Cartesian coordinates, but rather in a two dimensional manifold with new angular coordinates like latitude and longitude.

As hard as it is, apparently we have to give up our accustomed familiarity with the Euclidean space. By taking on the task of learning about Information Geometry, we must accommodate ourselves to the idea of the more abstract manifold and its attendant mathematical properties.

Learning to speak the language

Even the barest of introductions to just a few of the relevant concepts from Information Geometry quickly becomes a technical discussion. Important concepts such as *Riemannian manifolds, coordinate systems, inner products, metric tensors, Fisher information matrix, affine connections, flat spaces, projections*, and *divergence functions*, to name just a few, will be dragged out and used without any deep understanding. Basically, we just want to get these ideas down on paper so that we can proceed as quickly as possible to the more illuminating numerical examples.

Naturally, our intent is to show how these primitive concepts eventually get folded into a deeper explanation for how the IP can make optimal inferences. "Deeper" is somewhat of a misnomer in that we are just hoping that the mathematical treatment arising from Information Geometry might provide new insight, as it so often does. We always strive to interpret a result in Information Geometry into a probabilistic and inferential example as quickly as possible.

Information Geometry has borrowed liberally from the more mature mathematical disciplines of Linear Algebra, Vector Calculus, and Differential Geometry. Fortunately, not all of the advanced notions from these areas are needed for the task that Information Geometry has set for itself. Those concepts which allow a novel geometrical visualization of important ideas from logic, statistics, and information theory are emphasized here.

To repeat, our main emphasis in using Information Geometry is to quantitatively discuss and operate on the notion of information. Thus, the points in a manifold, defined by the models $\mathcal{M}_k$ that assigned numerical values to probabilities, can be looked at from a more visual geometric perspective. With new tools for operationally defining what information is, and how it can be inserted into probability distributions, we have a much better grasp of how to construct models for predictive probability distributions.

What to read?

It is a sad fact that there are NO GOOD BOOKS OR ARTICLES that easily explain inferencing, information, probability, entropy, and so on, from this advanced geometrical perspective. I am not going to claim that my efforts here are all that spectacular either.

All I am trying to accomplish is to provide a little more clarity and dispel ever so slightly the fog of mystery surrounding this topic. I attempt to hew closely to how Information Geometry offers a different perspective on core concepts in probability and information entropy.

The best current explanations, in my opinion, are offered in Sun-ichi Amari's corpus. I base most of my efforts here in Volume III on deciphering what Amari has presented. As I discussed in my *Apologia* to Volume I, I don't believe that Amari makes any *mathematical* errors. Nonetheless, his presentations are rife with *conceptual* errors. I take an inordinate delight in dissecting these conceptual errors at various points in the **Connections to the Literature** sections.

There do exist several books on these topics which I have consulted with varying degrees of success. In addition to Amari, I have read Kass and Vos [25], Murray and Rice [30], Frankel [11], Stoker [38], Kreyszig [27], Rao [33], and others listed in the Bibliography. I would recommend the recent book by Calin and Udriște [9] for a number of interesting mathematical expressions loosely tied together by an information processing approach that is similar to my own. All I can do is wish you the best of luck if you embark on the journey of trying to make sense of Information Geometry by consulting these various authors.

The final lesson from Information Geometry

In the end, after all the various esoteric mathematical paths we might pursue in Information Geometry, here is the final summary lesson that you should constantly bear in mind: **The two conceptual pillars discussed in Volumes I and II remain inviolable**.

Whatever we discover in Information Geometry will NOT impact one iota the principles contained in the formal rules for manipulating abstract probabilities. Nor will it negatively impact the utilization of the maximum entropy principle for assigning numerical values to probabilities.

Almost every Chapter in this Volume might serve as an extended example of these assertions. Information Geometry *does* at times provide one with an interesting alternative to the formal rules and the MEP, albeit at the cost of increased mathematical abstraction. But it does not contradict in any way the fundamental lessons about how to fold in the formal rules and the MEP to solve a problem that demands an inference for its solution.

Personally, I have no hang ups thinking about inferencing from the standpoint of *information, entropy, probability*, and so on. Apparently, though, there are some who remain quite uncomfortable adopting such a mental framework. For them, it may be more palatable to couch all of the information–centric concepts into an abstract geometric language. Whichever path you choose, the final destination remains the same.

Mathematica

When it gets down to the actual nuts and bolts of computation required for Information Geometry, *Mathematica* will be essential. I start off with a detailed excursion into *Mathematica*'s vector and matrix operations. I find that it is very helpful to carefully scan the symbolic output from a *Mathematica* evaluation to judge whether what you think is going on in some formula is actually the fact.

For example, I look at the symbolic evaluation of *dot products* involved in vector–matrix multiplications and compare them to the summation formulas I think are implementing the same operations. There are oftentimes surprises lurking whose resolution affords greater understanding.

Not much is made of this, although Wolfram has alluded to it from time to time, but *Mathematica* has primitive, but rather powerful, *theorem proving* capabilities. One version simply employs the built–in function **Equal[ ]** that I introduced in the first Volume, during an initial discussion of Boolean functions, to ask whether two different looking symbolic expressions happen to be the same. I use **Equal[ ]** to check whether the summation expressions and the dot product expressions are the same.

Since differentiation is obviously going to be one of the primary calculations within Information Geometry, the second appendix is devoted to an elementary introduction to how *Mathematica* does differentiation. I find it quite helpful to use examples from our previous exposure to the Maximum Entropy Principle. Once again, it is very instructive to have *Mathematica* evaluate things symbolically and compare this symbolic output to some formula we are relying upon.

For example, I examine one of Jaynes's formulas when he gets around to some technicalities in his explication of the MEP formalism. He presents a not very well–motivated formula involving the second derivative of the log of the partition function with respect to two model parameters. In order to satisfy myself of its correctness, I very carefully parsed *Mathematica*'s full symbolic output of Jaynes's formula. At first, this seemed to be somewhat daunting, but with a little practice very long symbolic expressions can be broken down into recognizable chunks and matched up very easily with their equivalent traditional mathematical forms.

Through trial and error, I stumbled upon some rather direct ways to calculate the metric tensor within *Mathematica*. Therefore, I devote some time and space in the third appendix to elucidating how that is done within simple inferencing problems involving dice tossing. I was pleased to integrate into this elementary tutorial about the metric tensor an extended discussion of the somewhat more complicated syntax of the "anonymous pure function," as Wolfram calls it, implemented in *Mathematica* through **Function[ ]**.

Contents

Preface   i

An Author's Apologia   iii

List of Figures   xv

List of Tables   xix

33 A New Language   **1**

    33.1 Introduction 1

    33.2 The Language of Information Geometry 1

    33.3 The Exponential Family 3

    33.4 Divergence Functions 5

    33.5 Replace Shannon Entropy with the Divergence Function? 6

    33.6 Speaking in Strange Tongues 7

    33.7 MEP Expressions Involving Differentiation 8

    33.8 Connections to the Literature 9

    33.9 Solved Exercises for Chapter Thirty Three 17

34 Introductory Concepts in Information Geometry   **45**

    34.1 Introduction 45

    34.2 A Manifold with Coordinate Systems 45

    34.3 Inner Product, Basis Vectors, and the Metric Tensor 47

    34.4 Norm, or Vector Length 48

    34.5 Probabilistic Interpretation of the Metric Tensor 49

	34.6 The Likelihood Equations	51
	34.7 Angles Between Vectors	53
	34.8 Connections to the Literature	54
	34.9 Solved Exercises for Chapter Thirty Four	57

35 Distance Between Probability Distributions — 75

	35.1 Introduction	75
	35.2 Length of a Curve and the Chain Rule	76
	35.3 Relative Entropy and Canonical Divergence	78
	35.4 Differential Relationships between Coordinate Systems	80
	35.5 Connections to the Literature	81
	35.6 Solved Exercises for Chapter Thirty Five	83

36 Inferences About Dice — 95

	36.1 Introduction	95
	36.2 Model for a Fair Die in IG	96
	36.3 Physical Nature of the Constraints	98
	36.4 The MEP and Linear Systems	107
	36.5 Calculating the Metric Tensor	111
	36.6 Distance to Nearby Points	112
	36.7 How the Dual Coordinate Systems Vary	113
	36.8 Connections to the Literature	118
	36.9 Solved Exercises for Chapter Thirty Six	122

37 Reasoning Logically about Kangaroos — 143

	37.1 Introduction	143
	37.2 Reviewing the Usual Details	145
	37.3 A Pure Deduction Problem	146
	37.4 Kangaroos from the Inferential Standpoint	149
	37.5 The Impact of any Data	157
	37.6 The Kangaroos and Information Geometry	158
	37.7 The Metric Tensor Revisited	161

CONTENTS

    37.8 Connections to the Literature 165

    37.9 Solved Exercises for Chapter Thirty Seven 169

38 Relative Entropy, Fisher Information, and the Metric Tensor   **199**

    38.1 Introduction 199

    38.2 Deconstructing Kullback's Argument 200

    38.3 Amplifying Remarks 206

    38.4 Length of Curves in a Manifold 208

    38.5 Related Concepts in Information Geometry 210

    38.6 Connections to the Literature 211

    38.7 Solved Exercises for Chapter Thirty Eight 218

39 Life on Mars   **225**

    39.1 Introduction 225

    39.2 Overview 226

    39.3 An Inferential Example about Life on Mars 227

    39.4 α–Project the Complicated Model 229

    39.5 Pythagorean Relationships 236

    39.6 Updating the IP's State of Knowledge 238

    39.7 Inference and Logic 239

    39.8 Connections to the Literature 240

    39.9 Solved Exercises for Chapter Thirty Nine 248

40 Triangular Relationships   **263**

    40.1 Introduction 263

    40.2 Law of Cosines 263

    40.3 Triangular Relationships in IG 267

    40.4 Amari's Theorem 3.7 269

    40.5 Divergence Notation 272

    40.6 Connections to the Literature 273

    40.7 Solved Exercises for Chapter Forty 276

41 Orthogonal Decompositions and Foliations — 285

- 41.1 Introduction 285
- 41.2 Mixed Coordinate System 286
- 41.3 Model Interactions and Foliations 291
- 41.4 Information Decomposition 293
- 41.5 Decomposition for a Larger State Space 296
- 41.6 IG's Reaction to the MEP 300
- 41.7 Connections to the Literature 304
- 41.8 Solved Exercises for Chapter Forty One 317

42 Boltzmann Machines — 343

- 42.1 Introduction 343
- 42.2 Preliminary Description 345
- 42.3 The Perspective from IG 347
- 42.4 Boltzmann Machines and Thermodynamics 348
- 42.5 The Kangaroos Meet the Boltzmann Machine 352
- 42.6 Prediction Using a Boltzmann Machine 354
- 42.7 Posterior Probability of Models 360
- 42.8 Connections to the Literature 361
- 42.9 Solved Exercises for Chapter Forty Two 365

43 Hidden Units for Boltzmann Machines — 383

- 43.1 Introduction 383
- 43.2 Various Inferencing Scenarios 386
- 43.3 The Formal Probabilistic Start 387
- 43.4 How to Avoid Probability 388
- 43.5 How Not to Avoid Probability 389
- 43.6 Probabilistic Boltzmann Machine 391
- 43.7 A Critical Assessment 394
- 43.8 Connections to the Literature 395
- 43.9 Solved Exercises for Chapter Forty Three 404

44 Canonical Inferences — 429
- 44.1 Introduction 429
- 44.2 Symbolic Expressions 430
- 44.3 Logistic Regression 435
- 44.4 Implications for More Realistic Scenarios 436
- 44.5 Relevance for Jeffreys's Prior? 438
- 44.6 Connections to the Literature 440
- 44.7 Solved Exercises for Chapter Forty Four 441

45 Variational Bayes — 463
- 45.1 Introduction 463
- 45.2 What Becomes Computationally Intractable? 465
- 45.3 Review of α–projections 468
- 45.4 Defining Free Energy 469
- 45.5 Connections to the Literature 472
- 45.6 Solved Exercises for Chapter Forty Five 474

46 Mean Field Theory — 499
- 46.1 Introduction 499
- 46.2 Binary Statements 501
- 46.3 The Ising Model 502
- 46.4 Adding Complexity 506
- 46.5 Comparing e-projection Notation 508
- 46.6 Connections to the Literature 509
- 46.7 Solved Exercises for Chapter Forty Six 513

47 Jeffreys's Prior — 527
- 47.1 Introduction 527
- 47.2 Back to Basics 529
- 47.3 What Upset Jeffreys? 532
- 47.4 Parametric Invariance 535
- 47.5 Geometrical Rationale for Jeffreys's Prior 536

47.6 Important Concepts	536
47.7 Connections to the Literature	538
47.8 Solved Exercises for Chapter Forty Seven	548

48 Harmony of the Spheres?   583

48.1 Introduction	583
48.2 Lengths and Volumes	584
48.3 Determinant of the Metric Tensor	585
48.4 Jeffreys's Prior for the Gaussian	587
48.5 Surface Areas and Volumes for Spheres	590
48.6 From Spheres to the Dirichlet Distribution	592
48.7 Amari's $\alpha = 0$ Representation	594
48.8 Connections to the Literature	594
48.9 Solved Exercises for Chapter Forty Eight	599

A Working with Vectors and Matrices in *Mathematica*   **615**

B Using *Mathematica* for Differentiation in IG   **623**

C *Mathematica* and the Metric Tensor   **629**

REFERENCES   **637**

List of Figures

34.1 *A schematic diagram illustrating the tangent space T_p at a point p in the sub–manifold S^2* 54

34.2 *A sketch of the (x,y) and (r,θ) coordinate systems* 57

34.3 *A* Mathematica *program for exploring constraint function assignments and model parameter adjustments* 67

34.4 *A sketch illustrating how a vector* **v** *is constructed from its coordinates and basis vectors* 73

36.1 *Output from* Mathematica *function* `chap36dice[ ]` 141

37.1 *A joint probability table for the kangaroo scenario under a model containing information about a logical implication* 147

37.2 *A generic joint probability table for the kangaroo scenario* 150

37.3 *A target joint probability table for the kangaroo scenario implementing the logic expression and containing no further information* 151

37.4 *A joint probability table for the kangaroo scenario under a model with three constraints* 153

37.5 *A joint probability table for the kangaroo scenario under a model with four constraints that induce a correlation between the genetic code and beer preference* 154

37.6 *A joint probability table for the kangaroo scenario under a model with five constraints inducing correlations between the genetic code and beer preference and beer preference and hand preference* 155

37.7 *A contingency table showing data supporting the implicational chain model. The normed frequency counts match the information in the five constraint functions of model $\mathcal{M}_3$ making it the maximum likelihood model* 158

37.8 *Diagram of the Rule 139 elementary cellular automaton* 173

37.9 *The "joint probability table" induced by a particular three variable Boolean function* $f(x, y, z) = axyz \bullet a'x'yz$ 175

37.10 *Wolfram's elementary cellular automaton following Rule 139* 198

39.1 *A sketch illustrating the Pythagorean relationships for four points in a Riemannian manifold* 236

39.2 *Six joint probability tables reflecting various models for the implication logic function* $A \to B$ 252

39.3 *A joint probability table with numerical assignments inspired by the logic expression for the life on Mars scenario* 255

39.4 *A joint probability table containing the numerical assignments when the complicated model of point p is α–projected down into a point q in $\mathcal{S}^4$* 259

39.5 *A joint probability table containing the numerical assignments when the complicated model of point p is $\mathbf{m}$–projected down into a point q in $\mathcal{S}^6$* 261

40.1 *The Law of Cosines for a triangle* 264

40.2 *The Law of Cosines for vectors* 265

40.3 *Finding $\cos \theta$ in the vector representation* 266

40.4 *Numerical example of the Law of Cosines to find the length of side b in triangle ABC* 276

41.1 *A sketch of how the intersecting sub–manifolds in a foliation determine the mixed coordinate system for any point. The four points, p, q, r, and s shown in the figure represent four different numerical assignments to the probability of the statements in the state space* . . 288

41.2 *A sketch of the Pythagorean relationships among the four probability distributions p, q, r, and s (not to scale). The $\alpha = -1$ and $\alpha = 1$ projections are shown* 290

41.3 *Mutually dual foliation of the Riemannian space designed to illustrate information decomposition* 293

41.4 *Another mutually dual foliation of the Riemannian space designed to illustrate information decomposition* 309

41.5 Mathematica *output showing the assigned probabilities, the Lagrange multipliers, the three constraint function averages, and the relative entropy between the complicated model of point p in the simple kangaroo scenario and the uniform distribution* 318

41.6 *The* Mathematica *output showing the assigned probabilities, the three Lagrange multipliers, the three constraint function averages, and the relative entropy between the less complicated model behind point q and the uniform distribution in the simple kangaroo scenario* 322

41.7 *The* Mathematica *output showing the assigned probabilities, the seven Lagrange multipliers, the seven constraint function averages, and finally the relative entropy between the complicated model behind point p and the uniform distribution behind point u for the $n = 8$ kangaroo scenario* 327

41.8 *How Information Geometry characterizes the MEP by visualizing points and distances within a Riemannian manifold. Point $q^\star$ is the maximum entropy distribution because it is as close as possible to point u and point p* 331

41.9 *The squared distances involved in the two triangles $\triangle pq^\star u$ and $\triangle rq^\star u$. Both triangles are right triangles so the Pythagorean theorem can be used* 336

42.1 *Sketch of a Boltzmann machine with five nodes. The overall state of the Boltzmann machine can be captured by a joint statement asserting which nodes are ON and which are OFF. The interconnection strength between nodes is indicated by a λ_j* 346

42.2 *The joint probability table for the Boltzmann machine example with five nodes. Each cell is labeled with the index i for the assignment Q_i under some model* 350

42.3 *The thirty two cell contingency table containing the observed frequency counts N_i in a sample of 640 kangaroos. The Boltzmann machine will "learn" from these data* 354

42.4 *The joint probability table for five traits of kangaroos as conditioned on the truth of a model that gives rise to strong deductions. This is a model behind a point p in the dually flat Riemannian manifold S^{32}* 370

43.1 *A restricted Boltzmann machine which will try to classify kangaroos through the assistance of hidden units acting as feature detectors* . . 385

43.2 *The* Mathematica *output for a restricted Boltzmann machine* 392

43.3 *Sketch of a multilayer perceptron consisting of an input layer, a hidden layer, and an output layer. Only a few of the weights are shown in order to keep the diagram uncluttered* 406

43.4 *Locating a cell in the joint probability table with a "dendrogram."* ... 419

43.5 *A joint probability table for three variables laid out with the C variable at the highest level* 421

43.6 *The joint probability table for all four variables of a Boltzmann machine following some logical constraints* 423

43.7 *The joint probability table for all four units of a simplified Boltzmann machine as the **m**-projection from a more complicated model* 428

44.1 *A fictitious contingency table to illustrate Jeffreys's "animals with feathers" problem* 452

44.2 *A binary predictor variable H has been added to Jeffreys's "animals with feathers" problem* 453

46.1 *The non–linear relationship between the dual parameters λ and $\langle F \rangle$ of the Ising model. A plot of $\tanh(\lambda)$ as a function of λ* 522

47.1 *A breakdown of all possibilities in sampling three balls from a total of four black balls and two red balls when drawing all three balls "in one grab."* 550

List of Tables

33.1 *A side by side comparison of the probability for a reaction time interval based on the correct Bayesian approach of averaging over the posterior probability of all models* versus *the maximum likelihood model based on the Gamma distribution* 39

36.1 *Comparison of the normed frequency counts, the MEP assignment, a linear assignment, and a superposed assignment on which Jaynes based his MEP assignment* 105

36.2 *The calculation of $g_{22}(p)$ using Equation (36.6)* 111

36.3 *A numerical example of the θ^j coordinate system changing in response to changes in the dual coordinate system η_j. The average values of the two constraint functions are approaching 4 and -1* . . . 115

36.4 *A continuation of the numerical example of how the θ^j coordinate system changes in response to changes in the dual coordinate system η_j. The numerical assignment to Q_4 is approaching 1* 117

37.1 *The computational details for $g_{21}(s)$, the covariance between the first and second constraint functions in model $\mathcal{M}_1$* 163

37.2 *Determining the location of 0s in a joint probability table using the DNF* 170

37.3 *The MEP calculations for a specific model inserting information about three marginal probabilities for the kangaroo scenario* 177

37.4 *The five constraint functions involved in model $\mathcal{M}_3$* 179

37.5 *The implicational chain model $\mathcal{M}_3$ inserts information about three marginal probabilities and two interactions* 179

39.1 *Locating the minimum value of a divergence function in order to find the α–projection of a point p representing a complicated model to a point q representing a simpler model* 231

39.2 *The minimum value of the divergence function when point p representing a complicated model is **m**–projected to a point q representing a simpler model* 235

39.3 *All fifteen dual coordinates for the point q **m**–projected from point p to a sub–manifold S^6* 235

39.4 *The functional assignments for Equation (39.5)* 249

39.5 *The dual coordinates for all fifteen constraints of the complicated model behind point p* 256

39.6 *The detailed listing of the mapping from statements to numbers as represented by the first ten constraint functions. The constraint function averages are shown in the last row* 257

39.7 *The detailed listing of the mapping from statements to numbers as represented by the final five constraint functions. The constraint function averages are shown in the last row* 258

41.1 *The m–affine, e–affine, and mixed coordinate systems for four points in a Riemannian manifold. The invariant probability distributions under any of these three coordinate systems are shown in the final column* 289

41.2 *The m–affine, e–affine, and mixed coordinate systems, together with the probability assignment for four points in a Riemannian manifold. Both points q and r live in the sub–manifold where there is no interaction* 292

41.3 *The m–affine, e–affine, and mixed coordinate systems, together with the probability assignment for seven points in a Riemannian manifold. Points q, r, and s live in the sub–manifold where there is no interaction. Points t and v live in the sub–manifold where there is no effect from the interaction or the second marginal probability. Point u contains the information from the fair model where there are no effects from the marginal probabilities or the interaction* 295

41.4 *A comparison of the η_i coordinates for the four points p, $q^\star$, r, and u. The three points p, $q^\star$, and r must share the same value for the first five coordinates. Point u does not share any common η_i values since it does not live in the $\boldsymbol{M}_5(p)$ sub–manifold* 332

42.1 *Evidence of an incipient combinatorial explosion for a full decomposition of a complicated model consisting of just eight nodes* 375

42.2 *All ten double interactions for a Boltzmann machine with 5 nodes* . 376

42.3 *A more detailed look at the five quadruple interactions of a complicated model behind some designed inferencing system* 379

42.4 *A summary of the categorizations, or pattern completions, made by a designed system for the intelligence and beer preference traits of kangaroos* 379

43.1 *All eight joint statements in the numerator of Bayes's Theorem used to calculate $P(A, B \,|\, C, D, E, \mathcal{M}_k)$* 412

43.2 *Summarizing the placement of the numerical assignments in the joint probability table based on a model that implements certain logical constraints* 422

44.1 *The posterior predictive probabilities for seeing future frequency counts of feathered animals with and without beaks in the remaining 5,000 that were left unsampled in the population* 460

47.1 *An exhaustive listing of all fifteen possible contingency tables in an $n = 3$ coin flip scenario involving $N = 4$ flips* 559

47.2 *The posterior probability for model $\mathcal{M}_A$ that specifies the assignment $Q_1 = Q_2 = 1/2$. These are Jeffreys's answers when the only other model is a model $\mathcal{M}_B$ that implements Laplace's position* 582

48.1 *The surface areas and volumes of n–balls with radius equal to 2* . . . 604

Chapter 33

A New Language

33.1 Introduction

As a beginning, I would like to introduce some of the special language that must be employed in crafting any discussion centered around Information Geometry (IG). Our intent here is a simple one. Let's practice speaking the language without seriously claiming that we understand either its grammar or its vocabulary.

We commence with some of IG's most basic vocabulary to get us off and running. The language, terms, and notation generally follow those presented in Amari [2].

33.2 The Language of Information Geometry

The most primitive notions of Information Geometry are that of a *manifold* and a *coordinate system*. A manifold S is an abstract collection of elements or points endowed with a coordinate system. Thankfully, Information Geometry is concerned with something more specific than this vague, abstract notion. IG identifies these *points* in the manifold S as entire probability distributions!

Each probability distribution in the manifold S is parameterized by a coordinate system of n real numbers given here by the notation of $\theta^0, \theta^1, \cdots, \theta^{n-1}$. However, we are free to choose other coordinate systems. As things develop, we will find ourselves employing a *dual coordinate system* expressed by $\eta_0, \eta_1, \cdots, \eta_{n-1}$.

Whether one uses subscripts as done above for $\eta_0, \eta_1, \cdots, \eta_{n-1}$, or superscripts as done for $\theta^0, \theta^1, \cdots, \theta^{n-1}$ for the coordinates is an example of one of the many notational peculiarities of Information Geometry. In this case, the notation for the coordinates takes it rationale from Differential Geometry, and the rules developed for Tensor Calculus. Additional specialized language refers to the θ coordinates as *covariant vectors*, while the η coordinates are called *contravariant vectors*.

The dimension of the manifold is then taken to be n and labeled as $\mathcal{S}^n$. This is the same as the dimension of the state space. Any causal modeling, however, takes place in a *sub-manifold* of dimension m where m is less than, and in applied problems, much, much less than n. Thus, one of the fundamental notions is the idea of *projecting* a point from a high dimensional manifold $\mathcal{S}^n$ to another point living at the intersection of two smaller orthogonal sub-manifolds.

Essential to all further discussions of probability within Information Geometry is the additional structure provided by Riemannian metrics. This is a mapping from the probability distributions, thought of as points in the manifold $\mathcal{S}^n$, to an inner product defined on another space, called the tangent space. Amari calls this tangent space a "local linearization" around any point p in the manifold.

It turns out that there is one Riemannian metric that has an intimate connection with a very well known statistical concept. This is the notion of information as introduced by Fisher in 1925 and called in his honor the Fisher information matrix.

With this additional structure of a Riemannian metric, traditionally given the notation of g, we begin to restrict our discussion to those Riemannian manifolds with the symbolic notation of $(\mathcal{S}, g)$. Furthermore, a Riemannian manifold is further qualified by calling it *smooth* which means that we are allowed to differentiate a function within the manifold any number of times that we please.

Another important geometrical concept in IG is the idea of a *curve*. These curves connect the points in the manifold and, as such, they can be "straight" or meander all over the place. We will be interested in finding the length of these curves, and especially those curves with the *minimum* length connecting two points in the manifold. A *coordinate curve* is simply a curve where only one of the coordinates θ^j is allowed to vary, while all of the other coordinates θ^i must remain fixed.

When delving into the esoterica of Information Geometry, we will constantly attempt to anchor ourselves by referring back to those ideas from probability theory and the Maximum Entropy Principle (MEP) that we will credit ourselves as being relatively well understood. As mentioned, IG alleviates some of the abstractness inherent in the mathematical definitions by encouraging us to imagine that the n-dimensional Riemannian manifold is a space consisting of probability distributions. In most of the literature, these probability distributions are said to belong to the *exponential family*.

The topic of exponential families has not been explicitly mentioned in either Volume I or Volume II. All we currently know about are the numerical assignments to probabilities afforded by the MEP. Nonetheless, our familiarity with the MEP permits us to state that any exponential family distribution is simply an MEP distribution. Any reference then to an exponential family in IG is a vital link to the entire development of the MEP as it was portrayed in Volume II.

33.3 The Exponential Family

Amari [2, pg. 33] provides this definition of an exponential family,

$$p(x;\theta) = \exp\left[C(x) + \sum_{i=1}^{n} \theta^i F_i(x) - \psi(\theta) \right] \tag{33.1}$$

Compare this to our MEP definition of a numerical assignment to a probability of a statement under the information contained in some model $\mathcal{M}_k$,

$$P(X = x_i \mid \mathcal{M}_k) = \frac{\exp\left[\sum_{j=1}^{m} \lambda_j F_j(X = x_i) \right]}{Z(\lambda_j)} \tag{33.2}$$

Pattern matching between the two expressions reveals the following equivalencies,

$$p(x;\theta) \equiv P(X = x_i \mid \mathcal{M}_k)$$

$$\theta^i \equiv \lambda_j$$

$$F_i(x) \equiv F_j(X = x_i)$$

$$\psi(\theta) \equiv \ln Z(\lambda_j)$$

$$C(x) \equiv 0$$

$$n \equiv m$$

Let's review the salient aspects of MEP distributions. Most importantly, any probability distribution generated by the MEP depends, in part, on m functions, $F_1(X = x_i), F_2(X = x_i), \cdots, F_m(X = x_i)$, together with their associated Lagrange multipliers $\lambda_1, \lambda_2, \cdots, \lambda_m$. These Lagrange multipliers are the parameters of some model $\mathcal{M}_k$. A model's job is to insert information into a probability distribution.

The Lagrange multipliers also happen to play the role of providing a coordinate system for sub–manifolds within $\mathcal{S}$. In fact, the θ coordinate system, which Amari labels as the one with the "natural" or "canonical" parameters, is interchangeable with the Lagrange multipliers.

All known probability distributions in statistics can be set up as examples of the exponential family when suitable functions,

$$F_1(X = x_i), F_2(X = x_i), \cdots, F_m(X = x_i)$$

together with their averages,

$$\langle F_1(x) \rangle, \langle F_2(x) \rangle, \cdots, \langle F_m(x) \rangle$$

are cleverly assigned. The η coordinate system is defined to be the same as these constraint function averages.

We studied a couple of examples of this approach in Volume II with the Gaussian and Cauchy distributions. We will expand our collection of MEP distributions by deriving the classic Gamma and Poisson distributions in exercises at the end of the Chapter.

The exponential family includes an extra feature in the $C(x)$ term sometimes called a "measure function," or a "prior probability." But does such a notion fit into the MEP formula? We equated $\exp[C(x)] = 1$ to finesse this issue about the exponential family definition and the MEP definition.

Amari presents the Gaussian distribution as an initial example of the exponential family. His characterization of the Gaussian, according to his definition of the exponential family in Equation (33.1), matches up exactly with my development of the Gaussian as an MEP distribution in Chapter Thirty of Volume II. We both have $F_1(x) = x$ and $F_2(x) = x^2$.

His canonical coordinates are,

$$\theta^1 = \frac{\mu}{\sigma^2}$$

$$\theta^2 = -\frac{1}{2\sigma^2}$$

which are exactly the same as the Lagrange multipliers λ_1 and λ_2 specified as the two parameters of my model.

His final term in Equation (33.1),

$$\psi(\theta) = \frac{\mu}{\sigma^2} + \log(\sqrt{2\pi}\sigma)$$

matches my,

$$Z(\lambda_1, \lambda_2) = \exp\left(\frac{\mu}{\sigma^2}\right) \sqrt{2\pi}\sigma$$

His $C(x) = 0$ matches the fact that there is no extraneous multiplying factor in the numerator of the Gaussian because $\exp[C(x)] = 1$. Amari's η parameters, η_1 and η_2, match up with $\eta_1 \equiv \langle F_1(x) \rangle = \mu$, and $\eta_2 \equiv \langle F_2(x) \rangle = \mu^2 + \sigma^2$.

It would be fair to say that we have, in fact, already exploited the exponential family of probability distributions merely because we have consistently employed the MEP algorithm. Constraint functions and their averages defined the active information inserted into a probability distribution. However, the complementary concept of information entropy gave us a way of profitably questioning what the term *missing information* might mean for a probability distribution.

The m functions, $F_1(X = x_i), F_2(X = x_i), \cdots, F_m(X = x_i)$, which mapped statements to real numbers were called *constraint functions* in Volume II. These constraint functions, and their associated averages, determine the simplicity or, if you prefer, the complexity of a model. A total of $m = n - 1$ constraint functions and coordinates would have to be used in a full decomposition of the manifold.

33.4 Divergence Functions

If information entropy is too squishy a concept for you, then perhaps the more concrete notion of distance between points is more to your liking. An elementary geometrical concept that plays a central role in defining models from the purview of Information Geometry is the notion of a distance between two points in the manifold. Alternatively, this distance can be thought of as some measure of the discrepancy between two probability distributions as indexed by the difference in their respective coordinates.

Information Geometry leverages the notion of what has been called a *divergence function* on the manifold $\mathcal{S}^n$. One very important divergence function stemming historically from the combined efforts of Shannon, Jaynes, and Kullback is the relative entropy function involving two points (distributions) p and q. The first expression on the left hand side is my notation followed by Amari's notation.

$$KL(p,q) = \sum_{i=1}^{n} p_i \ln\left(\frac{p_i}{q_i}\right) \xrightarrow{\text{Amari}} D^{(1)}(q \parallel p) = \int_{\mathcal{R}} p(x) \ln\left[\frac{p(x)}{q(x)}\right] dx \qquad (33.3)$$

This divergence function ties together fundamental notions about distance in an abstract space with information contained in the Fisher metric g. Equation (33.3) is one definition of an abstract "distance" between points p and q on the manifold. Or, as we prefer to express it in statistical inference, a measure of the amount of information that separates the two probability distributions $p(x)$ and $q(x)$,

$$p(x) \equiv P(X = x_i \,|\, \mathcal{M}_1) \text{ and } q(x) \equiv P(X = x_i \,|\, \mathcal{M}_2)$$

In Volume II, we took advantage of standard variational principles of calculus, in order to *maximize* an entropy function subject to the information inserted by a model in the form of constraint function averages. The divergence function may be treated in exactly the same fashion and *minimized* subject to the very same side constraints as in the MEP formalism.

After the variational mathematics takes place, the probability distribution $p(x)$ that results from the direct notion of the Kullback–Leibler relative entropy function is expressed in Amari's slightly different exponential family notation as,

$$p(x; \theta) = \exp\left[\ln q(x) + \sum_{i=1}^{n} \theta^i F_i(x) - \psi(\theta)\right] \qquad (33.4)$$

Re-write the exponential family in the preferred notation developed throughout Volumes I and II,

$$p(X = x_i \,|\, \mathcal{M}_k) = \frac{q(x_i) \exp\left[\sum_{j=1}^{m} \lambda_j F_j(x_i)\right]}{\sum_{i=1}^{n} q(x_i) \exp\left[\sum_{j=1}^{m} \lambda_j F_j(x_i)\right]} \qquad (33.5)$$

This different way of expressing the exponential family of probability distributions does tie into the MEP perspective in the following sense. We *are* trying to obtain a point p which is as close as possible to another point q in an abstract space. However, the "information" represented by constraint functions and their averages define q as that point (probability distribution) containing "no information."

If the point q represents the probability distribution with no information, and let us now relabel this point q as a point u, then its numerical assignment via the MEP is always the constant value $1/n$. Its influence cancels out in Equation (33.5), and we are back to the basic MEP formula.

We took some pains to emphasize an important concept in Volume II. Let me repeat it here. It is misguided to think of the divergence function as some sort of generalized replacement for Shannon entropy to implement the goal of assigning numerical values to probabilities of statements. Rather, the divergence function should be thought of as conceptually tied to Bayes's Theorem, and how it treats the updating of models when one piece is data is received. This is exactly how Kullback originally derived his measure.

33.5 Replace Shannon Entropy with the Divergence Function?

From the point of view of the MEP formalism, any two distributions p and q must be *first* assigned numerical values through the MEP formula before anything else takes place. As mentioned above, both p and q must be as close as possible to point u. After that much has been established, one can use the divergence function to talk about their separation, discriminability, distance, relative entropy, *etc*. It remains quite clear that p and q have the same status and the same origin. No mystery surrounds either of them.

On the other hand, those who want to replace the Shannon entropy function with the divergence function are faced with a dilemma. They also share the goal of assigning numerical values to distribution p. But the status of the distribution q then assumes mysterious proportions. It is on a completely different footing than p. It is talked about as if it served as some sort of baseline probability distribution, or perhaps as a reference distribution. Or, some speak of q as assuming the role of an initial "measure function" whatever that might be.

It is important to point out that we always indicate that information has been inserted into the probability density function by a model $\mathcal{M}_k$. These models $\mathcal{M}_k$, whether we embed them within the context of Information Geometry or not, serve as the operational definition of the type of information utilized for a probability assignment. No numerical value for a probability can be assigned in a vacuum; an information processor's uncertainty about the truth of any statement is conditioned on the information used. The probability must change as the information on which it is conditioned changes. This is axiomatic for the foundations of scientific inference.

In science, we use any observed data together with the current best explanatory models to reduce uncertainty. These $\mathcal{M}_k$ are tentative working hypotheses about the causal structure of the phenomena we are interested in. As scientists, we temporarily hold the position that the information resident in a model $\mathcal{M}_k$ *might be* true in order to see how closely it can match the observed data.

Only the information explicitly encoded in all of the constraint functions and their averages is included in the probability distribution. *All other information* not specified in the constraint functions is missing and excluded from the probability distribution. This goal is achieved when the missing information is maximized by treating Shannon's entropy function as the objective function in a constrained optimization problem.

The quantification of uncertainty as reflected by some probability distribution conditioned on $\mathcal{M}_k$ is an insurance policy. It protects us from including unknown or unwarranted information in a model that is considered to be some tentative working hypothesis about the causal structure of the phenomena under investigation. On the positive side, it also gives us a precise way of talking about the information that *has* been included in a model, and how alternative models differ in the amount of information they contain.

33.6 Speaking in Strange Tongues

In learning to speak this new language of Information Geometry, we have to allow ourselves to say the strangest things. And we must pretend that we really do understand this gibberish flowing so effortlessly from our tongues.

Much of the discussion in IG is couched, as we have intimated, in the jargon of several advanced mathematical disciplines. But at this elementary stage, the task is really just one of understanding what particular role a complicated sounding term plays in inferencing. Therefore, like everyone else, we find it not so hard to speak gibberish.

Tensor Calculus is an advanced mathematical discipline that defines objects called tensors in a very general and abstract way. As just one example, Physics has appropriated this tensor concept in the special and general theories of relativity.

Here is just one example of a difficult concept from the Tensor Calculus that plays a very specific role in inferencing. The expression $[T_p(S)]_r^q$ is called a "tensor field of type (q, r)". But as far as IG is concerned, only one particular type of tensor with the distinguishing feature that $q = 0$ and $r = 2$ is important. Furthermore, this tensor field $[T_p(S)]_2^0$ with $q = 0$ and $r = 2$ is called a "tensor field of covariant degree 2," or a "metric tensor of second order."

But this "tensor field of covariant degree 2" is simply the Fisher information matrix. It is the matrix $g_{rc}(p)$ defined at the point p which allows the manifold to be called a Riemannian manifold. I bring all of this up to emphasize that some of the effort in learning about IG is not in absorbing new advanced mathematical concepts, but rather in figuring out how some esoterically named object can be mapped to concepts we do care about when engaged in making inferences.

Thus, with a perfectly straight face, we can say things like: *I am going to investigate a family of exponential distributions as set of points in a smooth infinitely differentiable Riemannian manifold equipped with a metric tensor of second order as defined by the Fisher information matrix.* See how easy it all is!

But rather more prosaically and, perhaps more honestly, what we are really saying is something like: *I am going to look at relationships in my degrees of belief in the truth of statements under different information from many models.*

Here is an easier example where the new language, and its accompanying new notation, is widely used. It is borrowed from Differential Geometry. In contrast, though, to the Tensor Calculus this is a piece of cake. Whenever Amari needs to perform a summation involving a product where there is a repeated superscript and subscript as in,

$$\sum_{j=1}^{m} \theta^j \eta_j \equiv \theta \cdot \eta \tag{33.6}$$

the short form $\theta \cdot \eta$ with a "central dot" will be used. This is called "Einstein's summation convention." This new notation is mirrored in *Mathematica* expressions using the built–in function `Dot[arg1, arg2 ]`. For example, the dot product of two vectors θ and η is `Dot[theta, eta]`.

33.7 MEP Expressions Involving Differentiation

I suppose that it would not be too surprising, given the already mentioned influence from *Differential Geometry*, to learn that expressions involving differentiation play a significant role in *Information Geometry*. However, this reliance on differential expressions was a marked feature of the MEP formalism from the very beginning. Jaynes had already pointed out several important differential relationships within the MEP formalism as early as 1957.

Take note of Amari's notational abbreviations for differential operators,

$$\partial_i \equiv \frac{\partial}{\partial \theta^i} \tag{33.7}$$

$$\partial^j \equiv \frac{\partial}{\partial \eta_j} \tag{33.8}$$

in order to translate the highlighted MEP differential expressions into our new language.

Here is a partial listing of some notational equivalencies that exist between the MEP formalism on the left (my minor modification of Jaynes's notation) and Information Geometry on the right (Amari's notation),

$$\left\{ \frac{\partial \ln Z}{\partial \lambda_r} = \langle F_r \rangle \right\} \equiv \{ \partial_i \psi = \eta_i \} \tag{33.9}$$

$$\left\{ \frac{\partial H}{\partial \langle F_c \rangle} = -\lambda_c \right\} \equiv \{ \partial^j \varphi = \theta^j \} \tag{33.10}$$

$$\left\{ \frac{\partial \langle F_c \rangle}{\partial \lambda_r} = \frac{\partial^2 \ln Z}{\partial \lambda_r \, \partial \lambda_c} \right\} \equiv \left\{ \frac{\partial \eta_i}{\partial \theta^j} = g_{ij} \right\} \tag{33.11}$$

$$\left\{ \frac{\partial^2 \ln Z}{\partial \lambda_r \, \partial \lambda_c} = g_{rc}(p) \right\} \equiv \{ \partial_i \, \partial_j \, \psi = g_{ij} \} \tag{33.12}$$

$$\left\{ E_p \left[\frac{\partial \ln p}{\partial \lambda_r} \times \frac{\partial \ln p}{\partial \lambda_c} \right] = g_{rc}(p) \right\} \equiv \{ \langle \partial_i, \partial_j \rangle = g_{ij} \} \tag{33.13}$$

$$\left\{ -\frac{\partial \lambda_c}{\partial \langle F_r \rangle} = -\frac{\partial^2 H}{\partial \langle F_r \rangle \, \partial \langle F_c \rangle} \right\} \equiv \left\{ \frac{\partial \theta^i}{\partial \eta_j} = g^{ij} \right\} \tag{33.14}$$

$$\{ \mathbf{A} \mathbf{A}^{-1} = \mathbf{B}^{-1} \mathbf{B} = \mathbf{I} \} \equiv \{ g_{ij} \, g^{ij} = \delta_i^j \} \tag{33.15}$$

33.8 Connections to the Literature

Amari [3, pg. 1381] succinctly summarizes our current attitude about probability distributions when expressed in the new language of Information Geometry.

> The set S of the probability distributions can be regarded as an n–dimensional manifold (space), where $\boldsymbol{\theta}$ plays the role of a coordinate system introduced in S ... Any point (that is, any distribution) in S is specified by one $\boldsymbol{\theta}$. The $\boldsymbol{\theta}$ is called the natural or canonical parameter of the exponential family.

Bernardo & Smith [6, pg. 198] in their book **Bayesian Theory** present a more commonly seen definition for the exponential family. It is similar to Amari's definition as discussed in section 33.3. It looks like this,

$$p(x \,|\, \theta) = f(x) \, g(\theta) \, \exp\{ c \, \phi(\theta) \, h(x) \}$$

as compared to Amari's

$$p(x; \theta) = \exp \left[\, C(x) + \theta F(x) - \psi(\theta) \, \right]$$

I show the one parameter exponential family because I want to compare Bernardo and Smith's definition of the Gamma and Poisson distributions as members of the exponential family with my MEP derivation upcoming in Exercises 33.9.9 and 33.9.12.

They suggest [6, pp. 207–209] that the rationale for the exponential family of distributions can be founded on "information measures." They then proceed to *minimize* Kullback's discrepancy measure as the objective function using variational techniques similar to the MEP derivation which *maximized* Shannon's information entropy. A sketch of this approach was shown in section 33.4. This is the origin of the $f(x)$ term appearing in the exponential family definition above.

It is clear, however, that this $f(x)$ term ($C(x)$ for Amari) must be an already known probability distribution. As they say, $f(x)$ is an "approximation" to the desired "true" distribution $p(x \mid \theta)$. So the confusion begins with this proposal for an alternative to the standard MEP derivation.

We are left to grapple with these various attempts to label $f(x)$ either as an "approximation" to a "true" probability, or some sort of "measure function," or worse, as a "prior probability." But understandably, there is always a dispute as to what should be chosen as $f(x)$. So the very *raison d'être* of the MEP as an unambiguous way to assign numerical values to probabilities of joint statements disappears in a puff of smoke!

The MEP assignment $P(X = x_i \mid \mathcal{M}_k)$ requires no such thing as $f(x)$ because conceptually the MEP is not thought of as any kind of "approximation" to some "true" distribution. The MEP inserts "information" defined as averages of some number of constraint functions $F_j(X = x_i)$ into a distribution. At the same time, it makes sure that all of the missing information not explicitly mentioned as a constraint function average does not make it into the distribution. There is nothing mysterious and nothing ambiguous about the MEP. Everything is clearly defined, and the MEP procedure can be carried out without any misgivings as to whether some component like $f(x)$ is open to debate.

It is worthwhile to comment on the sometimes subtle differences in how people with similar world views think about fundamental issues. I have found that discussions about the exponential family and the MEP assignment can be quite revealing in this regard. I would like to take as my example here the opinions expressed by David J. C. MacKay in his wonderful book **Information Theory, Inference, and Learning Algorithms**.

I have learned a lot from MacKay and his book would be on any Bayesian's reading list. But look at how he takes a quite critical view of MEP assignments after having broached the issue by discussing "maximum likelihood fitting of exponential family models." [29, pp. 307–308]

Assume that a variable **x** comes from a probability distribution of the form

$$P(\mathbf{x} \mid \mathbf{w}) = \frac{1}{Z(\mathbf{w})} \exp\left(\sum_k w_k f_k(\mathbf{x}) \right)$$

where the functions $f_k(\mathbf{x})$ are given, and the parameters $\mathbf{w} = \{w_k\}$ are not known. A data set $\{\mathbf{x}^{(n)}\}$ of N points is supplied.

Show by differentiating the log likelihood that the maximum–likelihood parameters $\mathbf{w}_{\text{ML}}$ satisfy

$$\sum_{\mathbf{x}} P(\mathbf{x} \mid \mathbf{w}_{\text{ML}}) f_k(\mathbf{x}) = \frac{1}{N} \sum_n f_k(\mathbf{x}^{(n)}) \qquad (22.32)$$

where the left–hand sum is over *all* $\mathbf{x}$, and the right–hand sum is over the data points. A shorthand for this result is that each function–average under the fitted model must equal the function–average found in the data:

$$\langle f_k \rangle_{P(\mathbf{x} \mid \mathbf{w}_{\text{ML}})} = \langle f_k \rangle_{\text{Data}}$$

I would not really quibble over any of this development so far except to complain about "a variable $\mathbf{x}$ comes from a probability distribution ..." This is the egregious language that Jaynes lamented in his "mind projection fallacy" which essentially claims that an observation about the physical world is caused by the observer's state of knowledge.

We recapitulated Jaynes's demonstration in Volume II [8, pg. 418], within the context of MEP distributions, that maximum likelihood estimates of parameters are the same as taking the sample average of the constraint functions from the data, and using that average, after the fact, as if that sample average was just like any other information constructed as an expectation average with respect to the MEP distribution. This is exactly what MacKay's Equation (22.32) in the quote above is saying about the exponential family of distributions.

After he has set the scene for us, we arrive at MacKay's really interesting remarks which he introduces as "maximum entropy fitting of models to constraints."

> When confronted by a probability distribution $P(\mathbf{x})$ about which only a few facts are known, the *maximum entropy principle* (maxent) offers a rule for *choosing* a distribution that satisfies those constraints. According to maxent, you should select the $P(\mathbf{x})$ that maximizes the entropy
>
> $$H = \sum_{\mathbf{x}} P(\mathbf{x}) \log 1/P(\mathbf{x})$$
>
> subject to the constraints. Assuming the constraints assert that the *averages* of certain functions $f_k(\mathbf{x})$ are known, i.e.,
>
> $$\langle f_k \rangle_{P(\mathbf{x})} = F_k$$
>
> show by introducing Lagrange multipliers (one for each constraint, including normalization), that the maximum–entropy distribution has the form
>
> $$P(\mathbf{x})_{\text{Maxent}} = \frac{1}{Z} \exp\left(\sum_k w_k f_k(\mathbf{x}) \right)$$
>
> where the parameters Z and $\{w_k\}$ are set such that the constraints ... are satisfied.

> And hence the maximum entropy method gives identical results to maximum likelihood fitting of an exponential family model (previous exercise).
> The maximum entropy method has sometimes been recommended as a method for assigning prior distributions in Bayesian modeling. While the outcomes of the maximum entropy method are sometimes interesting, and thought–provoking, I do not advocate maxent as *the* approach to assigning priors.
>
> Maximum entropy is also sometimes proposed as a method for solving inference problems – for example, 'given that the mean of this unfair six–sided die is 2.5, what is its probability distribution $(p_1, p_2, p_3, p_4, p_5, p_6)$?' I think it is a bad idea to use maximum entropy in this way; it can give very silly answers. The correct way to solve inference problems is to use Bayes' theorem.
> [Emphasis in the original.]

This kind of characterization of the MEP demands a rebuttal. Let me reiterate at the outset that MacKay and I generally look at the world from a compatible viewpoint. But it remains a curious psychological phenomenon that a global world view does not necessarily translate into agreement on all of the low–level particulars. As they say, "The devil is in the details!"

We are not really "confronted by a probability distribution ... about which only a few facts are known ..." When we are at the beginning stages of the inferential process, and after we have managed to set up the state space, we are not confronted by *any* pre–existing probability distributions. None exist at this beginning stage.

The next stage in the inferential procedure is to *create* what did not exist before. What is incumbent on the IP is to formulate an epistemological state of knowledge where a vacuum previously existed. What we really are confronted with is the desire to assign numerical values to the abstract probabilities for the statements in the state space. As long as legitimate numerical values are assigned, one way is as good as the next (almost).

The MEP is simply *one* method with highly desirable features for making these numerical assignments. And most importantly, *all conceivable legitimate* numerical assignments may be made, and are made, through the auspices of the MEP.

It is not the case that the IP is trying to discover some one "true" probability assignment based on "only a few facts." During the course of making all conceivable assignments, the MEP will have to consider distributions about which nothing is known, very little is known, a substantial amount is known, quite a bit is known, and, finally, everything is known.

This core principle resides at the very heart of the MEP algorithm: Maximize the missing "background" information left over after taking account of all of the non–missing "foreground" information that has been supplied through some model.

For example, that foreground information, in the form of a constraint function average, may assign a numerical value to the probability for HEADS anywhere from 0 through 1. Clearly, the MEP is not restricted to the situation where "only a few facts are known." That is why the probability expression for an MEP assignment is always conditioned on the assumed truth of some model which may have supplied "nothing," or a "few facts," or "everything," as $P(X = x_i \,|\, \mathcal{M}_k)$ and not as $P(\mathbf{x})$.

In summary, it seems to me that MacKay gives the distinct impression that some sort of "choice" must be made for selecting "one" $P(\mathbf{x})$. Furthermore, the MEP is painted with the brush that it "chooses" one $P(\mathbf{x})$ based on a "few facts being known." Nothing could be further from the truth. No choices have to be made. All conceivable numerical assignments $P(X = x_i \,|\, \mathcal{M}_k)$ are implicitly made when contemplating the MEP in this early stage of the inferential process because every conceivable model $\mathcal{M}_k$ is considered.

Things become really objectionable when MacKay asserts that, "the maximum entropy method gives identical results to maximum likelihood fitting of an exponential family model ..." While true, the assertion misses the point completely and misleads the unwary.

The MEP assignment of numerical values to probabilities for statements in the state space is *logically prior* to any use of "maximum likelihood fitting." Any parameter $\langle F_j \rangle$ defining the information in some model *may* be set equal to the sample average $\overline{F}_j$ of that constraint function $F_j(X = x_i)$ based on the data. This represents just one MEP assignment $P(X = x_i \,|\, \mathcal{M}_{\mathrm{ML}})$ over the panoply of all conceivable assignments. But prior to the data, this fact about this particular MEP assignment is not known.

Once again, the MEP is NOT making a "choice" that happens to be exactly the same as a maximum likelihood fitting of an exponential family model. *After* the data have been gathered, this one particular MEP model $\mathcal{M}_{\mathrm{ML}}$ happens to possess the greatest posterior probability when compared to the posterior probability of any other MEP model $\mathcal{M}_k$, or $P(\mathcal{M}_{\mathrm{ML}} \,|\, \mathcal{D}) > P(\mathcal{M}_k \,|\, \mathcal{D})$.

A common mistake made by critics of the MEP is language that talks about the MEP being used to assign "priors." So far, the MEP has only been employed to assign numerical values to *joint* probabilities of statements in the state space conditioned on the specification of some model $\mathcal{M}_k$, and written as $P(X = x_i \,|\, \mathcal{M}_k)$. So far, it has NOT been called upon to assign a prior probability to models, written as $P(\mathcal{M}_k)$.

Furthermore, since the MEP assignment has absolutely nothing to do with any data, the notion that the MEP assignment is somehow contingent upon pre or post data considerations is absurd! Thus, the label *prior probability* should be reserved strictly for the "prior to the data" probability for models, $P(\mathcal{M}_k)$, and the label *posterior probability* should be reserved strictly for the "after the data is known" probability for models, $P(\mathcal{M}_k \,|\, \mathcal{D})$.

Any original, prior to the data, MEP assignment may however be updated after some data have been observed through the formal manipulation rules of probability theory, as in,

$$P(X_{N+1} = x_i \,|\, \mathcal{D}) = \sum_{k=1}^{\mathcal{M}} P(X_{N+1} = x_i \,|\, \mathcal{M}_k) \, P(\mathcal{M}_k \,|\, \mathcal{D}) \qquad (33.16)$$

Take care to note that the post–data revision of the prior probability of all the models, $P(\mathcal{M}_k \,|\, \mathcal{D})$, appearing as the second term on the right hand side, is also implemented through the formal manipulation rules,

$$P(\mathcal{M}_k \,|\, \mathcal{D}) = \frac{P(\mathcal{D} \,|\, \mathcal{M}_k)\, P(\mathcal{M}_k)}{P(\mathcal{D})} \qquad (33.17)$$

It is important to recognize that these formal rules have nothing to do with the MEP's original job of making numerical assignments to joint statements in the state space independently of any data.

As I have mentioned before, it is a helpful attitude to consider the MEP and Bayesian procedures as *complementary*, and not *antagonistic*! There should be no debate as to whether the MEP *or* Bayesian techniques are to be used in inferencing. *Both* should be applied in their respective domains of expertise.

For example, Bayes's Theorem knows nothing about how to assign numerical values to probabilities conditioned on information. In like manner, the MEP knows nothing about how to formally manipulate probability expressions.

The final sentence that I included in the quote from MacKay sums up this prevailing misconception that I encounter almost everywhere somebody is discussing the MEP. "The correct way to solve inference problems is to use Bayes' theorem." Yes and no.

The IP must rely upon the complementary and orthogonal contributions from both the formal manipulation rules (Bayes's theorem), AS WELL AS an algorithm for assigning numerical values to probabilities in consonance with the information implied by some model (the MEP). Neither one in isolation will accomplish the goal of reaching an inference; they must be used in concert.

I have no idea what MacKay is thinking of when he criticizes the MEP in the context of assigning probabilities to a die when the information in a model specifies that the mean should be, say, 2.5. It makes all the sense in the world to make the assignment provided by the MEP when this is, in fact, the information in a model. The justification is the very *raison d'être* for why we have the MEP in the first place. An IP wants an assignment that includes only the information that the mean is 2.5 and excludes all other information. I would like to know why the MEP provides a "very silly answer" for probability assignments to this die *under this model of causality*.

MacKay must be thinking that the MEP must be employed to make one choice of a "true" probability distribution for this die. Nothing could be further from the truth. The MEP will be used to make all conceivable legitimate probability assignments to the die faces. Whichever one of these MEP assignments happens to be supported by the data should be examined by the IP closely for some causal interpretation. A detailed example is given in Chapter Thirty Six when Jaynes's explanation is re–examined for the physical reasons, from the MEP perspective, as to why a die should show some particular frequency counts.

A NEW LANGUAGE

Returning now to the concerns of Information Geometry, Jaynes happens to be among the earliest sources for some of the more prominent IG expressions involving differentiation, despite later attributions to other authors that you sometimes find in the IG literature. These early expressions arose, as you might expect, in the context of Jaynes's explication of the MEP formalism.

Even though Jaynes's book was published right after the new millennium, much of the material originates half a Century earlier. This is true of these differential expressions which most likely stem from the 1950s. Unfortunately for us, these expressions, some of which were listed in Equations (33.9) through (33.15), were not motivated by Differential Geometry, but rather by analogy with similar equations that cropped up in statistical mechanics.

Jaynes encourages us to start thinking about these differential expressions by focusing on the Legendre transformation,

$$H_{\max}(Q_i) = \ln Z - \sum_{j=1}^{m} \lambda_j \langle F_j \rangle \qquad (33.18)$$

This is not a novel concept that the Legendre transformation should be applied to inferencing. It was dwelt on extensively in Volume II, and subsequently made the centerpiece of the *Mathematica* code to actually implement an MEP algorithm.

We now visit Chapter 11 of Jaynes's book [22, pg. 359] in order to highlight these relevant expressions. In Exercise 33.9.19, I present an essentially verbatim copy of Jaynes's explanation, as given below, in my notation with a few amplifications and *Mathematica* expressions.

Suppose we make a small change in one of the F_k; how does this change the maximum attainable H? We have from (11.59) [the Legendre transformation],

$$\frac{\partial S(F_1, \ldots, F_m)}{\partial F_k} = \sum_{j=1}^{m} \left[\frac{\partial \log Z(\lambda_1, \ldots, \lambda_m)}{\partial \lambda_j} \right] \left[\frac{\partial \lambda_j}{\partial F_k} \right] + \sum_{j=1}^{m} \frac{\partial \lambda_j}{\partial F_k} F_k + \lambda_k$$

which ... collapses to

$$\lambda_k = \frac{\partial S(F_1, \ldots, F_m)}{\partial F_k}$$

in which λ_k is given explicitly.

In Amari's rather spare notation, Jaynes's result is expressed as $\partial^j \varphi = \theta^j$.

Jaynes examined another expression involving differentiation emanating from within the MEP formalism. This eventually leads to the metric tensor.

We can derive some more interesting laws simply by differentiating ...

$$\frac{\partial F_k}{\partial \lambda_j} = \frac{\partial^2 \log Z(\lambda_1, \ldots, \lambda_m)}{\partial \lambda_j \, \partial \lambda_k} = \frac{\partial F_j}{\partial \lambda_k}$$

because the second cross–derivatives of $\log Z(\lambda_1, \ldots, \lambda_m)$ are symmetric in j and k. So, here is a general reciprocity law which will hold in any problem we do by maximizing the entropy. Likewise, if we differentiate ... we have

$$\frac{\partial \lambda_k}{\partial F_j} = \frac{\partial^2 S}{\partial F_j \partial F_k} = \frac{\partial \lambda_j}{\partial F_k}$$

another reciprocity law, which is, however, not independent of [the first differentiation above] because if we define the matrices $A_{jk} \equiv \partial \lambda_j / \partial F_k$, $B_{jk} \equiv \partial F_j / \partial \lambda_k$, we can easily see that they are inverse matrices: $A = B^{-1}$, $B = A^{-1}$. These reciprocity laws might appear trivial from the ease with which we derived them here; but when we get around to applications we'll see that they have highly nontrivial and nonobvious physical meanings. In the past, some of them were found by tedious means that made them seem mysterious and arcane.

It's a shame that in his book Jaynes never did get around to those applications where he would explain to us why the components of the metric tensor had these nontrivial and nonobvious physical meanings. Personally, I am very curious as to what he had in mind when he wrote this. I wonder how he envisioned the metric tensor fitting in with the Bayesian approach and the principle of maximum entropy.

33.9 Solved Exercises for Chapter Thirty Three

Exercise 33.9.1: Use the new language of Information Geometry to talk about five probability distributions. These distributions express degrees of belief about the truth of eight joint statements as they exist in some state space.

Solution to Exercise 33.9.1

The five probability distributions are synonymous with five points which might be arbitrarily labeled as p, q, r, s, and t. These five points are elements of a manifold with the notation of $\mathcal{S}^8$. In other words, the particular set of five points just mentioned, or the five probability distributions, is part of the space of potential points in the manifold, $\{p, q, r, s, t\} \in \mathcal{S}^8$.

The dimension of the state space is specified by $n = 8$, therefore the dimension of the manifold is represented by $\mathcal{S}^8$. By definition, the manifold must also possess a coordinate system for any point. When expressed in Amari's notation, this becomes $\varphi(p) = (\theta^0, \theta^1, \cdots, \theta^7)$. There also exists a dual coordinate system for any point in the manifold. This dual coordinate system is expressed in Amari's notation by $\phi(p) = (\eta_0, \eta_1, \cdots, \eta_7)$ such that any point in the manifold can be mapped to these coordinates.

Exercise 33.9.2: Connect the new notation for probability distributions with that used in the previous two Volumes.

Solution to Exercise 33.9.2

It is always a satisfying experience to be able to sketch out some connections linking our previous efforts with the new language of Information Geometry. Also, we need to disambiguate the conflicts in notation that inevitably crop up because eventually the storehouse of available symbols becomes exhausted.

We have just introduced, within the context of Information Geometry, about the simplest notation imaginable for entire probability distributions by designating probability distributions as *points*, p, q, r, $\ldots \in S^8$.

We have used Q_i when a short notation was required to indicate that a numerical assignment had been made to abstract probabilities. More importantly, we always tried to emphasize that the numerical assignment had been conditioned on the information resident in some model $\mathcal{M}_k$. These models are equivalent to specifying the parameters, or coordinates, in the n-dimensional Riemannian manifold. For example, at the top of the next page, three different models are specified by listing the coordinates in the form $\varphi(p) = (\theta^0, \cdots, \theta^{n-1})$. The last two points q and r represent models where some coordinates have collapsed back to 0.

$$p \equiv P(X = x_i \,|\, \mathcal{M}_1)$$

$$\equiv P(X = x_i \,|\, \theta^0, \theta^1, \cdots, \theta^j, \cdots, \theta^{n-1})$$

$$q \equiv P(X = x_i \,|\, \mathcal{M}_2)$$

$$\equiv P(X = x_i \,|\, \theta^0, \theta^1, \cdots, \theta^j, \cdots, \theta^{n-1} = 0)$$

$$r \equiv P(X = x_i \,|\, \mathcal{M}_3)$$

$$\equiv P(X = x_i \,|\, \theta^0, \theta^1, \cdots, \theta^{n-2} = 0, \theta^{n-1} = 0)$$

Exercise 33.9.3: Review the derivation given in Volume II relating the Lagrange parameter λ_0 to the partition function Z.

Solution to Exercise 33.9.3

During the course of the derivation of the MEP algorithm, it was shown that,

$$e^{\lambda_0 - 1} = \frac{1}{Z}$$

so that,

$$\ln\left(e^{\lambda_0 - 1}\right) = \ln\left(1/Z\right)$$

$$\lambda_0 - 1 = \ln 1 - \ln Z$$

$$\lambda_0 = 1 - \ln Z$$

The coordinate θ^0, or λ_0, the normalizing factor Z, represents the presence of the universal constraint function. The universal constraint reflects the fact that the numerical assignments to probabilities must sum to 1. Thus, there always implicitly existed a universal constraint function $F_0(X = x_i)$ with the constant value of 1, and accompanied by the constraint function average $\langle F_0 \rangle = \eta_0 = 1$. It is never explicitly shown because of the presence of Z.

Effectively, this meant that within the MEP algorithm only $n-1$ constraint functions with associated Lagrange multipliers λ_j had to be specified for a full decomposition of the state space. Any less complex model would then consist of $m < n-1$ constraint functions.

In our current context, we want to match up the Lagrange multipliers with the coordinate system θ^j. Thus, in the future, we need only specify $n-1$ coordinates in the vector $(\theta^1, \theta^2, \cdots, \theta^j, \cdots, \theta^{n-1})$ for an n dimensional manifold with an implicit θ^0 lurking in the background.

SOLVED EXERCISES FOR CHAPTER 33

Exercise 33.9.4: Suppose a Riemannian manifold $\mathcal{S}^4$ contains the point, or probability distribution, labeled as u. If the three Lagrange multipliers all happen to equal 0, then what are the coordinates for u?

Solution to Exercise 33.9.4

First, match up the Lagrange multipliers with the coordinates,

$$\lambda_1 \equiv \theta^1 = 0$$

$$\lambda_2 \equiv \theta^2 = 0$$

$$\lambda_3 \equiv \theta^3 = 0$$

The partition function Z must equal 4. Thus,

$$\theta^0 \equiv \lambda_0 = 1 - \ln Z = 1 - \ln 4 = -0.386294$$

The normalizing factor appearing in the denominator of the MEP algorithm has a value of $Z = 4$.

Moreover, since the remaining three coordinates are all 0, the numerator with $\lambda_1 = \lambda_2 = \lambda_3 = 0$ for any $F(X = x_i)$ will be $e^0 = 1$. The numerical assignment for the probability distribution u is then,

$$u \equiv P(X = x_i \,|\, \mathcal{M}_k)$$

$$\equiv P(X = x_i \,|\, \theta^0 = -0.386294,\ \theta^1 = 0,\ \theta^2 = 0,\ \theta^3 = 0)$$

$$= 1/4$$

$\mathcal{M}_k$ specifies the model parameters, the values for the coordinates.

Exercise 33.9.5: Recall the correlation model of the simple kangaroo scenario in Volume II. What was calculated there as the value for Fisher's information matrix?

Solution to Exercise 33.9.5

This exercise is in preparation for a numerical example of the metric tensor as expressed in Equations (33.11) and (33.12). As first presented in Chapter Twenty One of Volume II, this initial kangaroo scenario involved only four joint statements concerning beer and hand preference. Thus, the dimension of the state space was $n = 4$. A model with information about a correlation between beer preference and hand preference used three constraint functions and their averages, so $m = 3$. This correlational model resulted in a numerical assignment to the four probabilities with the $Q_i = (0.70, 0.05, 0.05, 0.20)$.

Fisher's information matrix, or the metric tensor $g_{rc}(p)$, was calculated as the 3×3 symmetric matrix,

$$g_{rc}(p) = \begin{pmatrix} 0.1875 & 0.1375 & 0.1750 \\ 0.1375 & 0.1875 & 0.1750 \\ 0.1750 & 0.1750 & 0.2100 \end{pmatrix}$$

Exercise 33.9.6: Pick a specific component in the metric tensor for the numerical example.

Solution to Exercise 33.9.6

Suppose that we select $r = 2$ and $c = 1$. This selects element $g_{21}(p) = 0.1375$ from the total of nine elements in the metric tensor. Thus, we have to show that according to Equations (33.11) and (33.12),

$$\frac{\partial \langle F_1 \rangle}{\partial \lambda_2} = \frac{\partial}{\partial \lambda_2} \frac{\partial \ln Z}{\partial \lambda_1} \equiv \frac{\partial^2 \ln Z}{\partial \lambda_2 \partial \lambda_1} = g_{21}(p) = 0.1375$$

Exercise 33.9.7: Employ *Mathematica* to help solve this problem.

Solution to Exercise 33.9.7

Referring back to Chapter Twenty One, section 21.3.3, under the correlation model these constraint functions and their expectations were given,

$$F_1(X = x_i) = (1, 1, 0, 0) \text{ with } \langle F_1 \rangle = 0.75$$

$$F_2(X = x_i) = (1, 0, 1, 0) \text{ with } \langle F_2 \rangle = 0.75$$

$$F_3(X = x_i) = (1, 0, 0, 0) \text{ with } \langle F_3 \rangle = 0.70$$

The constraint matrix for this problem is then,

```
cm = {{1, 1, 0, 0}, {1, 0, 1, 0}, {1, 0, 0, 0}}
```

Set up the symbolic expressions for the Lagrange multipliers with,

```
Map[Subscript[λ, #]&, Range[3]
```

so that the partition function Z can be formed as,

```
z = Total[Exp[Dot[Map[Subscript[λ, #]&, Range[3], cm]]]
```

This expression returns,

$$Z = e^{\lambda_1 + \lambda_2 + \lambda_3} + e^{\lambda_1} + e^{\lambda_2} + 1$$

Next, have *Mathematica* calculate the second partial derivative of the log of the partition function with respect to λ_2 and λ_1 in order to implement,

$$g_{21}(p) = \frac{\partial^2 \ln Z}{\partial \lambda_2 \, \partial \lambda_1}$$

This is accomplished through `D[Log[z], λ₂, λ₁]`

We would get back a symbolic answer from this expression, but, in fact, we prefer a numerical answer. So, we will replace the symbolic values of the Lagrange multipliers with their actual values from the MEP solution,

`ReplaceAll[D[Log[z], λ₂, λ₁],`
`    {λ₁ → -1.386294, λ₂ → -1.386294, λ₃ → 4.025352}]`

When the above differentiation was evaluated by *Mathematica*, the correct value was returned showing that $g_{21}(p) = 0.1375$.

Exercise 33.9.8: Check this result through a numerical differentiation formula that approximates the second partial derivative.

Solution to Exercise 33.9.8

There is a finite difference derivative formula that can be used to approximate a second partial derivative [32, pg. 174]. This formula is a generalization of the central difference approximation which was used in Volume II to check that the partial differentiation of the log of the partition function Z with respect to the coordinate λ_j was the jth constraint function average,

$$\frac{\partial \ln Z}{\partial \lambda_j} = \langle F_j \rangle$$

Now, we would like a comparable formula that approximates a second partial differentiation of log Z with respect to the two coordinates λ_2 and λ_1 in order to check the value returned by the above evaluation of $g_{21}(p)$,

$$\frac{\partial^2 \ln Z}{\partial \lambda_2 \, \partial \lambda_1} = g_{21}(p)$$

For a function $f(\mathbf{x})$ where $\mathbf{x}$ is a vector, we have the following approximation,

$$\frac{\partial^2 f(\mathbf{x})}{\partial x_i \, \partial x_j} \approx \frac{f(\mathbf{x} + \delta e_i + \delta e_j) - f(\mathbf{x} + \delta e_i) - f(\mathbf{x} + \delta e_j) + f(\mathbf{x})}{\delta^2}$$

with $f(\mathbf{x}) \equiv \ln Z(\lambda_1, \lambda_2, \lambda_3)$. Since we want to approximate $g_{21}(p)$, $r = 2, c = 1$, and we have $x_i \equiv \lambda_2$, $x_j \equiv \lambda_1$, $e_i \equiv (0, 1, 0)$, and $e_j \equiv (1, 0, 0)$.

Letting $\delta = 0.001$, we have for the approximation,

$$
\begin{aligned}
f(\mathbf{x} + \delta e_i + \delta e_j) &= \ln Z(\lambda_1 + 0.001, \lambda_2 + 0.001, \lambda_3) \\
&= 1.610938996 \\
f(\mathbf{x} + \delta e_i) &= \ln Z(\lambda_1, \lambda_2 + 0.001, \lambda_3) \\
&= 1.610188765 \\
f(\mathbf{x} + \delta e_j) &= \ln Z(\lambda_1 + 0.001, \lambda_2, \lambda_3) \\
&= 1.610188765 \\
f(\mathbf{x}) &= \ln Z(\lambda_1, \lambda_2, \lambda_3) \\
&= 1.609438671 \\
\frac{\partial^2 \ln Z}{\partial \lambda_2 \, \partial \lambda_1} &\approx \frac{1.610938996 - 1.610188765 - 1.610188765 + 1.609438671}{0.000001} \\
&\approx 0.137431
\end{aligned}
$$

Exercise 33.9.9: Derive the Gamma distribution as a typical example of the MEP formalism, and therefore as an exponential family distribution.

Solution to Exercise 33.9.9

The Gaussian and Cauchy distributions were proven to be examples of the MEP formalism in Chapters Thirty and Thirty One of Volume II. And, of course, every discrete distribution we have dealt with so far has also been an MEP distribution.

As asserted in this Chapter, every MEP distribution must also be an exponential family distribution. We are going to expand the list of well-known probability distributions derivable from the maximum entropy principle by now deriving the Gamma distribution. In this exercise, we shall derive the Gamma distribution as one way to make numerical assignments for the probability of statements referring to the real line from 0 to $+\infty$.

In my days as a research psychologist, the Gamma distribution was frequently mentioned as the probability distribution of choice for reaction times to respond correctly to a series of test questions. The idea was that a reaction time always had to be greater than 0. In addition, it would be nice not to have an arbitrary limit on how long that reaction time might be. Since reaction times by human test subjects are naturally skewed to show quite a few very long reaction times, the Gamma probability distribution was regarded as superior to the Gaussian distribution.

SOLVED EXERCISES FOR CHAPTER 33

Since the state space for the Gamma distribution extends over the real line from 0 to ∞, we would like to generate a continuous distribution, and therefore a probability density function through the auspices of the MEP. The standard Gamma distribution reflects information from a model with two parameters,

$$\text{pdf}\,(x\,|\,\mathcal{M}_k) \equiv \text{pdf}\,(x\,|\,\lambda_1, \lambda_2)$$

One of the nice features about the MEP approach is that we always have the generic template as a jumping off point. The generic MEP assignment for the probability density function of a continuous variable x is,

$$\text{pdf}\,(x\,|\,\mathcal{M}_k) = \frac{\exp\left[\,\lambda_1 F_1(x) + \lambda_2 F_2(x) + \cdots + \lambda_m F_m(x)\,\right]}{Z(\lambda_1, \lambda_2, \cdots, \lambda_m)}$$

The discrete statements indexed by $(X = x_i)$ about a measurement being in some interval on the real line have been replaced by the continuous analog $x\,dx$ as $dx \to 0$. The sum in the partition function is then replaced by an integral,

$$Z(\lambda_1, \lambda_2, \cdots, \lambda_m) = \int_R \exp\left[\,\lambda_1 F_1(x) + \lambda_2 F_2(x) + \cdots + \lambda_m F_m(x)\,\right] dx$$

Thus, for the specific case of the Gamma distribution that inserts information into the probability distribution with two parameters ($m = 2$),

$$\text{pdf}\,(x\,|\,\text{Gamma}) = \frac{\exp\left[\,\lambda_1 F_1(x) + \lambda_2 F_2(x)\,\right]}{Z}$$

$$Z = \int_0^\infty \exp\left[\,\lambda_1 F_1(x) + \lambda_2 F_2(x)\,\right] dx$$

The mapping takes the statements to real numbers through $F_1(x) = x$ and $F_2(x) = \ln x$. The information in a Gamma distribution can also be specified by the constraint function averages $\langle F_1 \rangle$ and $\langle F_2 \rangle$. Therefore, information about an average x and an average $\ln x$ gets inserted into a Gamma distribution.

With thus much determined, we now have the MEP template looking like,

$$\text{pdf}\,(x\,|\,\text{Gamma}) = \frac{\exp\left[\lambda_1 x + \lambda_2 \ln x\right]}{Z}$$

$$= \frac{\exp\left[\lambda_1 x\right] \exp\left[\lambda_2 \ln x\right]}{Z}$$

$$= \frac{\exp\left[\lambda_1 x\right] x^{\lambda_2}}{Z}$$

Relabel the parameters as $\lambda_1 \equiv -\beta$ and $\lambda_2 \equiv \alpha - 1$ to yield,

$$\text{pdf}\,(x\,|\,\text{Gamma}) = \frac{x^{\alpha-1} \exp\left[-\beta x\right]}{Z}$$

Every MEP assignment must satisfy the universal constraint. Thus,

$$Z = \int_0^\infty x^{\alpha-1} \exp(-\beta x)\, dx$$

$$= \frac{\Gamma(\alpha)}{\beta^\alpha}$$

The final form of the MEP assignment for the Gamma distribution is exactly the same as the exponential family definition,

$$\text{pdf}(x\,|\,\text{Gamma}) = \frac{x^{\alpha-1} \exp(-\beta x)}{Z}$$

$$= \frac{x^{\alpha-1} \exp(-\beta x)}{\frac{\Gamma(\alpha)}{\beta^\alpha}}$$

$$\text{pdf}(x\,|\,\alpha,\beta) = \frac{\beta^\alpha\, x^{\alpha-1}\, e^{-\beta x}}{\Gamma(\alpha)}$$

Having found $\text{pdf}(x\,|\,\text{Gamma})$, we can now calculate the constraint function averages. The universal constraint function average $\langle F_0 \rangle$ is,

$$\int_0^\infty \frac{\beta^\alpha\, x^{\alpha-1}\, e^{-\beta x}}{\Gamma(\alpha)}\, dx = 1$$

The first constraint function average $\langle F_1 \rangle$ is,

$$\int_0^\infty x\, \frac{\beta^\alpha\, x^{\alpha-1}\, e^{-\beta x}}{\Gamma(\alpha)}\, dx = \frac{\alpha}{\beta}$$

The second constraint function average $\langle F_2 \rangle$ is,

$$\int_0^\infty \ln(x)\, \frac{\beta^\alpha\, x^{\alpha-1}\, e^{-\beta x}}{\Gamma(\alpha)}\, dx = \psi(\alpha) - \ln(\beta)$$

where $\psi(\alpha)$ is the polygamma function that also appeared in Volume II during the MEP derivation of the Cauchy distribution.

Exercise 33.9.10: Have *Mathematica* verify the above integrations.

Solution to Exercise 33.9.10

First, let's find the partition function $Z(\alpha, \beta) = \int_0^\infty x^{\alpha-1} e^{-\beta x}\, dx$,

```
Integrate[x^(α - 1) Exp[-β x], {x, 0, ∞},
                        Assumptions → β > 0 && α > 0]
```

returns $\beta^{-\alpha}\, \Gamma(\alpha)$.

SOLVED EXERCISES FOR CHAPTER 33

Next, verify that the universal constraint is satisfied with,

`Integrate[`β`^`α `x^(`$\alpha - 1$`) Exp[-`β `x] / Gamma[`α`], {x, 0, `∞`},`
     `Assumptions` $\rightarrow$ $\beta > 0$ `&&` $\alpha > 0$`]`

returning 1.

Finally, verify that the first constraint function average,

`Integrate[x `β`^`α `x^(`$\alpha - 1$`) Exp[-`β `x] / Gamma[`α`], {x, 0, `∞`},`
     `Assumptions` $\rightarrow$ $\beta > 0$ `&&` $\alpha > 0$`]`

returns α/β, and that the second constraint function average,

`Integrate[Log[x] `β`^`α `x^(`$\alpha - 1$`) Exp[-`β `x] / Gamma[`α`], {x, 0, `∞`},`
     `Assumptions` $\rightarrow$ $\beta > 0$ `&&` $\alpha > 0$`]`

returns `- Log[`β`] + PolyGamma[0, `α`]`.

This last expression is a little less opaque if we wrap `TraditionalForm[ ]` around the result in order to see $\psi^{(0)}(\alpha) - \log(\beta)$ instead.

Of course, *Mathematica* is equipped with the built–in Gamma distribution,

`GammaDistribution[`α, β`]`

that could also be used to check the integrations. For example,

`Integrate[x PDF[GammaDistribution[`α`, 1/`β`], x], {x, 0, `∞`},`
     `Assumptions` $\rightarrow$ $\beta > 0$ `&&` $\alpha > 0$`]`

and,

`Integrate[Log[x] PDF[GammaDistribution[`α`, 1/`β`], x], {x, 0, `∞`},`
     `Assumptions` $\rightarrow$ $\beta > 0$ `&&` $\alpha > 0$`]`

return the same answers. *Mathematica* defines our β parameter as $1/\beta$.

Exercise 33.9.11: Rely on *Mathematica* to confirm the formula for the probability density function of the Gamma distribution as derived by the MEP in Exercise 39.9.9.

Solution to Exercise 33.9.11

This is the expression for the probability density function of the Gamma distribution as it was derived by the MEP,

$$P(X = x_i \mid \mathcal{M}_k) \rightarrow \text{pdf}(x \mid \alpha, \beta) = \frac{\beta^\alpha x^{\alpha-1} e^{-\beta x}}{\Gamma(\alpha)}$$

Evaluating `PDF[GammaDistribution[`α`, 1/`β`], x]` in `FullForm[ ]` returns an answer with many `Power[ ]` expressions that we might prefer to see in another format. `PowerExpand[ ]` is a very useful *Mathematica* function designed to turn expressions containing the function `Power[ ]` into their preferred form. With this in mind, evaluate instead,

`PowerExpand[PDF[GammaDistribution[`α`, 1/`β`], x]]`

which, after rearranging *Mathematica*'s idiosyncratic ordering, returns the desired answer of,

$$\frac{e^{-x\beta}\, x^{-1+\alpha}\, \beta^\alpha}{\text{Gamma}\,[\alpha]} \equiv \frac{\beta^\alpha\, x^{\alpha-1}\, e^{-\beta x}}{\Gamma(\alpha)}$$

Exercise 33.9.12: Derive the Poisson probability distribution through the MEP formalism, and therefore as an exponential family distribution.

Solution to Exercise 33.9.12

Deriving the classic Poisson distribution as an MEP distribution is a very interesting exercise because one cannot help but face up to the many conflicting conceptual approaches that one sees in the literature for this seemingly elementary distribution. I choose to emphasize the differences in my MEP derivation and one that Bernardo & Smith present for the Poisson distribution [6, pg. 199] because it seems to almost everyone that both stem from a common philosophical rationale. But there are traps for the unwary!

To begin, one immediately notices that the Poisson distribution is similar to the Gamma distribution just discussed. Instead of setting up a continuous state space stretching from 0 to ∞ as the Gamma does, the Poisson is restricted to integer values stretching from 0 to ∞.

Using my example from psychological testing once again, perhaps we now record the number of correct answers within a fixed time period instead of reaction times to individual questions. Even though, just like the Gamma distribution, it is absurd to admit that large numbers approaching infinity should be considered, it is easier mathematically to just go ahead and allow this possibility. Giving up on trying to decide on some arbitrary cut–off point on large reaction times, or total number correct, is an acceptable trade–off because we can live with the ridiculously low probabilities for these situations.

Just as before, start with the MEP template for the Poisson distribution with $m=2$ parameters and two constraint functions $F_1(X=x_i)$ and $F_2(X=x_i)$. The dimension of the state space n does go off to ∞, but one of the strengths of the MEP is for exactly this situation where $m <<< n$.

$$P(X=x_i\,|\,\lambda_1,\lambda_2) = \frac{\exp\left[\lambda_1\, F_1(X=x_i) + \lambda_2\, F_2(X=x_i)\right]}{Z}$$

Let $F_1(X=x_i) = x_i$ and $F_2(X=x_i) = \ln(x_i!)$ to arrive at,

$$P(X=x_i\,|\,\lambda_1,\lambda_2) = \frac{\exp\left[\lambda_1\, x_i + \lambda_2\, \ln(x_i!)\right]}{Z}$$

If we let the Lagrange multipliers take the form $\lambda_1 \equiv \ln \lambda$ and $\lambda_2 = -1$,

$$P(X = x_i \,|\, \lambda_1, \lambda_2) = \frac{\exp[\,x_i \ln \lambda - \ln(x_i!)\,]}{Z}$$

$$= \frac{\left(\frac{\lambda^{x_i}}{x_i!}\right)}{Z}$$

But the partition function is the sum over the state space of all the terms that could appear in the numerator, so

$$Z = \sum_{x_i=0}^{\infty} \frac{\lambda^{x_i}}{x_i!}$$

The right hand side is recognized as the definition of e^λ, and, as a consequence, we have arrived at the classic form of the Poisson distribution as derived via the MEP,

$$P(X = x_i \,|\, \text{Poisson}) = e^{-\lambda} \frac{\lambda^{x_i}}{x_i!}$$

Exercise 33.9.13: Have *Mathematica* verify the above sums.

Solution to Exercise 33.9.13

First of all, let's have *Mathematica* verify the expression for any individual i^{th} term appearing in the numerator,

$$e^{x_i \ln \lambda - \ln(x_i!)}$$

by evaluating,

$$\texttt{Exp[x Log[}\lambda\texttt{] - Log[x!]]}$$

The answer returned is indeed,

$$\frac{\lambda^x}{x!}$$

The sum of all these terms from 0 to ∞,

$$\texttt{Sum[}\lambda\texttt{\^{}x / x!, \{x, 0, }\infty\texttt{\}]}$$

evaluates to e^λ.

The average for the universal constraint function,

$$\texttt{Sum[Exp[-}\lambda\texttt{] }\lambda\texttt{\^{}x / x!, \{x, 0, }\infty\texttt{\}]}$$

correctly evaluates to 1, and the average for the first constraint function,

$$\texttt{Sum[x Exp[-}\lambda\texttt{] }\lambda\texttt{\^{}x / x!, \{x, 0, }\infty\texttt{\}]}$$

also correctly evaluates to λ.

There doesn't seem to be any simple closed form expression for the average of the second constraint function when $x \to \infty$, but a sum over a relatively large x at a given value for λ will yield an answer. For example, at $\langle F_1 \rangle \equiv \lambda = 2$, $\langle F_2 \rangle = 1.09118$ through,

```
Sum[Log[x!] Exp[-λ] λ^x / x!, {x, 0, 10000}] /. λ → 2 // N
```

As with the Gamma distribution, these sums could also be performed with the *Mathematica* built–in Poisson distribution. For example,

```
Sum[x PDF[PoissonDistribution[λ], x], {x, 0, Infinity}]
```

which also evaluates to λ and,

```
Sum[Log[x!] PDF[PoissonDistribution[λ], x], {x, 0, Infinity}]
                      /. λ → 2 // N
```

returns 1.09118.

Exercise 33.9.14: How do Bernardo & Smith match up the Poisson and Gamma distributions with the various terms in the exponential family of distributions?

Solution to Exercise 33.9.14

They write the one parameter exponential family as [6, pg. 198],

$$p(x \mid \theta) = f(x)\, g(\theta)\, \exp\{c\, \phi(\theta)\, h(x)\}$$

The Poisson distribution,

$$\text{Po}(x \mid \theta) = \frac{\theta^x e^{-\theta}}{x!}$$

is then translated by them as a typical member of the exponential family through the choices that $f(x) = (x!)^{(-1)}$, $g(\theta) = e^{-\theta}$, $h(x) = x$, $\phi(\theta) = \log \theta$, and $c = 1$.

These choices seem to me just as *ad hoc* as might be leveled against the MEP choices as to what information to insert into a distribution. How is the choice of $f(x) = 1/x!$ justified? Is $f(x)$ is supposed to be some "approximation" to the "true" distribution $p(x \mid \theta)$? Is it supposed to be some kind of "prior distribution?" Is it some kind of "measure?" The choices for $f(x)$, $g(\theta)$, $h(x)$, $\phi(\theta)$, and c are adjusted to whatever will mesh nicely into the generic exponential family template.

In my derivation strictly following the MEP template, there was no need for any $f(x)$. Instead of leaning upon some "approximation," or "prior," or "measure function" of dubious origin, another constraint function $F_2(X = x_i) = \ln x_i!$ and another Lagrange multiplier $\lambda_2 = -1$ were called upon. From the MEP view of things, it is clear that $g(\theta) \equiv 1/Z(\lambda_1, \cdots, \lambda_m)$. No mystery here as to how the terms in the exponential family arise when viewed through the lens of the MEP.

SOLVED EXERCISES FOR CHAPTER 33

They write the Gamma distribution with α already specified as [6, pp. 208–209],

$$\mathrm{Ga}(x \mid \alpha, \theta) = (x^{\alpha-1}/\Gamma(\alpha))\, \theta^\alpha \exp\{-\theta x\}$$

with $h(x) = x$, $\phi(\theta) = \theta$, $c = -1$, $g(\theta) = \theta^\alpha$, and $f(x) = x^{\alpha-1}/\Gamma(\alpha)$.

Compare this with the MEP Gamma distribution assignment of,

$$\mathrm{pdf}\,(x \mid \alpha, \beta) = \frac{\beta^\alpha\, x^{\alpha-1}\, e^{-\beta x}}{\Gamma(\alpha)}$$

The appearance of the $f(x)$ term in the exponential family definition of the Gamma distribution raises the same set of questions as for the Poisson distribution. The MEP derivation does not depend on any $f(x)$, and yet it still arrives at the correct expression. Moreover, if their one function $h(x)$ had been correctly identified as two constraint functions $F_1(x) = x$ and $F_2(x) = \ln x$, then their $g(\theta)$ as the inverse of the partition function would have been correctly found as $\theta^\alpha/\Gamma(\alpha)$. The ambiguous, mysterious $f(x)$ term in the exponential family definition has not been necessary to derive these two distributions.

Exercise 33.9.15: Find MacKay's "very silly answer" that MEP provides for the assignment of probabilities to the six faces of a die under given information about the mean.

Solution to Exercise 33.9.15

In MacKay's critical comments about the MEP, he mentions in passing that the MEP would provide a "very silly answer" to a probability assignment for the six faces of a biased die when the given information is that the mean of these six faces is not the 3.5 we would expect for a fair die, but rather 2.5. Let's examine in detail what assignment the MEP makes in this case. I leave to your judgment whether the resulting assignment meets MacKay's criterion of being a "very silly answer."

The state space for this conventional die scenario has dimension $n = 6$. However, only models with $m = 1$ constraint function averages will be considered. The mapping from the statements in the state space $(X = x_i)$ to real numbers is,

$$F(X = x_i) = (1, 2, 3, 4, 5, 6)$$

The IP is interested in the numerical assignment to the probabilities for the ONE face through the SIX face, $P(X = x_i \mid \mathcal{M}_k)$, under the information in a model that specifies that the constraint function average, $\langle F \rangle$, is equal to 2.5.

The MEP is exquisitely designed to handle this assignment problem. The MEP template for the assignment with just one piece of information is the familiar,

$$P(X = x_i \mid \mathcal{M}_k) = \frac{\exp\left[\,\lambda\, F(X = x_i)\,\right]}{\sum_{i=1}^{6} \exp\left[\,\lambda\, F(X = x_i)\,\right]}$$

Use the Legendre transformation as the basis for the MEP algorithm that will find the six numerical assignments to the die faces. It will maximize the resulting information entropy as well as possessing the mathematical expectation that,

$$\sum_{i=1}^{6} F(X = x_i)\, P(X = x_i \,|\, \mathcal{M}_k) = 2.5$$

Minimize the expression below by varying λ while keeping $\langle F \rangle$ fixed at 2.5,

$$H_{\max}(p) = \min_{\lambda} \left[\, \ln Z - \lambda \langle F \rangle \,\right]$$

At a value of $\lambda = -0.371049$, the information entropy $H_{\max}(p)$ reaches a maximum value of 1.61358. The resulting probability assignment to the six faces of the biased die is then,

$$Q_1 = 0.347494$$
$$Q_2 = 0.239774$$
$$Q_3 = 0.165447$$
$$Q_4 = 0.114160$$
$$Q_5 = 0.078772$$
$$Q_6 = 0.054353$$

For example, the assigned probability to the THREE face of the biased die is,

$$Q_3 \equiv P(\text{THREE} \,|\, \mathcal{M}_k)$$
$$= \frac{\exp[\,3\lambda\,]}{\exp[\,\lambda\,] + \exp[\,2\lambda\,] + \cdots + \exp[\,6\lambda\,]}$$
$$= \frac{\exp[\,3(-0.371049)\,]}{1.98567}$$
$$= 0.165447$$

The mathematical expectation of this probability assignment satisfies the one given piece of information,

$$(1 \times 0.347494) + (2 \times 0.239774) + \cdots + (6 \times 0.054353) = 2.5 = \langle F \rangle$$

and, in addition, possess no additional information that was not specified under the model. Any assignment other than this MEP assignment would either not satisfy the one piece of information, or would contain additional information not specified by the model, or both.

SOLVED EXERCISES FOR CHAPTER 33

It is an absolutely essential requirement that this MEP assignment is made *prior* to any observations on the die, or any frequency counts made by rolling the die. In addition, all possible legitimate probability assignments take place implicitly through the MEP.

It is only after the fact, that is, *after* the die has been physically examined, or *after* the die has been rolled some number of times, that contemplating $P(\mathcal{M}_k \mid \mathcal{D})$ will lead the IP to more closely inspect some restricted number of MEP inspired models for a causal explanation. *If* this particular MEP assignment just discussed happened to be supported by physical examination of the die, or by recording frequency counts, then, as Jaynes masterfully explained, there might be a physical reason like misplaced center of gravity for thinking that this MEP assignment was not so silly after all.

If this particular MEP assignment were NOT supported by physical examination of the die, or by the observed frequency counts from rolling the die a number of times, then it would be automatically rejected through $P(\mathcal{M}_k \mid \mathcal{D})$. The IP does NOT have to make any preliminary judgment as to whether that MEP assignment is a "very silly answer." Maybe it is or maybe it isn't. The rules of inferencing will tell us the answer to that question, not MacKay's *a priori* opinion of the efficacy of any MEP assignment.

Exercise 33.9.16: Show the correspondence, and the far more important lack of correspondence, between my explanation of maximum likelihood within the context of the MEP and MacKay's version.

Solution to Exercise 33.9.16

In his book [29, pg. 310], MacKay presents a solution to what he calls a "Maximum likelihood fitting of an exponential–family model." He begins with:

> Assume that a variable **x** comes from a probability distribution of the form
>
> $$P(\mathbf{x} \mid \mathbf{w}) = \frac{1}{Z(\mathbf{w})} \exp\left(\sum_k w_k f_k(\mathbf{x})\right) \qquad (22.31)$$
>
> where the functions $f_k(\mathbf{x})$ are given, and the parameters $\mathbf{w} = \{w_k\}$ are not known. A data set $\{\mathbf{x}^{(n)}\}$ of N points is supplied.

From the very beginning, MacKay uses the kind of language that is endemic throughout the literature. It encapsulates the fallacious concept that Jaynes has labeled as the *Mind Projection Fallacy*.

He railed against such entrenched language time and time again because it clearly indicates that the user of such phrases as, "Assume that a variable **x** comes from a probability distribution," believes that physical manifestations originate from *information about that object as possessed by a conscious entity*!

Once one grasps the simple conceptual fact that probability distributions are epistemological constructs and not ontological constructs, thinking that things in the real world arise from an IP's state of knowledge about such things is ludicrous. As expected, Jaynes's protests fell on deaf ears.

So we choose not to employ such language, but prefer to express things thusly:

A probability distribution captures an IP's degree of belief about whether some statement concerning the real world is true. Any such degree of belief is changeable, and it changes as the information in a model changes.

MacKay's variable **x** that comes from a probability distribution with unknown parameters **w** translates for us into a degree of belief in the truth of a statement $(X = x_i)$ conditioned on the information in some model,

$$P(\mathbf{x} \,|\, \mathbf{w}) \equiv P(X = x_i \,|\, \mathcal{M}_k)$$

On the right hand side of his Equation (22.31), MacKay is guilty of his own personal *deus ex machina*. This probability distribution,

$$P(\mathbf{x} \,|\, \mathbf{w}) = \frac{1}{Z(\mathbf{w})} \, \exp\left(\sum_k w_k \, f_k(\mathbf{x})\right) \qquad (22.31)$$

appears out of nowhere with no justification proffered. It is merely asserted to be an exponential family model.

We always, and in every case, invoke the MEP distribution for any numerical assignment to the probabilities in the state space because the MEP inserts the requisite information within some model, and no other information from some other model, into the probability distribution.

$$P(X = x_i \,|\, \mathcal{M}_k) = \frac{\exp\left[\sum_{j=1}^{m} \lambda_j \, F_j(X = x_i)\right]}{Z(\lambda_1, \lambda_2, \cdots, \lambda_m)}$$

With these considerations in mind, we can proceed to MacKay's solution, while taking care of the difference in notation, in order to see that the mathematical steps in our two approaches can not really differ that much.

The *likelihood* $\mathcal{L}$ of the data is,

$$\mathcal{L} \equiv P(\mathcal{D} \,|\, \mathcal{M}_k) = \prod_{t=1}^{N} P(X_t = x_i \,|\, \mathcal{M}_k)$$

SOLVED EXERCISES FOR CHAPTER 33

while the *log likelihood* $\ln \mathcal{L}$ becomes,

$$P(\mathcal{D} \mid \mathcal{M}_k) = W(N)\, Q_1^{N_1}\, Q_2^{N_2} \cdots Q_n^{N_n}$$

$$\ln P(\mathcal{D} \mid \mathcal{M}_k) = \ln W(N) + \sum_{i=1}^{n} N_i \ln Q_i$$

$$= \sum_{i=1}^{n} N_i \left(\sum_{j=1}^{m} \lambda_j\, F_j(X = x_i) - \ln Z \right)$$

$$= \sum_{i=1}^{n} N_i \sum_{j=1}^{m} \lambda_j\, F_j(X = x_i) - \sum_{i=1}^{n} N_i \ln Z$$

$$= \sum_{i=1}^{n} N_i \sum_{j=1}^{m} \lambda_j\, F_j(X = x_i) - N \ln Z$$

The so-called "maximum likelihood estimates" occur when the partial derivative of the log likelihood with respect to each of the m parameters in the model is set equal to 0,

$$\frac{\partial \ln \mathcal{L}}{\partial \lambda_j} \equiv \frac{\partial \ln P(\mathcal{D} \mid \mathcal{M}_k)}{\partial \lambda_j} = 0 \qquad \text{for } j = 1, 2, \cdots, m$$

Pressing forward with each differentiation yields,

$$\frac{\partial \ln \mathcal{L}}{\partial \lambda_j} = \frac{\partial \left[\sum_{i=1}^{n} N_i \sum_{j=1}^{m} \lambda_j\, F_j(X = x_i) - N \ln Z \right]}{\partial \lambda_j}$$

$$= \sum_{i=1}^{n} N_i\, F_j(X = x_i) - N \frac{\partial \ln Z}{\partial \lambda_j}$$

$$= \sum_{i=1}^{n} N_i\, F_j(X = x_i) - N \langle F_j \rangle$$

Now, divide through by N to form the *sample* averages $\overline{F}_j(X = x_i)$ from the data and the constraint function averages,

$$\frac{\sum_{i=1}^{n} N_i F_j(X = x_i) - N \langle F_j \rangle}{N} = \sum_{i=1}^{n} \frac{N_i}{N} F_j(X = x_i) - \langle F_j \rangle$$

$$= \overline{F}_j(X = x_i) - \langle F_j \rangle$$

Finally, equate each of the above m equations to 0 in order to see that finding "the maximum likelihood estimates" amounts to the same thing as setting the information in the constraint function averages under one particular model equal to the sample averages obtained from the data,

$$\overline{F}_j(X = x_i) - \langle F_j \rangle = 0 \quad \text{for } j = 1, 2, \cdots, m$$

$$\langle F_j \rangle = \overline{F}_j$$

$$\sum_{i=1}^{n} F_j(X = x_i)\, P(X = x_i \,|\, \mathcal{M}_{ML}) = \overline{F}_j \quad \text{for } j = 1, 2, \cdots, m$$

For MacKay, this all plays out as,

$$\ln P(\{\mathbf{x}^{(n)}\} \,|\, \mathbf{w}) = -N \ln Z(\mathbf{w}) + \sum_n \sum_k w_k f_k(\mathbf{x}^{(n)})$$

leading to,

$$\sum_{\mathbf{x}} P(\mathbf{x} \,|\, \mathbf{w}_{ML})\, f_k(\mathbf{x}) = \frac{1}{N} \sum_n f_k(x^{(n)})$$

While there is no disagreement between us at this level, there does exist a profound gap in how we interpret things. MacKay reaches the startling conclusion that,

> And hence the maximum entropy method gives identical results to maximum likelihood fitting of an exponential family model ...

This would lead one to immediately surmise that the data necessary in finding the sample averages are intimately involved in the procedures that the MEP uses to assign numerical values to probabilities. It suggests the completely erroneous impression that the MEP is a data based procedure thoroughly equivalent to making maximum likelihood estimates. Nothing could be further from the truth!

It was to counter such interpretations as the commonly accepted wisdom that I repeatedly and somewhat vociferously emphasized throughout Volume II that as far as the MEP is concerned,

DATA IS NOT INFORMATION

and,

INFORMATION IS NOT DATA

What is true is that the probability of the model which, after the fact, proves to exhibit the same constraint function average as the sample averages might be labeled as "the maximum likelihood model." It must have a greater posterior probability than any other model, that is,

$$P(\mathcal{M}_{ML} \,|\, \mathcal{D}) > P(\mathcal{M}_\star \,|\, \mathcal{D}) \text{ when } \langle F_j \rangle_{ML} = \sum_{i=1}^{n} \frac{N_i}{N} F_j(X = x_i)$$

When the MEP assignments are initially and implicitly constructed to cover all conceivable legitimate probability assignments, that is to say, before any data whatsoever has been observed, the sample averages obviously do not even exist yet! After data have been observed, some implicit MEP assignment with its particular information embedded within the $\langle F_j \rangle$ will inevitably match the $\overline{F}_j$. Then, *after the fact*, we may choose to call this model a maximum likelihood model.

This is arguing in the forward direction from the prior existence of the MEP assignment to the existence of the maximum likelihood model. But to argue in the reverse direction, as MacKay does, that a maximum likelihood model defines *the one true* MEP assignment is fallacious.

My final question: How did the Q_i in $\ln \mathcal{L}$ ever get there in the first place? Not through any "maximum likelihood fitting of an exponential family model."

There comes a point where piling more words on in an attempt to convince someone becomes a futile gesture. That is why the old maxim,

You can lead a horse to water, but you can't make it drink!

was invented.

Exercise 33.9.17: Examine another common situation, here specifically involving the Gamma distribution, where the conceptual gap between MacKay and me is enormous.

Solution to Exercise 33.9.17

MacKay poses the following exercise [29, pg. 318]. On the face of it, it seems quite innocuous. Moreover, you will find myriad variations on this same problem throughout all of statistics and probability. After putting down my general objections, I will show the way I would approach the problem solution.

> N data points $\{x_n\}$ are drawn from a Gamma distribution $P(x \mid s, c) = \Gamma(x; s, c)$ with unknown parameters s and c. What are the maximum likelihood parameters s and c?

There are so very few words in the statement of the problem; how could anything go so far wrong in such a limited time? But the language that is used reveals enormous conceptual differences between the conventional approach, even for a confirmed Bayesian like MacKay, and my approach.

Data are **NEVER** drawn from any probability distribution! This is Jaynes's *Mind Projection Fallacy* in its purest form. Observations about the real world **NEVER** arise from anyone's *degree of belief* in the truth of the observation. It's the old belabored distinction between epistemology and ontology.

The parameters involved in some model are **ALWAYS** known! That is why they always appear to the right of the conditioned upon symbol in expressions like $P(X = x_i \,|\, \mathcal{M}_k)$. The IP could never make a numerical assignment conditioned on the parameters of a model if they weren't known.

It is **NOT** maximum likelihood *parameters* that we seek. It is rather that one model possessing the largest posterior probability when compared to any other conceivable model.[1] And even after we find this model with the greatest probability after the data have been observed, we still have to average over *all* of the models. We do not base our prediction of future events on just one model.

Alternative language must be used. N data points are N determinations that some statement in the state space was true. The original state space for the Gamma distribution is the real line extending from 0 to ∞. But this original infinite state space must be redefined into discrete categories. It helps to think about these statements as intervals on the real line. An integration over the probability density function of the Gamma distribution will be used as a model to assign numerical values to these intervals.

In addition, the fact that the IP strictly relies only on the class of Gamma distributions to make these assignments has the consequence that a vast number of conceivable models are being excluded. Roughly speaking, only smooth, unimodal models are being used. It is the same restrictive situation when the class of Gaussian models is the only source for numerical assignments to intervals over the real line from $-\infty$ to $+\infty$.

The Gamma distribution is called a member of the exponential family, but it is also an MEP distribution. This means that the limited information from models with only two parameters is being incorporated into the probability density function to make assignments. Either two Lagrange multipliers, λ_1 and λ_2, or two constraint function averages, $\langle F_1 \rangle$ and $\langle F_2 \rangle$, can be utilized to insert information into the Gamma probability density function. They are the parameters for a Gamma model. One of these two sets of dual coordinates must be specified, that is, must be known, in order to insert information into the pdf under the model $\mathcal{M}_k$.

Now, that my objections are out front and center, it is time to see how I would approach this problem from the combined MEP/Bayesian approach. Suppose we are conducting a study on cockpit resource management for airline pilots. One of the goals is to obtain some figure of merit for how pilots handle minor emergencies that should typically be rectified in a matter of several seconds.

For a numerical example, image that the experiment records the pilot's reaction time in seconds, x_t, to each of twenty five minor cockpit emergencies. These reaction times serve as $N = 25$ data points. Suppose the actual data look something like this,
$$\mathcal{D} = \underbrace{(5.92, 5.19, 13.26, \cdots, 3.40)}_{25 \text{ data points}}$$

---

[1] It takes some degree of self–control to avoid indulging in some naughty puns involving current celebrities when Bayesians are forced to use language like this.

with a sample mean reaction time equal to,

$$\overline{F}_1 = \frac{1}{N} \sum_{t=1}^{25} x_t = 10.2045$$

and a sample mean log reaction time of,

$$\overline{F}_2 = \frac{1}{N} \sum_{t=1}^{25} \ln x_t = 2.1608$$

Since the first constraint function average for the Gamma distribution is,

$$\langle F_1 \rangle = \frac{\alpha}{\beta}$$

and the second constraint function average for the Gamma distribution is,

$$\langle F_2 \rangle = \psi(\alpha) - \ln(\beta)$$

when the sample mean reaction time of $\overline{F}_1 = 10.2045$ is substituted for $\langle F_1 \rangle$, and the sample mean log of reaction times of $\overline{F}_2 = 2.1608$ is substituted for $\langle F_2 \rangle$, the maximum likelihood model has parameters equal to,

$$\mathcal{M}_{\text{ML}} \equiv (\alpha = 3.24302, \beta = 0.317803)$$

It is important for me to emphasize here the obvious fact that these are *known* parameters, whether we imagined them as legitimate values for α and β before the data, or as we have done here, used the sample averages to find the model best supported by the data. In either case, the model parameters are inserting desired information, and it is irrelevant whether they arose from pre–data or post–data considerations. In either case, they most certainly do represent a legitimate MEP assignment.

Earlier, I mentioned that we were going to transform the infinite state space into one consisting of discrete categories. Let's divide the interval over the real line from 0 to $+\infty$ in the following manner. There will be seven discrete categories so that the new state space will have dimension $n = 7$. The first category is all reaction times greater than 0 seconds and less than or equal to 5 seconds, the second category all reaction times greater than 5 seconds and less than or equal to 10 seconds, and so on, up to the final category for all reaction times greater than 30 seconds.

The probability for any one of these categories is an integration of the probability density function of the Gamma distribution over the appropriate reaction time interval. For example, the probability of a reaction time between five and ten seconds under some Gamma distribution model would be,

$$P(5 \leq x_i < 10 \,|\, \text{Gamma}) = \int_5^{10} \frac{\beta^\alpha \, x^{\alpha-1} \, e^{-\beta x}}{\Gamma(\alpha)} \, dx$$

The maximum likelihood model finds this probability to be,

$$P(5 < x_i \leq 10 \,|\, \text{Gamma}) = \int_5^{10} \frac{(0.317803)^{3.24302} \, x^{2.24302} \, e^{-0.317803x}}{\Gamma(3.24302)} \, dx = 0.3895$$

The probabilities for all seven intervals, when doing the integrations just as above and adjusting the time intervals as needed, do add up to 1.

But this is not the important calculation from the Bayesian perspective. The most important calculation is the degree of belief in the truth of the ith reaction time interval for any *next* cockpit emergency not already covered in the data base. We recognize this as involving the average of the posterior probability over all models, expressed generically as,

$$P(X_{N+1} = x_i \,|\, \mathcal{D}) = \int \cdot \int P(X_{N+1} = x_i \,|\, \mathcal{M}_k) \, P(\mathcal{M}_k \,|\, \mathcal{D}) \, dq_i$$

We found the beautifully simple formula that answers this as,

$$P(X_{N+1} = x_i \,|\, \mathcal{D}) = \frac{N_i + 1}{N + n}$$

Examining the data base, we find that there were $N_2 = 12$ frequency counts in the range from 5 to 10 seconds. Compare this final correct answer that we are looking for with the maximum likelihood answer above,

$$P(X_{N+1} = x_2 \,|\, \mathcal{D}) = \frac{12 + 1}{25 + 7} = 0.4063 \text{ versus } 0.3895$$

The probability is about 41% that this pilot will react to some new minor cockpit emergency somewhere between 5 to 10 seconds.

Table 33.1 at the top of the next page lists the comparison between the Bayesian answer as the average over all models and the answer from the one maximum likelihood model for all seven discrete reaction time intervals.

Here is the point. The Bayesian calculation of,

$$P(X_{N+1} = x_i \,|\, \mathcal{D})$$

is an average over every conceivable numerical assignment *whether it came from any Gamma distribution or not*.

It's not that the maximum likelihood model based on the Gamma distribution is all that bad. As a matter of fact, after making our "black–box" final correct answer for a probability of some reaction time interval, we might want to revisit that particular MEP model of a Gamma distribution with parameters,

$$\mathcal{M}_{\text{ML}} \equiv (\alpha = 3.24302, \beta = 0.317803)$$

Now, the hard part becomes creating some psychological theory as to why the times to react to cockpit emergencies should be reflected by a Gamma distribution with these particular parameters.

Table 33.1: *A side by side comparison of the probability for a reaction time interval based on the correct Bayesian approach of averaging over the posterior probability of all models* versus *the maximum likelihood model based on the Gamma distribution.*

i	Interval	Probability of interval Maximum likelihood model	Probability of interval Bayesian
1	$0 < x_i \leq 5$	0.1702	$5/32 = 0.1563$
2	$5 < x_i \leq 10$	0.3895	$13/32 = 0.4063$
3	$10 < x_i \leq 15$	0.2619	$7/32 = 0.2187$
4	$15 < x_i \leq 20$	0.1166	$2/32 = 0.0625$
5	$20 < x_i \leq 25$	0.0424	$2/32 = 0.0625$
6	$25 < x_i \leq 30$	0.0137	$1/32 = 0.0312$
7	$x_i > 30$	0.0057	$2/32 = 0.0625$
	Sums	1.0000	1.0000

Exercise 33.9.18: Show some of the *Mathematica* code used to make the calculations in the previous exercise.

Solution to Exercise 33.9.18

In order to avoid collecting data from any actual experiment, the 25 data points were generated by,

```
data = RandomVariate[GammaDistribution[5, 2], 25]
```

Why am I being inconsistent when I *simulate* data in this fashion?

The two sample averages were found with,

```
Mean[data] and Total[Log[data]] / 25
```

The parameters for the maximum likelihood model were calculated with,

```
FindRoot[{α/β = = 10.2045, PolyGamma[0, α] - Log[β] = = 2.1608},
   {{α, 5}, {β, 1/2}}]
```

The values of 5 and 1/2 are the values that the user may specify for the *Mathematica* built–in function `FindRoot[ ]` to start its search before it zeroes in on the actual values of the maximum likelihood model of,

$$\mathcal{M}_{\text{ML}} \equiv (\alpha = 3.24302, \beta = 0.317803)$$

In order to calculate the probability for any of the seven reaction time intervals based on the maximum likelihood model, a numerical integration using the *Mathematica* built–in operation **NIntegrate[**f**, {**x, x_{min}, x_{max}**}]** is performed on the Gamma distribution. For example, the probability for the third discrete category,

$$P(10 < x_3 \leq 15 \,|\, \mathcal{M}_{\text{ML}}) = 0.2619$$

is calculated with,

NIntegrate[PDF[GammaDistribution[3.24302, 1 / 0.317803], x],
$\qquad\qquad\qquad\qquad\qquad\qquad\qquad\qquad\qquad$**{x, 10, 15}]**

Exercise 33.9.19: Show the equivalent expressions between my chosen MEP notation and Jaynes's notation. Illustrate the differentiation of the entropy function with respect to the constraint function averages using *Mathematica*.

Solution to Exercise 33.9.19

The equivalence between Jaynes's notation and mine for the information entropy when its arguments are the constraint function averages is given by,

$$S(F_1, \ldots, F_k, \ldots, F_m) \equiv H(\langle F_1 \rangle, \ldots, \langle F_j \rangle, \ldots, \langle F_m \rangle)$$

and the equivalence between the partial differentiations with respect to these same constraint function averages is,

$$\frac{\partial S(F_1, \ldots, F_m)}{\partial F_k} \equiv \frac{\partial H(\langle F_1 \rangle, \ldots, \langle F_m \rangle)}{\partial \langle F_j \rangle}$$

The equivalent expressions for the Legendre transformation are,

$$\left\{ S(F_1, \ldots, F_m) = \log Z(\lambda_1, \ldots, \lambda_m) + \sum_{k=1}^{m} \lambda_k F_k \right\} \equiv \left\{ H_{\max}(Q_i) = \ln Z - \sum_{j=1}^{m} \lambda_j \langle F_j \rangle \right\}$$

Keeping the expressions as generic as possible, let the list of three Lagrange multipliers be $\boldsymbol{\lambda}$ = **List[a, b, c]**, and similarly the list of three constraint function averages be **F = List[f, g, h]**, to form the information entropy as,

H = Log[z] - Sum[$\boldsymbol{\lambda}$**[[i]] F[[i]], {i, 1, 3}]**

or alternatively as,

H = Log[z] - Dot[$\boldsymbol{\lambda}$**, F]**

Then the partial differentiation of the information entropy leads to the result,

```
D[H, F[[1]]]  →  - a
D[H, F[[2]]]  →  - b
D[H, F[[3]]]  →  - c
```

This confirms Jaynes's result, when expressed in our notation, that the change in information entropy with respect to a change in the jth constraint function average is reflected by the jth Lagrange multiplier,

$$\frac{\partial H}{\partial \langle F_j \rangle} = -\lambda_j$$

Remember that my Lagrange multipliers are the negative of Jaynes's.

Begin with my version of the Legendre transformation formula,

$$H_{\max}(Q_i) = \ln Z - \sum_{j=1}^{m} \lambda_j \langle F_j \rangle$$

Carry out the partial differentiation of the maximum entropy attainable by this model, $H_{\max}(Q_i)$, as changes are made to the constraint function averages $\langle F_j \rangle$. Symbolically then, this leads us to consider the expression $\frac{\partial H}{\partial \langle F_j \rangle}$. Jaynes carries out these steps,

$$\frac{\partial H}{\partial \langle F_j \rangle} = \sum_{j=1}^{m} \frac{\partial \ln Z}{\partial \lambda_j} \frac{\partial \lambda_j}{\partial \langle F_j \rangle} - \sum_{j=1}^{m} \frac{\partial \lambda_j}{\partial \langle F_j \rangle} \langle F_j \rangle - \lambda_j$$

$$\frac{\partial \ln Z}{\partial \lambda_j} = \langle F_j \rangle$$

$$= \left[\sum_{j=1}^{m} \langle F_j \rangle \frac{\partial \lambda_j}{\partial \langle F_j \rangle} \right] - \left[\sum_{j=1}^{m} \frac{\partial \lambda_j}{\partial \langle F_j \rangle} \langle F_j \rangle \right] - \lambda_j$$

$$\frac{\partial H}{\partial \langle F_j \rangle} = -\lambda_j$$

Now, on to Jaynes's "reciprocity laws." He starts with the already well–established result for the constraint function average,

$$\langle F_c \rangle = \frac{\partial \ln Z}{\partial \lambda_c}$$

Then,

$$\frac{\partial \langle F_c \rangle}{\partial \lambda_r} = \frac{\partial^2 \ln Z}{\partial \lambda_r \, \partial \lambda_c}$$

which he labels as the matrix **B**.

B is the same as the metric tensor which, in our notation, is $g_{rc}(p)$, and becomes in Amari's IG notation,

$$\frac{\partial \eta_i}{\partial \theta^j} = g_{ij}$$

Since the metric tensor **B** is symmetric,
$$\frac{\partial \langle F_c \rangle}{\partial \lambda_r} = \frac{\partial \langle F_r \rangle}{\partial \lambda_c}$$

Jaynes had just proved that, (in our preferred notation above),
$$\frac{\partial H}{\partial \langle F_c \rangle} = -\lambda_c$$

Differentiating this expression with respect to a constraint function average $\langle F_r \rangle$ yields,
$$\frac{\partial \lambda_c}{\partial \langle F_r \rangle} = \left(\frac{\partial}{\partial \langle F_r \rangle}\right)\left(-\frac{\partial H}{\partial \langle F_c \rangle}\right) = -\frac{\partial^2 H}{\partial \langle F_r \rangle \partial \langle F_c \rangle}$$

which he labels as the matrix **A**.

He then asserts that **A** is the inverse of **B**, and **B** is the inverse of **A** so that,
$$\mathbf{AB} = \mathbf{AA^{-1}} = \mathbf{B^{-1}B} = \mathbf{I}$$

In the more cumbersome tensor notation, this becomes,
$$g_{ij}\, g^{ij} = \delta_i^j$$

of Equation (33.15).

Thus,
$$-\frac{\partial \lambda_c}{\partial \langle F_r \rangle} = -\frac{\partial^2 H}{\langle F_r \rangle \langle F_c \rangle} = g^{rc}(p)$$

For example, the specific component in the inverse of the metric tensor for the simple kangaroo scenario where $r = 2$ and $c = 3$ has the value of,
$$-\frac{\partial \lambda_3}{\partial \langle F_2 \rangle} = g^{23}(p) = -25$$

We will examine in later exercises how these components in the inverse of the metric tensor indicate how much, and in what direction, any Lagrange multiplier is expected to change in response to changes in any constraint function average.

Exercise 33.9.20: Confirm the value found above for one component of the inverse metric tensor with the same numerical differentiation formula from Exercise 33.9.8.

Solution to Exercise 33.9.20

Utilize the same generic template for the numerical approximation of the second order partial differentiation of the information entropy with respect to the constraint function averages,

SOLVED EXERCISES FOR CHAPTER 33

$$\frac{\partial^2 f(\mathbf{x})}{\partial x_i \, \partial x_j} \approx -\frac{\partial^2 H}{\langle F_r \rangle \langle F_c \rangle}$$

$$= \frac{f(\mathbf{x} + \delta e_i + \delta e_j) - f(\mathbf{x} + \delta e_i) - f(\mathbf{x} + \delta e_j) + f(\mathbf{x})}{\delta^2}$$

Now, insert the specific changes needed for this problem. Let,

$$f(\mathbf{x}) \equiv -H(\langle F_1 \rangle, \langle F_2 \rangle, \langle F_3 \rangle)$$

$$x_i \equiv \langle F_2 \rangle$$

$$x_j \equiv \langle F_3 \rangle$$

$$e_i = (0, 1, 0)$$

$$e_j = (0, 0, 1)$$

Let $\delta = 0.001$ as before. The term by term calculation of the varying information entropies with changing constraint function averages is then,

$$f(\mathbf{x} + \delta e_i + \delta e_j) = -H(0.75, 0.751, 0.701)$$

$$= -0.8684834$$

$$f(\mathbf{x} + \delta e_i) = -H(0.75, 0.751, 0.70)$$

$$= -0.872507$$

$$f(\mathbf{x} + \delta e_j) = -H(0.75, 0.75, 0.701)$$

$$= -0.8670846$$

$$f(\mathbf{x}) = -H(0.75, 0.75, 0.70)$$

$$= -0.8711333$$

The final calculation yields,

$$\frac{\partial^2 f(\mathbf{x})}{\partial x_i \, \partial x_j} = \frac{(-0.8684834) - (-0.872507) - (-0.8670846) + (-0.8711333)}{0.000001}$$

$$= -25$$

This numerical differentiation routine confirms, through a different approach, that the value of the component in the second row and third column of the inverse metric tensor is $g^{23}(p) = -25$. The use of the generic p as the argument to the metric tensor indicates that this is a point p living in a Riemannian manifold as well as the probability distribution p for a particular correlational model.

Chapter 34

Introductory Concepts in Information Geometry

34.1 Introduction

In this Chapter, we practice speaking in the strange and, at times, uncomfortable vocabulary of this new language that has been deemed appropriate for Information Geometry. The technical apparatus is slowly enlarged, primarily with the objective of simply getting down on paper some relevant formulas. Don't expect any kind of deep understanding at this juncture. Rather, the practice is more like a child parroting back phrases it overhears from adults as its facility for language evolves.

34.2 A Manifold with Coordinate Systems

Let $\mathcal{S}$ be a manifold consisting of any number of points labeled as $p, q, r, \cdots$, and so on. In other words, the manifold is taken to be the set $\mathcal{S} = \{p, q, r, \cdots\}$. A manifold is characterized not only by the set $\mathcal{S}$, but also by a coordinate system arbitrarily labeled perhaps by x, y, z, or in the case of IG by $\theta^0, \theta^1, \cdots, \theta^{n-1}$.

A manifold has a dimension indicated by a positive integer n and written as $\mathcal{S}^n$. Suppose that the dimension of $\mathcal{S}$ is $n = 3$, or $\mathcal{S}^3$. We might specify that a point p has these three coordinates ($x = 1, y = 2, z = 3$). For the special concerns of IG, a point p is a probability distribution with coordinates labeled instead by (θ^1, θ^2) with θ^0 implicit. The coordinate system is a function taking a point in the manifold to a vector with n components.

There are other allowable coordinate systems that could be associated with $\mathcal{S}^3$. In some alternative coordinate system, the point p might now have coordinates represented by ($\eta_1 = 3.5, \eta_2 = 0.0393$). Another nearby point q might have the coordinates ($\eta_1 = 3.51, \eta_2 = 0.0394$), and so on.

Probability distributions are special in that the universal constraint is always present. Therefore, probability distributions, as points in the manifold, can always be represented by $(n-1)$ coordinates. So we write (θ^1, θ^2), or (η_1, η_2) as coordinates for an $\mathcal{S}^3$ with a θ^0 or η_0 implicit. For probability distributions, the implicit θ^0 is a function of $\ln Z$ and $\eta_0 = 1$.

There is nothing mysterious about these coordinate assignments; but they must be thought of from the typically abstract mathematical perspective. That is, they are defined as a mapping from the manifold to the real numbers, or, in other words, a mapping from some point in $\mathcal{S}$ to a vector with n components.

Technically, the mapping is expressed in the standard functional notation as,

Definition 1 $\varphi \colon \mathcal{S} \to \mathbb{R}^n$

where this notation for a mapping is familiar to us from Volume I when Boolean functions were defined.

Abstractly, a point p belonging to a set S, $p \in S$, is mapped to n coordinates as in $\varphi(p) = (x_1, x_2, \cdots, x_n)$. How might we translate this definition over to our main concern of generalizing deduction using logic to inference with probability?

Making the language reflect its inferential intent, an IP might say that some probability distribution p exists in a sub–space of the Riemannian manifold S, and therefore is represented by m model parameters. These coordinates can be written as the traditional MEP model parameters $(\lambda_1, \lambda_2, \cdots, \lambda_m)$ or $(\langle F_1 \rangle, \langle F_2 \rangle, \cdots, \langle F_m \rangle)$. Within IG, and by adopting Amari's notation, these model parameters are expressed as either $(\theta^1, \theta^2, \cdots, \theta^m)$ or $(\eta_1, \eta_2, \cdots, \eta_m)$.

As pedantic as all this seems to be, going through the exercise has some merit. Contemplating either the abstract definition, or the semantics of the translation to probabilistic inference, it is evident that the notion of the *probability of the model parameters* makes no sense whatsoever. As the definition above indicates, the model parameters are the coordinates *of* the points; they themselves do NOT have a probability distribution. Thus, language like *the probability of the model parameters*, common throughout the literature, is completely nonsensical.

In addition, note carefully what is NOT mentioned in the definition. Cast aside, if possible, the comfortable Euclidean mind set of three orthogonal axes labeled x, y, and z. The manifold has no familiar *extrinsic* set of coordinate axes by which all points can be referenced. All we have is this austere representation of points mapped to a coordinate system.

Later, linear vector spaces will be constructed at *each point* of the manifold. These are called tangent spaces. Introducing this complication does not suffice as even further structure must be added to the manifold in order to *connect* the disparate points in the manifold.

TECHNICAL CONCEPTS

The creators of this geometry stripped away all of our old familiar notions and rebuilt things from the ground up. One of the things done away with was the extrinsic Euclidean view. It was replaced by more abstract *intrinsic* notions where seemingly intuitive ideas like distance between points had to be rethought.

Nonetheless, in my opinion, it remains an open question as to whether IG could just as easily be done from an *external, Euclidean* perspective. There is a technical theorem in Differential Geometry called the Whitney Embedding Theorem which seems to indicate that anything that could be done in the intrinsic non–Euclidean space could also be done in a higher dimensional extrinsic Euclidean space. There is certainly some appeal to this approach, but, to my knowledge, none of our experts in IG have chosen to explicate matters from this alternative perspective.

34.3 Inner Product, Basis Vectors, and the Metric Tensor

Information Geometry is based to some extent on trying to generalize calculus to an abstract space. For this, it borrows heavily from the very well developed field of *Differential Geometry*. For example, IG has many differential formulas involving the coordinates of the manifold as was illustrated in the opening Chapter.

IG also borrows notions from Linear Algebra. Some of the essential notions from Linear Algebra are the *scalar product*, it's generalization called the *inner product*, and a variation called the *outer product*.

The scalar product between vectors, or, as it is sometimes called, the *dot product*, is shown in Equation (34.1). The notation for the *dot product* is expressed on the left hand side as an angled bracket. The right hand side illustrates the fact that operationally the *dot product* is a summation involving the components v^j and w^j of the two vectors **v** and **w**. Notice the convention of using superscripts for the components of the two vectors.

$$\langle \mathbf{v}, \mathbf{w} \rangle = \sum_{j=1}^{n} v^j w^j \qquad (34.1)$$

This notational transition to coordinates with superscripts rather than the more common subscripted variety is dictated by Tensor Calculus.

This above equation for the *dot product* is implemented in *Mathematica* by the built–in function **Dot[ ]**. Mercifully, *Mathematica* does not burden us with distinctions about subscripts or superscripts. It simply takes two lists as arguments to **Dot[** *arg1, arg2* **]** and returns their linear combination. As explored more fully in the exercises, *Mathematica* will take the vector **v** and form the scalar product with another vector **w**, by evaluating,

Dot[vectorVsuper, vectorWsuper]

and return the linear combination $v^1 w^1 + v^2 w^2 + v^3 w^3$.

Think of the points in the manifold $\mathcal{S}^n$ as similar to vectors in an n–dimensional vector space in the sense that any arbitrary vector $\mathbf{v}$ can be expanded as,

$$\mathbf{v} = \sum_{j=1}^{n} v^j \, \mathbf{e}_j \tag{34.2}$$

where the $\{\mathbf{e}_i, \mathbf{e}_j, \cdots, \mathbf{e}_k\}$ is the common notation for the basis vectors. Note that each $\mathbf{e}_j$ is itself an n–dimensional vector. Neither normality nor orthogonality must be assumed for the basis vectors.

Now, write the inner product of the two vectors $\mathbf{v}$ and $\mathbf{w}$ in terms of their expansions,

$$\langle \mathbf{v}, \mathbf{w} \rangle = \left\langle \sum_{j=1}^{n} v^j \mathbf{e}_j, \sum_{k=1}^{n} w^k \mathbf{e}_k \right\rangle \tag{34.3}$$

This can be worked into,

$$\langle \mathbf{v}, \mathbf{w} \rangle = \sum_{j=1}^{n} \sum_{k=1}^{n} v^j \langle \mathbf{e}_j, \mathbf{e}_k \rangle w^k \tag{34.4}$$

The inner product of the basis vectors $\mathbf{e}_j$ and $\mathbf{e}_k$ on the right hand side of Equation (34.4) is the *metric tensor*, and given the notation,

$$\mathbf{G} \equiv \langle \mathbf{e}_j, \mathbf{e}_k \rangle \tag{34.5}$$

where $\mathbf{G}$ takes the form of an $n \times n$ symmetric matrix.

In a nice concise format used in Linear Algebra, we then have for the scalar product of two arbitrary vectors,

$$\langle \mathbf{v}, \mathbf{w} \rangle = \mathbf{v} \cdot \mathbf{G} \cdot \mathbf{w}^{\mathbf{T}} \tag{34.6}$$

If $\mathbf{v}$ is a $(1 \times m)$ row vector, $\mathbf{G}$ an $(m \times m)$ matrix, and $\mathbf{w}^{\mathbf{T}}$ an $(m \times 1)$ column vector, then the conformable multiplications $(1 \times m) \cdot (m \times m) \cdot (m \times 1)$ in $\mathbf{v} \cdot \mathbf{G} \cdot \mathbf{w}^{\mathbf{T}}$ results in a (1×1) scalar.

34.4 Norm, or Vector Length

Again, from Linear Algebra, the definition of the squared length of some arbitrary vector $\mathbf{v}$ involves the inner product,

$$\| \mathbf{v} \|^2 = \langle \mathbf{v}, \mathbf{v} \rangle \tag{34.7}$$

The *norm* of the vector $\mathbf{v}$ is then,

$$\| \mathbf{v} \| = \sqrt{\langle \mathbf{v}, \mathbf{v} \rangle} \tag{34.8}$$

TECHNICAL CONCEPTS

From Equation (34.6),

$$\| \mathbf{v} \| = \sqrt{\mathbf{v} \cdot \mathbf{G} \cdot \mathbf{v^T}} \tag{34.9}$$

The distance between the two points represented as vectors $\mathbf{v}$ and $\mathbf{w}$ is then,

$$\| \mathbf{v} - \mathbf{w} \| = \sqrt{(\mathbf{v} - \mathbf{w}) \cdot \mathbf{G} \cdot (\mathbf{v} - \mathbf{w})^\mathbf{T}} \tag{34.10}$$

Suppose that we want to approximate the distance between two points p and q, now thought of as two probability distributions whose coordinates in the manifold are very close together. By very close together, we mean that the model parameters for q are displaced ever so slightly from p's corresponding model parameters.

Thus, we might write out the coordinates for p as $\{\theta^1, \theta^2, \cdots, \}$. The coordinates for q then look like $\{\theta^1 + d\theta^1, \theta^2 + d\theta^2, \cdots, \}$ with any $d\theta^j$ signifying a small change in the coordinates, or model parameters.

The distance between points p and q is approximated by making use of the "linear" approximation afforded by the tangent space at point p,

$$\| \mathbf{p} - \mathbf{q} \| = \sqrt{(\mathbf{p} - \mathbf{q}) \cdot \mathbf{G} \cdot (\mathbf{p} - \mathbf{q})^\mathbf{T}} \tag{34.11}$$

or, more explicitly, by bringing into play the coordinates and the summations,

$$\| \mathbf{p} - \mathbf{q} \| = \sqrt{\sum_{r=1}^{m} \sum_{c=1}^{m} (\theta_p^r - \theta_q^r)\, g_{rc}(p)\, (\theta_p^c - \theta_q^c)^T} = \sqrt{\sum_{r=1}^{m} \sum_{c=1}^{m} g_{rc}(p)\, d\theta^r\, d\theta^c} \tag{34.12}$$

The switch in notation from an n, j, and k to an m, c, and r is undertaken to avoid confusion with these already appropriated symbols within the MEP formalism.

34.5 Probabilistic Interpretation of the Metric Tensor

How might an IP think about the metric tensor when it comes to embedding the concept within the inferential framework? The metric tensor is defined on the tangent space at a point p, so we will write it as $g_{rc}(p)$. Don't confuse the notation for the point p with the $P(\star \mid \star)$ of the probability operator.

The MEP algorithm provides us with a numerical assignment for the point p as a member of the exponential family,

$$p \equiv P(x_i \mid \mathcal{M}_k) = \frac{\exp\left[\lambda_1 F_1(x_i) + \lambda_2 F_2(x_i) + \cdots + \lambda_m F_m(x_i)\right]}{Z(\lambda_1, \lambda_2, \cdots, \lambda_m)} \tag{34.13}$$

where we have used the abbreviated form x_i for the statement $(X = x_i)$. The logarithmic transform of p is,

$$\ln p = \lambda_1 F_1(x_i) + \lambda_2 F_2(x_i) + \cdots + \lambda_m F_m(x_i) - \ln\left[Z(\lambda_1, \lambda_2, \cdots, \lambda_m) \right] \quad (34.14)$$

The basis vectors are given in various notations,

$$\mathbf{e}_r \equiv \partial_r \equiv \frac{\partial f}{\partial \theta^r} \quad (34.15)$$

and we have just seen in Equation (34.5) that the metric tensor is defined as the inner product of the basis vectors,

$$g_{rc}(p) \equiv \langle \mathbf{e}_r, \mathbf{e}_c \rangle$$

We have then,

$$g_{rc}(p) = \langle \partial_r, \partial_c \rangle = \left\langle \frac{\partial f}{\partial \theta^r}, \frac{\partial f}{\partial \theta^c} \right\rangle \quad (34.16)$$

or,

$$g_{rc}(p) = \left\langle \frac{\partial \ln p}{\partial \lambda_r}, \frac{\partial \ln p}{\partial \lambda_c} \right\rangle \quad (34.17)$$

The IG notation for the coordinates θ^i and θ^j is now translated into the equivalent MEP notation of the Lagrange multipliers λ_r and λ_c because eventually we will be dealing with a matrix consisting of r rows and c columns,

$$\frac{\partial \ln p}{\partial \lambda_r} = F_r(x_i) - \frac{\partial \ln Z}{\partial \lambda_r}$$

$$\frac{\partial \ln Z}{\partial \lambda_r} = \langle F_r \rangle$$

$$\frac{\partial \ln p}{\partial \lambda_r} = F_r(x_i) - \langle F_r \rangle$$

$$g_{rc}(p) = \left\langle \frac{\partial \ln p}{\partial \lambda_r}, \frac{\partial \ln p}{\partial \lambda_c} \right\rangle$$

$$= \langle F_r(x_i) - \langle F_r \rangle, F_c(x_i) - \langle F_c \rangle \rangle \quad (34.18)$$

Within probability theory (due to Rao and Fisher), the inner product is expressed as $\langle f_1(x), f_2(x) \rangle = E[f_1(x) \times f_2(x)]$ where $E[\cdots]$ is the expectation operator. The probability measure plays the role of the "weight" in the general theory of inner products on function spaces.

When the inner product is written in this form,

$$\langle f_1(x), f_2(x) \rangle = \int [f_1(x) \times f_2(x)] p(x) \, dx \quad (34.19)$$

TECHNICAL CONCEPTS

the metric tensor becomes,

$$g_{rc}(p) = \int_R \left[(F_r(x_i) - \langle F_r \rangle) \times (F_c(x_i) - \langle F_c \rangle) \right] p(x)\, dx \qquad (34.20)$$

However, this definition of a metric tensor happens to be the same as the definition of the **variance–covariance matrix** for the constraint functions $F_j(x_i)$.

Thus, we have arrived at an important statistical interpretation for the metric tensor $g_{rc}(p)$. It is seen to be the variance–covariance matrix for the constraint functions and, as such, is an $m \times m$ symmetric matrix.

Thus, for $m = 3$, the symbolic appearance of $g_{rc}(p)$ in matrix form would look like this,

$$g_{rc}(p) = \begin{pmatrix} g_{11} & g_{12} & g_{13} \\ g_{21} & g_{22} & g_{23} \\ g_{31} & g_{32} & g_{33} \end{pmatrix} \qquad (34.21)$$

Symmetry of the matrix requires that corresponding off–diagonal entries be the same. For example, $g_{13} = g_{31}$.

The metric tensor, when defined for the tangent space at every point in the manifold $\mathcal{S}$, is also known as the Fisher information matrix. The inner product is sometimes called the *scalar product* and, indeed,

$$\langle \mathbf{v}, \mathbf{w} \rangle = \mathbf{v}\,\mathbf{G}\,\mathbf{w}^{\mathbf{T}}$$

does produce a scalar after the vector and matrix multiplications have been carried out. This is shown in detail in Appendix A.

It so happens that the constraint function averages $\langle F_j \rangle$ are the dual coordinates given the notation η_i by Amari. This relationship will be illustrated many times in numerical examples upcoming in future Chapters.

The foregoing derivation of the metric tensor was given at a rather frenetic pace. As an apology, further Chapters do spend more time developing in–depth numerical examples for the metric tensor.

34.6 The Likelihood Equations

There is an orthodox statistical concept originating with Fisher, and subsequently used almost exclusively by his followers for the purposes of "parameter estimation." This is the maximum likelihood method together with the associated concept of the *score function*. Its introduction here will help pave the way for further discussion of the metric tensor.

Again, we remind the reader that the epithet "orthodox" is meant to refer to that very broadly based camp of statisticians that rejected the Bayesian approach.

The well–documented historical debate between Fisher and Jeffreys is the canonical example of the differences dividing the orthodox camp from the Bayesian camp.

The log–likelihood equations,

$$\frac{\partial \ln \mathcal{L}}{\partial \lambda_j} = 0 \qquad \text{for } j = 1 \text{ to } m \tag{34.22}$$

where $\mathcal{L}$ stands for the likelihood of the data, remains central to the orthodox development of parameter estimation. For our purposes here, the appearance of a partial differentiation is important as well because partial differentiations are pervasive operations throughout Information Geometry.

If we look at the exponential family resulting from the MEP algorithm, then the likelihood equation is interesting indeed. Let a generic numerical assignment from the MEP algorithm using m constraint functions, as specified under some model $\mathcal{M}_k$, be written as usual,

$$P(X = x_i \,|\, \mathcal{M}_k) = \frac{\exp\left[\,\lambda_1 F_1(X = x_i) + \lambda_2 F_2(X = x_i) + \cdots + \lambda_m F_m(X = x_i)\,\right]}{Z(\lambda_1, \lambda_2, \cdots, \lambda_m)}$$

The probability of the data under this model is,

$$\mathcal{L} = \prod_{t=1}^{N} P(X_t = x_i \,|\, \mathcal{M}_k) \tag{34.23}$$

with the log–likelihood then,

$$\ln \mathcal{L} = \sum_{t=1}^{N} \left[\sum_{j=1}^{m} \lambda_j F_j(X_t = x_i) - \ln Z \right] \tag{34.24}$$

We see that this can be turned into the sample averages from the data,

$$\ln \mathcal{L} = \sum_{j=1}^{m} \lambda_j \overline{F}_j - \ln Z \tag{34.25}$$

Taking the partial derivative of the log–likelihood with respect to the parameters in the model, we are left with a system of m equations looking like,

$$\ln \mathcal{L} = \sum_{j=1}^{m} \lambda_j \overline{F}_j - \ln Z$$

$$\frac{\partial \ln \mathcal{L}}{\partial \lambda_j} = \overline{F}_j - \frac{\partial \ln Z}{\partial \lambda_j}$$

$$= \overline{F}_j - \langle F_j \rangle \tag{34.26}$$

TECHNICAL CONCEPTS

After invoking the orthodox prescription that all of the partial differentiations should be set equal to 0, and with this taking care of the division by N to form the sample averages,

$$\frac{\partial \ln \mathcal{L}}{\partial \lambda_j} = 0$$

$$\overline{F}_j = \langle F_j \rangle \qquad \text{for } j = 1 \text{ to } m \qquad (34.27)$$

We might very well draw the following implication from the above: "Maximum likelihood" directs that the *expectation* of the constraint functions that existed within the MEP formalism, before any data, should be set equal, within some model, to the *sample averages* of the constraint functions as computed from the data. This gives rise to the assertion that the constraint functions $F_j(X = x_i)$ within the MEP formalism are the same as the "sufficient statistics" within the orthodox formalism.

34.7 Angles Between Vectors

Angles are obviously a fundamental geometrical notion and orthogonal angles are important in Information Geometry. Not only is a metric essential for defining lengths, it also comes into play for defining angles. Curves intersecting at 90° are called orthogonal and will be seen to be closely related to curves that are minimum distance curves between two points. Since the cosine of 90° is 0, checking for orthogonality implies the discovery of those conditions leading to $\cos(\phi) = 0$.

The definition of the cosine of an angle ϕ between two vectors $\mathbf{v}$ and $\mathbf{w}$ uses the already defined concepts of the inner product and lengths of vectors,

$$\cos(\phi) = \frac{\langle \mathbf{v}, \mathbf{w} \rangle}{\| \mathbf{v} \| \| \mathbf{w} \|} \qquad (34.28)$$

As a matter of fact, this definition is sometimes also used for the inner product,

$$\langle \mathbf{v}, \mathbf{w} \rangle = \| \mathbf{v} \| \| \mathbf{w} \| \cos(\phi) \qquad (34.29)$$

The metric tensor gets involved in this definition of the angle because the vectors $\mathbf{v}$ and $\mathbf{w}$ are expanded with respect to their natural basis vectors $\mathbf{e}_j$. Therefore,

$$\cos(\phi) = \frac{g_{jk} v^j w^k}{\sqrt{g_{jk} v^j v^j} \sqrt{g_{jk} w^k w^k}} \qquad (34.30)$$

If $\cos(\phi) = 0$, then $\langle \mathbf{v}, \mathbf{w} \rangle = g_{jk} v^j w^k = 0$.

34.8 Connections to the Literature

We can begin to decipher some of Amari's introductory material [1, pg. 26] with what we have developed so far. A good illustration of the idea of a tangent space at a point p is provided by considering the surface of a sphere as an example of a manifold. Refer to Figure 34.1 below showing two basis vectors,

$$\mathbf{e}_1 = \frac{\partial}{\partial \theta} \text{ and } \mathbf{e}_2 = \frac{\partial}{\partial \phi} \qquad (34.31)$$

tangent to the point p on the surface of the sphere and thus constituting the tangent space at that point.

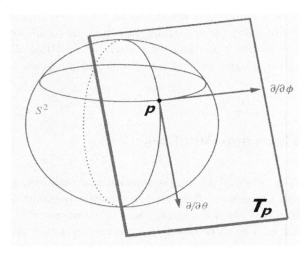

Figure 34.1: *A schematic diagram illustrating the tangent space T_p at a point p in the sub–manifold $\mathcal{S}^2$, the surface of a sphere (adapted from [11, pg. 26]).*

A fundamental concept in Linear Algebra is the notion that an arbitrary vector can be expanded with respect to some basis vectors. This idea is very close to one in Boolean Algebra presented early on in Volume I [7, pg. 90] where we discussed the expansion of a Boolean function $f(A, B)$ through orthonormal basis functions and written as,

$$f(A, B) = \sum_{i=1}^{k} c_i(A, B) \, \phi_i(A, B) \qquad (34.32)$$

The set of four basis functions $\phi_i(A, B)$ were AB, $A\overline{B}$, $\overline{A}B$, and $\overline{A}\,\overline{B}$. The four "coefficients" were $f(T, T)$, $f(T, F)$, $f(F, T)$, and $f(F, F)$.

The analog in Linear Algebra is,

$$\mathbf{v} = \sum_{j=1}^{n} v^j \, \mathbf{e}_j \qquad (34.33)$$

where the orthonormal basis functions are replaced by the basis vectors $\mathbf{e}_j$ and the coefficients are replaced by the components v^j. Amari's notation for the basis vector $\mathbf{e}_j$ is the tangent vector to the jth coordinate curve,

$$\partial_j = \frac{\partial}{\partial \theta^j}$$

The collection of all these tangent vectors constructed at each of the m coordinate curves for some model in the submanifold of dimension m span what is called the "tangent space" T_θ where Amari uses the subscript θ referring to the coordinate system.

Therefore, by definition, any vector A belonging to this tangent space must be a linear combination of the basis vectors ∂_j. Amari adopts the notation A^j for the components of this vector A, so that he writes,

$$\mathbf{v} = \sum_{j=1}^{m} v^j \mathbf{e}_j = \sum_{j=1}^{m} A^j \partial_j \equiv A^j \partial_j \equiv A \qquad (34.34)$$

where in the final step the Einstein summation convention can be employed because the upper and lower index are repeated in the summation.

This quote from Amari [1, pg. 26] was the inspiration for Exercise 34.9.9.

> The tangent space T_θ is a linearized version of a small neighborhood at θ of S, and an infinitesimal vector $d\theta = d\theta^i \partial_i$ denotes the vector connecting two neighboring points θ and $\theta + d\theta$ or two neighboring distributions $p(x, \theta)$ and $p(x, \theta + d\theta)$.

The material in section 34.5 is based on these remarks by Amari,

> Let us introduce a metric in the tangent space T_θ. It can be done by defining the inner product $g_{ij}(\theta) = \langle \partial_i, \partial_j \rangle$ of two basis vectors ∂_i and ∂_j at θ. To this end, we represent a vector $\partial_i \in T_\theta$ by a function $\partial_i l(x, \theta)$ in x, where $l(x, \theta) = \log p(x, \theta)$ and ∂_i (in $\partial_i l$) is the partial derivative $\partial/\partial \theta^i$. Then, it is natural to define the inner product by
>
> $$g_{ij}(\theta) = \langle \partial_i, \partial_j \rangle = E_\theta[\partial_i l(x,\theta) \, \partial_j l(x,\theta)],$$
>
> where E_θ denotes the expectation with respect to $p(x, \theta)$. This g_{ij} is the Fisher information matrix. Two vectors A and B are orthogonal when
>
> $$\langle A, B \rangle = \langle A^i \partial_i, B^j \partial_j \rangle = A^i B^j g_{ij} = 0$$

It is convenient after this quote to remark on the notational equivalencies. First, Amari's $p(x, \theta)$ is my $P(X = x_i \,|\, \mathcal{M}_k)$. Then, $l(x, \theta)$ is $\ln[P(X = x_i \,|\, \mathcal{M}_k)]$. Amari's rather cryptic $\partial_i \equiv \partial_i l(x, \theta)$ is,

$$\partial_r \equiv \frac{\partial \ln P(X = x_i \,|\, \mathcal{M}_k)}{\partial \lambda_r}$$

Then, Amari's
$$g_{ij}(\theta) = \langle \partial_i, \partial_j \rangle = E_\theta[\partial_i l(x, \theta) \, \partial_j l(x, \theta)]$$
gets translated into,
$$g_{rc}(p) = \sum_{i=1}^{n} \left[\frac{\partial \ln P(X = x_i \,|\, \mathcal{M}_k)}{\partial \lambda_r} \times \frac{\partial \ln P(X = x_i \,|\, \mathcal{M}_k)}{\partial \lambda_c} \right] P(X = x_i \,|\, \mathcal{M}_k)$$

and then into,
$$g_{rc}(p) = \sum_{i=1}^{n} \left[(F_r(x_i) - \langle F_r \rangle) \times (F_c(x_i) - \langle F_c \rangle) \right] \times Q_i$$

You have to get used to varying notation even with the same author. For example, here is Amari in [2, pg. 28] talking about the "Fisher metric." Even though we are talking about the same concepts here, the changing notation always causes one to pause and wonder whether, in fact, we *are* talking about the same thing. For example, Amari uses different notation for the coordinate system and the labels "Fisher metric" and "information metric" instead of the previously employed "Fisher information matrix" in the quote below.

> Let $S = \{p_\xi \,|\, \xi \in \Xi\}$ be an n–dimensional statistical model. Given a point $\xi(\in \Xi)$, the **Fisher information matrix** of S at ξ is the $n \times n$ matrix $G(\xi) = [g_{ij}(\xi)]$ where the $(i,j)^{th}$ element $g_{ij}(\xi)$ is defined by the equation below; in particular when $n = 1$, we call this the **Fisher information**.
> $$g_{ij}(\xi) \stackrel{\text{def}}{=} E_\xi[\partial_i l_\xi \, \partial_j l_\xi] = \int \partial_i \, l(x;\xi) \, \partial_j \, l(x;\xi) \, p(x;\xi) \, dx \quad (2.6)$$
> where $\partial_i \stackrel{\text{def}}{=} \frac{\partial}{\partial \xi^i}$,
> $$l_\xi(x) = l(x;\xi) = \log p(x;\xi), \quad (2.7)$$
> (log denotes the natural logarithm), and E_ξ denotes the expectation with respect to the distribution p_ξ, $E_\xi[f] \stackrel{\text{def}}{=} \int f(x) \, p(x;\xi) \, dx$. Although there are some models for which the integral in Equation (2.6) diverges, we assume in our discussion below that $g_{ij}(\xi)$ is finite for all ξ and all i, j, and that $g_{ij} : \Xi \to \mathbb{R}$ is C^∞.
>
> . . .
>
> Now suppose that the assumptions above hold, and define the inner product of the natural basis of the coordinate system $[\xi^i]$ by $g_{ij} = \langle \partial_i, \partial_j \rangle$. This uniquely determines a Riemannian metric $g = \langle \, , \, \rangle$. We call this the **Fisher metric**, or alternatively, the **information metric**. Since g_{ij} defined by Equation (2.6) behaves according to Equation (1.22) under coordinate transformations, we see that the Fisher metric is invariant over the choice of coordinate systems.

34.9 Solved Exercises for Chapter Thirty Four

Exercise 34.9.1: Illustrate the use of polar coordinates as an example of two different coordinate systems.

Solution to Exercise 34.9.1

This is the classic and beginning example of alternative coordinate systems. It has nothing at first glance to do with probability theory.

Consider a two dimensional Euclidean space. The Cartesian coordinate system is expressed in x and y coordinates, while the polar coordinate system is expressed by r and θ. Draw a right triangle inside a circle of radius r with the hypotenuse of length r making an angle θ with x–axis. The point p is located on the circumference of the circle at certain x and y coordinates, or alternatively, at certain r and θ coordinates. See Figure 34.2 below.

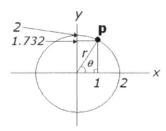

Figure 34.2: *A sketch of the (x, y) and (r, θ) coordinate systems.*

This avails the trigonometric definitions for $\cos\theta$ and $\sin\theta$ to be used as,

$$\cos\theta = \frac{\text{adjacent}}{\text{hypotenuse}}$$

$$\frac{x}{r} = \cos\theta$$

$$x = r\cos\theta$$

$$\sin\theta = \frac{\text{opposite}}{\text{hypotenuse}}$$

$$\frac{y}{r} = \sin\theta$$

$$y = r\sin\theta$$

For example, a point p on a circle of radius $r = 2$ where the hypotenuse makes an angle of $\theta = 60°$ has an x coordinate of $x = 1$ and a y coordinate of $y = 1.73205$.

Relying upon *Mathematica*, use either **N[2 Cos[60 Degree]]**, or alternatively, the equivalent **2 Cos[60 π / 180]** to calculate that $x = 1$.

Going in the other direction, we have,

$$r = \sqrt{x^2 + y^2}$$
$$= \sqrt{(1)^2 + (\sqrt{3})^2}$$
$$= 2$$
$$\tan\theta = \frac{y}{x}$$
$$= \frac{\sqrt{3}}{1}$$
$$\theta = \tan^{-1}(\sqrt{3})$$
$$= 60°$$

Relying on *Mathematica* to confirm this last result,

ArcTan[Sqrt[3]]

returns $\pi/3$ which we know is $60°$, but make *Mathematica* spit it out explicitly by complementing the above expression with,

N[ArcTan[Sqrt[3]] / Degree]

to see the answer of 60.

Any point p on the surface (circumference) of the circle (the manifold) can be expressed in (x, y) coordinates or in (r, θ) coordinates. Remember that a manifold is simply a collection of points together with some coordinate system. And, now we have the transformation between the Cartesian coordinate system and the polar coordinate system.

Exercise 34.9.2: In the future, something called the *Jacobian matrix* will have to be calculated. Illustrate the calculation of the Jacobian matrix for polar coordinates.

Solution to Exercise 34.9.2

The definition of the Jacobian matrix **J** involves the partial differentiation of the first coordinate system with respect to the second. Here, for the polar coordinate

SOLVED EXERCISES FOR CHAPTER 34

system, this is a matrix consisting of four elements,

$$\mathbf{J} = \begin{pmatrix} \frac{\partial x}{\partial r} & \frac{\partial x}{\partial \theta} \\ \frac{\partial y}{\partial r} & \frac{\partial y}{\partial \theta} \end{pmatrix}$$

After all four partial differentiations have been carried out, the Jacobian matrix **J** looks like this,

$$\mathbf{J} = \begin{pmatrix} \cos\theta & -r\sin\theta \\ \sin\theta & r\cos\theta \end{pmatrix}$$

Mathematica will calculate the Jacobian matrix for us with,

```
MatrixForm[D[{r Cos[θ], r Sin[θ]}, {{r, θ}}]]
```

Exercise 34.9.3: Also, the *determinant* of the Jacobian matrix will have to be calculated. Illustrate the calculation of the determinant of the Jacobian matrix for polar coordinates.

Solution to Exercise 34.9.3

A first pass with *Mathematica*,

```
Det[D[{r Cos[θ], r Sin[θ]}, {{r, θ}}]]
```

will produce the result,

$$r(\cos\theta)^2 + r(\sin\theta)^2$$

This can be reduced to the better answer r with,

```
Det[D[{r Cos[θ], r Sin[θ]}, {{r, θ}}]] // Simplify
```

We recognize that the trigonometric identity $(\cos\theta)^2 + (\sin\theta)^2 = 1$ has been applied.

Exercise 34.9.4: If the scalar product of two vectors **v** and **w** can be expressed in the simple format of the dot product, then what is the metric tensor?

Solution to Exercise 34.9.4

The dot product of two vectors **v** and **w** was defined earlier in Equation (34.1) as,

$$\langle \mathbf{v}, \mathbf{w} \rangle = \sum_{j=1}^{m} v^j w^j$$

Suppose further that the metric tensor is defined at a point p in a sub–manifold of dimensionality $m = 3$ where the basis vectors are the unit normal vectors. Then,

$$\mathbf{G} \equiv g_{rc}(p) = \begin{pmatrix} 1 & 0 & 0 \\ 0 & 1 & 0 \\ 0 & 0 & 1 \end{pmatrix}$$

The full expression for the inner product of $\mathbf{v}$ and $\mathbf{w}$ then becomes,

$$\begin{aligned} \langle \mathbf{v}, \mathbf{w} \rangle &= \mathbf{v}\,\mathbf{G}\,\mathbf{w}^{\mathbf{T}} \\ &= \begin{pmatrix} v^1 & v^2 & v^3 \end{pmatrix} \cdot \begin{pmatrix} 1 & 0 & 0 \\ 0 & 1 & 0 \\ 0 & 0 & 1 \end{pmatrix} \cdot \begin{pmatrix} w^1 \\ w^2 \\ w^3 \end{pmatrix} \\ &= v^1 w^1 + v^2 w^2 + v^3 w^3 \\ &= \sum_{j=1}^{3} v^j\, w^j \end{aligned}$$

Let's confirm this in various ways with *Mathematica*. For the first way, create a vector $\mathbf{v}$ by,

vectorV = Array[v, 3]

producing a list with the three components of the vector $\mathbf{v}$,

{v[1], v[2], v[3]}

followed by the vector $\mathbf{w}$,

vectorW = Array[w, 3]

also producing a list with the three components of the vector $\mathbf{w}$,

{w[1], w[2], w[3]}

The dot product **Dot[vectorV, vectorW]** forms the expected scalar,

v[1]w[1] + v[2]w[2] + v[3]w[3]

For a second way closer to the superscripted version in Equation (34.1), create the two vectors with,

vectorVsuper = Array[Function[x, Superscript[v, x]], 3]
vectorWsuper = Array[Function[x, Superscript[w, x]], 3]

then the dot product,

$$\text{Dot[vectorVsuper, vectorWsuper]}$$

will look like,

$$v^1 w^1 + v^2 w^2 + v^3 w^3$$

The same scalar product can be produced in a third way through a summation,

$$\text{Sum[v[j] w[j], \{j, 1, 3\}]}$$

If the metric tensor **G** is represented by the matrix,

$$\text{matrixG} = \{\{1, 0, 0\}, \{0, 1, 0\}, \{0, 0, 1\}\}$$

then,

$$\text{Dot[Dot[vectorV, matrixG], vectorW]}$$

also returns the same scalar, now looking like,

$$v[1]w[1] + v[2]w[2] + v[3]w[3]$$

Exercise 34.9.5: Show how *Mathematica* can create the metric tensor in the last exercise from the inner product of basis vectors.

Solution to Exercise 34.9.5

Applying Equation (34.5),

$$g_{rc}(p) = \langle \mathbf{e}_r, \mathbf{e}_c \rangle$$

where the $\mathbf{e}_r$ and the $\mathbf{e}_c$ are the generic r^{th} and c^{th} basis vectors. Then for a sub-manifold of dimension $m = 3$ the three basis vectors are,

$$\mathbf{e}_1 = \{1, 0, 0\}$$
$$\mathbf{e}_2 = \{0, 1, 0\}$$
$$\mathbf{e}_3 = \{0, 0, 1\}$$

These are seen to be equivalent to the basis vectors of a three-dimensional Euclidean space. As explained in Appendix A, *Mathematica* produces the metric tensor **G**,

$$\mathbf{G} = \begin{pmatrix} 1 & 0 & 0 \\ 0 & 1 & 0 \\ 0 & 0 & 1 \end{pmatrix}$$

with,

$$\text{Outer[Dot, \{\{1,0,0\}, \{0,1,0\}, \{0,0,1\}\},}$$
$$\text{\{\{1,0,0\}, \{0,1,0\}, \{0,0,1\}\}, 1]}$$

Exercise 34.9.6: Use *Mathematica* to illustrate Equations (34.2) through (34.6).

Solution to Exercise 34.9.6

For the numerical example, let $m = 3$. Construct a vector **v**, a vector **w**, and then form their dot product $\sum_{j=1}^{m} v^j w^j$ just as in the previous exercise.

```
vectorV = Array[v,3];
vectorW = Array[w,3];
Dot[vectorV, vectorW]
```

which produces the dot product `v[1]w[1] + v[2]w[2] + v[3]w[3]`.

Now, construct the basis functions as `basisF = Array[F, 3]`. Equation (34.2) is then $\sum_{j=1}^{3} v^j \mathbf{e}_j \equiv$ `Dot[vectorV, basisF]` for the decomposition of vector **v** and $\sum_{k=1}^{3} w^k \mathbf{e}_k \equiv$ `Dot[vectorW, basisF]` for the decomposition of vector **w**.

Equation (34.3) then tells us to construct the inner product according to the decomposition of each vector,

$$\langle \mathbf{v}, \mathbf{w} \rangle = \left\langle \sum_{j=1}^{n} v^j \mathbf{e}_j, \sum_{k=1}^{n} w^k \mathbf{e}_k \right\rangle$$

accomplished by,

```
Outer[Times, Dot[vectorV, basisF], Dot[vectorW, basisF]]
```

When we are engaged in a specific numerical example to confirm a general formula, we should examine the output in very close detail. The above expression evaluates to,

```
F[1] F[1] v[1] w[1] + F[1] F[2] v[2] w[1] + F[1] F[3] v[3] w[1] +
F[1] F[2] v[1] w[2] + F[2] F[2] v[2] w[2] + F[2] F[3] v[3] w[2] +
F[1] F[3] v[1] w[3] + F[2] F[3] v[2] w[3] + F[3] F[3] v[3] w[3]
```

If Equation (34.4) is correct, then the alternative way of forming the inner product from the decomposition of each vector,

$$\langle \mathbf{v}, \mathbf{w} \rangle = \sum_{j=1}^{n} \sum_{k=1}^{n} v^j \langle \mathbf{e}_j, \mathbf{e}_k \rangle w^k$$

expressed in *Mathematica* as,

```
Expand[Dot[Dot[vectorV, Outer[Times, basisF, basisF]], vectorW]]
```

should evaluate to exactly the same expression.

Rather than subjecting our eyeballs to excessive strain in order to confirm this equality, take advantage of the *Mathematica* built-in function **Equal[ ]**, as it was introduced in Volume I in connection with Boolean Algebra. Simply place both of these expressions as the arguments to **Equal[***arg1, arg2***]** as in,

```
Equal[
  Expand[
    Dot[Dot[vectorV,Outer[Times,basisF,basisF]],vectorW]],
    Outer[Times,Dot[vectorV,basisF],Dot[vectorW,basisF]]
                                              (* close Equal *) ]
```

which mercifully evaluates to **True**.

So, the metric tensor **G** is apparently, as Equation (34.5) asserts, the inner product of the basis vectors,

$$\mathbf{G} = \langle \mathbf{e}_j, \mathbf{e}_k \rangle$$

Exercise 34.9.7: What does parameter invariance mean?

Solution to Exercise 34.9.7

Let's start with the simple coin tossing scenario. If the IP fixes the constraint functions at our *de facto* setting of,

$$F(X = x_1) = 1 \text{ and } F(X = x_2) = 2$$

then the information supplied under a model $\mathcal{M}_k$ with the Lagrange multiplier, or model parameter, set at $\lambda = -1.0986$ results in a numerical assignment for the probability of HEADS of $P(\text{HEADS} \mid \mathcal{M}_k) = 0.75$.

However, the mapping from the statements in the state space to numbers was arbitrary. What is not arbitrary, and what must remain invariant, is not the model parameter λ, but the desired numerical assignment for the probability of HEADS under the model.

Therefore, if the constraint functions change to,

$$F(X = x_1) = 2 \text{ and } F(X = x_2) = 4$$

the invariant numerical assignment of 0.75 must be calculated with an appropriate change in the model parameter. It is easy here to see that if the constraint functions have been multiplied by 2, λ must be divided by 2. Now, the model parameter becomes $\lambda = -0.5493$, but the assignment for the probability of HEADS under the model remains the same at 0.75.

In a similar fashion, if the constraint functions were changed to,

$$F(X = x_1) = 3 \text{ and } F(X = x_2) = 6$$

the original model parameter $\lambda = -1.0986$ gets divided by 3 and the new model parameter is $\lambda = -0.3662$. The assignment for the probability of HEADS under the model remains the same at 0.75.

Now, of course, when the Lagrange multiplier λ changes, the dual parameter $\langle F \rangle$ must change as well. As the average of the constraint functions, it is being multiplied along with the constraint functions. Thus, the dual parameter changes from $\langle F \rangle = 1.25$ to $\langle F \rangle = 2.50$ to $\langle F \rangle = 3.75$.

What if, in addition to the multiplication, a constant is added to the constraint function? Suppose the original constraint functions are multiplied by 3 followed by adding a constant of 4. The new constraint functions are,

$$F(X = x_1) = 7 \text{ and } F(X = x_2) = 10$$

The new model parameter is $\lambda = -0.3662$ demonstrating that adding a constant did not change the model parameter from what we had found when we just multiplied by 3.

Given this finding, it is interesting to see what happens when the constraint function average $\langle F \rangle$ as a constant is subtracted from the constraint function as in $F(X = x_i) - \langle F \rangle$. When the arbitrary constraint function $F(X = x_1) = 7$ and $F(X = x_2) = 10$ was constructed, the dual parameter $\langle F \rangle$ for the model that established $P(\text{HEADS} | \mathcal{M}_k) = 0.75$ must be,

$$\langle F \rangle = F(X = x_1) Q_1 + F(X = x_2) Q_2 = (7 \times 3/4) + (10 \times 1/4) = 7.75$$

What then is the Lagrange multiplier for the new constraint function?

$$F^\star(X = x_1) \equiv F(X = x_1) - \langle F \rangle = -0.75$$

$$F^\star(X = x_2) \equiv F(X = x_2) - \langle F \rangle = +2.25$$

It is $\lambda = -0.3662$.

To summarize, whether the constraint function for the coin tossing scenario is constructed as,

$$F(X = x_1) = 1$$
$$F(X = x_2) = 2$$

or,

$$F(X = x_1) = 7$$
$$F(X = x_2) = 10$$

or,

$$F(X = x_1) = -0.75$$
$$F(X = x_2) = +2.25$$

is immaterial as long as the model parameter, or its dual, is adjusted accordingly so that the probability for the joint statements in the state space under the model remain invariant. Pursuing the notion of *parameter invariance* is a red herring which will distract you from the real goal of achieving invariance for the probability assignment to the statements in the state space.

Exercise 34.9.8: Write a *Mathematica* program to do the exploration of the model parameters as discussed in the previous exercise.

Solution to Exercise 34.9.8

I show you here a small *Mathematica* program I wrote to calculate the results discussed in the last exercise. I also would like to accomplish another objective with this exercise.

In Appendix D of Volume II, I made a brief plea asking authors of *Mathematica* programs to provide their readers with some sort of top–down deconstruction to aid in deciphering their programs. So I am going to provide another example of the kind of effort I am looking for in this regard. I am going to provide a sequence of meta–code that reveals more and more internal structure of my program.

At the very highest level, we have **Manipulate[** *expr, controls, options*] so that the user can interact with the program by changing key variables. The key variables here that a user needs to manipulate are the Lagrange multiplier λ and the mapping from the two statements in the state space, HEADS and TAILS, to numbers.

So we start off with,

Manipulate[*(a function to show how the probability changes with arguments of λ and mappings represented by constraint functions), controls, options* **]**

Next, after filling in this function with its three arguments, indicate where the controls and the initialization option will occur,

Manipulate[cointoss[λ, x1, x2],
     (controls for x1, x2, and λ *)*
Initialization:→(*(* define cointoss function which also present results *)* **)**
            (close Manipulate *)* **]**

Next, show the control for λ,

Manipulate[cointoss[λ, x1, x2],
     (controls for x1 and x2 *)*
     {{λ, 0, Style["λ", Bold, 28]}, -2, 2, .0001,
Appearance→"Labeled", ControlPlacement→Top, ImageSize→Large},
  Initialization:→ (*(* define cointoss function and present results *)* **)**
            (close Manipulate *)* **]**

Next, show the important part of the definition for the cointoss function within the Initialization option to the Manipulate[ ] command,

```
Manipulate[
  cointoss[λ, x1, x2],
                (* controls for x1 and x2 *)
  {{λ, 0, Style["λ",Bold, 28}, -2, 2, .0001,
  Appearance →"Labeled", ControlPlacement→Top, ImageSize→Large},
      Initialization:→ (cointoss[λ_, x1_, x2_] := Module[{q, avg},
      q = N[Exp[λ x1] / (Exp[λ x1] + Exp[λ x2])];
      avg = q x1 +((1 - q) x2);
                (* present results *)
                                     (* close Module *) ]
                                     (* close Initialization *) )
                                     (* close Manipulate *) ]
```

One could continue on with this enterprise to accommodate whatever level of pedagogy desired. But any help along these lines from the creator of a *Mathematica* program would certainly be appreciated. With the above as a guide, I will stop here and present the program in its entirety.

```
Manipulate[
  cointoss[λ, x1, x2],
              (* controls begin here *)
  Row[{
      Control[{{x1, 1, Style["x₁", Bold, 20]}, Range[10],
                                        ImageSize→Large}],
      Spacer[50],
      Control[{{x2, 2, Style["x₂", Bold, 20]}, Range[10],
                                        ImageSize→Large}]
                                                        }
                                        (* close Row *) ],
  Delimiter,
  {{λ, 0, Style["λ", Bold, 28]}, -2, 2, .0001,
              Appearance →"Labeled", ImageSize→Large},
              (* cointoss function defined here *)
  Initialization:→ (cointoss[λ_, x1_, x2_] := Module[{q, avg},
    q = N[Exp[λ x1] / (Exp[λ x1] + Exp[λ x2])];
    avg = q x1 +((1 - q) x2);
              (* present results *)
      Panel @
       Pane[
        Labeled[
          Grid[{{Style["P (HEADS)", Bold, Red, 20],
                  Style["Average", Bold, Blue, 20]},
                  {Style[q, Bold, Red, 18], Style[avg, Bold, Red, 18]}},
```

```
         Frame → All, Spacings→{2, 2}(* close Grid *)],
      Style["This calculation is for Exercise 34.9.8 in
            Chapter 34, Volume 3.", Bold, 16], Top
                                      (* Close Labeled *) ],
            ImageSize → {375,150}(* Close Pane *) ]
                                      (* close Module *) ]
                                      (* close Initialization *) )
                                      (* close Manipulate *) ]
```

With this program, the user can set up a mapping from HEADS and TAILS like $F(X = x_1) \to 7$ and $F(X = x_2) \to 10$. Then, the model parameter λ can be adjusted at these particular settings of the constraint function to find out where the invariant probability assignment exists. Here, $\lambda = -0.366204$.

The interactive nature of this *Mathematica* program and how the results are presented is shown below as Figure 34.3. When the controls for **x1** and **x2** were set to 7 and 10, the control for λ was manipulated until P(HEADS) showed an assignment of 0.75.

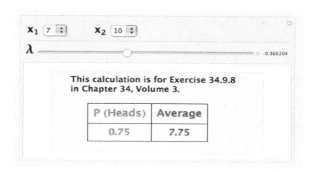

Figure 34.3: *A Mathematica program for exploring constraint function assignments and model parameter adjustments.*

Exercise 34.9.9: Illustrate the use of Equations (34.11) and (34.12) with the first kangaroo correlation model.

Solution to Exercise 34.9.9

Exercise 33.9.5 introduced the simpler kangaroo correlation model as a testbed for some numerical examples in IG. We will continue to rely upon this inferential scenario for the current numerical example for finding the distance between two close by models.

Let p and q be two points in a smooth Riemannian manifold of dimension $n = 4$ endowed with a metric tensor g defined at every point in the manifold. Technically, we have set up the structure $(\mathcal{S}, g)$.

Both points p and q can be mapped to a three dimensional vector of real numbers using $\varphi \colon \mathcal{S} \to \mathbb{R}^m$. The coordinates for p are,

$$\varphi(p) = (\theta_p^1, \theta_p^2, \theta_p^3) = (-1.38629, -1.38629, 4.02535) = \mathbf{p}$$

while the not very different coordinates for q are,

$$\varphi(q) = (\theta_q^1, \theta_q^2, \theta_q^3) = (-1.39629, -1.37629, 4.01535) = \mathbf{q}$$

It is instructive to review the numerical calculation using Equation (34.11) to see what is involved in the pattern of the vector and matrix multiplications.

$$\| \mathbf{p} - \mathbf{q} \| = \sqrt{(\mathbf{p} - \mathbf{q}) \cdot \mathbf{G} \cdot (\mathbf{p} - \mathbf{q})^{\mathbf{T}}}$$

First, we have the 1×3 row vector $(\mathbf{p} - \mathbf{q})$ multiplying the symmetric 3×3 matrix $\mathbf{G}$ from the left. $(\mathbf{p} - \mathbf{q}) \cdot \mathbf{G}$ will result in a 1×3 row vector. Next, multiply the 3×1 column vector $(\mathbf{p} - \mathbf{q})^{\mathbf{T}}$ from the left by the result for $(\mathbf{p} - \mathbf{q}) \cdot \mathbf{G}$ to obtain the scalar $(\mathbf{p} - \mathbf{q}) \cdot \mathbf{G} \cdot (\mathbf{p} - \mathbf{q})^{\mathbf{T}}$. This is exactly the scalar we want from this series of vector times matrix times vector operations because we then need to take the square root of this result.

Identify the vector $(\mathbf{p} - \mathbf{q})$ with the symbolic coordinate differences $(\theta_p^r - \theta_q^r)$. Now, bring in Equation (34.12) at this juncture for a more explicit spelling out of the pattern of vector-matrix-vector multiplications just described. The distance between the two probability distributions is,

$$\| \mathbf{p} - \mathbf{q} \| = \sqrt{\sum_{r=1}^{m} \sum_{c=1}^{m} (\theta_p^r - \theta_q^r) \, g_{rc}(p) \, (\theta_p^c - \theta_q^c)}$$

or, even better,

$$\| \mathbf{p} - \mathbf{q} \| = \sqrt{\sum_{r=1}^{m} \sum_{c=1}^{m} g_{rc}(p) \, d\theta^r \, d\theta^c}$$

The coordinate differences are evident, but the metric tensor $g_{rc}(p)$ needs to be identified before the calculation can begin. The nine components of the metric tensor at point p have already been calculated as the Fisher information matrix for this simple kangaroo correlation model,

$$g_{rc}(p) = \begin{pmatrix} 0.1875 & 0.1375 & 0.1750 \\ 0.1375 & 0.1875 & 0.1750 \\ 0.1750 & 0.1750 & 0.2100 \end{pmatrix}$$

SOLVED EXERCISES FOR CHAPTER 34

This is seen to be in the required form of a 3×3 symmetric matrix. After the vector–matrix–vector multiplications are performed (easily done by *Mathematica* as illustrated in the next Exercise), the distance separating the two probability distributions, as approximated by the distance between the vectors in the tangent space at p, is found to be,

$$\| \mathbf{p} - \mathbf{q} \| = 0.00556776$$

Exercise 34.9.10: Use *Mathematica* to perform the calculations in the previous exercise.

Solution to Exercise 34.9.10

After the values for the parameters of the two nearby models,

```
thetaP = {-1.38629, -1.38629, 4.02535}
thetaQ = {-1.39629, -1.37629, 4.01535}
```

and the Fisher information matrix,

```
G = {{0.1875, 0.1375, 0.1750},
     {0.1375, 0.1875, 0.1750},
     {0.1750, 0.1750, 0.2100}}
```

have been specified, the first vector–matrix multiplication is,

```
Dot[(thetaP - thetaQ), G]
```

while the entire vector–matrix–vector multiplication is wrapped up as,

```
Dot[Dot[(thetaP - thetaQ), G], (thetaP - thetaQ)]
```

At the final step,

```
Sqrt[Dot[Dot[(thetaP - thetaQ), G], (thetaP - thetaQ)]]
```

The answer is exactly the same if we implement the summation formula with,

```
Sqrt[Sum[G[[r, c]] (thetaP[[r]] - thetaQ[[r]])
         (thetaP[[c]] - thetaQ[[c]]), {r, 1, 3}, {c, 1, 3}]]
```

Exercise 34.9.11: What is the angle between two vectors v and w in the tangent space at some point p?

Solution to Exercise 34.9.11

Solve the problem for a Riemannian manifold $\mathcal{S}^3$ with a metric tensor $\mathbf{G}$ defined for every point in the manifold. The state space would then consist of the $n = 3$ joint statements $(X = x_i)$. The MEP algorithm will return numerical assignments for the probabilities $P(X = x_i \mid \mathcal{M}_k)$ under some model $\mathcal{M}_k$. Suppose that the information inserted by $\mathcal{M}_k$ defines the numerical assignment to the distribution p with the information consisting of two constraint functions, $F_1(X = x_i)$ and $F_2(X = x_i)$, and their averages, the dual coordinates η_1 and η_2.

The information in model $\mathcal{M}_k$ consists in the two constraint functions taking on the values,

$$F_1(X = x_i) = (1, 1, 0) \text{ and } F_2(X = x_i) = (0, 1, 0)$$

and with the constraint average for each of these constraint functions,

$$\eta_1 \equiv \langle F_1 \rangle = 0.40 \text{ and } \eta_2 \equiv \langle F_2 \rangle = 0.30$$

It is easy to see why the MEP algorithm assigns the following numerical values under this model,

$$P(X = x_i \mid \mathcal{M}_k) = (0.1, 0.3, 0.6)$$

The canonical basis vectors $\mathbf{e_1}$ and $\mathbf{e_2}$ are,

$$\mathbf{e_1} = F_1(X = x_i) - \langle F_1 \rangle = (1 - .4, 1 - .4, 0 - .4) = (0.6, 0.6, -0.4)$$

and,

$$\mathbf{e_2} = F_2(X = x_i) - \langle F_2 \rangle = (0 - .3, 1 - .3, 0 - .3) = (-0.3, 0.7, -0.3)$$

The metric tensor $g_{rc}(p)$ was defined as the Fisher information matrix,

$$g_{rc}(p) = \sum_{i=1}^{n} \left[(F_r(X = x_i) - \langle F_r \rangle) \times (F_c(X = x_i) - \langle F_c \rangle) \right] P(X = x_i \mid \mathcal{M}_k)$$

and this works out to be a 2×2 symmetric matrix,

$$g_{rc}(p) = \begin{pmatrix} 0.24 & 0.18 \\ 0.18 & 0.21 \end{pmatrix}$$

With the metric tensor calculated, we can now proceed to specify the components v^j and w^k of the two vectors $\mathbf{v}$ and $\mathbf{w}$. Then, the cosine of their angle can be calculated.

The MEP algorithm returned the two theta coordinates,

$$(\theta^1, \theta^2) = (-1.79176, 1.09861)$$

when it solved the numerical assignment for $P(X = x_i \mid \mathcal{M}_k)$. For the sake of this exercise, select the components for vectors **v** and **w** close to these values as $(v^1, v^2) = (-2, 1)$ and $(w^1, w^2) = (-1.5, 1.5)$.

So now it is just a matter of performing the vector–matrix calculations in,

$$\cos(\phi) = \frac{g_{rc} v^r w^c}{\sqrt{g_{rc} v^r v^r} \sqrt{g_{rc} w^c w^c}}$$

Appendix A shows how to make *Mathematica* carry out these computations with the result that,

$$\cos(\phi) = 0.745$$

and the angle between the two vectors is close to 42°.

Exercise 34.9.12: Change a given coordinate system for Jaynes's loaded die scenario.

Solution to Exercise 34.9.12

This numerical example will foreshadow another detailed discussion of Jaynes's loaded dice upcoming in Chapter Thirty Six. We first broached this inferential problem in Chapter Nineteen of Volume II when we were introducing the MEP as a way to assign numerical values to probabilities when rolling dice.

This inference involved a state space of dimension $n = 6$, but we restricted ourselves to Jaynes's model subspace where $m = 2$. This model subspace seemed appropriate for scenarios of a "loaded die" involving displacement in the center of gravity and unequal length along one of the die's axes.

The argument put forward is that the model parameters can flexibly adapt to changing constraint functions so long as the probability assignments must remain invariant. Subject Jaynes's original Lagrange multipliers to an arbitrary linear transformation. In order to motivate this discussion, remember that the two original constraint functions devised to capture the departure from the center of gravity and unequal length along the THREE–FOUR axis of the die, were,

$$F_1(X = x_i) = (1, 2, 3, 4, 5, 6)$$

$$F_2(X = x_i) = (1, 1, -2, -2, 1, 1)$$

Under Jaynes's maximum likelihood model based on the data from 20,000 rolls of the die, the sample constraint function averages $\overline{F}_j$ were set equal to a model's probabilistic expectation $\langle F_j \rangle$,

$$\langle F_1 \rangle = 3.5983$$

$$\langle F_2 \rangle = 0.1393$$

This specification of the information in a model led to finding that the Lagrange multipliers, our two coordinates, were equal to,

$$\lambda_1 = 0.0317244$$

$$\lambda_2 = 0.0717764$$

Now change the constraint functions to,

$$F_1(X = x_i) = (-2.5, -1.5, -0.5, +0.5, +1.5, +2.5)$$

$$F_2(X = x_i) = (1/2, 1/2, -1, -1, 1/2, 1/2)$$

This represents a transformation from the first original constraint functions by subtracting the constant 3.5, and multiplication by 1/2 for the second original constraint function. The constraint function averages must change to,

$$\langle F_1 \rangle = 0.09830$$

$$\langle F_2 \rangle = 0.06965$$

Given the discussion in this Chapter, the subtraction of a constant from the first constraint function should leave the first coordinate unchanged, while multiplying the second constraint function by 1/2 should double the second coordinate.

And, in fact, running the MEP algorithm to find the numerical assignments under a model with this "altered" information results in,

$$\lambda_1 = 0.0317244$$

$$\lambda_2 = 0.1435553$$

This information leads to exactly the same numerical assignments for the six probabilities as before. Note, however, that the metric tensor $\mathbf{G}$ is different under the two different coordinate systems.

Things that must remain invariant under coordinate transformations do remain invariant. For example, consider the two distances under each coordinate system. Taking $(\mathbf{p} - \mathbf{q})$ equal to $(0.01, -0.01)$, then the distance between the two points p and q in the vector space is calculated as,

$$\| \mathbf{p} - \mathbf{q} \| = \sqrt{(\mathbf{p} - \mathbf{q}) \, \mathbf{G} \, (\mathbf{p} - \mathbf{q})^\mathbf{T}} = 0.021863$$

Under the transformed coordinates, with $(\mathbf{p} - \mathbf{q})^\star$ equal to $(0.01, -0.02)$, and with the new metric tensor $\mathbf{G}^\star$, the calculation of the distance also results in,

$$\| \mathbf{p} - \mathbf{q} \|^\star = \sqrt{(\mathbf{p} - \mathbf{q})^\star \, \mathbf{G}^\star \, (\mathbf{p} - \mathbf{q})^{\star \, \mathbf{T}}} = 0.021863$$

Even though the choice of a particular mapping is not critical, *some* mapping from the statements to real numbers must nonetheless be selected. Then whatever selection has been made is adhered to for the duration.

SOLVED EXERCISES FOR CHAPTER 34

Exercise 34.9.13: Create an easy example that pays close attention to the language and resulting symbolic expressions involved in the concept of vector *coordinates* and *bases*.

Solution to Exercise 34.9.13

Disentangling the coordinates and bases can be a tricky enterprise if we don't pay careful attention. Let the tangent space T_p at some point p within a Riemannian manifold be a vector space of dimension $m = 3$. Let the *basis* of this vector space be designated as $(\mathbf{e}_1, \mathbf{e}_2, \mathbf{e}_3)$.

Then, any vector $\mathbf{v}$ in the tangent space can be written as a linear combination of the three elements in this basis,

$$\mathbf{v} = v^1 \mathbf{e}_1 + v^2 \mathbf{e}_2 + v^3 \mathbf{e}_3$$

where (v^1, v^2, v^3) are the *coordinates* of $\mathbf{v}$. We are following what was written before in Equation (34.2) as the expansion of a vector,

$$\mathbf{v} = \sum_{j=1}^{3} v^j \mathbf{e}_j$$

Figure 34.4 sketches out an expansion for a vector $\mathbf{v}$ in two dimensions. Add the two vectors $v^1 \mathbf{e}_1$ and $v^2 \mathbf{e}_2$ by the tactic of placing vectors tail to head.

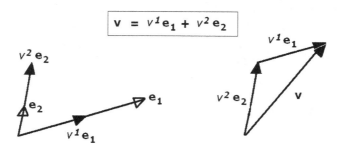

Figure 34.4: *A sketch illustrating how a vector $\mathbf{v}$ is constructed from its coordinates and basis vectors.*

To find the coordinates of a vector $\mathbf{v} = (0, 0, 1)$ with respect to these three basis vectors presented below,

$$\mathbf{e}_1 = (1, 1, 1)$$
$$\mathbf{e}_2 = (-1, 1, 0)$$
$$\mathbf{e}_3 = (1, 0, -1)$$

one must find the coordinates v^1, v^2, and v^3 such that,

$$v^1 \mathbf{e}_1 + v^2 \mathbf{e}_2 + v^3 \mathbf{e}_3 = \mathbf{v}$$

$$v^1 \mathbf{e}_1 + v^2 \mathbf{e}_2 + v^3 \mathbf{e}_3 = (0, 0, 1)$$

$$v^1 (1, 1, 1) + v^2 (-1, 1, 0) + v^3 (1, 0, -1) = (0, 0, 1)$$

This allows us to set up the three linear equations,

$$v^1 - v^2 + v^3 = 0$$

$$v^1 + v^2 + 0 = 0$$

$$v^1 + 0 - v^3 = 1$$

where we find that $v^1 = 1/3$, $v^2 = -1/3$, $v^3 = -2/3$.

Thus, the *coordinates* of the vector $\mathbf{v} = (0, 0, 1)$ with respect to the *basis* of $\mathbf{e}_1 = (1, 1, 1)$, $\mathbf{e}_2 = (-1, 1, 0)$, and $\mathbf{e}_3 = (0, 0, 1)$ are $(v^1, v^2, v^3) = (1/3, -1/3, -2/3)$.

The metric tensor $\mathbf{G}$ for the tangent space constructed at point p for these basis vectors would be found by *Mathematica* as,

$$\mathbf{G} = \begin{pmatrix} 3 & 0 & 1 \\ 0 & 2 & 0 \\ 1 & 0 & 1 \end{pmatrix}$$

through invocation of the **Outer[ ]** function,

```
Outer[Dot, {{1,1,1}, {-1,1,0}, {0,0,1}},
      {{1,1,1}, {-1,1,0}, {0,0,1}}, 1]
```

Chapter 35

Distance Between Probability Distributions

35.1 Introduction

One of the advantages of IG is that it allows you to conceptualize core inferencing concepts in a different way. As a primary example, substitute the geometric intuition concerning distances between points in a space as a replacement for the notion of the information entropy resident in probability distributions.

So far, we have emphasized the concept of a *manifold*, the *points* within the manifold, and finally the *coordinates* for any point in the manifold. As far as mathematical operations are concerned, differentiation of functions with respect to these coordinates is apparently very important. We begin to slowly enlarge this corpus of rather bare bones constructs.

It is quite natural to envision these points in the manifold as densely filling up the manifold. Then, the idea of curves connecting any number of points presents itself. What would be the length of such curves, and what is their analog for our ultimate objective of making inferences?

The points on a curve change because their coordinates are changing. From the perspective of Differential Geometry, the coordinates are changing as a function of another parameter, given the standard label of t. So, in Differential Geometry it makes sense to talk about the $\theta^i(t)$.

But we are immediately faced with a clash in the definition of core concepts. When we employ the MEP to assign numerical values to probabilities, the notion of a *parameter* has already been defined as the coordinates for a model. We can not now introduce this new notion involving t, and also call it a parameter as well. But such important distinctions seem to be generally ignored in the literature.

Moreover, from the standpoint of the IP and its desire to make inferences, it does seem to make more sense to think of the coordinates changing as a function of the changing information as the models $\mathcal{M}_k$ are changing. Thus, we may write equivalent versions for the probability of a statement conditioned on a model, or alternatively thinking in terms of coordinates,

$$P(X = x_i \,|\, \mathcal{M}_k) \equiv P(X = x_i \,|\, \theta^1, \theta^2, \cdots, \theta^m) \tag{35.1}$$

If we write a probability conditioned on the idea of coordinates changing as a function of t,

$$P(X = x_i \,|\, \theta^1(t), \theta^2(t), \cdots, \theta^m(t)) \tag{35.2}$$

then this means that the model $\mathcal{M}_k$ is no longer fixed. In fact, Equation (35.2) indicates that the model parameters are changing with t.

In Differential Geometry and applications to Physics, this variable t is thought of as something like a clock ticking away the time. Thus, the tendency is to formulate things in terms of some physical variable changing as time progresses. A bug is walking around a circle. Its position in terms of the $x(t)$ and $y(t)$ coordinates is dependent on the time because it is a fast or a slow bug. How is its position on the circle changing as time marches inexorably on?

But in inferencing, different models reflect different information. It's a stretch to try and shoehorn an IP's feelings about models and the quantification of different amounts of consciously inserted information, (or equivalently, differing amounts of residual missing information) with some time parameter t ticking away. As a matter of fact, an IP is forced as a matter of principle to consider all coordinate values as potential hypotheses. There doesn't seem to be any intrinsic motivation to discover model parameters that are a function of some other "parameter" t.

Despite my misgivings, the math already exists, and I have to accommodate myself to it. Thus, the formulas for the length of curves in the following section will, in fact, show the presence of a "parameter" t.

35.2   Length of a Curve and the Chain Rule

An important thing we would like to do in Information Geometry is to measure the length of a curve in a Riemannian manifold. Usually, we are trying to find curves with some minimum length. The curve is a succession of probability distributions starting out, say, from a point p and ending up at a point q.

In these early stages, we have no recourse but to pluck facts out of the air as needed. We have been doing this with complete abandon so far. Another one of these unmotivated facts that is now needed comes from multivariate calculus. I am referring to the *chain rule* for functions of several variables. The other formula appearing out of the blue is the integral representation for the length of a curve.

DISTANCE BETWEEN DISTRIBUTIONS

One of the simplest generic mathematical representations for the length of a curve between two points p and q is given by,

$$s(t) = \int_p^q \sqrt{x' \cdot x'} \, dt \tag{35.3}$$

where,

$$x' \equiv \frac{df(p)}{dt} \tag{35.4}$$

An application of the chain rule for this generic function $f(p)$ gives,

$$\frac{df(p)}{dt} = \sum_{r=1}^m \frac{\partial f(p)}{\partial \theta^r} \times \frac{d\theta^r}{dt} \tag{35.5}$$

Assume that this generic $f(p)$ stands for the log transform of the MEP formula when conditioned on some model, or equally well, the exponential family representation for the probability distribution at point p. The emphasis here is on the fact that the probability distribution is conditioned on the specification of the coordinates θ^r. The coordinates are changing as a function of t tracing out different probability distributions, or points, along the curve.

After substituting for $\sqrt{x' \cdot x'}$ in Equation (35.3), we have,

$$ds = \sqrt{\left\langle \sum_{r=1}^m \frac{\partial f(p)}{\partial \theta^r} \frac{d\theta^r}{dt}, \sum_{c=1}^m \frac{\partial f(p)}{\partial \theta^c} \frac{d\theta^c}{dt} \right\rangle} \tag{35.6}$$

Substituting the log transform of the probability assignment for $f(p)$,

$$ds = \sqrt{\left\langle \sum_{r=1}^m \frac{\partial \ln p}{\partial \theta^r} \frac{d\theta^r}{dt}, \sum_{c=1}^m \frac{\partial \ln p}{\partial \theta^c} \frac{d\theta^c}{dt} \right\rangle} \tag{35.7}$$

With,

$$\left\langle \frac{\partial \ln p}{\partial \theta^r}, \frac{\partial \ln p}{\partial \theta^c} \right\rangle \equiv \langle \mathbf{e}_r, \mathbf{e}_c \rangle \tag{35.8}$$

transition to the expressions,

$$ds = \sqrt{\langle \mathbf{e}_r, \mathbf{e}_c \rangle \frac{d\theta^r}{dt} \frac{d\theta^c}{dt}} \tag{35.9}$$

$$= \sqrt{g_{rc}(p) \frac{d\theta^r}{dt} \frac{d\theta^c}{dt}} \tag{35.10}$$

And, finally, after this string of questionable and dubious manipulations, we arrive at the summing up through an integration of each of these ds, the length of the curve between the two probability distributions p and q,

$$\int_p^q ds \, dt = \sqrt{g_{rc}(t) \, d\theta^r(t) \, d\theta^c(t)} \, dt \tag{35.11}$$

To find the length of a curve, adopt the usual calculus mentality of constructing a number of very short line segments along the curve and then add up the lengths of all the short segments. The short segments are allowed to get shorter and shorter as necessary.

From the last Chapter, the length of the short segment between the two very close points p_0 and p_1 in the vector space constructed at point p_0 would be,

$$\| \mathbf{p_0} - \mathbf{p_1} \| = \sqrt{(\mathbf{p_0} - \mathbf{p_1})\, \mathbf{G}\, (\mathbf{p_0} - \mathbf{p_1})^\mathbf{T}} \qquad (35.12)$$

In the alternative summation notation where the components of the vector $(\mathbf{p_0} - \mathbf{p_1})$ are shown by $d\theta^r$ and $d\theta^c$, this would look like,

$$\| \mathbf{p_0} - \mathbf{p_1} \| = \sqrt{\sum_{r=1}^{m}\sum_{c=1}^{m} g_{rc}(p_0)\, d\theta^r\, d\theta^c} \qquad (35.13)$$

Summing up some number N of these small segments leads to,

$$\sum_{t=0}^{N} \| \mathbf{p_t} - \mathbf{p_{t+1}} \| = \sum_{t=0}^{N} \sqrt{(\mathbf{p_t} - \mathbf{p_{t+1}})\, \mathbf{G(t)}\, (\mathbf{p_t} - \mathbf{p_{t+1}})^\mathbf{T}} \qquad (35.14)$$

Differential Geometry expresses this result concisely through use of the Einstein summation convention and the *differential* ds^2 on the left hand side for the so-called "squared distance of a small element of arc" as,

$$ds^2 = g_{rc}\, d\theta^r\, d\theta^c \qquad (35.15)$$

but personally I prefer to keep the summations around so it's harder for me to get confused later on.

35.3 Relative Entropy and Canonical Divergence

The development of a distance formula between points from Differential Geometry is nice, but we already have in our possession the relative entropy formula to quantify the separation between probability distributions. Is there a connection between the two formulas? Has the preceding discussion provided an alternative geometric flavor to satisfy those who are uncomfortable with our preferred informationally couched language?

The Kullback–Leibler relative entropy formula has the familiar ring of,

$$KL(p,q) = \int p(x) \ln\left[\frac{p(x)}{q(x)}\right] dx \qquad (35.16)$$

It turns out that this informational measure of the separation, or discriminability, between two close by probability distributions $p(x)$ and $q(x)$ is readily expressed in terms of the "distance of a small element of arc" as,

$$ds = \sqrt{2KL(p,q)} \qquad (35.17)$$

Even more interesting is that Amari presents a distance–like concept in IG called the *canonical divergence*. He defines the canonical divergence with this formula,

$$D(p \parallel q) \stackrel{\text{def}}{=} \psi(p) + \varphi(q) - \theta^i(p)\,\eta_i(q) \tag{35.18}$$

Unfortunately, everything is couched in IG jargon with little immediate relevance to inferencing. For example, there are two points p and q that belong to a manifold S. There are two coordinate systems for this manifold, the θ^i and the η_i. The Riemannian manifold is labeled as a "dually flat space" with the structural template of (S, g) augmented even further with a notation of (S, g, ∇, ∇^*).

If there can be just one goal in this book, I hope it can be to make such definitions more transparent. The way to proceed, I believe, is to build from our already solid base consisting of the formal manipulation rules of probability theory, and the principle of maximum entropy for assigning numerical values to probabilities.

In Volume II, while explaining the MEP algorithm, we derived the important relationship linking the information entropy of a probability distribution with the partition function, the Lagrange multipliers, and the constraint function averages. In that preferred notation of the MEP formalism, this fundamental relationship became the Legendre transformation,

$$H_{\max}(p) = \min_{\lambda_j} \left[\ln Z_p - \sum_{j=1}^{m} \lambda_j^p \times \langle F_j \rangle_p \right] \tag{35.19}$$

In much the same way, by invoking the relative entropy formula followed by the MEP formula for numerical assignment, Amari's canonical divergence definition is seen to be simply the relative entropy between two probability distributions p and q after the probability distribution q is expressed by its MEP assignment.

The first tricky thing is to reverse the arguments in the divergence function in the left hand side of Equation (35.16). The Kullback–Leibler relative entropy between these two probability distributions is expressed in terms of the canonical divergence function $D(\star \parallel \star)$,

$$KL(p, q) \equiv D(q \parallel p)$$

The above definition for the canonical divergence in Equation (35.18) is replaced by switching p and q everywhere,

$$D(q \parallel p) \equiv KL(p, q) = \psi(q) + \varphi(p) - \theta^i(q)\,\eta_i(p) \tag{35.20}$$

But what is the canonical divergence when expressed in informational terms like Equation (35.19)? As Exercise 35.6.3 will show in detail,

$$D(q \parallel p) = \ln Z_q - H_{\max}(p) - \sum_{j=1}^{m} \lambda_j^q \langle F_j \rangle_p \tag{35.21}$$

35.4 Differential Relationships between the Dual Coordinate Systems

We have already begun to freely use these two prescribed coordinate systems for the Riemannian manifolds. The points in these manifolds are envisioned to represent entire probability distributions.

The first coordinate system discussed was expressed as θ^j. We might think of this notation as a direct translation from the MEP perspective of the Lagrange multipliers λ_j. The second, so-called dual coordinate system, was expressed by η_j, and these coordinates were described as the direct translation of the constraint function expectation, $\langle F_j(X = x_i) \rangle$.

IG is party to a whole host of partial differential relationships between these so-called *dual* coordinate systems. One of the more interesting transformations between these dual coordinate systems from the perspective of the MEP are the *Legendre transformations*. Another question that IG is well positioned to answer, given its emphasis on the metric tensor, is how one set of coordinates will change in response to changes in the dual set of coordinates.

A basic question to ask in this context is how the constraint function averages will change with a change in the Lagrange multipliers. There are two differential relationships involving the metric tensor that answer this question.

$$\frac{\partial \eta_c}{\partial \theta^r} = g_{rc}(p) \tag{35.22}$$

and,

$$\frac{\partial \theta^r}{\partial \eta_c} = g^{rc}(p) \tag{35.23}$$

Equation (35.23) shows the inverse of the metric tensor with superscripts as opposed to the subscripts we have been using. This is the notation peculiar to Differential Geometry and Tensor Calculus. As mentioned, in this case, it is merely an indicator that $g^{rc}(p)$ is the inverse of $g_{rc}(p)$, and *vice versa*.

Drawing out these relationships symbolically is fairly easy. We will start with our old familiar notation from the MEP. The partial differentiation of the log of the partition function with respect to the jth Lagrange multiplier is equal to the jth constraint function average,

$$\frac{\partial \ln Z}{\partial \lambda_j} = \langle F_j \rangle \tag{35.24}$$

Performing a second partial differentiation, where now we switch over to the less ambiguous r and c subscripts,

$$\frac{\partial}{\partial \lambda_r}\left(\frac{\partial \ln Z}{\partial \lambda_c}\right) = \frac{\partial \langle F_c \rangle}{\partial \lambda_r} \tag{35.25}$$

DISTANCE BETWEEN DISTRIBUTIONS

But the second partial differentiation of the log of the partition function with respect to two Lagrange multipliers is defined as the metric tensor at point p.

$$\frac{\partial^2 \ln Z}{\partial \lambda_r \, \partial \lambda_c} = g_{rc}(p) \tag{35.26}$$

Putting this into Amari's IG notation, we have,

$$\frac{\partial \langle F_c \rangle}{\partial \lambda_r} = g_{rc}(p) \tag{35.27}$$

$$\frac{\partial \eta_k}{\partial \theta^j} = g_{jk} \tag{35.28}$$

The dual relationship is expressed through the inverse of the metric tensor as the second partial derivative of the information entropy function with respect to the dual coordinate system,

$$-\frac{\partial^2 H}{\partial \eta_c \, \partial \eta_r} = g^{rc}(p) \tag{35.29}$$

The metric tensor $g_{rc}(p)$ will be an $m \times m$ symmetric matrix as will its inverse, $g^{rc}(p)$. We are also using the linear algebra notation $\mathbf{G}^{-1}(p) \equiv g^{rc}(p)$. Of course, the matrix multiplication of the metric tensor with its inverse must result in the identity matrix $\mathbf{G}\,\mathbf{G}^{-1} = \mathbf{I}$ with 1s along the diagonal and 0s elsewhere.

This section serves the necessary function of getting these equations down in print. A better understanding will come later when all of these relationships will be elucidated through numerical examples.

35.5 Connections to the Literature

The definition of the canonical divergence was taken from Amari [2, pg. 61].

> Let (S, g, Δ, Δ^*) be a dually flat space, on which we are given mutually dual affine coordinate systems $\{[\theta^i], [\eta_i]\}$ and their potentials $\{\psi, \varphi\}$. Given two points $p, q \in S$, let
>
> $$D(p \parallel q) \stackrel{\text{def}}{=} \psi(p) + \varphi(q) - \theta^i(p)\,\eta_i(q)$$
>
> ...
>
> We call this D the **canonical divergence** ...

There are now large chunks of Amari's presentation that we can somewhat follow, if only in the sense that the language has a certain familiar ring to it. For example, here is a quote from [1, pp. 33–34],

We give an example of a 1–flat (*i.e.*, $\alpha = 1$) manifold S. The density function of exponential family $S = \{p(x, \theta)\}$ can be written as

$$\exp\{\theta^i x_i - \psi(\theta)\}$$

with respect to an appropriate measure, where $\theta = (\theta^i)$ is called the natural or canonical parameter. From

$$\partial_i l(x, \theta) = x_i - \partial_i \psi(\theta), \ \partial_i \partial_j l(x, \theta) = -\partial_i \partial_j \psi(\theta),$$

we easily have

$$g_{ij}(\theta) = \partial_i \partial_j \psi(\theta), \ \Gamma^{(\alpha)}_{ijk}(\theta) = \frac{1-\alpha}{2} \partial_i \partial_j \partial_k \psi$$

Hence, the 1–connection $\Gamma^{(1)}_{ijk}$ vanishes identically in the natural parameter, showing that θ gives a 1–affine coordinate system. A curve $\theta(t) = a^i(t) + b^i$, which is linear in the θ–coordinates, is a 1–geodesic, and conversely.

Since an α–flat manifold is $-\alpha$ flat, there exists a $-\alpha$ coordinate system $\eta = (\eta_i) = (\eta_1, \ldots, \eta_n)$ in an α–flat manifold S. Let $\partial^i = \partial/\partial \eta_i$ be the tangent vector of the coordinate curve η_i in the new coordinate system η. The vectors ∂^i form a basis of the tangent space T_η (*i.e.* at T_θ where $\theta = \theta(\eta)$) of S. When the two bases $\{\partial_i\}$ and $\{\partial^i\}$ of the tangent space T_θ satisfy

$$\langle \partial_i, \partial^j \rangle = \delta_i^j$$

at every point θ (or η), where δ_i^j is the Kronecker delta (denoting the unit matrix), the two coordinate systems θ and η are said to be mutually dual. (Nagaoka and Amari (1982)).

Theorem 2.2. When S is α–flat, there exists a pair of coordinate systems $\theta = (\theta^i)$ and $\eta = (\eta_i)$ such that i) θ is α–affine and η is $-\alpha$–affine, ii) θ and η are mutually dual, iii) there exist potential functions $\psi(\theta)$ and $\phi(\eta)$ such that the metric tensors are derived by differentiation as

$$g_{ij}(\theta) = \langle \partial_i, \partial_j \rangle = \partial_i \partial_j \psi(\theta)$$

$$g^{ij}(\eta) = \langle \partial^i, \partial^j \rangle = \partial^i \partial^j \phi(\eta)$$

where g_{ij} and g^{ij} are mutually inverse matrices so that

$$\partial_i = g_{ij} \partial^j, \ \partial^i = g^{ij} \partial_j$$

holds, iv) the coordinates are connected by the Legendre transformation

$$\theta^i = \partial^i \psi(\eta), \ \eta_i = \partial_i \psi(\theta) \qquad (2.8)$$

where the potentials satisfy the identity

$$\psi(\theta) + \phi(\eta) - \theta \cdot \eta = 0, \qquad (2.9)$$

where $\theta \cdot \eta = \theta^i \eta_i$.

In the case of an exponential family S, ψ becomes the cumulant generating function, the expectation parameter $\eta = (\eta_i)$

$$\eta_i = E_\theta[x_i] = \partial_i \psi(\theta)$$

is 1–affine, θ and η are mutually dual, and the dual potential $\phi(\eta)$ is given by the negative entropy,

$$\phi(\eta) = E[\log p],$$

where the expectation is taken with respect to the distribution specified by η.

35.6 Solved Exercises for Chapter Thirty Five

Exercise 35.6.1: Review the MEP derivation for the fair model and a nearby model in the simplest kangaroo scenario.

Solution to Ex. 35.6.1

This well–used example has a state space of dimension $n = 4$. All the models discussed will take place in the full Riemannian manifold with $m = 3$. Thus, all model parameters under the MEP perspective will consist of three Lagrange multipliers, $(\lambda_1, \lambda_2, \lambda_3)$.

For example, the correlated model first studied in Chapter Twenty One, Volume II, had the numerical assignment of [8, pg. 125],

$$Q_1 = 0.70$$
$$Q_2 = 0.05$$
$$Q_3 = 0.05$$
$$Q_4 = 0.20$$

under the information in a model $\mathcal{M}_k$ that $\langle F_1 \rangle = 3/4$, $\langle F_2 \rangle = 3/4$, and $\langle F_3 \rangle = 0.70$.

From the IG perspective, there is a mapping from a point p in the manifold $\mathcal{S}^4$ to four coordinates,

$$\varphi(p) = (\theta^0, \theta^1, \theta^2, \theta^3)$$

with the first coordinate θ^0 implicit because probability distributions must sum to 1. These three canonical coordinates for point p, $(\theta^1, \theta^2, \theta^3)$, are synonymous with the Lagrange multipliers $(\lambda_1, \lambda_2, \lambda_3)$. Under the dual coordinate system,

$$\phi(p) = (1, \eta_1, \eta_2, \eta_3)$$

with, once again, the first coordinate, which is always 1, implicit. These coordinates are synonymous with the constraint function averages, $\langle F_1 \rangle$, $\langle F_2 \rangle$, and $\langle F_3 \rangle$.

For the fair model, and let us label it as point p_0, the canonical coordinates are,

$$\varphi(p_0) = (\theta^1, \theta^2, \theta^3) = (0, 0, 0)$$

while the dual coordinates are,

$$\phi(p_0) = (\eta_1, \eta_2, \eta_3) = (1/2, 1/2, 1/4)$$

The MEP arrives at the numerical assignment of,

$$Q_1 = 0.25$$
$$Q_2 = 0.25$$
$$Q_3 = 0.25$$
$$Q_4 = 0.25$$

under the information in the fair model $\mathcal{M}_0$ that the constraint function averages are $\langle F_1 \rangle = 1/2$, $\langle F_2 \rangle = 1/2$, and $\langle F_3 \rangle = 1/4$.

The MEP formula provides the numerical assignment under any model with,

$$Q_i = \frac{\exp\left[\sum_{j=1}^{3} \lambda_j F_j(X = x_i)\right]}{\sum_{i=1}^{4} \exp\left[\sum_{j=1}^{3} \lambda_j F_j(X = x_i)\right]}$$

Since all three Lagrange multipliers are equal to zero under the given information of the fair model, the partition function $Z(\lambda_1, \lambda_2, \lambda_3)$ is equal to 4. Each numerator for Q_i is equal to 1, so all $Q_i = 1/4$ under the fair model.

The metric tensor $g_{rc}(p_0)$ constructed from the basis vectors at point p_0 must be a symmetric (3×3) matrix. *Mathematica* will calculate this metric tensor with,

matrixG = Outer[D[Log[z], #1, #2]&, {λ_1, λ_2, λ_3}, {λ_1, λ_2, λ_3}]
/. {$\lambda_1 \to 0$, $\lambda_2 \to 0$, $\lambda_3 \to 0$}

resulting in the (3×3) symmetric tensor,

$$g_{rc}(p_0) = \begin{pmatrix} 1/4 & 0 & 1/8 \\ 0 & 1/4 & 1/8 \\ 1/8 & 1/8 & 3/16 \end{pmatrix}$$

Let a nearby model be labeled as point p_1 with the changed coordinates of $\varphi(p_1) = (0.001, 0.002, -0.003)$. Since this is a review, we must delve into the details of the constraint functions. When we set up the matrix of constraint functions in *Mathematica* as,

cm = {{1, 1, 0, 0}, {1, 0, 1, 0}, {1, 0, 0, 0}}

it was meant that, for example, the mapping from statements about $(X = x_1)$ took the form,

$$F_1(X = x_1) = 1$$
$$F_2(X = x_1) = 1$$
$$F_3(X = x_1) = 1$$

or that the mapping from statements about $(X = x_2)$ took the form,

$$F_1(X = x_2) = 1$$
$$F_2(X = x_2) = 0$$
$$F_3(X = x_2) = 0$$

Note the distinction in the construction of the constraint function averages $\langle F_j \rangle$. For example, $\langle F_1 \rangle$ is found as,

$$\langle F_1 \rangle = \sum_{i=1}^{4} F_1(X = x_i) Q_i$$
$$= (1 \times Q_1) + (1 \times Q_2) + (0 \times Q_3) + (0 \times Q_4)$$
$$= Q_1 + Q_2$$

as compared to $\langle F_2 \rangle$,

$$\langle F_2 \rangle = \sum_{i=1}^{4} F_2(X = x_i) Q_i$$
$$= (1 \times Q_1) + (0 \times Q_2) + (1 \times Q_3) + (0 \times Q_4)$$
$$= Q_1 + Q_3$$

Thus, the numerator in the MEP formula has a fixed index i while the index j varies. For fixed index $i = 1$ with j varying from 1 to 3, the numerator for Q_1 becomes,

$$\text{numerator } Q_1 = \exp\left[\lambda_1 F_1(X = x_1) + \lambda_2 F_2(X = x_1) + \lambda_3 F_3(X = x_1)\right]$$
$$= \exp\left[\lambda_1(1) + \lambda_2(1) + \lambda_3(1)\right]$$
$$= \exp\left[\lambda_1 + \lambda_2 + \lambda_3\right]$$

In a similar manner, the numerator for Q_2 becomes,

$$\text{numerator } Q_2 = \exp\left[\lambda_1 F_1(X = x_2) + \lambda_2 F_2(X = x_2) + \lambda_3 F_3(X = x_2)\right]$$
$$= \exp\left[\lambda_1(1) + \lambda_2(0) + \lambda_3(0)\right]$$
$$= \exp\left[\lambda_1\right]$$

The four numerators are then,

$$\text{numerator } Q_1 = \exp[\lambda_1 + \lambda_2 + \lambda_3]$$

$$\text{numerator } Q_2 = \exp[\lambda_1]$$

$$\text{numerator } Q_3 = \exp[\lambda_2]$$

$$\text{numerator } Q_4 = \exp[0]$$

The partition function Z as the sum over all four numerators is then,

$$Z = \exp[\lambda_1 + \lambda_2 + \lambda_3] + \exp[\lambda_1] + \exp[\lambda_2] + 1$$

Of course, we never have to go down to this level of detail once we have understood the constrained optimization rationale behind the MEP formula. *Mathematica* will take care of the details through,

```
z = Total[Exp[Dot[Map[Subscript[λ, #]&, Range[3]], cm]]]
```

Calculate the numerical assignment to the third statement in the state space under the nearby model. The numerator for Q_3 will be,

$$\text{numerator } Q_3 = \exp[\lambda_2]$$

$$= \exp[0.002]$$

$$= 1.002002$$

The partition function $Z(\lambda_1, \lambda_2, \lambda_3) = 4.003003$ through the addition of all four numerators so that $Q_3 = 1.002002/4.003003 = 0.250313$.

The numerical assignment found by the MEP for all four statements in the state space in the simple kangaroo scenario under the information in a model "close by" to the fair model is then,

$$Q_1 = 0.249812$$

$$Q_2 = 0.250062$$

$$Q_3 = 0.250313$$

$$Q_4 = 0.249812$$

These assignments are calculated by *Mathematica* by defining a user supplied function **MEPprob[** *arg1, arg2* **]** with two arguments. The first argument is a list of the coordinates under a given model, and the second argument is the constraint matrix.

SOLVED EXERCISES FOR CHAPTER 35

```
MEPprob[theta_List, cm_List] := Module[{q, z, p},
            q = N[Exp[Dot[theta, cm]]];
            z = Total[q];
            p = q / z;
            {NumberForm[z, {8, 6}], p, Total[p]}]
```

`MEPprob[{0, 0, 0}, cm]` will return the fair model's numerical assignment of all $Q_i = 0.25$, while a call to `MEPprob[{.001, .002, -.003}, cm]` will return the nearby model's numerical assignment as listed above.

To conclude this review, let's include something new from IG. Check whether the "distance" between these two probability distributions for the fair model and the nearby model is approximately the same under the tangent space approximation and the relative entropy.

When the tangent space T_{p_0} is constructed at point p_0, the distance from p_0 to a nearby point p_1 *in the vector space* with the metric tensor $\mathbf{G}$, is calculated with the formula,

$$\| \mathbf{p_0} - \mathbf{p_1} \| = \sqrt{(\mathbf{p_0} - \mathbf{p_1}) \cdot \mathbf{G} \cdot (\mathbf{p_0} - \mathbf{p_1})^{\mathbf{T}}}$$

This separation between the fair model and a nearby model is implemented in *Mathematica* by,

`Sqrt[Dot[Dot[(thetap0 - thetap1), matrixG], (thetap0 - thetap1)]]`

The element of arc length returned has a value of 0.000829156.

The alternative summation formula,

$$\| \mathbf{p_0} - \mathbf{p_1} \| = \sqrt{\sum_{r=1}^{3} \sum_{c=1}^{3} g_{rc}(p_0) \, d\theta^r \, d\theta^c}$$

is implemented in *Mathematica* by,

`Sqrt[Sum[matrixG[[r, c]] (thetap0[[r]] - thetap1[[r]])`
`      (thetap0[[c]] - thetap1[[c]]), {r, 1, 3}, {c, 1, 3}]]`

which must return the same value of 0.000829156.

Now, the question is whether this definition of the distance between two points is approximately the same as the relative cross entropy between two probability distributions? The relationship between an entropy based view and the vector space based view is asserted to be,

$$\| \mathbf{p_0} - \mathbf{p_1} \| \approx \sqrt{2 \, KL(p_0, p_1)}$$

The relative entropy between the fair model probability distribution p_0 and the nearby model probability distribution p_1 is given by,

$$KL(p_0, p_1) = \sum_{i=1}^{4} p_{0,i} \ln\left(\frac{p_{0,i}}{p_{1,i}}\right)$$

This can be implemented in *Mathematica* with the aid of the previously defined **MEPprob[ ]** function. Write a new function called **reDistance[** *arg1, arg2, arg3* **]** with three arguments.

```
reDistance[theta1_List, theta2_List, cm_List] :=
        Sqrt[2 Total[MEPprob[theta1, cm]
        Log[MEPprob[theta1, cm] / MEPprob[theta2, cm]]]]
```

With a call to,

reDistance[{0, 0, 0}, {.001, .002, -.003}, cm]

the answer returned has the value of 0.000829213 as compared to 0.000829156, thus exhibiting the aforementioned close approximation to the vector space definition of the separation between the two probability distributions.

Exercise 35.6.2: Make up an example where the sum of the elements of arc length between several probability distributions approximates the integral representation for the length of a curve.

Solution to Ex. 35.6.2

With the first exercise as a backdrop, we can now calculate the element of arc length between any number of close by probability distributions. Imagine, then, that there is a digital clock ticking away that starts at $t = 0$ and ends at $t = 4$. There are thus five distinct probability distributions $\mathbf{p_0}$ through $\mathbf{p_4}$ for the four statements in the kangaroo scenario state space. Each one of these distinct probability distributions contains different information under a model, and therefore will possess different numerical assignments as dictated by these models.

The coordinates $\theta^j(t)$ are changing slowly as a function of t. In other words, the parameters $\lambda_j(t)$, as well as the information resident in these models, is changing as a function of t. We will start out as before with the fair model possessing coordinates $(\theta^1(t) = 0, \theta^2(t) = 0, \theta^3(t) = 0)$ at $t = 0$. $\theta^1(t)$ increases by 0.001, $\theta^2(t)$ increases by 0.002, and $\theta^3(t)$ decreases by 0.015 with each tick of the clock.

The distance from point $\mathbf{p_0}$ to point $\mathbf{p_4}$ in the integral representation,

$$s(t) = \int_{p_0}^{p_4} \sqrt{g_{rc}(t) \frac{d\theta^r}{dt} \frac{d\theta^c}{dt}} \, dt$$

should be approximated by this sum,

$$\sum_{j=0}^{3} \| \mathbf{p_j} - \mathbf{p_{j+1}} \| = \| \mathbf{p_0} - \mathbf{p_1} \| + \cdots + \| \mathbf{p_3} - \mathbf{p_4} \|$$

SOLVED EXERCISES FOR CHAPTER 35

A symbolic analogy between the integral representation containing the explicit occurrence of dt and the approximation by the sum with a Δt_j small enough is sometimes shown as, [25, pg. 303],

$$\sum_{j=0}^{3} \| \mathbf{p_j} - \mathbf{p_{j+1}} \| = \sum_{j=0}^{3} \frac{\| \mathbf{p_j} - \mathbf{p_{j+1}} \|}{\Delta t_j} \Delta t_j \approx \int_{p_0}^{p_4} \sqrt{g_{rc}(t) \frac{d\theta^r}{dt} \frac{d\theta^c}{dt}} \, dt$$

For us, this means that we recognize that both the coordinates and the metric tensor will be changing as a function of t when the summation takes place. Let θ^1 change by $+0.001$, θ^2 by $+0.002$, and θ^3 by -0.015.

With the accomplishments of the first exercise, we can calculate any $\| \mathbf{p_j} - \mathbf{p_{j+1}} \|$, that is, any term in the summation. But this requires four separate calculations of the metric tensor $\mathbf{G}(p_j)$ at the points $\mathbf{p_0}$ through $\mathbf{p_3}$. The metric tensor is also changing as a function of t.

Since we already have calculated the metric tensor for the fair model at point $\mathbf{p_0}$, look first at,

$$\| \mathbf{p_0} - \mathbf{p_1} \| = \sqrt{(\mathbf{p_0} - \mathbf{p_1}) \cdot \mathbf{G}(\mathbf{p_0}) \cdot (\mathbf{p_0} - \mathbf{p_1})^{\mathbf{T}}}$$

through a slight generalization of the,

`Sqrt[Dot[Dot[(thetap0 - thetap1), matrixG], (thetap0 - thetap1)]]`

of the first exercise.

Since we will want to calculate for changing coordinates and changing metric tensors, construct a function **arcLength[** *arg1, arg2, arg3* **]** with three arguments. The first two arguments are the lists containing the coordinates for point $\mathbf{p_j}$ and point $\mathbf{p_{j+1}}$. The final argument is the list of lists containing the metric tensor at point $\mathbf{p_j}$.

```
arcLength[thetapj_List, thetapj1_List, G_List] :=
   Sqrt[Dot[Dot[(thetapj - thetapj1), G], (thetapj - thetapj1)]]
```

A call to **arcLength[{0, 0, 0}, {.001, .002, -.015}, matrixGp0]** returns a value of 0.00567340. This makes sense because this new probability distribution at point $\mathbf{p_1}$ is "close" to the fair model because the coordinates have changed ever so slightly.

What is the new numerical assignment to the four probabilities of the statements in the state space under the changed information of this model that is very similar to the fair model? Use a slightly modified version of last exercise's **MEPprob[]** in order to find the constraint function averages for the given list of coordinates,

```
MEPprob1[theta_List, cm_List] := Module[{q, z, p, cfa},
            q = N[Exp[Dot[theta, cm]]];
            z = Total[q];
            p = q / z;
            cfa = Dot[cm, p];
            {NumberForm[z, {8, 6}], p, Total[p], cfa}]
```

A call to **MEPprob1[{.001, .002, -.015}, cm]** returns the new assignment,

$$Q_1 = 0.247570$$

$$Q_2 = 0.250810$$

$$Q_3 = 0.251061$$

$$Q_4 = 0.250559$$

It also returns the new constraint function averages $\langle F_1 \rangle$, $\langle F_2 \rangle$, and $\langle F_3 \rangle$, that is, the new dual coordinates $\phi(p_1) = (\eta_1, \eta_2, \eta_3) = (0.49838, 0.498631, 0.247570)$.

So this calculation represents the first term $\| \mathbf{p_0} - \mathbf{p_1} \| = 0.0056734$ in the overall summation of,

$$\sum_{j=0}^{3} \| \mathbf{p_j} - \mathbf{p_{j+1}} \|$$

We will also need to calculate the changed metric tensors $\mathbf{G(p_1)}$, $\mathbf{G(p_2)}$, and $\mathbf{G(p_3)}$ at points $\mathbf{p_1}$, $\mathbf{p_2}$, and $\mathbf{p_3}$ for the next three terms in the summation approximation. When this is done, the summation is equal to,

$$\sum_{j=0}^{3} \| \mathbf{p_j} - \mathbf{p_{j+1}} \| = 0.00567340 + 0.00565562 + 0.00563772 + 0.00561971 = 0.022586$$

This is the approximate value of the integral for the arc length between points p_0 and p_4,

$$s(t) = \int_{p_0}^{p_4} \sqrt{g_{rc}(t) \frac{d\theta^r}{dt} \frac{d\theta^c}{dt}} \, dt \approx 0.022586$$

Exercise 35.6.3: Derive the informationally inspired formula for the canonical divergence as shown in section 35.3.

Solution to Ex. 35.6.3

In contrast to Amari's IG inspired definition for the canonical divergence,

$$D(p \parallel q) \stackrel{\text{def}}{=} \psi(p) + \varphi(q) - \theta^i(p)\, \eta_i(q)$$

the information entropy inspired formula is derived below for its dual format,

$$D(q \parallel p) \stackrel{\text{def}}{=} \psi(q) + \varphi(p) - \theta^i(q)\, \eta_i(p)$$

As shown in the upcoming derivation,

$$D(q \parallel p) = \ln Z_q - H_{\max}(p) - \sum_{j=1}^{m} \lambda_j^q \langle F_j \rangle_p$$

Pay careful attention to the fact that the partition function Z_q and the Lagrange multipliers λ_j^q come from the probability distribution for q. However, the negative information entropy $-H_{\max}(p)$ and the constraint function averages $\langle F_j \rangle_p$ come from the probability distribution for p.

$$D(q \parallel p) = KL(p, q)$$

$$KL(p, q) = \sum_{i=1}^{n} p_i \ln\left(\frac{p_i}{q_i}\right)$$

$$= \sum_{i=1}^{n} p_i (\ln p_i - \ln q_i)$$

$$= \sum_{i=1}^{n} p_i \ln p_i - \sum_{i=1}^{n} p_i \ln q_i$$

$$= -H_{\max}(p) - E_p [\ln q_i]$$

$$= -H_{\max}(p) - E_p \left[\sum_{j=1}^{m} \lambda_j^q F(X = x_i) - \ln Z_q \right]$$

$$= -H_{\max}(p) - \sum_{j=1}^{m} \lambda_j^q E_p [F(X = x_i)] - E_p [\ln Z_q]$$

$$= \ln Z_q - H_{\max}(p) - \sum_{j=1}^{m} \lambda_j^q E_p [F(X = x_i)]$$

$$D(q \parallel p) = \ln Z_q - H_{\max}(p) - \sum_{j=1}^{m} \lambda_j^q \langle F_j \rangle_p$$

$$\equiv \psi(q) + \varphi(p) - \theta^i(q) \eta_i(p)$$

Exercise 35.6.4: What is the approximate change to be expected in the third constraint function average when the first Lagrange multiplier starts to change from its value at the fair model?

Solution to Ex. 35.6.4

Within the context of the simple kangaroo scenario, we are asking for,

$$\frac{\partial \langle F_3 \rangle}{\partial \lambda_1} \equiv \frac{\partial}{\partial \lambda_1} \left(\frac{\partial \ln Z}{\partial \lambda_3} \right) \equiv \frac{\partial^2 \ln Z}{\partial \lambda_1 \, \partial \lambda_3} \equiv \frac{\partial \eta_3}{\partial \theta^1} = g_{13}(p)$$

at point p, the probability distribution under the fair model. From the first exercise, we know that $g_{13}(p) = g_{31}(p) = 1/8$.

It is easy enough to see that $\langle F_3 \rangle$ is equal to 0.25 at the fair model. If λ_1 were to increase by 0.01, then this third constraint function average would change to 0.25125. So this change to $\langle F_3 \rangle$ also represents the changed information in the new model when compared to the fair model,

$$\langle F_3(\lambda_1 = 0.01, \lambda_2 = 0, \lambda_3 = 0)\rangle - \langle F_3(\lambda_1 = 0, \lambda_2 = 0, \lambda_3 = 0)\rangle = 0.25125 - 0.25$$

$$\Delta y = 0.00125$$

The approximate change to be expected in the constraint function averages based on the linear approximation inherent in the differential is,

$$\Delta y \approx f'(x)\,\Delta x$$

From above, $\Delta y = 0.00125$, and since the difference in the first Lagrange multiplier was 0.01, $\Delta x = 0.01$, and since $f'(x) = g_{13}(p) = 0.125$, the approximation is very good,

$$\Delta y \approx f'(x)\,\Delta x$$

$$f'(x)\,\Delta x = 0.125 \times 0.01$$

$$\Delta y \approx 0.00125$$

Exercise 35.6.5: What is the approximate change to be expected in the third constraint function average when the first Lagrange multiplier starts to change from its value at a highly correlated model?

Solution to Ex. 35.6.5

In the last exercise, we looked at a change in one of the parameters at the point in the manifold where the fair model existed. Now, look at the change in the same parameter for a point at a different place in the manifold where the coordinates are,

$$(\theta^1, \theta^2, \theta^3) = (-1.38629, -1.38629, 4.02535)$$

These are the coordinates for the correlated model and the numerical assignment,

$$Q_1 = 0.70$$

$$Q_2 = 0.05$$

$$Q_3 = 0.05$$

$$Q_4 = 0.20$$

The dual coordinates at this point are,

$$(\eta_1, \eta_2, \eta_3) = (0.75, 0.75, 0.70)$$

If the first Lagrange multiplier λ_1 changes by $+0.01$ to a close by model where,

$$(\theta^1, \theta^2, \theta^3) = (-1.37629, -1.38629, 4.02535)$$

SOLVED EXERCISES FOR CHAPTER 35

then η_3 increases to 0.701747. The difference Δy in the third constraint function average at these two models is then,

$$\Delta y = 0.701747 - 0.70 = 0.001747$$

The linear approximation to this exact difference afforded by the differential is,

$$\Delta y \approx f'(x)\,\Delta x$$

$$f'(x)\,\Delta x = 0.175 \times 0.01$$

$$\Delta y \approx 0.00175$$

The value of the metric tensor component representing $f'(x)$ where $r = 1$ and $c = 3$ is,

$$g_{13}(p) = \frac{\partial}{\partial \lambda_1}\left(\frac{\partial \ln Z}{\partial \lambda_3}\right) = \frac{\partial \eta_3}{\partial \theta^1} \equiv f'(x) = 0.175$$

where now point p is the correlated model. Thus, we would expect a change of about 0.00175 in the third constraint function average, or, in other words, in the value of $\langle F_3 \rangle \equiv Q_1$ as the first Lagrange multiplier is changed by +0.01.

Exercise 35.6.6: What is the approximate change to be expected in the first Lagrange multiplier if the third constraint function average starts to change from its value under the highly correlated model?

Solution to Ex. 35.6.6

This is the inverse of the problem solved in the last exercise. The IP is now asking about,

$$\frac{\partial \lambda_1}{\partial \langle F_3 \rangle} \equiv \frac{\partial}{\partial \langle F_3 \rangle}\left(-\frac{\partial H}{\partial \langle F_1 \rangle}\right) \equiv -\frac{\partial^2 H}{\partial \langle F_3 \rangle \partial \langle F_1 \rangle} \equiv \frac{\partial \theta^1}{\partial \eta_3} = g^{31}(p)$$

When the inverse of the metric tensor at the correlated model was calculated by *Mathematica*, $g^{31}(p) = -25$. Thus, the IP would expect, based on the linear approximation afforded by the differential, that $\Delta y \approx f'(x)\,\Delta x$.

But Δy is now the difference in the first Lagrange multiplier caused by changing the third constraint function average by some small amount, say, 0.001. Thus, the expected change in the first Lagrange multiplier λ_1 would be about,

$$\Delta y \approx f'(x)\,\Delta x$$

$$f'(x)\,\Delta x = -25 \times 0.001$$

$$\lambda_1^{\text{New}} - \lambda_1^{\text{Old}} \approx -0.025$$

Use the MEP algorithm to find that $\lambda_1^{\text{New}} = -1.41148$ when $\langle F_3 \rangle$ is changed to 0.701. Recall that $\lambda_1^{\text{Old}} = -1.38629$ when $\langle F_3 \rangle$ is set at 0.70 under the correlated model. Thus, $-0.025 \approx -1.41148 - (-1.38629) = -0.02519$.

Ask yourself an obvious question based on this result. Since the inverse metric tensor is also symmetric, does that mean that the change in the *third* Lagrange multiplier when the *first* constraint function average changes is still approximated by -0.025? That is,

$$\frac{\partial \lambda_3}{\partial \langle F_1 \rangle} = g^{13}(p) = g^{31}(p) = -25$$

The answer is, "Yes." In the original model, the third Lagrange multiplier was $\lambda_3^{\text{Old}} = +4.02535$. When the first constraint average is changed from $\langle F_1 \rangle = 0.75$ to 0.751, the third Lagrange multiplier changes to $\lambda_3^{\text{New}} = +4.00054$. The approximation still works,

$$-0.025 \approx \lambda_3^{\text{New}} - \lambda_3^{\text{Old}} = +4.00054 - (+4.02535) = -0.02481$$

Chapter 36

Inferences About Dice

36.1 Introduction

Jaynes's tutorial exposition describing how the Maximum Entropy Principle (MEP) can be applied to dice tossing is a classic. It is one of my all time favorite examples of how an inferential problem is analyzed in depth from an Information Processing perspective. I highly recommend Jaynes's original treatment of Wolf's dice data [17, pp. 258–273] to see a master at the peak of his powers.

With all due respect to Jaynes, we shall appropriate this problem, just as we did in Volume II, not only to recapitulate some of the important lessons about the MEP, but also to introduce some fundamental notions in Information Geometry. We will provide slight alterations to Jaynes's original presentation from time to time, all the while hoping that the spirit of Jaynes's tutorial has not been unduly mangled in the process.

Of course, we have been taking advantage of the dice tossing scenario for some time now. If we assume that we record the number of spots appearing on the turned up face of an ordinary die, this familiar game can illustrate a state space consisting of $n = 6$ statements. Now we will revisit it because of its familiarity as a comfortable anchor point. Information Geometry is a complex subject, and it would be preferable to refer back to things which we understand well.

Apparently, as Jaynes tells the story, in the latter half of the 19th Century, the Swiss astronomer Rudolph Wolf, now famous for "Wolf's sunspot data," conducted some dice rolling experiments in his laboratory. Jaynes advances the remarkable proposition that it should be possible to infer some physical characteristics of one of the actual die used in Wolf's experiments by analyzing the frequency results reported from a huge number of throws. He wanted to demonstrate that the MEP was not just some technique from statistical mechanics that physicists used to study molecules.

Jaynes conducts a long and very detailed explanation of how an IP might conduct such an inferential analysis using the MEP algorithm. There are lots of extraneous details in this explanation about things like the chi–square test on the frequency data which, by Jaynes's own admission, were he now alive, he would relegate to the rubbish bin. Nonetheless, his approach based on a common sense discussion of how a die is actually constructed, and how one ties in that knowledge with constraint functions and constraint function averages, is essential to our discussion.

36.2 Model for a Fair Die in IG

Before we listen to what Jaynes has to tell us about the physical nature of a die and how that might be translated into constraint functions, let's conduct a bit of a review on the whole dice tossing scenario. As this problem is so familiar, we attempt to make things less boring by employing the language of Information Geometry to talk about the model for a fair die.

Let a smooth Riemannian manifold $\mathcal{S}$ of dimension $n = 6$ be endowed with a metric tensor $g_{rc}(p)$ at each point p. A coordinate system maps any point p in the manifold to a vector of real numbers with five elements in the notation $\varphi(p) = (\theta^1, \cdots, \theta^5)$ with θ^0 implicit. The dual coordinate system maps the same point to a different vector with the notation $(\eta_1, \cdots, \eta_5)$ with η_0 implicit. The causal modeling of the die will take place in a sub–manifold of dimension $m < n - 1$. In this Chapter, all such models will reside in this sub–space of $m = 2$.

Thus, the IP will assign a numerical value to the probability that one of the six die faces will turn face up after being rolled conditioned on models of the form,

$$Q_i \equiv P(X = x_i \,|\, \mathcal{M}_k) \equiv P(X = x_i \,|\, \theta^1, \theta^2) \tag{36.1}$$

or, expressed equivalently in terms of the dual coordinate system,

$$Q_i \equiv P(X = x_i \,|\, \mathcal{M}_k) \equiv P(X = x_i \,|\, \eta_1, \eta_2) \tag{36.2}$$

Two constraint functions, $F_1(X = x_i)$ and $F_2(X = x_i)$, will have to be specified. Then, $F_1(X = x_i) - \langle F_1 \rangle$ and $F_2(X = x_i) - \langle F_2 \rangle$ will be the natural basis vectors $\mathbf{e_1}$ and $\mathbf{e_2}$ defining the two dimensional tangent space at any point p. The metric tensor $\mathbf{G}$ will be a 2×2 symmetric, positive definite matrix, of the form,

$$\mathbf{G} \equiv g_{rc}(p) = \begin{pmatrix} g_{11} & g_{12} \\ g_{21} & g_{22} \end{pmatrix} \tag{36.3}$$

The components of the metric tensor, $g_{rc}(p)$, are found by calculating the covariance between the constraint functions,

$$g_{rc}(p) = \sum_{i=1}^{6} \left[(F_r(X = x_i) - \langle F_r \rangle) \times (F_c(X = x_i) - \langle F_c \rangle) \right] Q_i \tag{36.4}$$

INFERENCES ABOUT DICE

The MEP formula tells us that the numerical assignments for the point in the manifold $\mathcal{S}^6$ for the fair model $\mathcal{M}_0$ will be computed by,

$$P(X = x_i \,|\, \mathcal{M}_0) = \frac{\exp\left[\,\lambda_1\, F_1(X = x_i) + \lambda_2\, F_2(X = x_i)\,\right]}{Z(\lambda_1, \lambda_2)} \tag{36.5}$$

Since we are discussing the model of a fair die, we know that both parameters of the fair model, that is both Lagrange multipliers, will be 0. The constraint functions, however they might ultimately be defined, simply do not come into play for the fair model.

The numerator for any of the six occurrences in the state space is computed to be the same value, $e^0 = 1$. The denominator, the partition function Z, is the sum over all six of these values, or,

$$Z(\lambda_1, \lambda_2) = \sum_{i=1}^{6} \exp\left[\,\lambda_1\, F_1(X = x_i) + \lambda_2\, F_2(X = x_i)\,\right] = 6$$

And, of course, we are led by the MEP formula to assign the numerical values,

$$P(X = x_i \,|\, \mathcal{M}_0) = \{1/6,\ 1/6,\ 1/6,\ 1/6,\ 1/6,\ 1/6\}$$

as the probability to each possible statement in the state space.

The presence of the universal constraint dictates,

$$\lambda_0 = 1 - \ln Z = -0.7918$$

so the coordinates for the fair model are,

$$\theta^0 = -0.7918$$

$$\theta^1 = 0$$

$$\theta^2 = 0$$

The dual coordinates will have the values,

$$\eta_0 = 1$$

$$\eta_1 = \sum_{i=1}^{6} F_1(X = x_i)\, P(X = x_i \,|\, \mathcal{M}_0)$$

$$\eta_2 = \sum_{i=1}^{6} F_2(X = x_i)\, P(X = x_i \,|\, \mathcal{M}_0)$$

These can be calculated as soon as we give explicit form to the constraint functions.

Also, we will wait to compute $g_{rc}(p)$ because both $F_j(X = x_i)$ and η_j are required. Later, we shall also calculate distances between the point p representing the fair model and other points in the manifold, $\{q, r, s, \cdots\}$, representing other models.

36.3 Physical Nature of the Constraints

Here we repeat Jaynes's rationale for constructing the constraint functions. For the sake of the numerical example, we are going to be content with two constraint functions. Both constraints are generated by thinking about the physical nature of the die, or as Jaynes puts it, "imperfections" or "deviations" from a perfect cube.

Jaynes examined the normed frequency counts for the six faces of the die over 20,000 throws and noticed a couple of curious features. There was a pattern of increasing frequency recorded in Wolf's data moving from the ONE to the SIX face. As a consequence, the average number of spots was ever so slightly greater than expected from a fair die.

Also, the frequencies of both the THREE and FOUR faces seemed somewhat more depressed than would be expected. Contemplating these experimental facts, Jaynes wondered what imperfections in the die might cause such discrepancies.

36.3.1 The first constraint function

Jaynes postulated two physical effects, the first related to the shift in the center of gravity, and the second related to the dimensions of a cube. Either, or both, of these physical effects could have an impact on the Q_i through the information in a model. Label the impact on the Q_i due to the shift in the center of gravity by δ_1, and the impact on the Q_i due to different lengths of the axes of the cube by δ_2.

Jaynes surmised that the shift in the center of gravity of the cube might very well be related to the difference in material excavated from opposite faces in order to indicate the spots.[1] Thus, there was a difference of 5 between the ONE spot and the SIX spot, a difference of 3 between the TWO spot and the FIVE spot, and a difference of 1 between the THREE spot and the FOUR spot.

List the linear impact of δ_1 on the fair model for all six individual Q_i as,

$$Q_1 = 1/6 - 2.5\,\delta_1$$

$$Q_2 = 1/6 - 1.5\,\delta_1$$

$$Q_3 = 1/6 - 0.5\,\delta_1$$

$$Q_4 = 1/6 + 0.5\,\delta_1$$

$$Q_5 = 1/6 + 1.5\,\delta_1$$

$$Q_6 = 1/6 + 2.5\,\delta_1$$

In summary, $Q_i = 1/6 + f_1(x_i)\,\delta_1$ where,

$$f_1(x_i) = (-2.5, -1.5, -0.5, +0.5, +1.5, +2.5)$$

The function $f_1(x_i)$ provides those differences between the spots on opposite faces.

---

[1] Jeffreys had advanced this idea earlier to explain frequency deviations in rolling dice.

INFERENCES ABOUT DICE

Of course, these effects in $f_1(x_i)$ have to add up to 1 because the Q_i, no matter what the impact of δ_1 or δ_2, must still add to 1. We observe that $f_1(x_i)\,\delta_1$ depresses Q_1, Q_2, and Q_3 by ever decreasing amounts, and that Q_4, Q_5 and Q_6 are boosted by ever increasing amounts.

Selecting then the first constraint function $F_1(X = x_i)$ as $f_1(x_i)$, form the constraint function average,

$$\langle F_1 \rangle = \sum_{i=1}^{6} F_1(X = x_i)\, Q_i$$

In detail this is spelled out as,

$$\sum_{i=1}^{6} F_1(X = x_i)\, Q_i =$$

$$[(-2.5) \times (1/6 - 2.5\,\delta_1)] + [(-1.5) \times (1/6 - 1.5\,\delta_1)] + [(-0.5) \times (1/6 - 0.5\,\delta_1)] +$$

$$[(+0.5) \times (1/6 + 0.5\,\delta_1)] + [(+1.5) \times (1/6 + 1.5\,\delta_1)] + [(+2.5) \times (1/6 + 2.5\,\delta_1)]$$

$$= 0 + 17.5\,\delta_1$$

Upon further reflection, we observe that this function of the observables,

$$F_1(X = x_i) = (-2.5, -1.5, -0.5, +0.5, +1.5, +2.5)$$

is simply the difference between the number of spots on the die face and 3.5, the average for a fair die.

$$F_1(X = x_i) = (1.0 - 3.5, 2.0 - 3.5, 3.0 - 3.5, 4.0 - 3.5, 5.0 - 3.5, 6.0 - 3.5)$$

If the first constraint function were changed to,

$$F_1^\star(X = x_i) = \{1, 2, 3, 4, 5, 6\}$$

it is easily seen that the average of this function for a fair die is 3.5.

Form the constraint function average for $F_1^\star(X = x_i)$ with the same Q_i showing the deviations from a perfect die,

$$\langle F_1^\star \rangle = \sum_{i=1}^{6} F_1^\star(X = x_i)\, Q_i$$

In detail this is,

$$\sum_{i=1}^{6} F_1^\star(X = x_i) Q_i =$$

$$[1 \times (1/6 - 2.5\,\delta_1)] + [2 \times (1/6 - 1.5\,\delta_1)] + [3 \times (1/6 - 0.5\,\delta_1)] +$$

$$[4 \times (1/6 + 0.5\,\delta_1)] + [5 \times (1/6 + 1.5\,\delta_1)] + [6 \times (1/6 + 2.5\,\delta_1)]$$

$$= 3.5 + 17.5\,\delta_1$$

Jaynes took the recorded average of the spots over the 20,000 rolls of the die, which turned out to be equal to 3.5983. He then subtracted the expected value for a fair die, $3.5983 - 3.5000 = 0.0983$, as an estimate for this particular physical effect $\delta_1 = 0.0983/17.5 = 0.0056$.

36.3.2 The second constraint function

The same approach can be taken with regard to the second physical imperfection related to the dimensions of a cube. The impact on the Q_i due to different lengths of the axes of the cube will be δ_2. If the length of the THREE–FOUR axis of the cube is longer than the ONE–SIX and TWO–FIVE axes, then the THREE and FOUR face will "feel" the impact of this imperfection in the dimensions of the cube. Once again, these effects in $f_2(x_i)$ on δ_2 must add up to 1 because the Q_i must still add to 1.

If we were to list the linear impact of δ_2 on the fair model for all six individual Q_i as,

$$Q_1 = 1/6 + 0.5\,\delta_2$$
$$Q_2 = 1/6 + 0.5\,\delta_2$$
$$Q_3 = 1/6 - 1.0\,\delta_2$$
$$Q_4 = 1/6 - 1.0\,\delta_2$$
$$Q_5 = 1/6 + 0.5\,\delta_2$$
$$Q_6 = 1/6 + 0.5\,\delta_2$$

Thus, $Q_i = 1/6 + 0.5 f_2(x_i)\,\delta_2$ where,

$$f_2(x_i) = (+1, +1, -2, -2, +1, +1)$$

We observe that δ_2 increases Q_1, Q_2, Q_5, and Q_6 by the same amount. Both Q_3 and Q_4 are decreased by the same amount.

INFERENCES ABOUT DICE

Go ahead and add a second constraint in order to reflect this suggested physical imperfection in the cube. Jaynes straightforwardly suggests this constraint function,

$$F_2(X = x_i) = \{+1, +1, -2, -2, +1, +1\}$$

to reflect the negative impact on the THREE and FOUR spots. The other four possibilities in the state space are not affected, and the overall impact must balance out in the end. Wolf's actual data were not close to an expectation of 0 for this function (considering 20,000 throws were made), but rather had a value of 0.1393.

Selecting then the second constraint function $F_2(X = x_i)$ as $f_2(x_i)$, form the constraint function average,

$$\langle F_2 \rangle = \sum_{i=1}^{6} F_2(X = x_i) Q_i$$

In detail this is,

$$\sum_{i=1}^{6} F_2(X = x_i) Q_i =$$

$$1 \times (1/6 + 0.5\,\delta_2) + 1 \times (1/6 + 0.5\,\delta_2) + (-2) \times (1/6 - 1.0\,\delta_2) +$$

$$(-2) \times (1/6 - 1.0\,\delta_2) + 1 \times (1/6 + 0.5\,\delta_2) + 1 \times (1/6 + 0.5\,\delta_2)$$

$$= 0 + 6\,\delta_2$$

An estimate of the effect of this second physical imperfection is then,

$$\delta_2 = \frac{0.1393}{6} = 0.0232$$

36.3.3 A supply of tentative models

Jaynes has creatively supplied us with a physical model of the die incorporating two imperfections. The first is an alteration to the center of gravity caused by unequal amounts of material removed from each face, while the second is a deviation from equality of all three axes of the cube caused by how the die was cut out from the supplied material.

The first constraint causes greater frequencies as the spots go from ONE to SIX, while the second constraint causes lower frequencies for the THREE and FOUR spots, while the ONE, TWO, FIVE and SIX spots are higher but equal among themselves. The overall impact from the action of both constraints could be a complicated interaction, but to a first approximation can be modeled as a combined linear effect.

We found that the linear impact on the Q_i from the displaced center of gravity was,
$$Q_i = 1/6\,[\,1 + 6\,\delta_1\,f_1(x_i)\,]$$
and the linear impact on the Q_i from the longer length of the THREE–FOUR axis was,
$$Q_i = 1/6\,[\,1 + 3\,\delta_2\,f_2(x_i)\,]$$
The combined effect from both imperfections is approximated by,
$$Q_i = 1/6\,[\,1 + 6\,\delta_1\,f_1(x_i)\,] \times [\,1 + 3\,\delta_2\,f_2(x_i)\,]$$
But Jaynes tells us that this expression is hardly different from,
$$Q_i = 1/6\,\exp\,[\,6\,\delta_1\,f_1(x_i) + 3\,\delta_2\,f_2(x_i)\,]$$
Pattern matching this expression with the standard MEP formula,
$$Q_i = \frac{e^{\lambda_1\,F_1(X=x_i) + \lambda_2\,F_2(X=x_i)}}{Z(\lambda_1, \lambda_2)}$$
the following analogies leap out at us,
$$\lambda_1 \approx 6\delta_1$$
$$\lambda_2 \approx 3\delta_2$$
$$F_1(X = x_i) \equiv f_1(x_i)$$
$$F_2(X = x_i) \equiv f_2(x_i)$$
$$Z(\lambda_1, \lambda_2) \approx 6$$

Qualitatively, having proposed this class of tentative models based on physical imperfections of a cube, we might reason about the normed frequencies expected in a large number of dice rolls: The ONE and TWO spots should be lower than 1/6, the THREE and FOUR spots should be right around 1/6 due to the first constraint, but if the effect of the second constraint is fairly strong could be lower than 1/6, and conceivably might even be lower than the ONE and TWO spots. Meanwhile, the frequency for the FIVE and SIX spots, being the beneficiary of both constraints, should be higher than 1/6, with the SIX spot having the largest frequency of all.

Having taken us this far with physical reasoning about possible imperfections in a cube, Jaynes completes the circle by telling us, in effect, that *information* about these physical effects can be inserted by models where the MEP will provide the appropriate numerical assignments to the probabilities for the six faces of the die.

INFERENCES ABOUT DICE

36.3.4 The MEP solution

We can now proceed to use the MEP algorithm to find the numerical assignments to the probabilities of the six faces based on this class of models with two constraint functions.[2] The information inserted into the distribution are the values of the constraint functions as given in the last section and the averages of these constraint functions.

Jaynes used the experimental data gathered over the 20,000 tosses of the die for the constraint averages. The information in Jaynes's model is then,

$$\langle F_1 \rangle = \overline{F}_1 = 0.0983 \text{ and } \langle F_2 \rangle = \overline{F}_2 = 0.1393$$

Thus, Jaynes is actually using the maximum likelihood model based on the sample averages for these two constraint functions over Wolf's 20,000 recorded dice rolls.

The MEP algorithm yields the following solution for the Lagrange multipliers: $\lambda_1 = 0.0317244$ and $\lambda_2 = 0.0717764$. This result indicates that the two constraints are indeed operating to cause deviations from the perfect cube of the fair die. The numerical assignment to the probability of the six faces based on this model was,

$$Q_i \equiv P(X = x_i \mid \mathcal{M}_k) = (0.16433, 0.16963, 0.14118, 0.14572, 0.18656, 0.19258)$$

The normed frequencies in Wolf's data are relatively close to these numerical assignments, so this model would have strong support from the data. Jaynes did a χ^2 test to reach the same conclusion. He decided to stop adding any more constraint functions when he thought this test indicated no further "signal" could be extracted from the data.[3]

Recall a very similar rationale employed when discussing logistic regression in Chapter Twenty Three of Volume II [8] where sample averages were also used as the information provided by a model. This model will always turn out to be the one most supported by the data, and is the Bayesian equivalent of maximum likelihood.

The qualitative reasoning carried out previously can now be bounced off these MEP numerical assignments. The steady increase in probability as the spots progress from ONE to SIX is evident from the first constraint. However, the effect from the second constraint is so strong that the probabilities of the THREE and FOUR spots are lowered significantly. As surmised, the probabilities for the FIVE and SIX spots are increased, with the SIX spot possessing the highest probability of all.

In fact, this kind of qualitative argument about which way the probabilities should move is supported by the linear adjustments that were made to the fair model probabilities based on the supposed imperfections in the cube. And it was this linear approach that Jaynes showed tied in neatly with the MEP assignment.

---

[2] The results reported here were calculated independently from Jaynes, but confirm his published answers. My answers to the Lagrange multipliers are defined to be the negative of Jaynes's calculation.

[3] This is the part of his analysis of Wolf's dice data that Jaynes later regretted.

For example, consider the numerical assignment Q_1 to the probability for the ONE face. The linear adjustments to the fair model probability of 1/6 to the ONE face consisted, first of all, of a displacement of the center of gravity due to the difference of five spots in excavated material between the ONE and SIX face. Secondly, the ONE–SIX axis was the same length as the TWO–FIVE axis, but shorter than the THREE–FOUR axis.

Thus, the linear argument produced,

$$Q_1 = 1/6 + \delta_1 f_1(x_1) + \delta_2 f_2(x_1) = 1/6 + (-2.5)(0.0056) + (1/2)(0.0232) = 0.16427$$

On the other hand, the adjustments to the fair model probability of 1/6 to the FOUR face consisted of a small center of gravity increase, and a lengthened THREE–FOUR axis. Thus, the linear argument here produced,

$$Q_4 = 1/6 + \delta_1 f_1(x_4) + \delta_2 f_2(x_4) = 1/6 + (+0.5)(0.0056) + (-1)(0.0232) = 0.14627$$

The first Lagrange multiplier is approximately equal to the strength of the first physical effect,

$$\lambda_1 \approx 6\delta_1 = 6(0.0056) = 0.0336$$

and the second Lagrange multiplier is approximately equal to the strength of the second physical effect,

$$\lambda_2 \approx 3\delta_2 = 3(0.0232) = 0.0696$$

The partition function for this model of $Z = 6.04$ was also close to the 6 appearing as a denominator in the linear approximation.

Table 36.1 at the top of the next page presents a comparison of the actual data, that is, the normed frequency counts, with the MEP assignment, the linear effects assignment, and the multiplicative or superposed assignment that led to the MEP. On the face of it, there doesn't seem much to distinguish these three assignments. We can't look at each assignment's information entropy measure and choose the one with the lowest $H(Q_i)$ because they don't all have the same constraint function averages.

However, what we might do is to look at the relative entropy of each of these three assignments compared to the actual data. The Kullback–Leibler information measure $\sqrt{2\,KL(p,q)}$ was calculated for each of the three assignments as shown in the final row of Table 36.1. The actual normed frequency counts played the role of point p, and each of the three assignments assumed the role, in turn, of point q. The MEP assignment is "closest" to the data, with the superposed assignment second, and the linear effects assignment the furthest away from the data.

Table 36.1: *Comparison of the normed frequency counts, the MEP assignment, a linear assignment, and a superposed assignment on which Jaynes based his MEP assignment.*

Face	N_i/N	MEP	Linear	Superposed
ONE	0.16230	0.164330	0.164232	0.163255
TWO	0.17245	0.169627	0.169849	0.169263
THREE	0.14485	0.141175	0.140641	0.141032
FOUR	0.14205	0.145725	0.146259	0.145867
FIVE	0.18175	0.186564	0.186701	0.187288
SIX	0.19660	0.192578	0.192318	0.193296
Sums	1.00000	1.000000	1.000000	1.000000
Distance from data		0.021701	0.023112	0.022090

36.3.5 The proper treatment of the data

However, Jaynes's interweaving of the model development and the observed results for Wolf's dice data, as just described, leaves a lot to be desired. What is the proper way to think about prediction based on all 20,000 throws of Wolf's dice?

In other words, what is an IP's degree of belief for the die to come up on any face on the *next* toss when it already has the results from 20,000 previous tosses? Because it is instructive to do so, let's recapitulate all of the stages we would pass through in reasoning this out.

Suppose the IP wants to assess its degree of belief that the next roll of the die will result in a THREE after 20,000 previous rolls. The beginning stage would pose two statements: (1) statement $A \equiv$ "A THREE will appear uppermost after the die has been tossed for the twentieth thousand and first time." and (2) statement $B \equiv$"A FOUR appeared on the first throw, a SIX appeared on the second throw, ..., a ONE appeared on 20,000th and final throw." The IP's degree of belief in statement A when conditioned on the truth of statement B is $P(A \mid B)$.

The known data are,

$$\mathcal{D} \equiv (X_1 = x_4) \text{ and } (X_2 = x_6) \text{ and } \ldots \text{ and } (X_{20,000} = x_1)$$

So at the next stage we write,

$$P(A \mid B) \equiv P(X_{20,001} = x_3 \mid \mathcal{D})$$

or since $N = 20,000$,

$$P(A \mid B) \equiv P(X_{N+1} = x_3 \mid \mathcal{D})$$

We have spent a lot of time in both Volumes I and II developing the general prediction formula based on the formal manipulation rules of probability. This is the formula for the probability of any number of future frequency counts conditioned on all known past frequency counts.

We now write out this general posterior prediction formula,

$$P(M_1, M_2, \cdots, M_n \mid N_1, N_2, \cdots, N_n)$$

Take the data to be the observed frequency counts $N_1, N_2, \cdots, N_n$ collected for each statement in the state space to arrive at an equivalent expression,

$$P(M_1, M_2, \cdots, M_n \mid \mathcal{D})$$

When the IP is interested in the ith statement at the *next* $(N+1)^{st}$ trial, this turns into,

$$P(M_1 = 0, M_2 = 0, \cdots, M_i = 1, \cdots, M_n = 0 \mid \mathcal{D})$$

For our current question of interest, we then have,

$$P(M_1 = 0, M_2 = 0, M_3 = 1, M_4 = 0, M_5 = 0, M_6 = 0 \mid \mathcal{D})$$

By Laplace's *Rule of Succession*, this has the answer of,

$$P(X_{N+1} = \text{THREE} \mid \mathcal{D}) = \frac{N_3 + 1}{N + n} = \frac{2897 + 1}{20000 + 6} = 0.144856543$$

Jaynes presents the raw frequency data in [22, pg. 327, Table 1].

Of course, this correct answer for the prediction of a future THREE cannot be much different than the normed frequency count for a THREE,

$$g_3 = \frac{N_3}{N} = \frac{2897}{20000} = 0.144850000$$

The maximum likelihood model based on the sample of $N = 20000$ with constraint function averages $\langle F_1 \rangle$ and $\langle F_2 \rangle$ set equal to the sample averages $\overline{F}_1$ and $\overline{F}_2$ returned $Q_3 = 0.14117$.

At this juncture, it important to emphasize that the correct answer of,

$$P(X_{N+1} = \text{THREE} \mid \mathcal{D}) = 0.14486$$

was an average over ALL models for the NEXT trial whereas,

$$Q_3 \equiv P(X_t = \text{THREE} \mid \mathcal{M}_{ML}) = 0.14117$$

is the prediction by ONE model on ANY trial.

In my opinion, Jaynes's presentation of the Wolf's dice data scenario is *backwards* from the way it should have gone. Having *first* calculated the "black–box" solution above for all $P(X_{N+1} = x_i \mid \mathcal{D})$, he should have then started investigating plausible

INFERENCES ABOUT DICE

physical models that might lay behind this result. This is the stage at which the MEP becomes really valuable. The MEP is superfluous when all models are being averaged over because you don't need any explicit breakdown for the q_i for the integration to take place successfully.

However, when you are seeking a restricted class of causal models that underpins the black–box prediction, the MEP does come into its own. This is where physical insight, buttressed by reliance on the MEP formula for the Q_i, helps to discover that restricted class of good causal models. This is exactly what Jaynes did so brilliantly when analyzing Wolf's dice data. He invented the constraint functions $F_1(X = x_i)$ and $F_2(X = x_i)$ to operationally implement within the MEP formula a way to capture physical deviations on a perfect cube like misplaced center of gravity and lengthened dimensions.

My only quibble is that the role of the data in forming the maximum likelihood model and the role of the data in the formal manipulation rules for deriving the prediction formula were jumbled up. Jaynes should have done a better job at distinguishing these concepts, but this criticism on my part is just a re–play of the whole conceptual confusion pounced on with a vengeance in Volume II.

DATA IS NOT INFORMATION
AND
INFORMATION IS NOT DATA

36.4 The MEP and Linear Systems

Jaynes identified three general things that might happen in the MEP formula when information is introduced through some model. In the first case, every thing goes swimmingly because every time the IP adds a new constraint function, together with its associated average, the MEP algorithm produces completely sensible output.

The information entropy must decrease with each additional constraint given the assumption that these constraints are not redundant. The information entropy, the quantitative measure of any missing information, is decreasing with each new model. The constraint functions are linearly independent, there is no redundancy, and no contradictory information has been introduced.

For example, the model with $F_1(X = x_i)$ and $\langle F_1 \rangle = 0.0983$ inserted more information than the fair model. The information entropy under this new model had to drop from the absolute maximum of $\ln n = \ln 6 = 1.7918$. This value is the quantitative measure of the amount of *missing* information under the fair model, and since no information was inserted by the fair model, the missing information must be at its maximum. Adding more information in the form of $F_2(X = x_i)$ and $\langle F_2 \rangle = 0.1393$ lowered the amount of missing information even further.

In the second case, suppose that the IP attempts to add *redundant* information through some model. The IP is unaware that some supposedly new constraint function and its average are not implementing any new additional information above and beyond what is already there. The MEP algorithm still functions properly, but reports back exactly the same numerical assignments Q_i. In addition, the MEP algorithm signals a redundant piece of information by telling us that the information entropy has not changed. The missing information has not been reduced, so no new information has been inserted.

For example, if the IP were to try a new model with the supposedly new information of $F_3(X = x_i) = (1, 2, 3, 4, 5, 6)$ and constraint function average of $\langle F_3 \rangle = 3.5983$, the numerical assignments Q_i are exactly the same as the previous model. The MEP algorithm works perfectly fine in this case.

We notice, however, that the information entropy has not budged at all when compared to the previous model with just $\langle F_1 \rangle$ and $\langle F_2 \rangle$. Thus, the IP has tried to insert redundant information. m is still at 2, and not at 3 as the IP might have supposed when it tried to add $\langle F_3 \rangle$. In the language of Linear Algebra, the matrix of constraint functions with a redundant constraint function is not linearly independent.

Even worse, in the third case, the IP attempts to insert additional information that is *contradictory* to the information already inserted via the existing functions and their averages. The MEP algorithm will NOT work properly, and error messages will proliferate. No legitimate assignments can be produced.

For example, suppose a model already exists with the above $F_1(X = x_i)$ and $F_2(X = x_i)$ and their averages. Then, the IP decides it wants to look at a new model with information $F_3(X = x_i) = (1, 0, 0, 0, 0, 0)$ and $\langle F_3 \rangle = 0.9$. This is incompatible with the information stipulated in $\langle F_1 \rangle$ and $\langle F_2 \rangle$. If $Q_1 = 0.9$, which by itself is legitimate, then no possible assignment could ever satisfy $\langle F_2 \rangle = 0.0983$.

It is a worthwhile exercise to examine the constraint functions and their averages simply as *linear systems*. This linear system is the set of $j = 1$ to m equations,

$$\sum_{i=1}^{n} F_j(X = x_i) Q_i = \langle F_j \rangle$$

or, in the general notation for linear systems,

$$\mathbf{A}\mathbf{x} = \mathbf{b}$$

where $\mathbf{A}$ is a given $(m \times n)$ matrix, $\mathbf{b}$ is a given length m vector, and $\mathbf{x}$ is the length n vector of unknowns to be solved for. Thus, matrix $\mathbf{A}_{m \times n}$ is the matrix of m constraint functions $F_j(X = X_i)$, the column vector $\mathbf{x}_{n \times 1}$ contains the n numerical assignments Q_i, and the column vector $\mathbf{b}_{m \times 1}$ contains the m constraint function averages $\langle F_j \rangle$ where j runs from 1 to m.

In a linear system, the vector **x** of unknowns is solved for through,

$$\mathbf{x} = \mathbf{A}^{-1}\mathbf{b}$$

We can find Q_i satisfying a *linear* system by multiplying the constraint function average vector by the inverse of the matrix of constraint functions. This solution might be surprisingly close to the Q_i found by the MEP algorithm.

We thought it important in Volume II, when dissecting the MEP formula as a Lagrange multiplier method variational problem, to emphasize the information entropy function as the objective function to be maximized, while the constraint function averages were mere side conditions. But consider reversing this attitude.

Now, the foremost goal of numerical assignment is to insert information from constraint function averages into a probability distribution. After this much has been accomplished, try to achieve a secondary goal of maximizing the information entropy. But the way the MEP is usually presented, one gets the impression that it is the other way around. We can use a linear systems analysis to shed some light on reversing the importance of constraint function averages *vis-à-vis* entropy maximization.

In saying all of this, I do not want to dismiss the absolutely essential role that maximizing the information entropy plays in the MEP formalism. It acts as a vital regularizer, or smoothing operation, to provide probabilities that (1) are not negative and (2) are as spread out as possible. A purely linear system for finding the Q_i will not accomplish this goal.

The MEP does an excellent job of handling the fact that we always have an *underdetermined* system when $m \ll n - 1$. In other words, our proposed models will attempt to insert information for m as small as possible when compared to n. In this Chapter, Jaynes's class of models inserted information from $m = 2$ physical imperfections of a cube when the state space of the die tossing scenario was $n = 6$.

The advantage of the MEP solution over a linear system solution is that it resolves any ambiguity a linear system might have for an underdetermined system by accepting that solution to the linear system that maximizes the information entropy. In this way, the entropy acts as a kind of *regularizer* to find acceptable solutions that might escape a purely linear systems approach.

In all of this discussion of linear systems and the MEP algorithm, one question must remain uppermost. What must remain *invariant*?

The numerical assignments Q_i are the only entities which must remain unaltered despite any transformations to the constraint functions and their averages. Thus, the $F_j(X = x_i)$ and their associated $\langle F_j \rangle$ need not be fixed in stone. Even though Jaynes said, in commenting on the constraint functions, that "here is where the physics comes in," there still remains a great deal of flexibility in the choice of the $F_j(X = x_i)$.

The $F_j(X = x_i)$ can be transformed if the appropriate transformation on the $\langle F_j \rangle$ also takes place. This means that any new $\langle F_j^\star \rangle$ must be,

$$\sum_{i=1}^{n} F_j^\star(X = x_i)\, Q_i = \langle F_j^\star \rangle$$

where the Q_i have not changed.

We have seen examples of this with Wolf's dice. It doesn't make any difference if the IP uses a model with the information,

$$F_1(X = x_i) = (-2.5, -1.5, -0.5, 0.5, 1.5, 2.5) \text{ with } \langle F_1 \rangle = 0.0983$$

or a model with the information,

$$F_1^\star(X = x_i) = (1, 2, 3, 4, 5, 6) \text{ with } \langle F_1^\star \rangle = 3.5983$$

because the Q_i which determine the constraint function averages are the same in both cases. If the coordinates $\eta_j \equiv \langle F_j \rangle$ must transform in a certain way, then since the Lagrange multipliers are the dual coordinates, they also must transform in a certain way.

I would recommend a linear system analysis of the constraint functions (**A**), assigned probabilities (**x**), and constraint function averages (**b**),

$$\mathbf{A}\,\mathbf{x} = \mathbf{b}$$

as a *diagnostic* tool.

As a diagnostic tool, a linear systems analysis could be viewed as a helpful adjunct to any MEP algorithm in order to help identify those main characteristics of information that Jaynes often highlighted and were the recent objects of our discussion. It is helpful to find out through the linear system analysis, prior to invoking any MEP algorithm, whether the IP's initial thoughts about introducing information via the models are well–behaved, or whether there is some unsuspected redundant information, or, even worse, contradictory information that the IP has tried to insert through a model.

If there is a problem that arises from the linear system analysis, it probably would be a good idea for the IP to re–think its approach as to how it might want to incorporate information through the models.

See Exercises 36.9.2, 36.9.5, 36.9.6, and 36.9.7 for further numerical examples of the ideas presented in this section. These exercises also highlight a few pertinent *Mathematica* built–in functions for performing linear systems analysis as a prelude to a standard MEP analysis.

36.5 Calculating the Metric Tensor

It is now time to return to those things that IG is concerned with. The constraint functions and their averages for Wolf's dice data have been thoroughly discussed, so let's calculate the metric tensor at the tangent space of any point in the manifold. We labeled p the assignment for the fair model, so calculate the metric tensor at this point first.

Use the definition for the metric tensor ensuing from the inner product of the natural basis vectors $\mathbf{e_r}$ and $\mathbf{e_c}$ given earlier where a fixed r and c indicates the component $g_{rc}(p)$ of the matrix $\mathbf{G}$,

$$\langle \mathbf{e_r}, \mathbf{e_c} \rangle = \sum_{i=1}^{6} (F_r(X = x_i) - \langle F_r \rangle) \times (F_c(X = x_i) - \langle F_c \rangle) \, P(X = x_i \mid \mathcal{M}_p) \quad (36.6)$$

The constraint averages under the fair model $\mathcal{M}_p$ are,

$$\langle F_1 \rangle = 3.5$$

$$\langle F_2 \rangle = 0$$

The metric tensor at point p is found to be,

$$\mathbf{G} \equiv g_{rc}(p) = \begin{pmatrix} 2.9167 & 0 \\ 0 & 2 \end{pmatrix}$$

This matrix is symmetric and positive definite. The variance of $F_1(x_i) = 2.9167$, while the variance of $F_2(x_i) = 2$. The covariance between the first and second constraint functions is 0. The easy details involved in determining $g_{22}(p) = 2$ for $r = 2$ and $c = 2$ are shown below in Table 36.2. The answer is the sum over the final column.

Table 36.2: *The calculation of $g_{22}(p)$ using Equation (36.6).*

C1	C2	C3	C4	C5	C6
i	$F_2(x_i) - \langle F_2 \rangle$	$F_2(x_i) - \langle F_2 \rangle$	C2 x C3	$P(x_i \mid \mathcal{M}_p)$	C5 x C4
1	+1	+1	1	1/6	1/6
2	+1	+1	1	1/6	1/6
3	−2	−2	4	1/6	4/6
4	−2	−2	4	1/6	4/6
5	+1	+1	1	1/6	1/6
6	+1	+1	1	1/6	1/6
	Sums			1	2

Let point q be Jaynes's MEP solution for Wolf's data. q has different coordinates than p, different basis vectors, and a different metric tensor. The new coordinates in both coordinate systems for q are,

$$\theta^1 = 0.0317$$

$$\theta^2 = 0.0718$$

$$\eta_1 = 0.0983$$

$$\eta_2 = 0.1393$$

The metric tensor at point q is,

$$g_{rc}(q) = \begin{pmatrix} 3.0942 & 0.0778 \\ 0.0778 & 1.8413 \end{pmatrix}$$

Some numerical examples showing how *Mathematica* makes the calculations for this metric tensor are part of the Exercises and Appendix C.

36.6 Distance to Nearby Points

Consider another point s very close by to point p. In other words, we are thinking about another model that assigns numerical values to the six faces of the die that are very similar to those assigned by the fair model. What is the distance between these two nearby points, or, what is the separation between these two probability distributions? What information separates the two distributions?

The coordinates for the fair model were $\{\theta^1 = 0, \theta^2 = 0\}$. Set new coordinates for point s with a $d\theta = 0.001$, or $\{\theta^1 = 0.001, \theta^2 = 0.001\}$. The physical effects from the two constraints are now going to impact the frequencies recorded for the die, but these parameters from the new model indicate that they are very weak. They shouldn't cause much perturbation to the fair model's all $Q_i = 1/6$.

The new values of the dual coordinates for s will be calculated as the expectation of the constraint functions after the new numerical assignments to the probabilities are found. Of course, the IP could have specified the dual coordinates first as the constraint averages and then found the resulting values of the Lagrange multipliers.

Given the new parameters under model $\mathcal{M}_s$, the IP finds that $Z_s = 6.02105$ as compared to $Z_p = 6$. The numerical assignments under the new model differ only slightly from the fair model. More importantly, they differ in the expected direction,

$$Q_i = P(X = x_i \,|\, \mathcal{M}_s) = \{0.1664, 0.1666, 0.1663, 0.1664, 0.1671, 0.1672\}$$

There is an upward trend in the assigned probabilities, interrupted by the depressed values for the THREE and FOUR spots, with the SIX spot having the largest value.

INFERENCES ABOUT DICE

With $P(X = x_i \mid \mathcal{M}_s)$ known, the expectations of the two constraint functions, and thus the dual coordinates η_1 and η_2 are found to be,

$$\langle F_1 \rangle_s = \eta_1 = 3.5029$$

and

$$\langle F_2 \rangle_s = \eta_2 = 0.0020$$

The metric tensor at point p has already been calculated,

$$g_{rc}(p) = \begin{pmatrix} 2.91667 & 0 \\ 0 & 2 \end{pmatrix}$$

The distance between the two points as calculated by the local linearization in the tangent space at p where $m = 2$,

$$\| \mathbf{p} - \mathbf{s} \| = \sqrt{(\mathbf{p} - \mathbf{s}) \cdot \mathbf{G}(\mathbf{p}) \cdot (\mathbf{p} - \mathbf{s})^{\mathbf{T}}}$$
$$= 0.00221736$$

The distance between these very close points as calculated by a function of the relative entropy $KL(p,s)$ is closely approximated by the local linearization in the tangent space as just found,

$$KL(p, s) = \sum_{i=1}^{6} p_i \ln\left(\frac{p_i}{s_i}\right)$$

$$\| \mathbf{p} - \mathbf{s} \| \approx \sqrt{2 \times KL(p,s)}$$

$$\sqrt{2 \times KL(p,s)} = 0.00221781$$

36.7 How the Dual Coordinate Systems Vary

As a prelude to a more formal study of the joint variation of the dual coordinate systems, let's examine here, while still remaining within the context of Jaynes's analysis of Wolf's dice data, how one set of coordinates change when responding to changes to its dual coordinates.

We will begin in a numerical fashion by examining how the θ coordinates vary when the model space changes in response to changes in the η coordinates. In other words, if a model's parameters are changed through the constraint function averages, how are the Lagrange multipliers changing in response?

For the remainder of this numerical example, let the first constraint function be defined as,

$$F_1(X = x_i) = (1, 2, 3, 4, 5, 6)$$

It is easiest to begin with the observation that η_1 can only take on values between 1 and 6. As the average of the first constraint function, these extremes are easily identified. $\eta_1 = 1$ is reached when the numerical assignment by some model makes the ONE face a certainty,

$$Q_i = P(X = x_i \mid \mathcal{M}_k) = (1, 0, 0, 0, 0, 0)$$

and likewise, $\eta_1 = 6$ is reached when the numerical assignment by some model makes the SIX face a certainty,

$$Q_i = P(X = x_i \mid \mathcal{M}_k) = (0, 0, 0, 0, 0, 1)$$

For all other numerical assignments to the probabilities, $1 < \eta_1 < 6$. For example, Jaynes set $\eta_1 = 3.5983$ for Wolf's dice data, picking this model, without explicitly saying so, because it was the maximum likelihood model.

For the remainder of this numerical example, let the second constraint function be defined as,

$$F_2(X = x_i) = (1/2, 1/2, -1, -1, 1/2, 1/2)$$

Through a similar analysis, η_2 must vary between a minimum of -1 and a maximum of $1/2$. The minimum of $\eta_2 = -1$ is reached, for example, when the numerical assignment makes $Q_1 = Q_2 = Q_5 = Q_6 = 0$ and $Q_3 = Q_4 = 1/2$.

What happens to θ^1 and θ^2 as the model space parameters expressed in terms of η_1 and η_2 are changing? We've already seen that when the model $\mathcal{M}_p$ is specified through $\eta_1 = 3.5$, and $\eta_2 = 0$, both θ^1 and θ^2 take on the value 0.

Suppose, for the sake of a numerical example, that we want to investigate what happens to the θ coordinates when η_1 approaches 4, and η_2 approaches -1. In other words, the numerical assignment to the probabilities for the six faces of the die is approaching that case just mentioned where the probability of the FOUR face is nearly 1 and the probabilities for the other five faces are all approaching 0.

Before the quantitative calculations take place, it is always a good strategy to conduct a qualitative argument based on the physics of a cube just as Jaynes did for Wolf's die. This helps prevent egregious blunders in blindly accepting computer output. It's obvious before we even start off on this enterprise that the physical characteristics of such a "die" that is almost certain to show a FOUR on every toss must be pretty bizarre.

If we want the FOUR face to have a very high probability, and the other five faces combined an extremely low probability, then this is asking, in some sense, for the first constraint function to still operate in somewhat the same way as it did in explaining Wolf's dice data. Thus, the "strength" of the first constraint function postulating some displacement in the center of gravity toward the higher number faces should still be present. But the face with larger number of spots is much more highly favored than before. The THREE face has a chunk of steel close to it. The first Lagrange multiplier λ_1 should be much larger than a real die.

However, this first physical effect is overwhelmed by the second physical effect of the length of the THREE–FOUR axis of the cube. The die becomes a very thin plate with an extremely short THREE–FOUR dimension and relatively larger ONE–SIX and TWO–FIVE dimensions. Thus, when this extremely flat "die" is "tossed" it is almost certainly going to land with either the THREE or the FOUR face showing. The "strength" of this second constraint, reflected in the second Lagrange multiplier λ_2, must be even greater than λ_1. But the THREE face is very heavy and always lands face down. Thus, the FOUR face with physical characteristics of this die is almost certainly going to land with the FOUR face up.

These kinds of qualitative arguments are also valuable because they highlight the central role of the constraint functions. The two constraint functions were initially thought up because of the pressures brought to bear by Wolf's extensive data. Now, however, invoking them for the current scenario seems a bit tortured. Jaynes once issued the cryptic comment, "The constraint functions are where you put in the *physics*." Personally, I wish that he had spent a lot more time elaborating on that particular topic.

Now on to the actual computations for the numerical example. Careful study of Table 36.3 below and Table 36.4 two pages over will be worthwhile.

As known reference points, we begin in row 1 with the fair die model, and row 2 with Jaynes's model for Wolf's imperfect die. The fair model is specified by setting the parameters $\eta_1 = 3.5$ and $\eta_2 = 0$. The MEP algorithm finds that dual coordinate system parameters are $\theta^1 = \theta^2 = 0$ with, of course $Q_4 = 1/6$.

Table 36.3: *A numerical example of the θ^j coordinate system changing in response to changes in the dual coordinate system η_j. The average values of the two constraint functions are approaching* 4 *and* -1.

η_1	η_2	θ^1	θ^2	Q_4
3.5000	+0.0000	0.0000	0.0000	0.1667
3.5983	+0.0697	+0.0317	+0.1436	0.1457
3.7000	−0.1000	+0.0760	−0.1995	0.2076
3.8000	−0.6000	+0.2421	−1.2118	0.4108
3.9000	−0.8000	+0.6874	−2.2057	0.5767
3.9900	−0.9000	+1.4984	−3.7589	0.7628
3.9900	−0.9500	+2.4886	−5.5631	0.8926

Jaynes's model must have the different η_j and θ^j parameters shown in the second row. Mainly because of the influence of the lengthened THREE–FOUR axis, the assigned value of $Q_4 = 0.1457$ is lower than the fair model's assigned value of $1/6$. The next five rows gradually adjust the model parameters so that the assigned probability to the FOUR face is increasing and approaching 1.

The numerator in the MEP formula for the FOUR face will always be,

$$\exp\left[\lambda_1 F_1(X = x_4) + \lambda_2 F_2(X = x_4)\right] = \exp\left[4\lambda_1 - \lambda_2\right]$$

when the mappings $F_1(X = x_4) = 4$ and $F_2(X = x_4) = -1$ are substituted for the statement,

$$(X = x_4) \equiv \text{"The die showed the FOUR face."}$$

For example, in row 3, the numerator for Q_4 is,

$$\exp\left[(4 \times 0.0760) + 0.1995\right] = 1.6545$$

The partition function Z, found as the sum over all six numerators, is,

$$Z = \sum_{i=1}^{6} \exp\left[\lambda_1 F_1(X = x_i) + \lambda_2 F_2(X = x_i)\right] = 7.9695$$

resulting in an assigned probability to the FOUR face under this model of,

$$Q_4 = P(X = x_4 \mid \mathcal{M}_k) = \frac{1.6545}{7.9695} = 0.2076$$

As we let the model parameters η_1 and η_2 approach 4 and -1, we can watch the concomitant change in the dual parameters θ^1 and θ^2. The value of θ^1 does, as argued, become larger to enforce the effect of the change in the center of gravity depending on the difference in the amount of material excavated for the number of spots on each face. This physical effect, in and of itself, would also be raising the probability for the FIVE and SIX faces.

However, the second constraint function reflecting the shortened dimension of the THREE–FOUR axis is being enforced even more strongly. As shown in the final row, when the model parameters have reached $\eta_1 = 3.99$ and $\eta_2 = -0.95$, the numerical assignment to the probability for the FOUR face has climbed drastically to $Q_4 = 0.8926$. The FIVE and SIX faces have only a combined probability of about 0.033, while the THREE face has a probability of about 0.07. The combined probability for the ONE and TWO faces is close to zero.

Table 36.4 at the top of the next page takes this example one step further by continuing on the path begun in the previous table by focusing in on models with an even closer approach to 4 and -1. We examine models where η_1 goes from 3.9900 to 3.9990 and η_2 from -0.9900 to -0.9992. In the final row, and with the final model listed, we are very close to reaching $Q_4 = 1$ with $Q_4 = 0.9974$.

At these rather extreme numbers, the non–linearity between the coordinates η_j and θ^j becomes more evident. η_1 only has to increase by 0.001 to reach 4, and η_2 only has to decrease by 0.0008 to reach -1 to achieve $Q_4 = 1$. But θ^1 must increase all the way to $+\infty$ and θ^2 must decrease all the way to $-\infty$ during these small η intervals for a comparable effect.

INFERENCES ABOUT DICE

Table 36.4: *A continuation of the numerical example of how the θ^j coordinate system changes in response to changes in the dual coordinate system η_j. The numerical assignment to Q_4 is approaching 1.*

η_1	η_2	θ^1	θ^2	Q_4
3.9900	−0.9900	3.7343	−8.3150	0.9702
3.9990	−0.9990	6.0581	−12.9525	0.9970
3.9990	−0.9991	6.1171	−13.1015	0.9972
3.9990	−0.9992	6.1798	−13.2636	0.9974

Computationally, we also start to see some extreme numbers. For the model in the final row, Q_1 is on the order of 10^{-17} and Z is on the order of 10^{16}. Of course, given that Q_4 is very close to 1 and Z is on the order of 10^{16}, the numerator of Q_4,

$$\exp\left[\,(4 \times 6.1798) + 13.2636\,\right]$$

must also be on the order of 10^{16}.

36.7.1 The metric tensor sheds some light on this example

The changes in one set of coordinates is related differentially to the changes in the dual set through,

$$\frac{\partial \eta_r}{\partial \theta^c} = g_{rc}(p)$$

and,

$$\frac{\partial \theta^r}{\partial \eta_c} = g^{rc}(p)$$

In section 36.5, the metric tensor $\mathbf{G} = g_{rc}(p)$ at point p was shown as,

$$\mathbf{G} = \begin{pmatrix} 2.9167 & 0 \\ 0 & 2 \end{pmatrix}$$

The inverse of the metric tensor at the same point is,

$$\mathbf{G^{-1}} = \begin{pmatrix} 0.342857 & 0 \\ 0 & 0.5 \end{pmatrix}$$

With matrix notation it is obvious that,

$$\mathbf{G}\mathbf{G^{-1}} = \mathbf{I}$$

where $\mathbf{I}$ is the identity matrix. The matrix notation is easier on the eye than the equivalent subscripted superscripted notation together with the Einstein summation convention,

$$\{\mathbf{G}\mathbf{G^{-1}} = \mathbf{I}\} \equiv \left\{ g_{jk}\, g^{jk} = \delta^j_k \right\}$$

We can exploit the metric tensor to discover how changes to one parameter will affect its dual. For example, we might be interested in investigating some of the more extreme models in Table 36.4. Arbitrarily labeling as point s where the parameters are $\eta_1 = 3.9990$ and $\eta_2 = -0.9991$, we can ask the question: how much will a slight change to η_2 impact θ_1? If η_2 is decreased by 0.0001, approximately how much will θ_1 increase?

Calculate the inverse of the metric tensor at point s to find the component in the first row and second column,

$$\frac{\partial \theta_1}{\partial \eta_2} = g^{12}(s) = -607.39739$$

A decrease of 0.0001 from $\eta_2 = -0.9991$ to $\eta_2 = -0.9992$ will cause a change to $\theta_1 = 6.1171$ of approximately,

$$0.0001 \times (-607.39739) = -0.06074 \quad \rightarrow \quad 6.1171 - \theta_1 = -0.0607$$

$$\rightarrow \quad \theta_1 \approx 6.1778$$

as compared to the actual value of $\theta_1 = 6.1798$ at $\eta_2 = -0.9992$.

The metric tensor and its inverse also highlight the numerical instability that will ensue when the coordinates are approaching some singular value. The components of the metric tensor become very small as the models start to approach $\eta_1 \rightarrow 4$ and $\eta_2 \rightarrow -1$. The consequence is that the entries for the inverse of the metric tensor, $g^{rc}(s)$, become very large. The upshot is that small changes to the constraint function averages,

$$\frac{\partial \theta^r}{\partial \eta_c} = g^{rc}(s)$$

are going to result in relatively large changes in the Lagrange multipliers. This effect was becoming evident for the last few models investigated in Table 36.4.

36.8 Connections to the Literature

Who better to have the final word in this Chapter than Jaynes himself? Here are some quotes from various papers where Jaynes addresses conceptual issues arising from his analysis of various dice tossing scenarios.

As early as 1962, Jaynes [15, pp. 41–45] had been using dice to illustrate the maximum entropy approach. He gave a detailed mathematical account of the 1962 scenario later on in 1978 [17, pg. 244] illustrating how one could use MEP to assign probabilities to the six faces of a die based on the one constraint that the average number of spots differed from the value of 3.5 that one would expect from a perfect die. Further on in that same article [17, pp. 258–268], he presents the example that we have adapted for use in this Chapter, analyzing 20,000 throws of one of two dice as recorded by the 19th Century Swiss astronomer Rudolph Wolf.

In analyzing data from dice tossing, Jaynes is trying to tell us that probability theory is concerned with making logical inferences based on incomplete information. It is in this sense that we have used the adjective "epistemological" to describe probability theory as logic. It is for the very same reason that we constantly refer back to an "information processor" and its "state of knowledge."

Probability is not something attached to physical objects which causes them to act "randomly." (Yes, we are talking about you, quantum mechanics.) Because probability is concerned with logical inferences made by information processors, inferences may transcend the physical constraints of space, time, and matter.

The same general techniques that require both the application of the formal manipulation rules from probability theory and a numerical assignment algorithm like the MEP can be employed to study disparate topics such as image restoration, statistical mechanics, and dice tossing with equal aplomb! Jaynes [18, pg. 328] explains it thusly,

> Today, another sixty years have passed, [since an unsatisfactory analysis of the Wolf dice data by the economist John Maynard Keynes] and to the best of the writer's knowledge no statistician has ever attempted to draw any specific inferences about the imperfections in Wolf's dice from these data. Yet to a physicist those data tell us something very clear and simple about his dice; information that can be extracted by a straightforward entropy analysis that does not require us to go into complicated mechanical details. ... it still seems generally believed that the principles of statistical mechanics apply only to molecules; and not to dice.

Jaynes exploits the MEP for hypothesis testing in earnest [18, pp. 320–321],

> ... in a real application one will wish, if possible, to choose the constraint matrix [...] so that the resulting quantities [...] represent systematic physical influences, real or conjectured, (for example, eccentric position of the center of gravity of a die), which constrain the frequencies to be different from the uniform distribution of absolute maximum entropy [...]. In using entropy analysis for hypothesis testing, the mathematical relations are used in the other direction, considering the [frequencies] as known experimentally. A successful hypothesis about the systematic influences is then one for which the experimentally observed entropy [...] is sufficiently close to the maximum [...] permitted by the assumed constraints ...

One could imagine conducting a very sophisticated, albeit a very complicated, simulation of this die tossing problem according to the known laws of physics. The shift in the center of gravity, unequal dimensions of the cube, the height at which the die was tossed, the properties of the surface of the table, the exact method by which the die was tossed, the temperature of the room, and so on, *ad infinitum*, (or to the exhaustion of the physical insight of the experimenter) could be specified to determine which die face turns up on a given trial.

Then, all the factors which could not be physically controlled from trial to trial would have to be varied and the next toss simulated. In this way we could tabulate the frequencies of the die face over a large number of trials. But Jaynes [17, pp. 266–267] tells us that such a Herculean effort is not necessary. Judicious employment of the maximum entropy principle allows us to avoid the inconsequential details that would just cancel out over the many trials of the simulation.

> Far from "ignoring the physics," it [the maximum entropy principle] leads us to concentrate our attention on the part of the physics that is *relevant*. Success in using it does not require that we take into account all of the dynamical details; it is enough if we can recognize, whether by common-sense analysis or by inspection of the data, *what are the systematic influences at work, that represent the "physical constraints?"* If by any means we can recognize these, maximum entropy then takes over and supplies the rest of the solution, which does not depend on dynamical details but only on counting the possibilities.

Physicists invoked a similar rationale when trying to explain statistical mechanics. For example, to determine the frequencies with which molecules occupy various energy levels, it is not necessary to know the exact dynamical history of every molecule in a gas; it is sufficient to know the relevant physical constraints, for example, the temperature, the pressure, and the volume.

Jaynes [17, pg. 270] goes on to mention some of the important characteristics of the constraint functions from the standpoint of Linear Algebra. For example, the MEP prefers that the set of constraint functions $F_j(X = x_i)$ be linear independent vectors. It is a curious feature of the MEP algorithm that it automatically ignores *redundant* constraint functions. Jaynes makes note of a mathematical point that the metric tensor is positive definite if the constraint functions *are* linearly independent.

> If the [constraint functions] are linearly independent, the manifold ... has dimensionality m. Otherwise, [the manifold] is of some dimensionality $m' < m$; the set of [constraint] functions ... is then redundant, in the sense that at least one of them could be removed ... While the presence of redundant [constraint] functions ... proves to be harmless in that it does not affect the actual results of entropy maximization, ... it is a nuisance for the present purposes .... In the following we assume that any redundant functions have been removed, so that $m' = m$.

Reflecting on Jaynes's analysis of Wolf's dice was my personal motivation for the "Signals plus Noise" characterization of models as introduced in Chapter Thirty Two of Volume II [8, pg. 521]. Jaynes advocated an intuitively compelling strategy of "subtracting out" from the data whatever new contributions there might be from ever increasingly complicated models. He suggested to keep on doing this until you reach a point where you have extracted a "signal," and all that's left is the "noise."

For Wolf's dice data, Jaynes first subtracted out the signal due to a displaced center of gravity, but discovered that there was still a recognizable pattern to the residue. He then subtracted out the effects due to a lengthened THREE–FOUR axis and deemed that at this point all that was left was noise.

Jaynes reluctantly settled on a chi–square criterion for judging whether he had, in fact, reached the pure noise stage. He expressed regrets about this choice later on. Ironically, in the mid 1950s (Jaynes's Mobil–Socony lectures), at a much earlier date than his discussion of Wolf's dice data, he had already hinted that an alternative criterion based on relative entropy might be more appropriate.

Despite the delight we experience in observing Jaynes's thought processes as he went about analyzing Wolf's dice from the physical perspective of defects to a perfect cube, we do, in the end, require something more definitive than Jaynes's intuition. Nonetheless, Jaynes's breezy excursions into what constitutes a "signal" and the "noise" for Wolf's dice are certainly an interesting beginning rationale.

It's our job to accept his insight about the physical characteristics of dice as they relate to the MEP. Then, we have the interesting task of discovering how well this mode of analysis might translate over to other problems. In particular, how *does* one capture relevant causal variables within the MEP constraint functions $F_j(X = x_i)$ in problems other than the rolling of dice?

36.9 Solved Exercises for Chapter Thirty Six

Exercise 36.9.1: Find Q_3 as a superposition of two linear effects and compare to the Q_3 from the MEP formula.

Solution to Exercise 36.9.1

In section 36.3.3, a formula was derived for the superposition of the linear impact of two physical imperfections in a die. For the specific case of the departure from a fair die for the THREE face,

$$Q_3 = 1/6\,[\,1 + 6\,\delta_1\,f_1(x_3)\,][\,1 + 3\,\delta_2\,f_2(x_3)\,]$$

$$f_1(x_3) = -0.5$$

$$f_2(x_3) = -2.0$$

$$\delta_1 = \frac{0.0983}{17.5}$$

$$= 0.005617$$

$$\delta_2 = \frac{0.1393}{6}$$

$$= 0.023217$$

$$Q_3 = 1/6\,[\,1 + 6\,(0.005617 \times (-0.5))\,][\,1 + 3\,(0.023217 \times (-2))\,]$$

$$= 0.141032$$

Calculating Q_3 from the MEP formula with the same constraint functions and constraint function averages yields,

$$Q_3 = \frac{e^{\lambda_1\,F_1(X=x_3) + \lambda_2\,F_2(X=x_3)}}{Z(\lambda_1, \lambda_2)}$$

$$= \frac{e^{(0.0317244 \times (-0.5)) + (0.0717764 \times (-2))}}{6.03962}$$

$$= 0.141175$$

Exercise 36.9.2: Use *Mathematica* to solve a set of linear equations in order to find the Q_i.

Solution to Exercise 36.9.2

Mathematica possesses a built–in function **LinearSolve[** *arg1, arg2* **]** with two arguments. Matrix **A** is *arg1*, and vector **b** is *arg2*. This function finds the answer

SOLVED EXERCISES FOR CHAPTER 36

to the set of linear system equations $\mathbf{A}\mathbf{x} = \mathbf{b}$ by returning the vector $\mathbf{x}$. For our situation, the matrix $\mathbf{A}$ will be the matrix of constraint functions **cm**, the vector **b** will be the vector of constraint function averages **cfa**, while the vector $\mathbf{x}$, the sought–for solutions to the linear system, will be the Q_i.

For the first numerical example let,

$$\mathtt{cm1 = \{\{1,\ 1,\ 1,\ 1,\ 1,\ 1\},\ \{1,\ 2,\ 3,\ 4,\ 5,\ 6\}\}}$$
$$\mathtt{cfa1 = \{1,\ 3.5\}}$$

This would be the linear system equivalent to the fair model. Then, evaluate,

$$\mathtt{LinearSolve[cm1,\ cfa1]}$$

which returns the Q_i as (1/6, 1/6, 1/6, 1/6, 1/6, 1/6), obviously the same result that the MEP formula would return for this model.

What about the reformulation of the first constraint function to,

$$F_1(X = x_i) = (-2.5, -1.5, -0.5, +0.5, +1.5, +2.5)$$

as it was discussed for Wolf's dice data? Then we would have,

$$\mathtt{cm2 = \{\{1,\ 1,\ 1,\ 1,\ 1,\ 1\},\ \{-2.5,\ -1.5,\ -0.5,\ 0.5,\ 1.5,\ 2.5\}\}}$$
$$\mathtt{cfa2 = \{1,\ 0\}}$$

This would also be the same linear system equivalent to the fair model with,

$$\mathtt{LinearSolve[cm2,\ cfa2]}$$

also returning the Q_i as (1/6, 1/6, 1/6, 1/6, 1/6, 1/6).

How long can these solutions to the linear system go on to duplicate the MEP results? Consider next that model where the information about the displaced center of gravity and longer length of the THREE–FOUR dimension was included. Now we have,

$$\mathtt{cm3 = \{\{1,\ 1,\ 1,\ 1,\ 1,\ 1\},}$$
$$\mathtt{\{-2.5,\ -1.5,\ -0.5,\ 0.5,\ 1.5,\ 2.5\},}$$
$$\mathtt{\{1,\ 1,\ -2,\ -2,\ 1,\ 1\}\}}$$
$$\mathtt{cfa3 = \{1,\ 0.0983,\ 0.1393\}}$$

The linear system assignment,

$$\mathtt{LinearSolve[cm3,\ cfa3]}$$

returns the Q_i as,

$$Q_i = (0.164232, 0.169849, 0.140641, 0.146259, 0.186701, 0.192318)$$

which are barely different than the Q_i from the MEP formula listed in Table 36.1.

The information entropy of this linear system assignment must be nearly the same as the MEP assignment. Using the following *Mathematica* expression for the information entropy,

```
- Total[p Log[p]]
```

the linear system assignment evaluates to 1.7852223, while the information entropy from the MEP algorithm is 1.7852240. This difference in the information entropy between the linear assignment and the MEP assignment is only 0.0000017.

This result highlights the fact that including the constraint average information into the numerical assignment is the primary goal, while maximizing the entropy is, in some sense, secondary. Although, it *is* interesting to take note that the linear solution, as any non–MEP assignment must do, has included ever so slightly more unwanted information than the MEP solution which, by definition, must possess the maximum value of the information entropy.

Exercise 36.9.3: Compare two MEP solutions for the Q_i in Jaynes dice scenario which contain "different" information from two models.

Solution to Exercise 36.9.3

Compare the MEP solution as just calculated in the last exercise for the Q_i in an $m = 2$ model where the information is,

$$F_1(X = x_i) = (1, 2, 3, 4, 5, 6)$$

$$\langle F_1 \rangle = 3.5983$$

$$F_2(X = x_i) = (1/2, 1/2, -1, -1, 1/2, 1/2)$$

$$\langle F_2 \rangle = 0.06965$$

with the $m = 2$ model Jaynes used in the solution to Wolf's dice data where the information is,

$$F_1^\star(X = x_i) = (-2.5, -1.5, -0.5, 0.5, 1.5, 2.5)$$

$$\langle F_1^\star \rangle = 0.0983$$

$$F_2^\star(X = x_i) = (1, 1, -2, -2, 1, 1)$$

$$\langle F_2^\star \rangle = 0.1393$$

The Q_i produced from these two "different" models are exactly the same. The information in the two models is exactly the same. The Lagrange multipliers for the first model are,

$$\lambda_1 = 0.031724$$

$$\lambda_2 = 0.143553$$

while the Lagrange multipliers for Jaynes's model are,

$$\lambda_1^\star = 0.031724$$

$$\lambda_2^\star = 0.071776$$

We observe that λ_1 is exactly the same as $\lambda_1^\star$, and λ_2 is exactly twice $\lambda_2^\star$. $\langle F_1^\star \rangle$ is $\langle F_1 \rangle + 3.5$, while $\langle F_2 \rangle$ is one–half of $\langle F_2^\star \rangle$. This observation is motivated by Jaynes telling us that transformations on the $F_j(X = x_i)$ and the λ_j do not necessarily result in different information.

Exercise 36.9.4: Demonstrate numerically with the MEP formula that $Q_4^\star$ is equal to Q_4.

Solution to Exercise 36.9.4

In the previous exercise, it was asserted that the Q_i from two supposedly different models are, in fact, exactly the same if the proper transformations are carried out on the constraint functions and Lagrange multipliers. Calculate the numerical assignment $Q_4^\star$ for the probability of seeing the FOUR face on a throw of Wolf's die under Jaynes's specification for the model, and Q_4 under a "different" model with different constraint functions and constraint function averages.

For this example, we are going to define the two constraint functions and their averages under Jaynes's information as,

$$F_1^\star(X = x_i) = (-2.5, -1.5, -0.5, +0.5, +1.5, +2.5)$$

$$\langle F_1^\star \rangle = 0.0983$$

$$F_2^\star(X = x_i) = (+1, +1, -2, -2, +1, +1)$$

$$\langle F_2^\star \rangle = 0.1393$$

The MEP formula calculates $Q_4^\star$ under Jaynes's information as,

$$Q_4^\star = \frac{\exp\left[\lambda_1^\star F_1^\star(X = x_4) + \lambda_2^\star F_2^\star(X = x_4)\right]}{Z(\lambda_1^\star, \lambda_2^\star)}$$

$$= \frac{\exp\left[0.031724\,(0.5) + 0.071776\,(-2)\right]}{6.0396238}$$

$$= 0.1457$$

The information under a "different" model is defined by,

$$F_1(X = x_i) = (1, 2, 3, 4, 5, 6)$$

$$\langle F_1 \rangle = 3.5983$$

$$F_2(X = x_i) = (+1/2, +1/2, -1, -1, +1/2, +1/2)$$

$$\langle F_2 \rangle = 0.06965$$

The MEP formula calculates the same Q_4 under this "different" model,

$$\begin{aligned} Q_4 &= \frac{\exp\left[\lambda_1 F_1(X = x_4) + \lambda_2 F_2(X = x_4)\right]}{Z(\lambda_1, \lambda_2)} \\ &= \frac{\exp\left[0.031724\,(4) + 0.143553\,(-1)\right]}{6.7488847} \\ &= 0.1457 \end{aligned}$$

Alternatively,

$$\begin{aligned} \ln\left(\frac{Q_4^\star}{Q_4}\right) &= \ln Q_4^\star - \ln Q_4 \\ &= (\lambda_1^\star F_1^\star(X = x_4) + \lambda_2^\star F_2^\star(X = x_4) - \ln Z^\star) - \\ &\quad (\lambda_1 F_1(X = x_4) + \lambda_2 F_2(X = x_4) - \ln Z) \\ &= \ln\left(\frac{Z}{Z^\star}\right) + [\lambda_1^\star F_1^\star(X = x_4) + \lambda_2^\star F_2^\star(X = x_4)] - \\ &\quad [\lambda_1 F_1(X = x_4) + \lambda_2 F_2(X = x_4)] \\ &= \ln\left(\frac{6.7488847}{6.0396238}\right) + [0.0317247(0.5) + 0.071776(-2)] - \\ &\quad [0.0317247(4) + 0.143531(-1)] \\ &= \ln\left(\frac{6.7488847}{6.0396238}\right) + [0.0317247(0.5 - 4)] \\ &= 0 \end{aligned}$$

$$\frac{Q_4^\star}{Q_4} = 1$$

SOLVED EXERCISES FOR CHAPTER 36

Exercise 36.9.5: Use *Mathematica* to verify that the matrix of some proposed constraint functions is linearly independent.

Solution to Exercise 36.9.5

Mathematica has the built–in function **MatrixRank[ ]** to find the rank of a matrix appearing as the argument to the function. Set up the constraint matrix for Jaynes's dice scenario as,

```
cm = {{-2.5, -1.5, -0.5, 0.5, 1.5, 2.5}, {1, 1, -2, -2, 1, 1}}
```

Then **MatrixRank[cm]** evaluates to 2, verifying that the constraint functions are linearly independent.

Exercise 36.9.6: Show where a linear systems solution breaks down, but still provides us with an important clue.

Solution to Exercise 36.9.6

Suppose we continue to use **LinearSolve[ ]** to find numerical assignments Q_i. We have the matrix **A** and the vector **b** as,

```
cm4 = {{1, 1, 1, 1, 1, 1},
       {1, 2, 3, 4, 5, 6},
       {1, 1, -2, -2, 1, 1},
       {-2.5, -1.5, -0.5, 0.5, 1.5, 2.5},
       {1, 0, 0, 0, 0, -1}}
cfa4 = {1, 3.5983, 0.1393, 0.0983, 1}
```

The linear system solution **LinearSolve[cm4, cfa4]** returns the Q_i as,

$$Q_i = (0.678275, -0.601215, -0.11638, 0.40328, 0.957765, -0.321725)$$

As a solution to the linear system,

$$\mathbf{A}\mathbf{x} = \mathbf{b}$$

this result is perfectly acceptable. For example,

$$x_1 - x_6 = 0.678275 - (-0.321725) = 1$$

As a solution for the Q_i, it is obviously totally unacceptable. If fed this information, the MEP algorithm would crash. The IP tried to insert *contradictory* information in the form of $Q_1 - Q_6 = 1$ and the average of the die faces as 3.5983. Although, as mentioned, the appearance of negative probabilities in the linear systems solution is a clue that, *before* we employ the MEP algorithm, we had better clean up the source of the contradictory information. Notice that the IP also inserted *redundant* information as well.

Exercise 36.9.7: How do we become aware of a redundancy in the matrix of constraint functions?

Solution to Exercise 36.9.7

There are $m = 5$ rows and $n = 6$ columns in matrix **A** which is our matrix of constraint functions `cm4`. However, `MatrixRank[cm4]` results in 4 telling us that a linear dependency exists in the constraint function matrix.

We might have a suspicion that the two vectors of constraint functions involving the center of gravity, namely,

$$\{1, 2, 3, 4, 5, 6\} \text{ and } \{-2.5, -1.5, -0.5, 0.5, 1.5, 2.5\}$$

are involved in a linear dependency. Formally, the requirement for linear dependence is that there exist coefficients c_i, not all zero, such that,

$$c_1 \mathbf{x_1} + c_2 \mathbf{x_2} + \cdots + c_n \mathbf{x_n} = \mathbf{0}$$

Examining the three vectors $\mathbf{x_1}$, $\mathbf{x_2}$, and $\mathbf{x_3}$, where $c_1 = -3.5$, $c_2 = 1$, and $c_3 = -1$, the set of linear equations is,

$$(-3.5 \times 1) + (1 \times 1) + (-1 \times -2.5) = 0$$
$$(-3.5 \times 1) + (1 \times 2) + (-1 \times -1.5) = 0$$
$$(-3.5 \times 1) + (1 \times 3) + (-1 \times -0.5) = 0$$
$$(-3.5 \times 1) + (1 \times 4) + (-1 \times +0.5) = 0$$
$$(-3.5 \times 1) + (1 \times 5) + (-1 \times +1.5) = 0$$
$$(-3.5 \times 1) + (1 \times 6) + (-1 \times +2.5) = 0$$

This is where the linear dependency exists and where redundant information was inserted by the model.

Exercise 36.9.8: Explain why, in some sense, it is misleading to focus exclusively on Jaynes's maximum likelihood model.

Solution to Exercise 36.9.8

Every single conceivable model is making some contribution to the probability of a die face appearing on the next throw. It is true that an enormous number of models make an utterly negligible contribution to the average,

$$P(X_{N+1} \mid \mathcal{D}) = \sum_{k=1}^{M} P(X_{N+1} \mid \mathcal{M}_k) \, P(\mathcal{M}_k \mid \mathcal{D})$$

because they have no support from the data.

SOLVED EXERCISES FOR CHAPTER 36

On the other hand, it is also true that an enormous number of models, similar to the maximum likelihood model, all make roughly equal contributions to the average. For an example, consider an alternative model that differs in the third decimal place in both Lagrange multipliers.

We have from Exercise 27.7.20 in Volume II the formula derived from the MEP for the relative weighting of any two models after having seen the data,

$$\frac{P(\mathcal{M}_\star \mid \mathcal{D})}{P(\mathcal{M}_J \mid \mathcal{D})} = \exp\left\{ N \times \left[\sum_{j=1}^{m} (\lambda_j^\star - \lambda_j^J)\, \overline{F}_j + \ln\left(\frac{Z_J}{Z_\star}\right) \right] \right\}$$

Here, $\mathcal{M}_J$ refers to Jaynes's maximum likelihood based model, and $\mathcal{M}_\star$ to a very similar model with a difference of 0.001 in the third decimal place of both Lagrange multipliers.

The calculation for the relative weighting of this alternative model as compared to Jaynes's model shows that,

$$\lambda_1^J = 0.0317247$$

$$\lambda_1^\star = 0.0307247$$

$$\lambda_2^J = 0.0717764$$

$$\lambda_2^\star = 0.0727764$$

$$\overline{F}_1 = 0.0983$$

$$\overline{F}_2 = 0.1393$$

$$Z_J = 6.0396328$$

$$Z_\star = 6.0398858$$

$$N = 20{,}000$$

$$\frac{P(\mathcal{M}_\star \mid \mathcal{D})}{P(\mathcal{M}_J \mid \mathcal{D})} = \exp\left\{ 20000 \times \left[(-0.001 \times 0.0983) + (0.001 \times 0.1393) + \ln\left(\frac{6.0396328}{6.0398858}\right) \right] \right\}$$

$$= 0.9824$$

So, in the end, the predictions from this alternative model, and its numerous ilk, are all contributing to the average with about the same weight as the maximum likelihood model. The contribution to the average from the maximum likelihood model must, of course, be the most highly weighted, but it is simply ONE value amongst very, very many other values from highly weighted models.

This is the reason why it would have been better, as was done in section 36.3.5, to report the prediction on the next toss of Wolf's die as an average over all models, and not the prediction from the single most highly weighted model.

Exercise 36.9.9: Use *Mathematica* to calculate the metric tensor for Jaynes's solution.

Solution to Exercise 36.9.9

The answer reported in section 36.5 as the metric tensor for Jaynes's solution at point q was,

$$\mathbf{G} \equiv g_{rc}(q) = \begin{pmatrix} 3.0942 & 0.0778 \\ 0.0778 & 1.8413 \end{pmatrix}$$

Mathematica will compute this metric tensor for us according to the formula in Equation (36.6) with the following short code. This code is already part of my existing MEP algorithm. Therefore, various relevant items such as the $Q_i \equiv$ `qi` have already been computed when it comes time to compute the metric tensor. We don't attempt to explain this code in detail here. That task will be relegated to Appendix C.

First, establish the list of tangent vectors as,

```
tangentvec = MapThread[#1 - #2 &, {cm, cfa}, 1]
```

This returns the two vectors,

{{-2.5983, -1.5983, -0.5983, .4017, 1.4017, 2.4017},
{.8607, .8607, -2.1393, -2.1393, .8607, .8607}}

Next, form the outer product of these tangent vectors to produce the metric tensor as,

```
metrictensor = Outer[Total[(qi #1 #2)] &,
                     tangentvec, tangentvec, 1]
```

metrictensor will then be a 2×2 symmetric matrix represented as always in *Mathematica* as two lists with each list containing two components of the metric tensor.

Exercise 36.9.10: How well does the local linearization in the tangent space at point p work to find the distance to Jaynes's model at point q?

Solution to Exercise 36.9.10

In section 36.6, the distance between the two nearby points p and s was approximated by calculating their separation distance in the tangent space at p. This was an approximation to the relative entropy measure of separation between two probability distributions dictated by the fair model p and a model s very close to the fair model. The approximation was good as the difference was only about 10^{-7}.

Jaynes's point q is farther away from point p than point s was. How well does the approximation hold up? In the tangent space defined at point p the distance is given by the formula,

$$\| \mathbf{p} - \mathbf{q} \| = \sqrt{(\mathbf{p} - \mathbf{q}) \cdot \mathbf{G}(\mathbf{p}) \cdot (\mathbf{p} - \mathbf{q})^{\mathbf{T}}}$$

Mathematica evaluates,

```
Sqrt[Dot[Dot[{-.0317244, -.0717764}, {{2.91667, 0}, {0, 2}}],
    {-.0317244, -.0717764}]]
```

as 0.115061. The relative entropy between these two distributions p and q is,

$$KL(p, q) = \sum_{i=1}^{6} p_i \ln\left(\frac{p_i}{q_i}\right)$$

$$\sqrt{2 \times KL(p, q)} = 0.115203$$

So, the approximation of 0.115061 is not quite so good as before with a difference of about 10^{-4}. But both measures are telling us that point q is definitely farther from the fair model than point s at distances of about 0.0022 vs. 0.1152.

Exercise 36.9.11: Use the second version described in Appendix C for computing the metric tensor for the fair model of Wolf's dice data.

Solution to Exercise 36.9.11

Two alternative versions were described in detail in Appendix C for computing the metric tensor at any point in the Riemannian sub–manifold $\mathcal{S}^m$. The second version relied upon the second derivative of the log of the partition function with respect to the Lagrange multipliers,

$$g_{rc}(p) = \frac{\partial^2 \ln Z}{\partial \lambda_r \, \partial \lambda_c}$$

The four components for the $m = 2$ model with $r = 1$ to 2 and $c = 1$ to 2 are calculated individually by,

```
D[Log[z], λr, λc] /. {λ1 → 0, λ2 → 0}]
```

Before this can be evaluated, the partition function Z must first be found. The matrix of constraint functions for Jaynes's solution of Wolf's dice data was,

```
cm = {{-2.5, -1.5, -0.5, 0.5, 1.5, 2.5}, {1, 1, -2, -2, 1, 1}}
```

Specifying the information in a model for a fair die results in the constraint function averages,

$$\text{cfa} = \{0, 0\}$$

Mathematica will show us the symbolic version of the partition function Z for any two Lagrange multipliers as,

```
z = Total[Exp[Dot[Map[Subscript[λ, #]&, Range[2]], cm]]
```

This returns the symbolic result for Z as,

$$Z = e^{-2.5\lambda_1+\lambda_2} + e^{-1.5\lambda_1+\lambda_2} + e^{-0.5\lambda_1-2\lambda_2} + e^{0.5\lambda_1-2\lambda_2} + e^{1.5\lambda_1+\lambda_2} + e^{2.5\lambda_1+\lambda_2}$$

Evaluating,

$$\text{D[Log[z], } \lambda_1, \lambda_1\text{] /. } \{\lambda_1 \to 0, \lambda_2 \to 0\}]$$

yields $g_{11}(p) = 2.91667$. In the same manner,

$$\text{D[Log[z], } \lambda_2, \lambda_2\text{] /. } \{\lambda_1 \to 0, \lambda_2 \to 0\}]$$

yields $g_{22}(p) = 2$ while $g_{12}(p) = g_{21}(p) = 0$. Recall that $g_{22}(p) = 2$ was confirmed independently using the alternative version in Table 36.2 in section 36.5. All four separate computations can be wrapped up as one overall result for the metric tensor as shown in Appendix C with,

```
Outer[Function[{x, y}, D[Log[z], x, y]], {λ₁, λ₂}, {λ₁, λ₂}]
                                                /. {λ₁ → 0, λ₂ →0}
```

Exercise 36.9.12: Illustrate the power of *Mathematica* for doing purely symbolic computation by verifying Jaynes's Equation 11.73 appearing in Chapter 11 of his book.

Solution to Exercise 36.9.12

Jaynes tells us that his covariance expression,

$$\langle f_j f_k \rangle - \langle f_j \rangle \langle f_k \rangle = \frac{1}{Z}\frac{\partial^2 Z}{\partial \lambda_j \partial \lambda_k} - \frac{1}{Z^2}\frac{\partial Z}{\partial \lambda_j}\frac{\partial Z}{\partial \lambda_k}$$

is, in fact, equal to,

$$\frac{\partial^2 \log Z}{\partial \lambda_j \partial \lambda_k}$$

Translate Jaynes's assertion into *Mathematica* with this very literal symbolic expression,

```
(D[z, λ₁, λ₂])/z - ((D[z, λ₁] D[z, λ₂])/z^2) === D[Log[z],λ₁, λ₂]
```

The evaluation yields the answer **True**, a truly remarkable feat in my estimation.

SOLVED EXERCISES FOR CHAPTER 36

Exercise 36.9.13: Begin to examine the plausibility of Jaynes's Equation (11.73) with a simple exercise using the coin flip scenario.

Solution to Exercise 36.9.13

I began the introduction to the MEP formalism in Volume II with about the simplest scenario imaginable, a flip of an ordinary coin landing HEADS or TAILS. The MEP formalism for assigning numerical values to $P(\text{HEADS})$ and $P(\text{TAILS})$ is grounded on the idea of information inserted into a probability distribution by some model. A model inserts information by establishing constraint functions $F_j(X = x_i)$ mapping the statements in the state space, $(X = x_i)$, to real numbers, together with the averages $\langle F_j \rangle$ of these constraint functions.

For the coin flip problem, since the dimension of the state space is $n = 2$, only one ($m = n - 1 = 1$) constraint function and its average are required. We have already been exposed to the idea that the $F_j(X = x_i)$ can be transformed as long as the accompanying constraint function averages undergo an analogous transformation that keeps the Q_i invariant. Therefore, with this degree of freedom, pick any convenient constraint function such as the arbitrary mapping,

$$F(X = x_1) \equiv F(\text{``HEADS''}) = 1 \text{ and } F(X = x_2) \equiv F(\text{``TAILS''}) = 2$$

The MEP assigns numerical values to the probabilities for HEADS and TAILS under the information $\mathcal{I}$ from some model $\mathcal{M}_k$,

$$Q_1 \equiv P(\text{``HEADS''} \mid \mathcal{I}) \longrightarrow P(X = x_1 \mid \mathcal{M}_k)$$

$$Q_2 \equiv P(\text{``TAILS''} \mid \mathcal{I}) \longrightarrow P(X = x_2 \mid \mathcal{M}_k)$$

through the familiar formula,

$$P(X = x_i \mid \mathcal{M}_k) \equiv Q_i = \frac{\exp\left[\lambda F(X = x_i)\right]}{\sum_{i=1}^{2} \exp\left[\lambda F(X = x_i)\right]}$$

Some model $\mathcal{M}_k$ will specify information via the parameter $1 \leq \langle F \rangle \leq 2$. The Lagrange multiplier λ will adjust accordingly. Or, because they are dual parameters, the Lagrange multiplier parameter could have been specified, and the constraint function average would have adjusted to produce the same Q_i. Suppose that, for the sake of this example, the information inserted by the model $\mathcal{M}_k$ is that $\langle F \rangle = 1.75$.

With the given constraint functions, the partition function $Z(\lambda)$ is,

$$Z = \exp(\lambda) + \exp(2\lambda)$$

We have the differential relationship connecting the log of the partition function to the constraint function average,

$$\frac{\partial \ln Z}{\partial \lambda} = \langle F \rangle$$

However, we also have Jaynes's derivation pointing to this alternative relationship,

$$\frac{\partial \ln Z}{\partial \lambda} = \langle F \rangle$$

$$\frac{\partial \ln Z}{\partial \lambda} = \frac{1}{Z}\frac{\partial Z}{\partial \lambda}$$

$$\frac{1}{Z}\frac{\partial Z}{\partial \lambda} = \langle F \rangle$$

$$Z = \exp(\lambda) + \exp(2\lambda)$$

$$\frac{\partial Z}{\partial \lambda} = \exp(\lambda) + 2\exp(2\lambda)$$

$$\frac{\partial \ln Z}{\partial \lambda} = \frac{\exp(\lambda) + 2\exp(2\lambda)}{\exp(\lambda) + \exp(2\lambda)}$$

$$\frac{\exp(\lambda) + 2\exp(2\lambda)}{\exp(\lambda) + \exp(2\lambda)} = \left[\frac{1}{\exp(\lambda) + \exp(2\lambda)}\right] \times [\exp(\lambda) + 2\exp(2\lambda)]$$

$$= \frac{1}{Z}\frac{\partial Z}{\partial \lambda}$$

$$\langle F \rangle = \frac{1 \times \exp[(1 \cdot \lambda)]}{\exp(\lambda) + \exp(2\lambda)} + \frac{2 \times \exp[(2 \cdot \lambda)]}{\exp(\lambda) + \exp(2\lambda)}$$

$$= \frac{\exp(\lambda) + 2\exp(2\lambda)}{\exp(\lambda) + \exp(2\lambda)}$$

$$= \frac{1}{Z}\frac{\partial Z}{\partial \lambda}$$

Since we know that the information in the model is $\langle F \rangle = 1.75$, solve for the value of λ with,

$$\texttt{Solve[D[Log[z], } \lambda \texttt{]==1.75]}$$

after first establishing that,

$$\texttt{z = Exp[}\lambda\texttt{] + Exp[2 }\lambda\texttt{]}$$

all of which returns with $\lambda = 1.09861$. Therefore, the MEP formula finds the assigned probability for HEADS under this model as,

$$P(X = x_1 \mid \mathcal{M}_k) = \frac{\exp(1.09861 \times 1)}{\exp(1.09861 \times 1) + \exp(1.09861 \times 2)} = 0.25$$

In addition, we can confirm that the constraint function average is indeed,

$$\langle F \rangle = \sum_{i=1}^{2} F(X = x_i) \, Q_i = (1 \times 0.25) + (2 \times 0.75) = 1.75$$

Continue to check that *Mathematica* confirms Jaynes's symbolic expression in this particular numerical example by evaluating the left hand side,

$$\frac{\partial^2 \log Z}{\partial \lambda_j \, \partial \lambda_k}$$

where now there is only one model parameter λ,

`D[Log[z], λ, λ] /. λ → 1.09861`

to yield the answer of 0.1875.

The right hand side,

$$\frac{1}{Z} \frac{\partial^2 Z}{\partial \lambda_j \, \partial \lambda_k} - \frac{1}{Z^2} \frac{\partial Z}{\partial \lambda_j} \frac{\partial Z}{\partial \lambda_k}$$

`D[z, λ, λ] / z - ((D[z, λ] (D[z, λ]) / z^2) /. λ → 1.09861`

also evaluates to 0.1875.

This is the value of the "Fisher information" as a variance,

$$E_p \left[F(X = x_i) - \langle F \rangle \times F(X = x_i) - \langle F \rangle \right]$$

and, in general, as a covariance. It is very easy to verify the above value of 0.1875 as a variance,

$$E_p \left[F(X = x_i) - \langle F \rangle \times F(X = x_i) - \langle F \rangle \right] = \left[\left(-\frac{3}{4} \times -\frac{3}{4} \right) \times \frac{1}{4} \right] + \left[\left(\frac{1}{4} \times \frac{1}{4} \right) \times \frac{3}{4} \right]$$

$$= \frac{3}{16}$$

Contrast "Fisher information" with the MEP definition of information as,

$$\langle F \rangle = E_p \left[F(X = x_i) \right]$$

The Fisher information works out to the correct value only when the appropriate value of $\langle F \rangle$ is specified. Then, the Fisher information and the MEP information are equivalent.

Exercise 36.9.14: Did Jaynes follow the standard definitions given for the variance and covariance in arriving at his Equation (11.70)?

Solution to Exercise 36.9.14

Most statistics textbooks will define the variance of a "random variable" $\mathbf{X}$ as,

$$E[(\mathbf{X} - \mu)^2]$$

which, upon expanding the term in parentheses, works out to,

$$\begin{aligned}
E[(\mathbf{X} - \mu)^2] &= E[(\mathbf{X}^2 - \mu\mathbf{X} - \mu\mathbf{X} + \mu^2)] \\
&= E[\mathbf{X}^2] - \mu E[\mathbf{X}] - \mu E[\mathbf{X}] + E[\mu^2] \\
&= E[\mathbf{X}^2] - \mu\mu - \mu\mu + \mu^2 \\
&= E[\mathbf{X}^2] - \mu^2
\end{aligned}$$

and the covariance between random variables $\mathbf{X}$ and $\mathbf{Y}$ as,

$$E[(\mathbf{X} - \mu_X)(\mathbf{Y} - \mu_Y)]$$

which works out to,

$$\begin{aligned}
E[(\mathbf{X} - \mu_X)(\mathbf{Y} - \mu_Y)] &= E[(\mathbf{XY} - \mu_X\mathbf{Y} - \mu_Y\mathbf{X} + \mu_X\mu_Y)] \\
&= E[\mathbf{XY}] - E[\mu_X\mathbf{Y}] - E[\mu_Y\mathbf{X}] + E[\mu_X\mu_Y] \\
&= E[\mathbf{XY}] - \mu_X\mu_Y - \mu_Y\mu_X + \mu_X\mu_Y \\
&= E[\mathbf{XY}] - \mu_X\mu_Y
\end{aligned}$$

Matching up with Jaynes's notation in Equation (11.70), the "random variables" $\mathbf{X}$ and $\mathbf{Y}$ are his constraint functions f_j and f_k. Furthermore, $f_j - \langle f_j \rangle \equiv \mathbf{X} - E[\mathbf{X}]$, $f_k - \langle f_k \rangle \equiv \mathbf{Y} - E[\mathbf{Y}]$, so that,

$$\begin{aligned}
\langle (f_j - \langle f_j \rangle)(f_k - \langle f_k \rangle) \rangle &= E[(\mathbf{X} - E[\mathbf{X}])(\mathbf{Y} - E[\mathbf{Y}])] \\
&= E[\mathbf{XY} - \mathbf{Y}E[\mathbf{X}] - \mathbf{X}E[\mathbf{Y}] + E[\mathbf{X}]E[\mathbf{Y}]] \\
&= E[\mathbf{XY}] - E[\mathbf{Y}]E[\mathbf{X}] - E[\mathbf{X}]E[\mathbf{Y}] + E[\mathbf{X}]E[\mathbf{Y}] \\
&= E[\mathbf{XY}] - E[\mathbf{X}]E[\mathbf{Y}] \\
&= E[\mathbf{XY}] - \mu_\mathbf{X}\mu_\mathbf{Y} \\
&= \langle f_j f_k \rangle - \langle f_j \rangle \langle f_k \rangle
\end{aligned}$$

Exercise 36.9.15: Use *Mathematica* for help in finding the differential expressions that Jaynes used in deriving his "reciprocity laws."

Solution to Exercise 36.9.15

With the development of the last exercise, two constraint function averages, $\langle f_j \rangle$ and $\langle f_k \rangle$ appear in the last term. We have employed the familiar,

$$\langle f_k \rangle = \frac{\partial \ln Z}{\partial \lambda_k}$$

many times, but Jaynes used another expression. It comes from a basic result concerning the differentiation of the log of a function,

$$\frac{\partial \ln f(x)}{\partial x} = \frac{1}{f(x)} \times \frac{\partial f(x)}{\partial x}$$

Applying this to our situation of interest, we have,

$$\langle f_k \rangle = \frac{\partial \ln Z}{\partial \lambda_k} = \frac{1}{Z} \frac{\partial Z}{\partial \lambda_k}$$

Therefore, $\langle f_j \rangle \langle f_k \rangle$ turns into,

$$\langle f_j \rangle \langle f_k \rangle = \frac{1}{Z} \frac{\partial Z}{\partial \lambda_j} \times \frac{1}{Z} \frac{\partial Z}{\partial \lambda_k} = \frac{1}{Z^2} \frac{\partial Z}{\partial \lambda_j} \frac{\partial Z}{\partial \lambda_k}$$

Mathematica confirms this finding when we ask it to differentiate the log of some function $f(x)$ with **D[Log[f[x]], x]**, by returning,

$$\frac{f'(x)}{f(x)}$$

For an easy example, set the partition function Z as $g(x)$ in the standard coin toss scenario of Exercise 36.9.13,

g[x] = Exp[x] + Exp[2 x]

Then, **D[g[x], x]** evaluates to,

$$e^x + 2\,e^{2x}$$

But **D[Log[g[x]], x]** evaluates to,

$$\frac{e^x + 2\,e^{2x}}{e^x + e^{2x}} \equiv \frac{g'(x)}{g(x)}$$

Taking this one step further, evaluate the second partial derivative of the log of function $f(x, y)$ with respect to arguments x and y,

D[Log[f[x, y]], x, y]

Translating the resulting *Mathematica* expression into a generic answer yields,

$$\frac{\partial^2 \ln f(x,y)}{\partial x\, \partial y} = \frac{1}{f(x,y)} \frac{\partial^2 f(x,y)}{\partial x\, \partial y} - \left[\frac{1}{f(x,y)} \frac{\partial f(x,y)}{\partial x} \times \frac{1}{f(x,y)} \frac{\partial f(x,y)}{\partial y} \right]$$

After some pattern matching, the term in the brackets is something we have just found above,

$$\langle f_j \rangle \langle f_k \rangle = \frac{1}{Z^2} \frac{\partial Z}{\partial \lambda_j} \frac{\partial Z}{\partial \lambda_k}$$

while the first term is,

$$\frac{1}{f(x,y)} \frac{\partial^2 f(x,y)}{\partial x\, \partial y} = \frac{1}{Z} \frac{\partial^2 Z}{\partial \lambda_j\, \partial \lambda_k}$$

Putting everything back together,

$$\frac{\partial^2 \ln Z}{\partial \lambda_j\, \partial \lambda_k} = \frac{1}{Z} \frac{\partial^2 Z}{\partial \lambda_j\, \partial \lambda_k} - \frac{1}{Z^2} \frac{\partial Z}{\partial \lambda_j} \frac{\partial Z}{\partial \lambda_k}$$

the same result that was asserted at the end of Exercise 36.9.12. See the final section of Appendix C for further details.

Exercise 36.9.16: What do you think Jaynes's objective was in developing his Equations (11.72) and (11.73)?

Solution to Exercise 36.9.16

The previous exercises now makes Jaynes's Equation (11.73) more transparent. He shows that the covariance between the constraint functions is,

$$\langle f_j f_k \rangle - \langle f_j \rangle \langle f_k \rangle = \frac{1}{Z} \frac{\partial^2 Z}{\partial \lambda_j\, \partial \lambda_k} - \frac{1}{Z^2} \frac{\partial Z}{\partial \lambda_j} \frac{\partial Z}{\partial \lambda_k}$$

But we have just shown what the expression on the right hand side is equal to. Thus, the covariance between the constraint functions is the same as Fisher's information matrix and the metric tensor. Mixing up our indices,

$$\langle f_j f_k \rangle - \langle f_j \rangle \langle f_k \rangle \equiv \frac{\partial}{\partial \lambda_r} \left(\frac{\partial \ln Z}{\partial \lambda_c} \right) \equiv \frac{\partial^2 \ln Z}{\partial \lambda_r\, \partial \lambda_c} = g_{rc}(p)$$

I suspect though that Jaynes's end goal was to bring out in the open some of his thoughts on these so-called "reciprocity laws" involving the MEP formalism,

$$g_{rc}(p) = \frac{\partial \langle F_c \rangle}{\partial \lambda_r} \text{ and } g^{rc}(p) = \frac{\partial \lambda_c}{\partial \langle F_r \rangle}$$

for their presumed impact on a deeper understanding of physical laws. However, I don't believe that he ever found that really revolutionary result that he felt he was close to with the reciprocity laws. I have not been able to discover anything Jaynes wrote shedding additional light on the reciprocity laws, despite suggestive hints on his part that hidden treasures had been uncovered. It remains to this day a tantalizing topic which is in need of further study.

SOLVED EXERCISES FOR CHAPTER 36

Exercise 36.9.17: What is Amari's very concise notation for the variance–covariance matrix of the constraint functions?

Solution to Exercise 36.9.17

Amari's equivalent IG flavored notation for Jaynes's notation and my adaptations to it, which we have been using exclusively so far in the exercises, looks like this,

$$\{\partial_i\, \partial_j\, \psi = g_{ij}\} \equiv \left\{ \frac{\partial^2 \ln Z}{\partial \lambda_r\, \partial \lambda_c} \equiv g_{rc}(p) \right\}$$

This is Amari's Equation (3.34) [2, pg. 59] which was shown earlier in section 33.7 as Equation (33.12).

Exercise 36.9.18: Write a *Mathematica* function to handle some of the computations discussed in this Chapter.

Solution to Exercise 36.9.18

We will document a function called **chap36dice[cfa_List]**. A list of constraint function averages is indicated as its one argument.

This is a strictly utilitarian function implementing the MEP algorithm in order to find numerical assignments for the probabilities in Jaynes's dice problem. With a little more work, it could be embedded within a **Manipulate[ ]** to facilitate looking at lots of things of interest in this problem. But for our present purposes we desire that the focus remain on the core MEP algorithm along with the additional tasks involved in calculating the metric tensor.

Following my own plea for some meta–code to aid the reader before he or she tries to decipher a complete *Mathematica* program, please first consider this overview.

We define a function with just one argument; a list that contains two constraint function averages in order to implement the information in Jaynes's $m = 2$ models for a "loaded" die. As we have discussed at length in several places, and most extensively in this Chapter, these models involve the physics of a displaced center of gravity, and the unequal length of one axis of a perfect cube compared with the other two axes.

```
chap36dice[cfa_List] := Module[
           (* my standard MEP algorithm *)
           (* definitions for tangent vectors *)
           (* definitions for metric tensor *)
           (* calculate inverse of the metric tensor *)
                    Column[{(* presentation of the calculations *) }]
                                            (* close Module *) ]
```

Let me first fit in the new code for the metric tensor into this skeletal outline. This code has already been addressed in Exercise 36.9.9.

```
chap36dice[cfa_List] := Module[
        (* my standard MEP algorithm *)
tangentvec = MapThread[#1 - #2&, {constraintmatrix, cfa}];
metrictensor = Outer[Total[(qi #1 #2)]&, tangentvec, tangentvec, 1];
inversemetrictensor = Inverse[metrictensor];
Column[{(* presentation of the calculations *) }]
                        (* close Module *) ]
```

Next, I show all of the local variables defined in the **Module[ ]**; the matrix of constraint functions being the most significant. Notice that I have selected two discussed constraint functions, the first for the displaced center of gravity, and the second for unequal length of the THREE–FOUR axis of the die.

```
chap36dice[cfa_List] := Module [{lambda, z, solution, entropy,
znew, lnew, numerator, qi, tangentvec, metrictensor,
inversemetrictensor,
constraintmatrix = {{1, 2, 3, 4, 5, 6}, {1, 1, -2, -2, 1, 1}}},
        (* my standard MEP algorithm *)
tangentvec = MapThread[#1 - #2&, {constraintmatrix, cfa}];
metrictensor = Outer[Total[(qi #1 #2)]&, tangentvec, tangentvec, 1];
inversemetrictensor = Inverse[metrictensor];
Column[{(* presentation of the calculations *) }]
                        (* close Module *) ]
```

At this juncture, I fill in the MEP algorithm, essentially the same code based on the Legendre transformation as presented before.

```
chap36dice[cfa_List] := Module [{lambda, z, solution, entropy,
znew, lnew, numerator, qi, tangentvec, metrictensor,
inversemetrictensor,
constraintmatrix = {{1, 2, 3, 4, 5, 6}}, {1, 1, -2, -2, 1, 1}}},
lambda = Map[Subscript[λ, #]&, Range[2]];
z = Total[Exp[Dot[lambda, constraintmatrix]]];
solution = NMinimize[Log[z] - Dot[lambda, cfa], lambda];
entropy = First[solution];
znew = First[z /. Rest[solution]];
lnew = First[lambda /. Rest[solution]];
numerator = N[Exp[Dot[lnew, constraintmatrix]]];
qi = numerator / znew;
tangentvec = MapThread[#1 - #2&, {constraintmatrix, cfa}];
metrictensor = Outer[Total[(qi #1 #2)]&, tangentvec, tangentvec, 1];
inversemetrictensor = Inverse[metrictensor];
Column[{(* presentation of the calculations *) }]
                        (* close Module *) ]
```

And, finally, I show a representative **Grid[ ]** from the section of the code that presents the results on the Lagrange multipliers.

```
Column[{Style["Lagrange multipliers",
              {FontFamily → "Verdana", Red, FontSize → 20}],
    Grid[{{λ₁, λ₂}, Chop[lnew]}, Frame → All, Spacings → {1, 1},
    BaseStyle → {FontFamily → "ArialBold", FontSize →18},
    Background → {None, {LightGreen, LightBlue}} (* close Grid *)],
                                    (* more results presented *)
                                       (* close Column list *) }
                                          (* close Column function *) ]
                                              (* close Module *) ]
```

A call to **chap36dice[{3.5983, 0.1393}]** results in the output shown in Figure 36.1 below.

Assigned Probabilities

Q_1	Q_2	Q_3	Q_4	Q_5	Q_6
0.16433	0.169627	0.141175	0.145725	0.186564	0.192578

Lagrange multipliers

λ_1	λ_2
0.0317244	0.0717764

Partition Function and Information Entropy

Z	Entropy
6.7488845	1.7852248

Metric Tensor

3.09417	0.07778
0.07778	1.84130

Inverse Metric Tensor

0.32353	−0.01367
−0.01367	0.54367

Figure 36.1: *Output from* Mathematica *function* **chap36dice[ ]**.

Chapter 37

Reasoning Logically about Kangaroos

37.1 Introduction

There is no argument that learning about Information Geometry is made easier by re-examining already understood inference problems. This Chapter follows up on our initial selection of an easy inference, namely, Jaynes's analysis of Wolf's dice counts, by returning to our faithful companions of Volumes I and II, the beer drinking kangaroos.

Again, we should always try to think about these kind of problems as essentially centered around an inference. To force attention on this issue, reasoning about the kangaroos is first introduced as a pure *deduction* problem before making the transition to an *inference* problem.

It is also imperative, from time to time, to re-orient ourselves about the really important issues in inferencing. It is easy to fall into the trap, especially when immersing oneself in the esoterica of IG, of neglecting the forest for the trees.

One of the most fundamental concepts that I have tried to constantly reinforce is the notion that probability theory *generalizes* logic. We would all prefer to reach deductive certainty in our reasoning by using logic, but, more often than not, that preferred path is closed off to us.

For example, logic deserts us when we would like to assert B and $A \to B$ as premises, and then reach some conclusion about A. Logic rightly informs us that such a desire is an invalid form of reasoning if one mistakenly tries to reason backward from B's happening to knowledge about A.

All is not lost, however. An IP might want to generalize logical reasoning by taking advantage of probability theory.

Fortunately, probability theory doesn't break down from the brittleness inherent in deductive reasoning. An IP can quite legitimately reason backwards from B's occurrence to an updated state of knowledge about A. An introduction to some of the details involved in various generalizations of Classical Logic was provided in Chapter Seven of Volume I [7, pp. 153–200].

And, of course, more complicated logic expressions other than the *modus ponens* example just given can be handled within the general framework of probability theory. Consider this chain of implications represented by the logic expression,

$$(A \to B) \land (B \to C)$$

The IP would like to reason logically about A through this chain of implications after C has occurred. I like to refer to such situations as examples of "circumstantial evidence."

Logic and deduction once again fail us if we foolishly attempt to establish the truth of A given C's occurrence. Logic capitulates and throws in the towel at the very beginning of the problem. It informs us rather impolitely that our attempts at this kind of reasoning are "invalid."

Alas, this type of reliance upon a deductive proof, to the exclusion of generalized probabilistic degree of belief, is also at the heart of much of judicial reasoning. It brings my blood to a boil when "it's only circumstantial evidence" is trotted out to argue that the lack of a deductive proof means that we cannot change our state of knowledge about whether a criminal event occurred.

I would like to examine in detail how inferencing and probability deal with this situation involving the pejorative phrase, "circumstantial evidence." Despite the judicial prejudice, we find that we are able to ascertain just how much we might believe in the truth of A given C's occurrence, despite the fact that we can never deductively prove "beyond a shadow of a doubt" that A is absolutely true or false.

Logic expressions, like the chain of implications used above, are the inspiration for models used to assign numerical values to probabilities. The MEP is relied upon to assign probabilities to joint statements about kangaroos.

I shall defer any further diatribes about the "logic" of the judicial system in order to explore the consequences of an implicational model that asserts a genetic influence on hand and beer preference. In consonance with the remarks made above about generalized reasoning, we wonder about a state of knowledge for the genetic influence after having observed, say, a kangaroo's hand or beer preference.

We realize full well that it is impossible to prove anything "beyond a shadow of a doubt" by reasoning backwards to the genetic influence from the occurrence of what is implied by the model. The critical realization is that, despite this negative lesson from deductive reasoning, an IP can still improve its degree of belief in the truth of a statement concerning the presence of the genetic influence.

We will not neglect advancing our understanding of Information Geometry: the notions of coordinate systems, the metric tensor as embodied within the Fisher information matrix, and the inverse of the metric tensor are brought up near the end. Since my hallmark is repetition, we also revisit the feature, given the duality of the two coordinate systems, of figuring out how much a Lagrange multiplier will change when a new model is proposed through the mechanism of changing a constraint function average.

37.2 Reviewing the Usual Details

The state space for this problem consists of joint statements about the genetic code (G), beer preference (B), and hand preference (H) of a kangaroo. Each one of the three statements, G, B, and H, can take on only two values, with the consequence that the dimension of the state space is $n = 8$. Every kangaroo can be placed into one, and only one, of eight mutually exclusive categories such as: (1) a right–handed Foster's drinker with the normal genetic code (GBH), or, (2) a left–handed Corona drinker with a mutated genetic code ($\overline{G}\,\overline{B}\,\overline{H}$).

Joint statements may be indicated in the usual way by, say, ($\overline{G}\,\overline{B}H$) for the statement, "The kangaroo is a right–handed Corona drinker with a mutated genetic code." Even more abstractly, the joint statements may be referenced with ($X = x_i$) where i runs from 1 through 8, the dimension of the state space. Thus, the joint statement ($\overline{G}\,\overline{B}H$) is equivalent to ($X = x_6$) when the joint probability table is constructed appropriately.

The degree of belief in any joint statement, say, $P(\overline{G}\,\overline{B}H)$ or $P(X = x_6)$, is written first as an abstract probability. Such probabilities can be assigned numerical values by conditioning on the truth of some model statement. When numerical assignments are eventually made to the joint probabilities, as conditioned on the information resident in some given model $\mathcal{M}_k$, the expressions must be properly written as $P(\overline{G}\,\overline{B}H \,|\, \mathcal{M}_k)$ or $P(X = x_6 \,|\, \mathcal{M}_k)$.

In order to show that the model has been influenced by a logic expression, write it out explicitly to the right of the conditioned upon symbol as, for example,

$$P(G \,|\, H, \mathcal{M}_k \equiv [\,G \to B\,] \wedge [\,B \to H\,]\,)$$

As mentioned in the **Introduction**, with these kinds of models, the IP is seen to be interested in the degree of belief about the presence of the normal genetic code given the presence of a known physical or personality characteristic of the kangaroo. This is circumstantial evidence held together by an implicational model thought to reflect actual cause and effect. No certain deduction can be carried out concerning the genetic code based on observing hand preference; nonetheless, probability theory will at least permit the IP to make a defensible inference.

Numerical values assigned through the implicational model will appear in an eight cell table of joint probabilities. Construct two tables, each consisting of two columns for the B characteristic and two rows for the H characteristic. Then, each one of these two sub–tables indexes the two G characteristics.

As always, a Q_1 appearing in cell 1 of the joint probability table will refer to an assigned numerical value for the joint probability, $P(GBH \mid \mathcal{M}_k)$. Likewise, a Q_2 appearing in cell 2 of the table will refer to an assigned numerical value for the joint probability, $P(G\overline{B}H \mid \mathcal{M}_k)$, and so on, up to a Q_8 which refers to an assigned numerical value for the joint probability, $P(\overline{G}\,\overline{B}\,\overline{H} \mid \mathcal{M}_k)$, appearing in cell 8.

The MEP formula will be used to assign these numerical values. Each model entertained will consist of $m \leq n - 1$ constraint functions, $F_j(X = x_i)$, and their associated constraint function averages, $\langle F_j \rangle$. In the last Chapter, simple models were constructed with only $m = 2$ constraint functions. Here, we will construct more complex models whose main intent is to exhibit a causal relationship between the genetic code and a kangaroo's physical and personality traits. These models will be guided by the information inherent in the implicational chain $(G \to B) \wedge (B \to H)$.

In the last Chapter, we admired Jaynes's way of creating constraint functions for dice based on physical considerations. In contrast, for the kangaroos we will employ the more prosaic, but generic technique of constructing constraint functions by looking at marginal sums within the joint probability table. This is less creative than Jaynes, but at least it is easy and generally applicable. Furthermore, this technique will not in any way hinder our attempts to come up with models that implement the essence of the implicational chain.

37.3 A Pure Deduction Problem

As mentioned in the **Introduction**, let's first look at this current kangaroo scenario as one of pure deduction as opposed to a problem in inference. Therefore, we can begin to discuss it from this perspective prior to delving into models and the MEP algorithm. One of the most crucial conceptual notions must be to reiterate Jaynes's insistence that probability theory generalizes Classical Logic. As a generalization, probability theory must return the same answers that Classical Logic provides us.

Suppose that, in a slight extension of the basic *modus ponens* syllogism, we hypothesize that the normal genetic code implies a preference for Foster's beer. Also, a preference for Foster's beer implies a right hand preference. Thus, we write the logic expression $(G \to B) \wedge (B \to H)$. Now, if G is TRUE, then B is TRUE, and it then follows that H is TRUE. This is a classical deduction.

Probability theory arrives at the same answer in the following standard way. Set up Bayes's Theorem asking for the probability of H predicated on assuming that G is true, and the model $(G \to B) \wedge (B \to H)$ is also true,

$$P(H \mid G, \mathcal{M}_k \equiv [\,G \to B\,] \wedge [\,B \to H\,]\,)$$

An application of Bayes's Theorem must return an answer of 1 if probability theory is a generalization of Classical Logic.

KANGAROOS AND LOGIC

The details are provided in Exercise 37.9.1 with the expected outcome that the probability of a right handed kangaroo is indeed 1 if it is true that the kangaroo possess the normal genetic code. This result is not only predicated on the knowledge that the kangaroo does possess the normal genetic code, but as well on the causal implication furnished through the information in the model.

Happily, probability theory reproduces the result already obtained by logical deduction. Of course, it does much more than that. It also solves any inference problem where, by definition, reasoning by logic is invalid. That problem appears soon when the IP desires to make the "inverse" inference about G when observations about B and/or H become available.

In addition, further insight can be extracted from the purely deductive argument which helps in thinking about the transition to an inferential attack. Back in Volume I, the implication logic function, $A \to B$, was defined by the functional assignment of F only if A assumed the value T and B assumed the value F. In the notation of Chapter Two, Volume I, $f_{13}(T, F) = F$. In the other three cases, the functional assignment was T.

So, when it comes time to think about numerical assignments via some model for an inferential solution, the deductive result tells us that the cell of a joint probability table where G is TRUE and where B and H are also both TRUE should have a non–zero probability.

Complementary to this fact, the cells of a joint probability table where G is TRUE and where either or both B and H are FALSE should contain 0s, or values close to 0. However, all the cells where G is FALSE can legitimately hold non–zero values. This is no problem because, after all, the functional assignment of the implication function is T when A assumes the value F.

Examine the joint probability table in Figure 37.1 for the ramifications of the generic implicational chain model. There must be, as we know quite well by know, $n = 8$ cells each holding an assigned numerical value Q_i for the probability of the joint statement indexed by that cell.

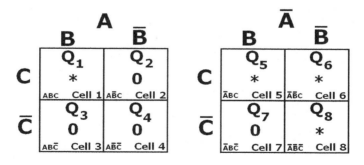

Figure 37.1: *A joint probability table for the kangaroo scenario under a model containing information about a logical implication.*

But Q_2, Q_3, and Q_4 in the sub–table under A must contain 0s due to the information in the model $(A \to B) \land (B \to C)$. If A is TRUE, then B cannot be FALSE leading to the 0s in cells 2 and 4. Likewise, if B is TRUE, then C cannot be FALSE leading to the 0 in cell 3.

There is only one zero in the sub–table under $\overline{A}$ and that is $Q_7 = 0$. Again, C cannot be FALSE if B is TRUE. Just as for the implication logic function $A \to B$, A may be FALSE with no requirement that B cannot be FALSE. Furthermore, B may be FALSE with no requirement that C cannot be FALSE. Thus, Q_5, Q_6, and Q_8 may contain non–zero probabilities. Of course, Q_1 must be a non–zero probability as well.

As explained in Chapters Two and Three of Volume I [7], any logic expression can be decomposed into *canonical forms*, one of which, called the disjunctive normal form (DNF), we found especially valuable. Thus the logic expression,

$$\mathcal{M}_k \equiv (A \to B) \land (B \to C)$$

acting as a reference model for the current kangaroo scenario has as its DNF,

$$ABC \lor \overline{A}BC \lor \overline{A}\,\overline{B}C \lor \overline{A}\,\overline{B}\,\overline{C}$$

The DNF serves as a useful heuristic for the location of the 0s in the joint probability table. The above DNF for the logic expression shows where the three variable Boolean function $f(\star, \star, \star) = T$. Wherever the function assigned an F we must place a 0. Thus, for example, $Q_7 = 0$ because $\overline{A}B\overline{C}$ is NOT one of the terms in the DNF. All of this is explained in great detail in Exercises 37.9.2 through 37.9.7.

From the logic function serving as the model, and reasoning via a deduction, we know that C must be TRUE given that A is TRUE. Probability theory as a generalization of logic must also reach the same conclusion.

Setting up Bayes's Theorem,

$$
\begin{aligned}
P(C \mid A, \mathcal{M}_k) &= \frac{P(AC \mid \mathcal{M}_k)}{P(A \mid \mathcal{M}_k)} \\
&= \frac{P(ABC \mid \mathcal{M}_k) + P(A\overline{B}C \mid \mathcal{M}_k)}{P(ABC \mid \mathcal{M}_k) + P(A\overline{B}C \mid \mathcal{M}_k) + P(AB\overline{C} \mid \mathcal{M}_k) + P(A\overline{B}\,\overline{C} \mid \mathcal{M}_k)} \\
&= \frac{Q_1 + Q_2}{Q_1 + Q_2 + Q_3 + Q_4} \\
&= \frac{Q_1 + 0}{Q_1 + 0 + 0 + 0} \\
&= 1
\end{aligned}
$$

It would naturally come to mind to consider other models close to $\mathcal{M}_k$ in order to capture some notion of an association, relationship, or correlation existing among the genetic code, beer preference, and hand preference variables. The implicational chain model emphasized so far serves mainly as a reference model. It clearly provides a jumping off point for other similar models where we wouldn't necessarily believe in some unbending direct causal connection between the genetic code and these characteristics of the kangaroos.

Rather, we would be more inclined to hypothesize that some strong correlations might be present. In that case, we would relax the imposition of the 0s in the joint probability table and replace them with small probabilities.

37.4 Kangaroos from the Inferential Standpoint

37.4.1 The MEP formula

Begin the transition from exposure to the problem as a pure deduction problem to a quantitative analysis as an inferential problem. To accomplish this, the IP will require a numerical assignment to the probabilities for all of the joint statements concerning G, B, and H.

As always, any numerical assignment must be conditioned on the truth of some model $\mathcal{M}_k$. We have already mentioned that these models will be inspired by the causal model, $\mathcal{M}_k \equiv (G \to B) \land (B \to H)$, where a normal genetic code induces right–handed Foster's drinking behavior.

The MEP formula for numerical assignments to the probability of any joint statement in the kangaroo state space is,

$$P(X = x_i \mid \mathcal{M}_k) \equiv Q_i = \frac{\exp\left[\sum_{j=1}^{m} \lambda_j F_j(X = x_i)\right]}{Z(\lambda_1, \lambda_2, \cdots, \lambda_m)} \quad (37.1)$$

where the partition function is defined, as always, as the sum over all n statements in the state space.

$$Z(\lambda_1, \lambda_2, \cdots, \lambda_m) = \sum_{i=1}^{n} \exp\left[\sum_{j=1}^{m} \lambda_j F_j(X = x_i)\right] \quad (37.2)$$

This issue of how to make numerical assignment to probabilities is, of course, quite familiar to us from Volume II and the examples presented so far here in Volume III.

37.4.2 A joint probability table

The promised illustration of the joint probability table for the kangaroo problem is shown in Figure 37.2 at the top of the next page. The table must consist of $n = 8$ cells where each cell indexes a joint statement.

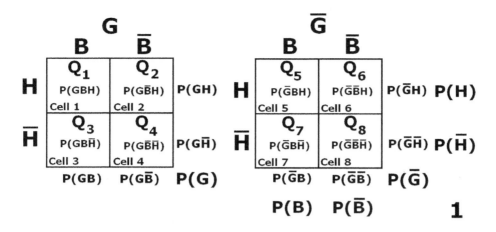

Figure 37.2: *A generic joint probability table for the kangaroo scenario.*

Cell 1 indexes joint statement GBH, cell 2 indexes joint statement $G\overline{B}H$, and so on, up to cell 8 which indexes the joint statement $\overline{G}\,\overline{B}\,\overline{H}$. The table is broken down into two sub-tables of four cells each. The first sub-table on the left is for G, and the second sub-table on the right is for $\overline{G}$. The two rows in each sub-table are for H and $\overline{H}$, while the two columns are for B and $\overline{B}$.

The cell numbers, the Q_i notation, and the marginal sums in symbolic form are presented. No numerical values are shown in the cells as yet. Those will be filled in when discussing various models. The one numerical value that can be written down at the very outset is the value 1 for the sum of the assigned numerical values over all eight cells.

37.4.3 A target joint probability table

What would a filled-in joint probability table that implements a version of the logic expression model look like? First of all, such a table should place four zeroes as the numerical assignments for Q_2, Q_3, Q_4, and Q_7. Furthermore, there is maximal uncertainty about the presence of the mutated genetic code. If the IP desired no further information than this in a model, the remaining probabilities should be spread out as evenly as possible. Consider Figure 37.3 at the top of the next page.

Everyone experiences a bit of discomfiture on their first exposure to the classical implication logic function $A \to B$. And textbooks go to quite a bit of trouble to alleviate the head scratching when it is explained that although if A is TRUE, B cannot be FALSE, it is allowed that A can be FALSE, and B can be TRUE. The abstract approach through the different functional assignments for the logic functions causes no problems because it is simply the definition of implication that at variable settings of $A = F$ and $B = T$, $f_{13}(A, B) = T$.

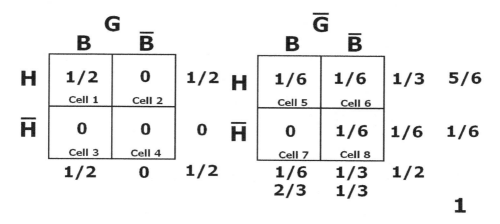

Figure 37.3: *A target joint probability table for the kangaroo scenario implementing the logic expression and containing no further information.*

Let's return to our little interlude with Sir Harold Jeffreys in Volume I about animals with feathers and their beaks. If A is the statement, "An animal has feathers." and B is the statement, "An animal possess a beak.", then $A \to B$ is interpreted as, "If an animal has feathers, then it must also possess a beak."

A logical deduction about ducks would be, "A duck is an animal with feathers, therefore it must also possess a beak." But as just explained, A can be FALSE and B TRUE. Thus, under the implication logic function, it is still perfectly acceptable to assert that, "A duckbill platypus does NOT have feathers, but it still possesses a beak."

That is why non–zero probabilities are placed into the joint probability table for the joint statements $\overline{G}BH$, $\overline{G}\,\overline{B}H$, and $\overline{G}\,\overline{B}\,\overline{H}$. A kangaroo may have a mutated genetic code, but it is still allowed to prefer Foster's and be right–handed. Also, it may have a mutated genetic code, prefer Corona and be right–handed, as well as have a mutated genetic code, prefer Corona and be left–handed. The kangaroo cannot, however, have a mutated genetic code, prefer Foster's and be left–handed.

The target joint probability table shown above in Table 37.3 captures all of these ramifications of the logic expression $(G \to B) \wedge (B \to H)$ serving as the reference model $\mathcal{M}_k$.

37.4.4 MEP formula with three constraint functions

To get our feet wet, before developing more complex models, here is an example where three pieces of information are inserted into the probability distribution. Thus, there will be three constraint functions, three associated constraint averages, together with the three Lagrange multipliers. The models will be labeled in the order in which we discuss them, so this model is model $\mathcal{M}_1$.

Set $m = 3$, so that the MEP formula looks like,

$$P(X = x_i \,|\, \mathcal{M}_1) = \frac{\exp\left[\,\lambda_1\, F_1(X = x_i) + \lambda_2\, F_2(X = x_i) + \lambda_3\, F_3(X = x_i)\,\right]}{Z(\lambda_1, \lambda_2, \lambda_3)} \qquad (37.3)$$

with the partition function $Z(\lambda_1, \lambda_2, \lambda_3)$,

$$Z(\lambda_1, \lambda_2, \lambda_3) = \sum_{i=1}^{8} \exp\left[\,\lambda_1\, F_1(X = x_i) + \lambda_2\, F_2(X = x_i) + \lambda_3\, F_3(X = x_i)\,\right] \qquad (37.4)$$

What do the first three constraint functions look like? The information takes the form of marginal probabilities about $G, B,$ and H considered by themselves. It is interesting and quite helpful that simply setting the constraint functions at appropriate values of 0s and 1s allows the IP to insert information about these marginal probabilities into the joint probability table.

The marginal probability for $P(G\,|\,\mathcal{M}_1)$, that is, the probability for the normal genetic code, is easily read straight from the joint probability table as,

$$P(G\,|\,\mathcal{M}_1) = Q_1 + Q_2 + Q_3 + Q_4$$

Likewise, the marginal probability for $P(B\,|\,\mathcal{M}_1)$, that is, the probability for being a Fosters drinker, is read from the joint probability table as,

$$P(B\,|\,\mathcal{M}_1) = Q_1 + Q_3 + Q_5 + Q_7$$

Finally, the marginal probability for $P(H\,|\,\mathcal{M}_1)$, that is, the probability for being right–handed, is just as easily read from the joint probability table as,

$$P(H\,|\,\mathcal{M}_1) = Q_1 + Q_2 + Q_5 + Q_6$$

The first constraint function is set up as $F_1(X = x_i) = (1, 1, 1, 1, 0, 0, 0, 0)$, and,

$$\langle F_1 \rangle = \sum_{i=1}^{8} F_1(X = x_i)\, Q_i = (1 \times Q_1) + (1 \times Q_2) + \cdots + (0 \times Q_8) = Q_1 + Q_2 + Q_3 + Q_4$$

to capture the desired information about the genetic code.

The second constraint function is set up as $F_2(X = x_i) = (1, 0, 1, 0, 1, 0, 1, 0)$, and,

$$\langle F_2 \rangle = \sum_{i=1}^{8} F_2(X = x_i)\, Q_i = (1 \times Q_1) + (0 \times Q_2) + \cdots + (0 \times Q_8) = Q_1 + Q_3 + Q_5 + Q_7$$

to capture the desired information about beer preference.

The third constraint function is set up as $F_3(X = x_i) = (1, 1, 0, 0, 1, 1, 0, 0)$, and,

$$\langle F_3 \rangle = \sum_{i=1}^{8} F_3(X = x_i) Q_i = (1 \times Q_1) + (1 \times Q_2) + \cdots + (0 \times Q_8) = Q_1 + Q_2 + Q_5 + Q_6$$

to capture the desired information about hand preference.

To complete the information that is inserted into a probability distribution under this model, the IP specifies that $\langle F_1 \rangle = 1/2$, $\langle F_2 \rangle = 2/3$, and $\langle F_3 \rangle = 5/6$. In other words, the IP has set up a tentative working hypothesis that half of the kangaroos possess the normal genetic code, two–thirds prefer to drink Foster's, and five–sixths are right–handed.

This identifies one specific model $\mathcal{M}_1$, or alternatively, the parameters λ_1, λ_2, and λ_3 which will make the proper numerical assignments to Q_1 through Q_8. The dual parameters, the constraint function averages, have been given the specific values listed above as the information because we are trying to approach the target joint probability table of Figure 37.3.

All the numerical details are spelled out in Exercise 37.9.9. The numerical values assigned to the eight cells of the joint probability table under this model are shown in Figure 37.4 below. Notice that all three constraint averages, shown boxed, are satisfied by this assignment.

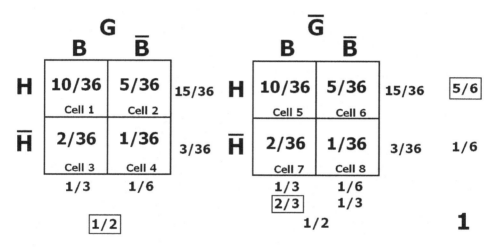

Figure 37.4: *A joint probability table for the kangaroo scenario under a model with three constraints.*

37.4.5 Another model with additional information

Although the numerical values assigned to the probabilities under model $\mathcal{M}_1$ are legitimate probabilities, it is hard to discern that they are in accord with the general direction we wanted from discussion of the purely deductive model and its associated target joint probability table. Moreover, under this model the probability of the genetic code does not change given knowledge of, say, hand preference. The IP would like to somehow modify this original model to bring in dependencies of hand and beer preference on the genetic code.

The IP can progress towards this goal by adding one constraint function so that m increases from $m = 3$ to $m = 4$. This new constraint function will allow for some association between genetic code and beer preference.

As was discussed in Volume II, this is an example of an "interaction constraint." They can be generated automatically by simply multiplying the constraint function for G with the constraint function for B to produce a new interaction constraint GB. This new constraint function vector is explicitly,

$$GB = F_4(X = x_i) = (1, 0, 1, 0, 0, 0, 0, 0)$$

By referring back to the joint probability table, one can easily see that cells 1 and 3 represent the marginal sum for GB. Of course, the constraint average $\langle F_4 \rangle$ must be part of the information in the new model $\mathcal{M}_2$. In order to more closely approach the assignment under the deductive model, the information in $\langle F_4 \rangle = Q_1 + Q_3$ is set to $1/2$. This will force Q_2 and Q_4 to zero.

As always, the MEP formula will provide the IP with numerical assignments to the eight cells of the joint probability table for this new model. Figure 37.5 shows these new assignments and how they do satisfy all four constraints.

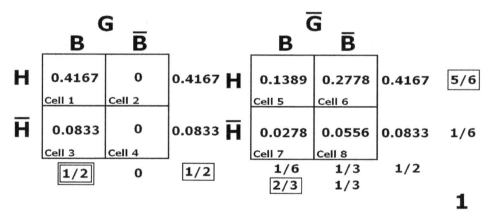

Figure 37.5: *A joint probability table for the kangaroo scenario under a model with four constraints that induce a correlation between the genetic code and beer preference.*

The same three constraint averages as used in the first model are shown with a single box, while the new double interaction constraint average GB is emphasized by drawing two boxes around it. The important point is that there now exists a dependency between the genetic code and beer preference. As mentioned, model $\mathcal{M}_2$ has located two of the targeted four zeroes in the proper place.

37.4.6 A third model

It would be fair to speculate that, perhaps by adding one more constraint enforcing a correlation between beer preference and hand preference, we might get even closer to the targeted joint probability table. It turns out that a model $\mathcal{M}_3$ with $m = 5$ constraint functions containing information about the BH double interaction makes numerical assignments exactly matching our requirements.

The additional constraint function vector is,

$$BH = F_5(X = x_i) = (1, 0, 0, 0, 1, 0, 0, 0)$$

Specify the information as $\langle F_5 \rangle = Q_1 + Q_5 = 2/3$.

Figure 37.6 shows the new assignments under the current model $\mathcal{M}_3$ satisfying all five constraints. The same three constraint averages as used in the first model are shown singly boxed, while the two double interaction constraint averages enforcing the correlations are emphasized by two boxes. The important point is that there now exists not only a dependency between the genetic code and beer preference, but also a new dependency between beer preference and hand preference. The details are explored in Exercise 37.9.10.

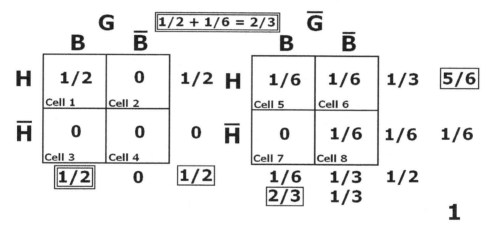

Figure 37.6: *A joint probability table for the kangaroo scenario under a model with five constraints inducing correlations between the genetic code and beer preference as well as between beer preference and hand preference.*

37.4.7 Bayes's Theorem calculations

The sole motivation for constructing these joint probability tables is to conduct an *inference* about G when B and/or H are known. The IP can no longer reason deductively by assuming G, and then asking about B or H. This attempt at reverse reasoning is what logic labels as "invalid."

So, first look at the state of knowledge concerning G when H is known, and when conditioned on the truth of the implicational chain model $\mathcal{M}_3$,

$$(G \to B) \wedge (B \to H)$$

This model was implemented via the MEP with five constraint functions and their averages. The first three specified marginal distributions for G, B, and H, while the last two specified the double interactions for GB and BH.

$$\begin{aligned}
P(G \,|\, H, \mathcal{M}_3) &= \frac{P(GH \,|\, \mathcal{M}_3)}{P(H \,|\, \mathcal{M}_3)} \\
&= \frac{P(GBH \,|\, \mathcal{M}_3) + P(G\overline{B}H \,|\, \mathcal{M}_3)}{P(H \,|\, \mathcal{M}_3)} \\
&= \frac{Q_1 + Q_2}{Q_1 + Q_2 + Q_5 + Q_6} \\
&= \frac{1/2 + 0}{1/2 + 0 + 1/6 + 1/6} \\
&= 0.60
\end{aligned}$$

The IP's degree of belief that G is TRUE has been raised from $1/2$ to 0.60 given the fact that H was observed.

Next, look at the state of knowledge concerning G when B is known, and when conditioned on the truth of the implicational chain model $\mathcal{M}_3$.

$$\begin{aligned}
P(G \,|\, B, \mathcal{M}_3) &= \frac{P(GB \,|\, \mathcal{M}_3)}{P(B \,|\, \mathcal{M}_3)} \\
&= \frac{P(GBH \,|\, \mathcal{M}_3) + P(GB\overline{H} \,|\, \mathcal{M}_3)}{P(B \,|\, \mathcal{M}_3)} \\
&= \frac{Q_1 + Q_3}{Q_1 + Q_3 + Q_5 + Q_7} \\
&= \frac{1/2 + 0}{1/2 + 0 + 1/6 + 0} \\
&= 0.75
\end{aligned}$$

The IP's degree of belief that G is TRUE has been raised from $1/2$ to $3/4$ given the fact that B was observed.

Finally, look at the state of knowledge concerning G when both B and H are known, and when conditioned on the truth of the implicational chain model $\mathcal{M}_3$.

$$
\begin{aligned}
P(G \mid B, H, \mathcal{M}_3) &= \frac{P(GBH \mid \mathcal{M}_3)}{P(BH \mid \mathcal{M}_3)} \\
&= \frac{P(GBH \mid \mathcal{M}_3)}{P(GBH \mid \mathcal{M}_3) + P(\overline{G}BH \mid \mathcal{M}_3)} \\
&= \frac{Q_1}{Q_1 + Q_5} \\
&= \frac{1/2}{1/2 + 1/6} \\
&= 0.75
\end{aligned}
$$

The IP's degree of belief that G is TRUE has been raised from 1/2 to 3/4 given the facts that both B and H were observed.

It makes perfect sense, given the implicational model, that knowing H is TRUE in addition to knowing that B is TRUE does not increase the IP's degree of belief in G compared to when it knew only that B was TRUE.

All of the actual numerical values for the Q_i in the Bayes's Theorem calculations come from the information in model $\mathcal{M}_3$. Recall that this information consisted of the marginal probabilities for G, B, and H, and the marginal probabilities for the interactions GB and BH. These numerical values for the Q_i were shown in the joint probability table of Table 37.6.

37.5 The Impact of any Data

Of course, all of this previous discussion has taken place within the hypothetical world of tentatively entertained models. The data will tell the IP how to update its belief in any model relative to all other possible models.

Suppose, for the sake of argument, a limited amount of data have been collected about the genetic code, beer preference, and hand preference of kangaroos. With this limited sample size, we have that $N = 12$ with the individual frequency counts N_i shown in the contingency table of Figure 37.7 at the top of the next page.

The normed frequency counts $f_i = N_i/N$ are shown at the margins of the table. These data can be seen to support the implicational chain model.

In the last section, the final model considered, model $\mathcal{M}_3$, permitted the IP a degree of belief equal to 3/4 in the truth of the statement that a kangaroo had the normal genetic code if it was a right–handed Foster's drinker,

$$P(G \mid B, H, \mathcal{M}_3) = 0.75$$

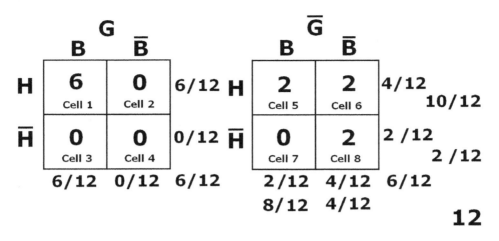

Figure 37.7: *A contingency table showing data supporting the implicational chain model. The normed frequency counts match the information in the five constraint functions of model $\mathcal{M}_3$ making it the maximum likelihood model.*

However, if the IP averages over all possible models, then, conditioned on the existing data of the twelve kangaroos, the degree of belief that the next kangaroo will possess the normal genetic code given that it was a right–handed Foster's drinker is reduced from 0.75 to 0.70.

$$P(G_{N+1} \,|\, B_{N+1}, H_{N+1}, \mathcal{D}) = \frac{N_1 + 1}{(N_1 + 1) + (N_5 + 1)} = \frac{7}{10}$$

The degree of belief in the statement of interest must be reduced when averaging over all models when compared to the prediction from just one model. Nonetheless, this limited amount of data can be seen to support our tentatively entertained model fairly well.

37.6 The Kangaroos and Information Geometry

Now that the fundamentals of the inferential problem involving the kangaroos have been discussed in some detail, make the transition to the specialized language of Information Geometry. The structure with the notation $(\mathcal{S}, g)$ has been defined as a smooth n–dimensional Riemannian manifold $\mathcal{S}^8$ with a metric tensor $g_{rc}(p)$ defined at each point p in the manifold. In other words, a tangent space can be constructed at each point.

For this problem, $\mathcal{S}^8$ is a eight dimensional manifold possessing a dual coordinate system with the notation of $(\theta^1, \theta^2, \cdots, \theta^7)$ and $(\eta_1, \eta_2, \cdots, \eta_7)$ with θ^0 and η_0 being implicitly present. The points $\{p, q, r, \cdots\} \in \mathcal{S}^8$ all refer to probability distributions with different coordinate values.

For example, we might call p the distribution under model $\mathcal{M}_3$, q the distribution under model $\mathcal{M}_2$, and s the distribution under model $\mathcal{M}_1$. The θ^i coordinates are matched up with the Lagrange multipliers λ_j, and the η_i coordinates with the constraint averages $\langle F_j \rangle$ from the MEP algorithm.

37.6.1 Partial differentiations

Some commentators have characterized Differential Geometry as the ability to do calculus on arbitrary spaces. Information Geometry has borrowed heavily from Differential Geometry, and, not surprisingly, there are many relationships involving partial differentiation. The most prominent differential relationships are reviewed here by looking at some numerical examples drawn from the kangaroo scenario.

The simplest important differential relationship in Information Geometry is the change in the partition function with respect to changes in the λ_j coordinates,

$$\frac{\partial \ln Z}{\partial \lambda_j} = \langle F_j \rangle$$

This result comes in handy because the natural basis vector $\mathbf{e}_r$ at point p is defined by,

$$\mathbf{e}_r = \frac{\partial \ln p}{\partial \lambda_r} = F_r(X = x_i) - \frac{\partial \ln Z}{\partial \lambda_r} = F_r(X = x_i) - \langle F_r \rangle$$

For example, the change in the logarithmic transform of the partition function as the parameter λ_4 in model $\mathcal{M}_3$ changes is,

$$\frac{\partial \ln Z}{\partial \lambda_4} = \langle F_4 \rangle = 1/2$$

A numerical example is presented in Exercise 37.9.13.

With the above differential relationship involving the partition function and the Lagrange multiplier, it becomes very easy to prove the lemma given below that Amari needs to demonstrate "flatness,"

$$\int \frac{\partial \ln p}{\partial \lambda_j} p(x) \, \mathrm{d}x = 0 \tag{37.5}$$

A few easy steps takes us to this result,

$$\begin{aligned}
\int \frac{\partial \ln p}{\partial \lambda_j} p(x) \, \mathrm{d}x &= E_p \left[\frac{\partial \ln p}{\partial \lambda_j} \right] \\
&= E_p [F_j(X = x_i) - \langle F_j \rangle] \\
&= \langle F_j \rangle_p - \langle F_j \rangle_p \\
&= 0
\end{aligned}$$

Another important differential relationship is reflected in the metric tensor as defined by the partial differentiations in the inner product involving the basis vectors $\mathbf{e}_r$ and $\mathbf{e}_c$, and then switching over to Amari's notation,

$$g_{rc}(p) = \langle \mathbf{e}_r, \mathbf{e}_c \rangle \equiv \left\langle \frac{\partial \ln p}{\partial \lambda_r}, \frac{\partial \ln p}{\partial \lambda_c} \right\rangle$$

$$g_{ij} = \langle \partial_i, \partial_j \rangle \equiv \left\langle \frac{\partial \ln p}{\partial \theta^i}, \frac{\partial \ln p}{\partial \theta^j} \right\rangle$$

We won't discuss examples of the metric tensor for the kangaroos right now, but will reserve them for the next section.

The metric tensor crops up in another way in addition to the distance and angular measurements. If the IP is curious about the relative changes in the dual coordinate system, these relationships are expressed as,

$$\frac{\partial \eta_r}{\partial \theta^c} = g_{rc}(p) \text{ and } \frac{\partial \theta^r}{\partial \eta_c} = g^{rc}(p)$$

where $g^{rc}(p)$ is the inverse matrix to $g_{rc}(p)$.

As we shall discover shortly in the next section, the off–diagonal elements in $g_{rc}(r)$, (i.e., model $\mathcal{M}_1$ with no correlations among the three constraint functions), are all 0. The off–diagonal elements in the inverse matrix $g^{rc}(r)$ are also 0. Do not confuse the point r with the row r in the metric tensor. This means that θ^r will not change if η_c changes. For example, in model $\mathcal{M}_1$, both θ^2 and θ^3 remain at a value of 0.619039, even if $\eta_1 \equiv \langle F_1 \rangle$ changes from 1/2.

Moreover, if a component in $g^{rc}(p)$ is large, then θ^r must increase or decrease rapidly for a very small change in η_c. In model $\mathcal{M}_2$, the fourth diagonal element $g^{44}(q) \approx 100$ is large. This means that as $\langle F_4 \rangle$, the constraint average on the genetic code and beer preference interaction, tries to approach 1/2 from 0.49, the Lagrange multiplier θ^4 enforcing this constraint must increase rapidly.

The off–diagonal elements in the inverse metric tensor are no longer 0 under model $\mathcal{M}_2$ as well. There are correlations built into this second model. This has the consequence that other Lagrange multipliers λ_j enforcing other constraints must also shift in a wild fashion.

37.6.2 Entropy and differentiation

The information entropy function,

$$H(p) = -\sum_{i=1}^{n} p_i \ln p_i$$

which was so prominently featured in Volume II, is also subject to differentiation in Information Geometry. In Volume II, we proved an important relationship, called the *Legendre transformation*. This expression is vitally important because it relates information entropy to the partition function, the Lagrange multipliers, and the constraint function averages,

$$H(p) = \ln Z - \sum_{j=1}^{m} \lambda_j \langle F_j \rangle \qquad (37.6)$$

The change in negative entropy $(-H)$ with a change in $\langle F_j \rangle$, the jth constraint function average, is the Lagrange multiplier of the jth constraint function, λ_j, and expressed in MEP notation as,

$$\frac{\mathrm{d}(-H)}{\mathrm{d}\langle F_j \rangle} = \lambda_j \qquad (37.7)$$

or, expressed in Amari's IG notation as,

$$\partial^i \varphi = \theta^i \qquad (37.8)$$

See Exercise 37.9.14 for a numerical example of Equation (37.7), followed by an interesting derivation of Equation (37.8) in Exercise 37.9.15 that closely follows Amari's derivation. Amari presents a proof relying on the *total differential*. That is why the notation above changed from the notation of the partial differentiation to the notation appropriate for manipulations with the total differential.

37.7 The Metric Tensor Revisited

A tangent space can be established at each point in the manifold with the basis vectors defined as $F_j(x_i) - \langle F_j \rangle$. These basis vectors are seen to consist of the MEP's constraint functions and constraint function averages. Any vector **v** in the tangent space is the linear combination of coefficients v^j and the basis vectors $\mathbf{e}_j$ expressed as $\sum_{j=1}^{m} v^j \, \mathbf{e}_j$.

The metric tensor at the point p is thought of from a statistical perspective as the Fisher information matrix,

$$g_{rc}(p) = E_p \left[\frac{\partial \ln p}{\partial \lambda_r} \times \frac{\partial \ln p}{\partial \lambda_c} \right]$$

where $E_p[\cdots]$ is the Expectation operator. The exponential family, or the MEP assignment under some model, looks like,

$$p \equiv P(X = x_i \,|\, \mathcal{M}_k) = \frac{\exp\left[\sum_{j=1}^{m} \lambda_j \, F_j(X = x_i)\right]}{Z(\lambda_1, \cdots, \lambda_m)}$$

To find the metric tensor, we first take the log transform of the MEP assignment followed by its partial derivative with respect to each Lagrange multiplier. The log transform is,

$$\ln p = \lambda_1 F_1(X = x_i) + \lambda_2 F_2(X = x_i) + \cdots + \lambda_m F_m(X = x_i) - \ln Z$$

while the partial derivatives are,

$$\frac{\partial \ln p}{\partial \lambda_j} = F_j(X = x_i) - \frac{\partial \ln Z}{\partial \lambda_j}$$

Since

$$\frac{\partial \ln Z}{\partial \lambda_j} = \langle F_j \rangle$$

we see that the metric tensor is, from this statistical perspective, the covariance matrix of the constraint functions, defined at each different point in the manifold,

$$g_{rc}(p) = E_p\left[(F_r(X = x_i) - \langle F_r \rangle) \times (F_c(X = x_i) - \langle F_c \rangle)\right]$$

In the notation of Information Geometry, the metric tensor is expressed as,

$$g_{ij}(\xi) = E_\xi\left[\partial_i l_\xi \partial_j l_\xi\right]$$

where,

$$l_\xi \equiv \log p(x; \xi)$$

37.7.1 Numerical example of the metric tensor at point s

We will present a numerical example of the metric tensor for the current inferential scenario involving the kangaroos. Under the first model considered, model $\mathcal{M}_1$, $m = 3$ constraint functions were employed. This was labeled as point s in the manifold. No information from any interactions were introduced by this model. Therefore, we would not expect any covariance components to appear in the matrix. Since $m = 3$, the Fisher information matrix at point s will be a 3×3 matrix.

The diagonal entries in the matrix will consist of the variances for each of the three constraint functions, while the off-diagonal entries will consist of the three possible covariances between the constraint functions. Instead of computing all nine values, only six values in the matrix have to be computed since a covariance matrix is symmetric.

Thus, the six entries $g_{11}, g_{21}, g_{22}, g_{31}, g_{32}, g_{33}$ will be computed with the three diagonal entries g_{11}, g_{22}, and g_{33} representing the three variances, and the three off-diagonal entries g_{21}, g_{31}, and g_{32} representing the three covariances.

The detailed computation of g_{21}, the covariance between the first and second constraint function, is shown in Table 37.1 at the top of the next page. Keep in mind that this is a covariance between constraint functions, not between some data. Compute one of the entries in the 3×3 matrix $g_{rc}(s)$, namely $g_{21}(s)$. The subscript $r = 2$ indexes the row, and the subscript $c = 1$ indexes the column.

Table 37.1: *The computational details for $g_{21}(s)$, the covariance between the first and second constraint functions in model $\mathcal{M}_1$.*

$F_2(x) - \langle F_2 \rangle$	$F_1(x) - \langle F_1 \rangle$	Col 1 × Col 2	Q_i	Col 3 × Col 4
$1 - 2/3$	$1 - 1/2$	$+1/6$	0.2778	$+0.0463$
$0 - 2/3$	$1 - 1/2$	$-1/3$	0.1389	-0.0463
$1 - 2/3$	$1 - 1/2$	$+1/6$	0.0556	$+0.0093$
$0 - 2/3$	$1 - 1/2$	$-1/3$	0.0278	-0.0093
$1 - 2/3$	$0 - 1/2$	$-1/6$	0.2278	-0.0463
$0 - 2/3$	$0 - 1/2$	$+1/3$	0.1389	$+0.0463$
$1 - 2/3$	$0 - 1/2$	$-1/6$	0.0556	-0.0093
$0 - 2/3$	$0 - 1/2$	$+1/3$	0.0278	$+0.0093$
	Sums		1.0000	0.0000

This easy calculation shows in detail that the covariance between $F_1(X = x_i)$ and $F_2(X = x_i)$ is 0,

$$g_{21}(s) = E_r\left[(F_1(X = x_i) - \langle F_1 \rangle) \times (F_2(X = x_i) - \langle F_2 \rangle)\right]$$

$$= 0$$

The complete metric tensor at the point s is shown as the matrix $g_{rc}(s)$,

$$g_{rc}(s) = \begin{pmatrix} 0.25 & 0 & 0 \\ \boxed{0} & 0.2222 & 0 \\ 0 & 0 & 0.1389 \end{pmatrix}$$

where the just calculated value for the covariance $g_{21} = 0$ is shown boxed.

If $\mathbf{G} \equiv g_{rc}(s)$, then the inverse of the metric tensor, $\mathbf{G}^{-1} \equiv g^{rc}(s)$, can be calculated as,

$$g^{rc}(s) = \begin{pmatrix} 4 & 0 & 0 \\ 0 & 4.5 & 0 \\ 0 & 0 & 7.2 \end{pmatrix}$$

37.7.2 Changes in dual coordinates

What kind of change is induced in an arbitrary constraint function average when an arbitrary Lagrange multiplier is changed? In other words, how responsive is, say, $\langle F_4 \rangle$ when λ_2 is changed, and all the other Lagrange multipliers are held constant? This is answered for this specific case by,

$$\frac{\partial \langle F_4 \rangle}{\partial \lambda_2}$$

and, in general by,
$$\frac{\partial \langle F_c \rangle}{\partial \lambda_r}$$

Other ways of writing this relationship would be,
$$\frac{\partial \eta_c}{\partial \theta^r} \equiv \partial_r \eta_c$$

But more revealing is,
$$\frac{\partial \ln Z}{\partial \lambda_c} = \langle F_c \rangle$$
$$\frac{\partial}{\partial \lambda_r}\left(\frac{\partial \ln Z}{\partial \lambda_c}\right) = \frac{\partial \langle F_c \rangle}{\partial \lambda_r}$$
$$\frac{\partial^2 \ln Z}{\partial \lambda_r \partial \lambda_c} = \frac{\partial \langle F_c \rangle}{\partial \lambda_r}$$
$$\frac{\partial \langle F_c \rangle}{\partial \lambda_r} = g_{rc}(p)$$

So, the metric tensor will tell us the answer to the question of how responsive a constraint function average is when some Lagrange multiplier is changed. See Exercise 37.9.20 for an in–depth analysis.

But we can also answer the inverse question of how responsive the Lagrange multiplier will be when some constraint function average is changed. In this case, we have,
$$\frac{\partial \lambda_c}{\partial \langle F_r \rangle} \equiv \frac{\partial \theta^c}{\partial \eta_r}$$

In Jaynes's notation,
$$\frac{\partial S}{\partial F_r} = \lambda_r$$
$$\frac{\partial}{\partial F_c}\left(\frac{\partial S}{\partial F_r}\right) = \frac{\partial^2 S}{\partial F_c \partial F_r}$$
$$\frac{\partial \lambda_r}{\partial F_c} = g^{rc}(p)$$

So, for example, if the IP were curious about how λ_5 must change for some small change in $\langle F_2 \rangle$, the component of the inverse metric tensor, $g^{25}(p) = g^{52}(p)$, will provide the answer.

We won't get bogged down in all of the details right now. Exercise 37.9.21 will illustrate the general idea that, if, say, $g^{52}(p) = -50$, there must be roughly a unit change in λ_5 in the negative direction as the average for constraint function $F_2(X = x_i)$ moves from, say, $\langle F_2 \rangle = 0.65$ to $\langle F_2 \rangle = 0.67$.

37.8 Connections to the Literature

One of the reasons why I admire Jaynes is that he is the only writer I know of who has gone on record to forcefully and explicitly state that measure theory is not needed in any way, shape, or form in order to understand probability. I have followed in his footsteps in my presentation of probability theory, also not discovering any inferential problems solvable only by resorting to measure theory. If Jaynes were wrong in this supposition about the necessity for measure theory, why have its proponents not immediately pounced on inferential scenarios where the solution did, in fact, depend critically on using measure theory?

Instead, Jaynes [22, pg. 664] says that he will bypass measure theory and still get the results he wants. In my take on the same issue, I have emphasized that an IP may rely heavily upon probability theory *as a generalization of logic* in order to obtain the results it wants. And it was more mathematically interesting to Jaynes, and for me as well, to examine the Boolean function expansions that underlie this generalization.

> We note also that measure theory is not always applicable, because not all sets that arise in real problems are measurable. For example, in many applications we want to assign probabilities to functions that we know in advance are continuous. ...
>
> Our value judgment is just the opposite; being concerned with the real world, we are willing to sacrifice preconceptions about measurable classes in order to preserve the aspects of the real world that are important in our problem. In this case, a form of our cautious approach policy will always be able to bypass measure theory in order to get the useful results we seek; for example, (1) expand the continuous functions in a finite–number n of orthogonal functions, (2) assign probabilities to the expansion coefficients in a finite–dimensional space R_n; (3) do the probability calculation; (4) pass to the limit $n \to \infty$ at the end. In a real problem we find that increasing n beyond a certain value makes a numerically negligible change in our conclusions ... So we need never depart from finite sets after all. Useful results, in various applications from statistical mechanics to radar detection, are found in this way.

Look at how we started off here in this Chapter in setting up a model inspired by a chain of implications. We established a Boolean function of three variables $f(x, y, z)$,

$$f \colon \mathbf{B}^3 \to \mathbf{B}$$

expanded canonically via Boole's Expansion Theorem with a finite number $n = 8$ of orthogonal Boolean functions and accompanying function expansion coefficients $f(T, T, T)$ through $f(F, F, F)$,

$$f(x, y, z) = [\, f(T,T,T)\, x \circ y \circ z\,] \bullet \cdots \bullet [\, f(F,F,F)\, x' \circ y' \circ z'\,]$$

Specialize to a logic function with three variables,

$$f_{139}(A, B, C) = (A \to B) \wedge (B \to C)$$

It too has its own DNF expansion involving a finite number of orthogonal logic functions, and accompanying function expansion coefficients $f(\star, \star, \star)$,

$$f_{139}(A, B, C) = f(\text{TRUE}, \text{TRUE}, \text{TRUE})\, ABC \vee \cdots \vee f(\text{FALSE}, \text{FALSE}, \text{FALSE})\, \overline{A}\,\overline{B}\,\overline{C}$$

In both cases, the function expansion coefficients, $f(\text{TRUE}, \text{TRUE}, \text{TRUE})$ through $f(\text{FALSE}, \text{FALSE}, \text{FALSE})$, were the important determiners of a model assigning numerical values to probabilities.

For example, the expansion coefficient $f(\text{TRUE}, \text{TRUE}, \text{FALSE})$ accompanying the orthogonal basis function $AB\overline{C}$ had the value FALSE. Therefore, the IP assigned a numerical value of 0 to cell 3 of the joint probability table under this implicational chain model $f_{139}(A, B, C)$. [7, Chapters One, Two, and Three]

The bane of trying to learn anything in IG is the lack of any standardized notation. Here is just one example of the difficulty of reading works in IG due to the prevalence of a non–standardized notation. Jaynes [22, pg. 359, Eq. 11.62] had already mentioned a differential relationship involving the change of information entropy with a change in constraint function average. He wrote it,

$$\lambda_k = \frac{\partial S(F_1, \ldots, F_m)}{\partial F_k}$$

Amari [2, pg. 59] expressed the same relationship as,

$$\partial^i \varphi = \theta^i$$

Amari employs an idiosyncratic and condensed notation in his proofs, but one that is nonetheless fully acceptable to mathematicians. For example, his proof (Refer to Exercise 37.9.15) of the above relationship starts and ends as follows:

$$\begin{aligned} \varphi &= \theta^i \eta_i - \psi \\ \partial^i \varphi &= \theta^i \end{aligned}$$

Sometimes, this kind of notation succeeds in enabling you to grasp proofs more quickly as it does in this example, but there are many other occasions where you are left scratching your head with a lot of notational macros left for you to unwind. Exercises 37.9.15 and 37.9.16 go through one of these notational unwinding tasks to give you a flavor of what is involved when you want to match up with, say, Jaynes's notation.

Whenever I talk about models in terms of the dual parameters λ_j and $\langle F_j \rangle$ within the MEP formalism, the correspondence with Amari when he talks about the dualistic structure of the exponential family is exact, even though all of the notation is different. The following quote from Amari [3, pp. 65–66] will make this clear.

We showed ... that with respect to an exponential family,

$$p(x;\theta) = \exp\left[\, C(x) + \theta^i\, F_i(x) - \psi(\theta)\,\right] \qquad (3.55)$$

the natural parameters $[\theta^i]$ form a 1–affine coordinate system. Now if we define

$$\eta_i = \eta_i(\theta) \stackrel{\text{def}}{=} E_\theta[F_i] = \int F_i(x)\, p(x;\theta)\, \mathrm{d}x \qquad (3.56)$$

then from Equations (2.33) and (2.9) we obtain $\eta_i = \partial_i \psi$. Furthermore, from Equations (2.34) and (2.8) we obtain $\partial_i \partial_j \psi = g_{ij}$. Hence $[\eta_i]$ is a (-1)–affine coordinate system dual to $[\theta^i]$, and ψ is the potential of a Legendre transformation. We call this $[\eta_i]$ the **expectation parameters** or the **dual parameters**.

The dual potential φ in Equation (3.36) is then given by

$$\begin{aligned}
\varphi(\theta) &= \theta^i\, \eta_i(\theta) - \psi(\theta) \\
&= E_\theta[\log p_\theta - C] \\
&= -H(p_\theta) - E_\theta[C]
\end{aligned} \qquad (3.57)$$

where H is the **entropy**: $H(p) \stackrel{\text{def}}{=} -\int p(x) \log p(x)\, \mathrm{d}x$. In addition, from Equation (3.38) we have

$$\varphi(\theta) = \max_{\theta'}\{\theta'^i\, \eta_i(\theta) - \psi(\theta')\} \qquad (3.58)$$

where the maximum is attained by $\theta' = \theta$.

From the definition of the Fisher information matrix ... we have

$$g_{ij}(\theta) = E_\theta[(F_i(x) - \eta_i)(F_j(x) - \eta_j)] \qquad (3.59)$$

Unfortunately, the overlap in agreement abruptly ends here when Amari begins to make remarks germane to orthodox statistics.

Some of the notational equivalencies between my MEP–inspired notation and Amari's IG–inspired notation are summarized below. Listed are the probability of a statement conditioned on a model, the MEP formula and the exponential family, the constraint function average, and the metric tensor (Fisher information matrix),

$$P(X = x_i \,|\, \mathcal{M}_k) \equiv p(x;\theta)$$

$$\frac{\exp\left[\sum_{j=1}^{m} \lambda_j F_j(X = x_i)\right]}{Z(\lambda_1, \lambda_2, \cdots, \lambda_m)} \equiv \exp\left[C(x) + \theta^i F_i(x) - \psi(\theta)\right]$$

$$\langle F_j \rangle \equiv \eta_i$$

$$\sum_{i=1}^{n} F_j(X = x_i)\, Q_i \equiv E_\theta[F_i]$$

$$\frac{\partial \ln Z}{\partial \lambda_j} \equiv \partial_i \psi$$

$$g_{rc}(p) \equiv g_{ij}(\xi)$$

$$\frac{\partial^2 \ln Z}{\partial \lambda_r \, \partial \lambda_c} \equiv \partial_i \partial_j \psi$$

$$\equiv \frac{\partial \langle F_c \rangle}{\partial \lambda_r}$$

$$\equiv \sum_{i=1}^{n} (F_r(X = x_i) - \langle F_r \rangle) \times (F_c(X = x_i) - \langle F_c \rangle) \times Q_i$$

and between the potentials of the Legendre transformation,

$$H_{\max}(Q_i) = \min_{\lambda_j}\left\{\ln Z - \sum_{j=1}^{m} \lambda_j \langle F_j \rangle\right\}$$

$$\varphi(\theta) = \max_{\theta'}\left\{\theta'^i \, \eta_i(\theta) - \psi(\theta')\right\}$$

37.9 Solved Exercises for Chapter Thirty Seven

Exercise 37.9.1: Carry out the formal logic operations to determine the numerator and denominator in Bayes's Theorem that generalize the purely deductive result found in section 37.3.

Solution to Exercise 37.9.1

Given the premises G and $(G \to B) \wedge (B \to H)$, the logical consequence is H. The formal manipulation rules for probability theory arrive at the same result.

Set up Bayes's Theorem for this problem as,

$$P(H \,|\, G, [G \to B] \wedge [B \to H]) = \frac{P(H \wedge G \wedge [G \to B] \wedge [B \to H])}{P(G \wedge [G \to B] \wedge [B \to H])}$$

See the next Exercise showing how the DNF expansion for the implication chain model can be found as,

$$GBH \vee \overline{G}BH \vee \overline{G}\,\overline{B}H \vee \overline{G}\,\overline{B}\,\overline{H}$$

Now employ the standard Boolean operations on the expressions within the probability operator in both the numerator and denominator. Start off with the denominator,

$$P(G \wedge [G \to B] \wedge [B \to H]) \;=\; P(G \wedge [GBH \vee \overline{G}BH \vee \overline{G}\,\overline{B}H \vee \overline{G}\,\overline{B}\,\overline{H}])$$

$$=\; P([GGBH \vee G\overline{G}BH \vee G\overline{G}\,\overline{B}H \vee G\overline{G}\,\overline{B}\,\overline{H}])$$

$$=\; P(GBH)$$

The numerator is easily found as well,

$$P(H \wedge G \wedge [G \to B] \wedge [B \to H]) \;=\; P(H \wedge [GBH])$$

$$=\; P(GBH)$$

Substituting these results back into Bayes's Theorem yields the expected answer,

$$P(H \,|\, G, [G \to B] \wedge [B \to H]) = \frac{P(GBH)}{P(GBH)} = 1$$

It is certain that H is TRUE given G is TRUE together with the implicational model linking G with B and B with H. There is no uncertainty that a kangaroo is right–handed given that the IP knows that the kangaroo possesses the standard genetic code, and given that the causal model captured by the logic function is correct.

Exercise 37.9.2: Rely on *Mathematica* to help you find the DNF for the logic expression $(A \to B) \wedge (B \to C)$.

Solution to Exercise 37.9.2

Mathematica has two built-in Boolean functions,

                      `BooleanConvert[ ]` and `BooleanTable[ ]`

which we can use to find the fully expanded DNF for a logic expression. First, form the logic expression $(A \to B) \wedge (B \to C)$ in *Mathematica*,

                      `And[Implies[A, B], Implies[B, C]]`

Follow this with an application of `BooleanTable[ ]` after first processing with a `BooleanConvert[ ]` in order to obtain the list of `True` and `False`. This list will tell us how to construct the fully expanded DNF. So, after evaluating,

`BooleanTable[BooleanConvert[And[Implies[A,B], Implies[B,C]]]]`

Mathematica returns the list,

        {`True, False, False, False, True, False, True, True`}

which will pinpoint the locations of the 0s in the joint probability table.

Table 37.2 below deciphers the heuristic by which we determine the locations of the 0s in the joint probability table. The first column is the standard ordering of the three arguments to a Boolean function. The next column shows this in the notation of logic variables.

Table 37.2: *Determining the location of 0s in a joint probability table using the DNF.*

$(\star,\star,\star)$	Logic variables	$f(\star,\star,\star)$	Cell	0 assigned?
TTT	ABC	TRUE	Cell 1	No
TTF	$AB\overline{C}$	FALSE	Cell 3	Yes
TFT	$A\overline{B}C$	FALSE	Cell 2	Yes
TFF	$A\overline{B}\,\overline{C}$	FALSE	Cell 4	Yes
FTT	$\overline{A}BC$	TRUE	Cell 5	No
FTF	$\overline{A}B\overline{C}$	FALSE	Cell 7	Yes
FFT	$\overline{A}\,\overline{B}C$	TRUE	Cell 6	No
FFF	$\overline{A}\,\overline{B}\,\overline{C}$	TRUE	Cell 8	No

The third column matches up the results of the **True** and **False** list as just evaluated by *Mathematica*. The fourth column matches up the location of the joint statements with the cells of the joint probability table.

One must be careful here because, for example, $TTF \equiv AB\overline{C}$ is actually Cell 3 given the way the joint probability table was constructed. Refer back to Figure 37.1 for the generic joint probability table. The final column tells us whether we should place a 0 in a particular cell to conform to the model of the logic expression $(A \rightarrow B) \wedge (B \rightarrow C)$.

Exercise 37.9.3: Based on the results in the previous exercise, what is the fully expanded DNF?

Solution to Exercise 37.9.3

Connect disjunctively with the $\vee$ operator those four terms in Table 37.2 with an assignment of TRUE to arrive at the DNF for $(A \rightarrow B) \wedge (B \rightarrow C)$,

$$ABC \vee \overline{A}BC \vee \overline{A}\,\overline{B}C \vee \overline{A}\,\overline{B}\,\overline{C}$$

Exercise 37.9.4: What does *Mathematica* report back as the DNF?

Solution to Exercise 37.9.4

After evaluating,

```
BooleanConvert[And[Implies[A, B], Implies[B, C]]]
```

Mathematica simply returns $\overline{A}\,\overline{B} \vee BC$.

Exercise 37.9.5: Is this result the same as the fully expanded DNF as found in Exercise 37.9.3?

Solution to Exercise 37.9.5

Yes, it is. Applying standard Boolean operations to the full DNF expression reduces it to the shorter expression returned by *Mathematica*. Examine the first two terms, extract the common factor of BC, and recognize an expression for T,

$$ABC \vee \overline{A}BC = BC(A \vee \overline{A})$$

$$A \vee \overline{A} = T$$

$$B \wedge C \wedge T = BC$$

Do exactly the same thing for the final two terms,

$$\overline{A}\,\overline{B}C \vee \overline{A}\,\overline{B}\,\overline{C} = \overline{A}\,\overline{B}(C \vee \overline{C})$$

$$C \vee \overline{C} = T$$

$$\overline{A} \wedge \overline{B} \wedge T = \overline{A}\,\overline{B}$$

For the final step, use **Commutativity** to see that,

$$ABC \vee \overline{A}BC \vee \overline{A}\,\overline{B}C \vee \overline{A}\,\overline{B}\,\overline{C} = \overline{A}\,\overline{B} \vee BC$$

Exercise 37.9.6: Use Wolfram's numbering system to discover the Boolean function of three variables that matches the logic expression.

Solution to Exercise 37.9.6

We'll make *Mathematica* do all of the hard work before checking the result by hand.

**FromDigits[Boole[BooleanTable[BooleanConvert[
            And[Implies[A, B], Implies[B, C]]]]], 2]**

returns the number 139. Create the three variable Boolean function,

f139 = BooleanFunction[139, 3]

The DNF of this Boolean function, $\overline{A}\,\overline{B} \vee BC$,

BooleanConvert[f139[A, B, C]]//TraditionalForm

is exactly the same as our original logic expression. To clinch the matter,

BooleanTable[f139]

returns the same list of functional assignments as the original logic expression,

{**True, False, False, False, True, False, True, True**}

Checking by hand, match up each **True** with 1 and each **False** with 0 for the binary expansion,

$$(1 \times 128) + (0 \times 64) + (0 \times 32) + (0 \times 16) + (1 \times 8) + (0 \times 4) + (1 \times 2) + (1 \times 1) = 139$$

SOLVED EXERCISES FOR CHAPTER 37

Exercise 37.9.7: Discuss the model in terms of Wolfram's elementary cellular automata.

Solution to Exercise 37.9.7

$f_{139}(A, B, C)$ is the three variable Boolean function $f(x, y, z)$ at the heart of one of Wolfram's 256 elementary CA. Which elementary CA is that? Why, Rule 139 of course.

Figure 37.8 shows Rule 139 as Wolfram pictures it. This diagram is always in terms of three cells A_N, B_N, C_N, colored white or black at time step N, which produce at time step $N + 1$ a black or white cell B_{N+1}. For the first part of the deterministic CA rule, the color of the three cells in eight sets of three cells matches up with the domain of $\mathbf{B}^3 \equiv \{\{T, T, T\}, \{T, T, F\}, \cdots, \{F, F, F\}\}$ where T is a black cell and F is a white cell.

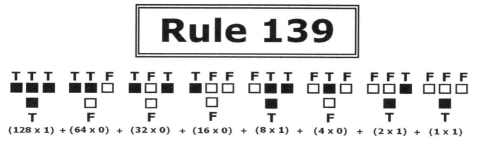

Figure 37.8: *Diagram of the Rule 139 elementary cellular automaton.*

For the second part of the deterministic CA rule, the color of the cell produced by these three cells is determined by a particular three variable Boolean function whose expansion via Boole's Expansion Theorem is,

$$f: \mathbf{B}^3 \to \mathbf{B} \equiv f_{139}(x, y, z)$$
$$= f(T,T,T) \circ xyz \bullet f(T,T,F) \circ xyz' \bullet \cdots \bullet f(F,F,F) \circ x'y'z'$$
$$= (T \circ ABC) \bullet (F \circ AB\overline{C}) \bullet \cdots \bullet (T \circ \overline{A}\,\overline{B}\,\overline{C})$$
$$f_{139}(A, B, C) = ABC \vee \overline{A}BC \vee \overline{A}\,\overline{B}C \vee \overline{A}\,\overline{B}\,\overline{C}$$

As we discovered in previous exercises, translating over into logic expressions, the value of TRUE or FALSE for the coefficient function $f(\star, \star, \star)$ of the eight orthogonal basis functions $ABC, AB\overline{C}, \cdots, \overline{A}\,\overline{B}\,\overline{C}$ tells us whether the cell is colored white or black. This coefficient function is the analog to the probability placed into a cell of the joint probability table.

Exercise 37.9.8: Elaborate on Jaynes's comment quoted in section 37.8 about probabilities and the expansion coefficients of basis functions.

Solution to Exercise 37.9.8

I first presented Boole's Expansion Theorem for a Boolean function in Chapter One of Volume I [7, pg. 9]. To me, this is a subtle, but often unrecognized, feature underlying probability theory. Jaynes is the only writer I know of who mentions the potential significance of expansion coefficients of functions for probability theory.

Here, I want to show an example using a three variable Boolean function where we get away from using just T and F as assignments for the expansion coefficients $f(\star, \star, \star)$. We are familiar with the canonical expansion of an arbitrary three variable Boolean function explicitly showing all of the orthogonal basis functions together with their coefficient functions,

$$f(x,y,z) = f(T,T,T)\,xyz \bullet f(T,T,F)\,xyz' \bullet \cdots \bullet f(F,F,F)\,x'y'z'$$

In this Chapter, we examined the consequences of a three variable logic function where any expansion coefficient equal to F told us where to place zeroes into a joint probability table. But it may have caused some head scratching as to why the other assignments all had expansion coefficients equal to T.

For a general Boolean Algebra, the assignments to the coefficient functions don't all necessarily have to be either T or F. For example, if the carrier set consists of the four elements,

$$\mathbf{B} = \{a, a', F, T\}$$

then an arbitrary expansion coefficient $f(\star, \star, \star)$ could take on the value of a or a' as well as T or F.

For an easy example, suppose that $f(T,T,T) = a$ and $f(F,T,T) = a'$ with all of the other coefficients equal to F. The function $f(x, y, z)$ now looks like this,

$$f(x,y,z) = axyz \bullet a'x'yz$$

What makes all of this work in the context of Classical Logic is the requirement that all of our statements must be TRUE or FALSE. There is no middle ground. Even though it *is* permissible with this carrier set to make assignments other than T or F, the actual values of the variables must be T or F to match up with probability as a generalization of logic. This can be illustrated quite nicely with the above function expansion,

$$f(x,y,z) = axyz \bullet a'x'yz$$

Let the variables take on the values $x = T$, $y = T$, and $z = T$, then,

$$f(x = T, y = T, z = T) = (a \circ T \circ T \circ T) \bullet (a' \circ T' \circ T \circ T)$$

$$= a \bullet (a' \circ F \circ T \circ T)$$

$$= a \bullet F$$

$$= a$$

Likewise, let the variables take on the values $x = F$, $y = T$, and $z = T$, then,

$$f(x = F, y = T, z = T) = (a \circ F \circ T \circ T) \bullet (a' \circ F' \circ T \circ T)$$

$$= F \bullet (a' \circ T \circ T \circ T)$$

$$= F \bullet a'$$

$$= a'$$

So, in the end, the coefficients of the orthogonal basis functions do, in fact, provide the probabilities. Figure 37.9 demonstrates what kind of joint probability table this would produce.

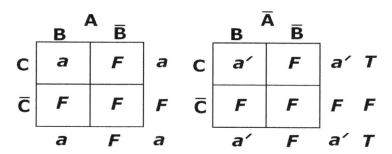

Figure 37.9: *The "joint probability table" induced by a particular three variable Boolean function $f(x, y, z) = axyz \bullet a'x'yz$.*

Mixing our metaphors, the variable labels in the joint probability table are the same three logic variables used in this Chapter. The joint probability for $P(ABC)$ would be its expansion coefficient $f(T,T,T) = a$. The joint probability for $P(\overline{A}BC)$, would be its expansion coefficient $f(F,T,T) = a'$. All of the other joint probabilities, say, $P(A\overline{B}C)$ would be F.

The beauty of Boolean Algebra is that it needs neither numbers nor arithmetical operators like $+$ or $\times$. However, we do need numbers and arithmetic operations for probability theory, so let $a = Q_1$ and $a' = Q_5$. The DNF for the model $\mathcal{M}_k$ of the joint probability table would be $ABC \vee \overline{A}BC = BC$. And,

$$P(B) = a \bullet F \bullet a' \bullet F = a \bullet a' = T = 1$$

Let's examine whether everything we've constructed so far is consistent with the formal rules of probability manipulation. We have this model based on a logic function BC, $\mathcal{M}_k \equiv BC$. If B is TRUE, what does it say about A?,

$$P(A \mid B, \mathcal{M}_k) = \frac{P(AB \mid \mathcal{M}_k)}{P(B \mid \mathcal{M}_k)} = \frac{P(ABC \mid \mathcal{M}_k) + P(AB\overline{C} \mid \mathcal{M}_k)}{P(B \mid \mathcal{M}_k)} = \frac{Q_1 + Q_3}{1} = Q_1$$

But $P(A) = Q_1$ before we knew that B was TRUE. Therefore, knowing that B is TRUE adds nothing more to what we know about A.

If C is TRUE, what does it say about A?,

$$P(A \mid C, \mathcal{M}_k) = \frac{P(AC \mid \mathcal{M}_k)}{P(C \mid \mathcal{M}_k)} = \frac{P(ABC \mid \mathcal{M}_k) + P(A\overline{B}C \mid \mathcal{M}_k)}{P(C \mid \mathcal{M}_k)} = \frac{Q_1 + Q_2}{1} = Q_1$$

The same situation as before. If C is TRUE, it adds nothing more to what we know about A.

But if B is TRUE, what does it say about C?

$$P(C \mid B, \mathcal{M}_k) = \frac{P(BC \mid \mathcal{M}_k)}{P(B \mid \mathcal{M}_k)} = \frac{P(ABC \mid \mathcal{M}_k) + P(\overline{A}BC \mid \mathcal{M}_k)}{P(B \mid \mathcal{M}_k)} = \frac{Q_1 + Q_5}{1} = 1$$

If B is TRUE, it says everything about C, and *vice versa*. It says that C is certain to happen. Everything is consistent. We conditioned on the truth of the model that said that both B and C were TRUE. If B is TRUE, then C must also be TRUE. Probability theory as a generalization of logic must confirm these specific cases.

Finally, notice that we are not allowed by Bayes's Theorem to condition on a statement and a model that are logically incompatible. Suppose that the IP asks about C given the model $\mathcal{M}_k$ and the truth of the statement that B is FALSE.

$$P(C \mid \overline{B}, \mathcal{M}_k) = \frac{P(\overline{B}C \mid \mathcal{M}_k)}{P(\overline{B} \mid \mathcal{M}_k)} = \frac{P(\overline{B}C \mid \mathcal{M}_k)}{0} \qquad \text{INVALID}$$

The model states that both B and C are TRUE, and the conditioning statement says that, at the same time, B must be FALSE. This is logically incompatible and trying to use Bayes's Theorem would be invalid.

SOLVED EXERCISES FOR CHAPTER 37

Exercise 37.9.9: Use the MEP formula to find the numerical assignments to all eight cells of the joint probability table when only three constraint functions specify the model.

Solution to Exercise 37.9.9

Bring in the MEP formula in Equations (37.3) and (37.4) to find the numerical assignment for all eight cells in the joint probability table under this particular model $\mathcal{M}_1$. Model $\mathcal{M}_1$ inserts information about the marginal probabilities for the genetic code, beer preference, and hand preference, and that is all the information that it provides.

The calculations for all eight numerical assignments Q_i under this model are summarized below in Table 37.3. The final row presents the relevant sums. The

Table 37.3: *The MEP calculations for a specific model inserting information about three marginal probabilities for the kangaroo scenario. The Lagrange multipliers for this model are $\lambda_1 = 0$, $\lambda_2 = 0.69315$, and $\lambda_3 = 1.60944$.*

Cell i	$F_1(x_i)$	$F_2(x_i)$	$F_3(x_i)$	Numerator	Q_i
1	1	1	1	$\exp[0 + 0.69315 + 1.60944] = 10$	10/36
2	1	0	1	$\exp[0 + 1.60944] = 5$	5/36
3	1	1	0	$\exp[0 + 0.69315] = 2$	2/36
4	1	0	0	$\exp[0] = 1$	1/36
5	0	1	1	$\exp[0.69315 + 1.60944] = 10$	10/36
6	0	0	1	$\exp[1.60944] = 5$	5/36
7	0	1	0	$\exp[0.69315] = 2$	2/36
8	0	0	0	$\exp[0] = 1$	1/36
Sums				$Z = 36$	1

first sum shown is the partition function,

$$Z(\lambda_1, \lambda_2, \lambda_3) = \sum_{i=1}^{8} \exp[\lambda_1 F_1(x_i) + \lambda_2 F_2(x_i) + \lambda_3 F_3(x_i)] = 36$$

The sum of all eight assignments, $\sum_{i=1}^{8} Q_i$, must, of course, equal 1. I refer to this as the universal constraint. It is written in the same manner as any other constraint, $F_0(x_i) = \{1, 1, 1, 1, 1, 1, 1, 1\}$, and $\langle F_0 \rangle = 1$. The satisfaction of any constraint may be checked at your leisure. For example,

$$\langle F_1 \rangle = \frac{10}{36} + \frac{5}{36} + \frac{2}{36} + \frac{1}{36} = \frac{18}{36} = \frac{1}{2}$$

In detail, consider the calculation of the numerical assignment to the seventh cell, $Q_7 \equiv P(\overline{G}, B, \overline{H} \mid \mathcal{M}_1)$,

$$Q_7 = \frac{\exp\left[\lambda_1 F_1(x_7) + \lambda_2 F_2(x_7) + \lambda_3 F_3(x_7)\right]}{Z(\lambda_1, \lambda_2, \lambda_3)}$$

$$= \frac{\exp\left[(\lambda_1 \times 0) + (\lambda_2 \times 1) + (\lambda_3 \times 0)\right]}{Z(\lambda_1, \lambda_2, \lambda_3)}$$

$$= \frac{\exp\left[\lambda_2\right]}{Z(\lambda_1, \lambda_2, \lambda_3)}$$

In order to satisfy the constraints, the MEP algorithm found that the Lagrange multipliers equaled,

$$\lambda_1 = 0$$

$$\lambda_2 = 0.69315$$

$$\lambda_3 = 1.60944$$

The numerator for Q_7 is then calculated as,

$$\text{Numerator of } Q_7 = \exp\left[\lambda_2\right]$$

$$= \exp\left[0.69315\right]$$

$$= 2$$

The partition function, $Z(\lambda_1, \lambda_2, \lambda_3)$, is the sum over all eight numerators calculated in the same way as just demonstrated, so $Z = 36$. The numerical value assigned to the probability for the joint statement in cell 7 of the joint probability table by model $\mathcal{M}_1$ is therefore $2/36 = 0.0556$. Refer back to Figure 37.4 to see all of the numerical assignments calculated in this exercise placed into their proper locations.

Exercise 37.9.10: Apply the MEP formula as above to find the numerical assignments to all eight cells of the joint probability table for the final implicational chain model $\mathcal{M}_3$.

Solution to Exercise 37.9.10

This is carried out in exactly the same manner as the previous exercise. Since the information in model $\mathcal{M}_3$ now includes two more constraint functions over and above the three constraint functions in model $\mathcal{M}_1$, m has increased from 3 to 5.

Table 37.4 is a precursor table showing all 5 constraint functions. $F_1(x_i)$, $F_2(x_i)$, and $F_3(x_i)$ are the same as for model $\mathcal{M}_1$ as they code for the marginal probabilities of G, B, and H. $F_4(x_i)$ implements the GB interaction, while $F_5(x_i)$ does the BH interaction. Consult this table by reading across any row to discern which Lagrange multipliers will be included in the numerator.

SOLVED EXERCISES FOR CHAPTER 37

Table 37.4: *The five constraint functions involved in model $\mathcal{M}_3$.*

Cell i	$F_1(x_i)$	$F_2(x_i)$	$F_3(x_i)$	$F_4(x_i)$	$F_5(x_i)$
1	1	1	1	1	1
2	1	0	1	0	0
3	1	1	0	1	0
4	1	0	0	0	0
5	0	1	1	0	1
6	0	0	1	0	0
7	0	1	0	0	0
8	0	0	0	0	0

With this much accomplished, the second table, Table 37.5, can be constructed showing the details of the numerator, the calculation of Z, and the actual numerical assignments to the probability for each cell of the joint probability table.

Table 37.5: *The implicational chain model $\mathcal{M}_3$ inserts information about three marginal probabilities and two interactions. The Lagrange multipliers for this model are $\lambda_1 = -35.9314$, $\lambda_2 = -28.1773$, $\lambda_3 = 0$, $\lambda_4 = 37.03$, and $\lambda_5 = 28.1773$.*

Cell i	Numerator	Q_i
1	$\exp[-35.9314 - 28.1773 + 0 + 37.03 + 28.1773] = 3$	3/6
2	$\exp[-35.9314 + 0] \approx 0$	0
3	$\exp[-35.9314 - 28.1773 + 37.03] \approx 0$	0
4	$\exp[-35.9314] \approx 0$	0
5	$\exp[-28.1773 + 0 + 28.1773] = 1$	1/6
6	$\exp[0] = 1$	1/6
7	$\exp[-28.1773] \approx 0$	0
8	$\exp[0] = 1$	1/6
Sums	$Z = 6$	1

Exercise 37.9.11: What must happen to the information entropy in these three models?

Solution to Exercise 37.9.11

Each succeeding model incorporates more information. Therefore, the information entropy $H(Q_i)$, as a quantitative measure of the amount of missing information, must decrease. Start with the fair model with no information, and correspondingly the maximum amount of missing information of $\ln n$. Then, continue on to $\mathcal{M}_1$ with $m = 3$, $\mathcal{M}_2$ with $m = 4$, and finally $\mathcal{M}_3$ with $m = 5$. The information entropy decreases in the expected fashion under each of these four models as,

$$H_0(Q_i) = 2.07944$$

$$H_1(Q_i) = 1.78022$$

$$H_2(Q_i) = 1.46197$$

$$H_3(Q_i) = 1.24245$$

Exercise 37.9.12: Use Bayes's Theorem to calculate the probability that a kangaroo drinks Fosters given that the IP knows it prefers to use its right hand. Assume the truth of the numerical assignments Q_i under model $\mathcal{M}_3$.

Solution to Exercise 37.9.12

The numerical assignment for the probability that a kangaroo prefers Foster's was $P(B) = 2/3$ under model $\mathcal{M}_3$. This particular piece of information was specified in the second constraint function average. The updated probability that the kangaroo prefers Foster's given that the IP knows that the kangaroo is right–handed is,

$$\begin{aligned}
P(B\,|\,H,\mathcal{M}_k) &= \frac{P(B,H\,|\,\mathcal{M}_k)}{P(H\,|\,\mathcal{M}_k)} \\[4pt]
&= \frac{P(G,B,H\,|\,\mathcal{M}_k) + P(\overline{G},B,H\,|\,\mathcal{M}_k)}{P(G,B,H\,|\,\mathcal{M}_k) + P(\overline{G},B,H\,|\,\mathcal{M}_k) + P(G,\overline{B},H\,|\,\mathcal{M}_k) + P(\overline{G},\overline{B},H\,|\,\mathcal{M}_k)} \\[4pt]
&= \frac{\text{Cell 1} + \text{Cell 5}}{\text{Cell 1} + \text{Cell 5} + \text{Cell 2} + \text{Cell 6}} \\[4pt]
&= \frac{1/2 + 1/6}{1/2 + 1/6 + 0 + 1/6} \\[4pt]
&= 4/5
\end{aligned}$$

The IP's degree of belief that the kangaroo drinks Foster's beer was raised from 2/3 to 4/5 when conditioned on the observation that the kangaroo uses its right hand and the information about the various marginal probabilities inserted under model $\mathcal{M}_3$.

Exercise 37.9.13: Illustrate numerically that the partial differentiation of the log of the partition function ψ with respect to a θ coordinate yields the η coordinate in model $\mathcal{M}_3$.

Solution to Exercise 37.9.13

Obviously, we are going to show the same thing as,

$$\frac{\partial \ln Z}{\partial \lambda_j} = \langle F_j \rangle$$

but now in Amari's notation. Once again, rely upon the crude, but effective, central difference formula as a numerical approximation for the derivative,

$$f'(x) = \frac{df(x)}{dx} \approx \frac{f(x+h) - f(x-h)}{2h}$$

where $f(x) = \psi$ and the arguments x are the θ^i. Thus, we are interested in finding out whether,

$$\partial_i \psi = \eta_i$$

The partition function under model $\mathcal{M}_3$ was $Z = 6$ for the five model parameter values of,

$$\theta^1 = -35.9314$$

$$\theta^2 = -28.1773$$

$$\theta^3 = 0$$

$$\theta^4 = +37.0300$$

$$\theta^5 = +28.1773$$

The log of the partition function is then,

$$\psi = \ln 6 = 1.791759$$

If we differentiate with respect to θ^4, and take $h = 0.01$,

$$\frac{\partial \psi}{\partial \theta^4} \approx \frac{\psi(\theta^1, \cdots, \theta^4 + h, \theta^5) - \psi(\theta^1, \cdots, \theta^4 - h, \theta^5)}{0.02}$$

$$\approx \frac{1.79677 - 1.78677}{0.02}$$

$$\approx 0.499997$$

The constraint function average for the GB interaction was specified as,

$$\langle F_4 \rangle \equiv \eta_4 = 0.50$$

Exercise 37.9.14: Illustrate numerically that the partial differentiation involving the change in the negative information entropy φ with respect to a constraint function average η yields the θ coordinate in model $\mathcal{M}_2$.

Solution to Exercise 37.9.14

For the example, look at how the negative of the information entropy changes under model $\mathcal{M}_2$ when the second constraint function average changes. Amari's generic notation for this case would be,

$$\partial^i \varphi = \theta^i \rightarrow \frac{\partial \varphi}{\partial \eta_2} = \theta^2$$

Since $\theta^2 = -0.6931$ under model $\mathcal{M}_2$, the numerical differentiation routine should approximate this value as $\langle F_2 \rangle = \eta_2 = 2/3$ undergoes small changes. The generic function $f(x)$ is the negative of the information entropy with four arguments η_i. In this case, the negative information entropy is changing only as a function of η_2. Change $\eta_2 = 0.66667$ by an $h = 0.01$ so that,

$$f(x+h) \equiv -H(1/2, 0.67667, 5/6, 1/2) = -1.46845$$

$$f(x-h) \equiv -H(1/2, 0.65667, 5/6, 1/2) = -1.45458$$

$$\frac{\partial \varphi}{\partial \eta_2} \approx \frac{-1.46845 - (-1.45458)}{0.02} = -0.6935$$

Exercise 37.9.15: Compare Amari's notation as he used it to prove the above differentiation to my amended notation from Jaynes.

Solution to Exercise 37.9.15

As mentioned already, Amari wrote out the partial differentiation of the negative information entropy with respect to a constraint function average as,

$$\partial^i \varphi = \theta^i$$

Jaynes wrote it as,

$$\frac{\partial S}{\partial F_k} = \lambda_k$$

whereas for me this expression becomes,

$$\frac{\partial(-H)}{\partial \langle F_j \rangle} = \lambda_j$$

Amari presented a very succinct and abstract proof for $\partial^i \varphi = \theta^i$ that relied on the total differential.

$$\varphi = \theta^i \, \eta_i - \psi$$

$$\mathrm{d}\varphi = \theta^i \, \mathrm{d}\eta_i + \eta_i \, \mathrm{d}\theta^i - \mathrm{d}\psi$$

$$\frac{\mathrm{d}\psi}{\mathrm{d}\theta^i} = \eta_i$$

$$\mathrm{d}\psi = \eta_i \, \mathrm{d}\theta^i$$

$$\mathrm{d}\varphi = \theta^i \, \mathrm{d}\eta_i + \eta_i \, \mathrm{d}\theta^i - \eta_i \, \mathrm{d}\theta^i$$

$$\mathrm{d}\varphi = \theta^i \, \mathrm{d}\eta_i$$

$$\frac{\mathrm{d}\varphi}{\mathrm{d}\eta_i} = \theta^i$$

$$\partial^i \equiv \frac{\partial}{\partial \eta_i}$$

$$\partial^i \varphi = \theta^i$$

Here is my translation in the more familiar language of the MEP notation. Start out with the Legendre transformation relating the information entropy appearing on the left hand side of the equation to the partition function, the Lagrange multipliers, and the constraint function averages on the right hand side,

$$H(p) = \ln Z - \sum_{j=1}^{m} \lambda_j \langle F_j \rangle$$

Amari uses φ to stand for the negative of the information entropy and then employs the Einstein summation convention for the sum,

$$-H(p) = \sum_{j=1}^{m} \lambda_j \langle F_j \rangle - \ln Z$$

$$\mathrm{d}(-H) = \sum_{j=1}^{m} \lambda_j \, \mathrm{d}\langle F_j \rangle + \sum_{j=1}^{m} \langle F_j \rangle \, \mathrm{d}\lambda_j - \mathrm{d}\ln Z$$

$$\mathrm{d}\ln Z = \sum_{j=1}^{m} \langle F_j \rangle \, \mathrm{d}\lambda_j$$

$$\mathrm{d}(-H) = \sum_{j=1}^{m} \lambda_j \, \mathrm{d}\langle F_j \rangle$$

$$\frac{\partial(-H)}{\partial \langle F_j \rangle} = \lambda_j$$

Exercise 37.9.16: Show in more detail the tricky step in the previous derivation.

Solution to Exercise 37.9.16

The tricky step is the one that asserts,

$$d \ln Z = \sum_{j=1}^{m} \langle F_j \rangle \, d\lambda_j$$

Let's take an $n = 4$ state space, say the simple kangaroo scenario most recently looked at in Exercise 35.6.1, with its first two standard constraint functions,

$$F_1(X = x_i) = (1, 1, 0, 0) \text{ and } F_2(X = x_i) = (1, 0, 1, 0)$$

for the $m = 2$ class of models. The MEP assignments are then,

$$Q_1 = \frac{e^{\lambda_1 + \lambda_2}}{Z}$$

$$Q_2 = \frac{e^{\lambda_1}}{Z}$$

$$Q_3 = \frac{e^{\lambda_2}}{Z}$$

$$Q_4 = \frac{1}{Z}$$

$$Z = e^{\lambda_1 + \lambda_2} + e^{\lambda_1} + e^{\lambda_2} + 1$$

The total differential of the log of the partition function is then,

$$d \ln Z = d[\ln(e^{\lambda_1 + \lambda_2} + e^{\lambda_1} + e^{\lambda_2} + 1)]$$

$$= \frac{e^{\lambda_1} \, d\lambda_1 + e^{\lambda_2} \, d\lambda_2 + e^{\lambda_1 + \lambda_2} \, d(\lambda_1 + \lambda_2)}{e^{\lambda_1 + \lambda_2} + e^{\lambda_2} + e^{\lambda_1} + 1}$$

$$= \frac{e^{\lambda_1 + \lambda_2} \, d\lambda_1}{Z} + \frac{e^{\lambda_1} \, d\lambda_1}{Z} + \frac{e^{\lambda_1 + \lambda_2} \, d\lambda_2}{Z} + \frac{e^{\lambda_2} \, d\lambda_2}{Z}$$

Having gotten this far in the algebraic transformation, we see that each term is in fact a Q_i, so that,

$$d \ln Z = (Q_1 + Q_2) \, d\lambda_1 + (Q_1 + Q_3) \, d\lambda_2$$

$$= \langle F_1 \rangle \, d\lambda_1 + \langle F_2 \rangle \, d\lambda_2$$

$$= \sum_{j=1}^{2} \langle F_j \rangle \, d\lambda_j$$

SOLVED EXERCISES FOR CHAPTER 37

Exercise 37.9.17: Show how Amari's notation would indicate the inverse metric tensor.

Solution to Exercise 37.9.17

Just through pure symbol shuffling, and not even really knowing what it all means, we can start with the expression reached at the end of Exercise 37.9.15,

$$\partial^i \varphi = \theta^i$$

Now apply Amari's differential operator ∂^j where,

$$\partial^j \equiv \frac{\partial}{\partial \eta_j}$$

to arrive at,

$$\partial^j \partial^i \varphi = \frac{\partial \theta^i}{\partial \eta_j}$$

But this expression is the definition of the inverse metric tensor,

$$\partial^j \partial^i \varphi = \frac{\partial \theta^i}{\partial \eta_j} = g^{ij}$$

Amari's notational expression for the inverse metric tensor is what Jaynes wrote as,

$$\frac{\partial \lambda_j}{\partial F_k} = \frac{\partial^2 S}{\partial F_k \, \partial F_j} = \mathbf{A}$$

where the matrix $\mathbf{A}$ was the inverse of the matrix $\mathbf{B}$, $\mathbf{A} = \mathbf{B}^{-1}$,

$$\frac{\partial F_j}{\partial \lambda_k} = \frac{\partial^2 \ln Z}{\partial \lambda_k \, \partial \lambda_j} = \mathbf{B}$$

Exercise 37.9.18: Show symbolically that differentiating the log of the partition function with respect to the Lagrange multiplier results in the constraint function average.

Solution to Exercise 37.9.18

See what happens for a very small state space with $n = 3$. For this symbolic illustration, select models where the constraint function vector is,

$$F(X = x_i) = (1, 2, 3)$$

For a model with $m = 1$, the log of the partition function is,

$$\ln Z = \ln \left[e^{(\lambda \times 1)} + e^{(\lambda \times 2)} + e^{(\lambda \times 3)} \right]$$

Have *Mathematica* carry out the differentiation of $\ln Z$ with respect to the Lagrange multiplier by,

```
D[Log[Exp[λ] + Exp[2λ] + Exp[3λ]], λ]
```

This evaluates to,
$$\frac{\partial \ln Z}{\partial \lambda} = \frac{e^\lambda + 2e^{2\lambda} + 3e^{3\lambda}}{e^\lambda + e^{2\lambda} + e^{3\lambda}}$$

Working directly from first principles we have that,

$$\langle F \rangle = \sum_{i=1}^{3} F(X = x_i) Q_i$$

$$= \left(1 \times \frac{e^\lambda}{e^\lambda + e^{2\lambda} + e^{3\lambda}}\right) + \left(2 \times \frac{e^{2\lambda}}{e^\lambda + e^{2\lambda} + e^{3\lambda}}\right) + \left(3 \times \frac{e^{3\lambda}}{e^\lambda + e^{2\lambda} + e^{3\lambda}}\right)$$

$$\frac{\partial \ln Z}{\partial \lambda} = \langle F \rangle$$

Exercise 37.9.19: Examine an expanded coin toss scenario with the help of *Mathematica* to illustrate how the total differential was used in the derivation of Amari's $\partial^i \varphi = \theta^i$.

Solution to Exercise 37.9.19

Expand the state space of the coin toss scenario from $n = 2$ to $n = 3$. Perhaps the IP has added EDGE as a possible observation to HEADS or TAILS. Trying to use as simple an expression as possible, evaluate the total differential of the negative information entropy expressed as the Legendre transformation by letting **F** stand for the constraint function average,

```
Dt[λ F - Log[z]]
```

This returns the answer for the total differential of the negative information entropy as,

```
λ Dt[F] - Dt[z]/z + F Dt[λ]
```

Now, substitute the actual partition function for **z** of the expanded coin toss scenario with the constraint functions $F(X = x_1) = 1$, $F(X = x_2) = 2$, and $F(X = x_3) = 3$ for `Dt[z]/z`,

```
Dt[Exp[λ] + Exp[2 λ] + Exp[3 λ]] / (Exp[λ] + Exp[2 λ] + Exp[3 λ])
```

to see that `Dt[z]/z` returns,

$$\frac{e^\lambda \, d\lambda + 2e^{2\lambda} \, d\lambda + 3e^{3\lambda} \, d\lambda}{e^\lambda + e^{2\lambda} + e^{3\lambda}} \equiv \frac{e^\lambda + 2e^{2\lambda} + 3e^{3\lambda}}{e^\lambda + e^{2\lambda} + e^{3\lambda}} \, d\lambda$$

SOLVED EXERCISES FOR CHAPTER 37

But this is exactly **F Dt[λ]** from the previous Exercise. Therefore, the last two terms cancel out, leaving us with the total differential of the negative information entropy as **λ Dt[F]** or, as Amari would have it,

$$\{d\varphi = \theta^i d\eta_i\} \equiv \{\partial^i \varphi = \theta^i\}$$

Exercise 37.9.20: Show the details of the differential relationship involved in the change of any constraint function average when any Lagrange multiplier is changed.

Solution to Exercise 37.9.20

Models from the class where $m = 5$, that is, those models calling on the double interactions involving GB and BH serve as the milieu for the numerical examples in this exercise and the next. One such model from this class, namely the distribution as reflected by point p, was found to implement the logic function for the kangaroo scenario used in this Chapter. The models in this example are generally very similar to the target joint probability table of Figure 37.3.

Suppose that the IP is curious about how the fourth constraint function average $\langle F_4 \rangle$ would change if various models are changing the information represented in the first Lagrange multiplier λ_1. In other words, the IP would like to compute,

$$g_{rc}(p) = \frac{\partial}{\partial \lambda_r}\left(\frac{\partial \ln Z}{\partial \lambda_c}\right) = \frac{\partial^2 \ln Z}{\partial \lambda_r \, \partial \lambda_c} = \frac{\partial \langle F_c \rangle}{\partial \lambda_r}$$

In this example then, $r = 1$ and $c = 4$. We can have *Mathematica* compute $g_{14}(p) = g_{41}(p) = 0.25$ with the details of how this is accomplished given in the next exercise.

The differential dy is a linear approximation for the difference Δy. In this case, this is the difference between two constraint function averages when the Lagrange multiplier λ_1 changes slightly. Generically, the differential dy is expressed as a linear approximation by,

$$dy = f'(x)\,\Delta x$$

where,

$$f'(x) \equiv \frac{\partial \langle F_c \rangle}{\partial \lambda_r}$$

and Δx would be some small change in the relevant Lagrange multiplier, say a difference of 0.02 for λ_1. Then, dy computed as,

$$dy = f'(x)\,\Delta x \equiv g_{14}(p)\,\Delta\lambda_1 = 0.25 \times 0.02 = 0.005$$

should be approximately equal to the difference in the fourth constraint function average calculated at two different models $\Delta x = \lambda_1^\star - \lambda_1^{\star\star} = 0.02$.

It turned out that $\lambda_1^p = -35.9314$ (along with the appropriate values for the other four Lagrange multipliers) was able to reproduce the desired probabilities in the target joint probability table. The motivation for this model was to assign probabilities of 0 where the logic function had indicated where the zeroes should be located.

For two models, where λ_1 was incremented by 0.01 and decremented by 0.01, we have the difference in their constraint function averages,

$$\Delta y = \langle F_4(\lambda_1 = -35.921365) \rangle - \langle F_4(\lambda_1 = -35.941365) \rangle$$

$$= 0.5025 - 0.4975$$

$$= 0.005$$

Thus, we find, in fact, that dy closely approximates the Δy change to $\langle F_4 \rangle$ by $g_{rc}(p) \times \Delta \lambda_1$ when the information in various models is changing λ_1 by small amounts.

Operationally, this is how the above plays out when the computations are done in *Mathematica*. This is strictly utilitarian, that is, non–elegant code, which has the redeeming virtue of being easy to follow.

First, create a function f[ ] to compute a constraint function average. The arguments to f[ ] are a list of the Lagrange multipliers. This function specifically returns $\langle F_4 \rangle$.

```
f[lambda_List] := Module[{num, z, cfa, qi,
                 cm = {{1, 1, 1, 1, 0, 0, 0, 0},
                       {1, 0, 1, 0, 1, 0, 1, 0},
                       {1, 1, 0, 0, 1, 1, 0, 0},
                       {1, 0, 1, 0, 0, 0, 0, 0},
                       {1, 0, 0, 0, 1, 0, 0, 0}}},
         num = Exp[Dot[lambda, cm]];
         z = Total[num];
         qi = num / z;
         cfa = Dot[cm, qi];
         cfa[[4]]]
```

A call to,

```
f[{-35.921365, -28.177282, 0, 37.029977, 28.177282}]
```

returns $\langle F_4 \rangle = 0.5025$, while a call to,

```
f[{-35.941365, -28.177282, 0, 37.029977, 28.177282}]
```

returns $\langle F_4 \rangle = 0.4975$.

SOLVED EXERCISES FOR CHAPTER 37   189

Exercise 37.9.21: Show the details of the differential relationship involved in the change of any Lagrange multiplier when any constraint function average is changed.

Solution to Exercise 37.9.21

First of all, to orient ourselves, we are aware that the dual coordinates λ_r and $\langle F_c \rangle$ are linked when the parameters of a model change. Changing the information resident in $\langle F_c \rangle$ might also change λ_r. But what is that quantitative relationship? It is captured in the differential relationship,

$$\frac{\partial \lambda_r}{\partial \langle F_c \rangle} = g^{rc}(p)$$

where $g^{rc}(p)$ is the inverse of the metric tensor $g_{rc}(p)$.

For the numerical example, we will suppose that the IP is interested in how λ_5 must change when the information in $\langle F_2 \rangle$ changes. We are still working within the context of the kangaroo inferential scenario where we focus in on the class of $m = 5$ models implementing slight variations of the implicational chain model. Suppose that the IP is interested in these models because they have been supported by data like that in Figure 37.7.

As a matter of fact, the IP wouldn't want to place all of its bets in the rigid demands of the pure implicational model, as it was highlighted in this Chapter, with all of its zeroes. The very first observation of a kangaroo in say cell 3, that is, a left–handed Foster's drinking kangaroo with the normal genetic code, would destroy model $\mathcal{M}_3$ immediately. As an alternative, the IP would prefer to keep around those models that offer slight deviations from the zeroes, so that one errant observation is not a cause for disaster.

Consider then three variations on model $\mathcal{M}_3$ where the zeroes of model $\mathcal{M}_3$ are replaced with small assigned probabilities. The constraint average $\langle F_2 \rangle$ for,

$$F_2(X = x_i) = (1, 0, 1, 0, 1, 0, 1, 0)$$

that is, the marginal probability for beer preference, might be easily changed from $\langle F_2 \rangle = 2/3$ to three close variations on model $\mathcal{M}_3$, models $\mathcal{M}_4$, $\mathcal{M}_5$, and $\mathcal{M}_6$,

$$\langle F_2 \rangle_{\mathcal{M}_4} = 0.66$$

$$\langle F_2 \rangle_{\mathcal{M}_5} = 0.65$$

$$\langle F_2 \rangle_{\mathcal{M}_6} = 0.67$$

The four cells in the target joint probability table that previously held 0s under model $\mathcal{M}_3$ now hold small assigned probabilities close to 0.01 under these three revised models. Label the new probability distribution under the model information $\langle F_2 \rangle_{\mathcal{M}_4} = 0.66$ as point p in the Riemannian manifold. (If you try to work this problem on your own, you will notice that other constraint function averages will have to change as well.)

We know how to compute the metric tensor $g_{rc}(p)$ at point p. We have provided several examples already starting from the definition of the metric tensor as Fisher information,

$$g_{rc}(p) = E_p\left[(F_r(X = x_i) - \langle F_r \rangle) \times (F_c(X = x_i) - \langle F_c \rangle)\right]$$

For a different attack, but one which must arrive at the same components of the metric tensor, try computing,

$$\frac{\partial^2 \ln Z}{\partial \lambda_r \, \partial \lambda_c} = g_{rc}(p)$$

with *Mathematica*.

Set up the constraint matrix for the $m = 5$ models for the three main effects of G, B, and H, together with the two double interactions of GB and BH.

```
cm = {{1, 1, 1, 1, 0, 0, 0, 0}, {1, 0, 1, 0, 1, 0, 1, 0},
      {1, 1, 0, 0, 1, 1, 0, 0}, {1, 0, 1, 0, 0, 0, 0, 0},
      {1, 0, 0, 0, 1, 0, 0, 0}}
```

Then create the symbolic expression for the partition function Z as,

```
z = Total[Exp[Dot[Map[Subscript[λ,#]&, Range[5]], cm]]]
```

which evaluates correctly to,

$$Z = e^{\lambda_1+\lambda_2+\lambda_3+\lambda_4+\lambda_5} + e^{\lambda_1+\lambda_3} + e^{\lambda_1+\lambda_2+\lambda_4} + e^{\lambda_1} + e^{\lambda_2+\lambda_3+\lambda_5} + e^{\lambda_3} + e^{\lambda_2} + 1$$

Finally, take the second partial derivative of $\ln Z$ with respect to λ_5 and λ_2,

```
D[Log[z], λ5, λ2] /. {λ1 → -2.77259, λ2 → -2.97326,
                      λ3 → 0, λ4 → 3.75342, λ5 → 3.04452}
```

which evaluates to $g_{52}(p) = g_{25}(p) = 0.2142$. The *Mathematica* code shows that the values for the five Lagrange multipliers of the model where $\langle F_2 \rangle_{\mathcal{M}_4} = 0.66$ have been substituted for the evaluation.

The complete metric tensor $\mathbf{G}$ as a 5×5 symmetric matrix is listed below,

$$\mathbf{G} = g_{rc}(p) = \begin{pmatrix} 0.2500 & 0.1500 & 0.0682 & 0.2400 & 0.1432 \\ 0.1500 & 0.2244 & 0.1020 & 0.1632 & \boxed{0.2142} \\ 0.0682 & 0.1020 & 0.1600 & 0.0742 & 0.1260 \\ 0.2400 & 0.1632 & 0.0742 & 0.2496 & 0.1558 \\ 0.1432 & \boxed{0.2142} & 0.1260 & 0.1558 & 0.2331 \end{pmatrix}$$

However, this is not what we want. What we have computed above is the change in the constraint function average with changes in the Lagrange multiplier,

$$g_{52}(p) = \frac{\partial^2 \ln Z}{\partial \lambda_5 \, \partial \lambda_2} = \frac{\partial}{\partial \lambda_5}\left(\frac{\partial \ln Z}{\partial \lambda_2}\right) = \frac{\partial \langle F_2 \rangle}{\partial \lambda_5}$$

But by asking *Mathematica* to take the inverse of the metric tensor $\mathbf{G}^{-1}$ as,

SOLVED EXERCISES FOR CHAPTER 37

```
Inverse[metrictensor]
```

we find that,
$$\frac{\partial \lambda_5}{\partial \langle F_2 \rangle} = g^{25}(p) = g^{52}(p) = -39.2157$$

From this result we now know that the approximate change of λ_5 with a change of $\langle F_2 \rangle$ from 0.67 to 0.65 must be $0.02 \times (-39.2157) \approx -0.8$. In fact, as was mentioned at the end of section 37.7, λ_5 changes from 2.70 to 3.51 as the information for these two models changes in the prescribed fashion from $\langle F_2 \rangle_{\mathcal{M}_6} = 0.67$ to $\langle F_2 \rangle_{\mathcal{M}_5} = 0.65$.

This is a nice example from Calculus showing that the *differential*, dy, serves as a linear function approximation to Δy. The dominant linear part is,
$$dy = df(x) = f'(x)\,\Delta x$$

So, for this particular example,
$$dy \approx \Delta y = \lambda_5(\langle F_2 \rangle = 0.67) - \lambda_5(\langle F_2 \rangle = 0.65) = (-39.2157) \times 0.02 = -0.7843$$

where the differential dy does closely approximate Δy,
$$dy = -0.7843 \approx \Delta y = 2.69622 - 3.50715 = -0.8109$$

The inverse of the metric tensor matrix $\mathbf{G}^{-1}$ at the point where $\langle F_2 \rangle = 0.66$, as calculated by *Mathematica*, is also a symmetric 5×5 matrix and listed below,

$$\mathbf{G}^{-1} = g^{rc}(p) = \begin{pmatrix} 53.1250 & 3.1250 & 0 & -53.1250 & 0 \\ 3.1250 & 43.4399 & 5.8824 & -8.6806 & \boxed{-39.2157} \\ 0 & 5.8824 & 11.7647 & 0 & -11.7647 \\ -53.1250 & -8.6806 & 0 & 60.7639 & 0 \\ 0 & \boxed{-39.2157} & -11.7647 & 0 & 46.6853 \end{pmatrix}$$

We notice that there are several zeroes scattered throughout this matrix. For example, $g^{51}(p) = 0$. This must mean that λ_5 does NOT change at all when $\langle F_1 \rangle$ is changing. Sure enough, making small informational changes to some model $\mathcal{M}_k$ where $\langle F_1 \rangle = 1/2$ to close by models where $\langle F_1 \rangle = 0.49$ and $\langle F_1 \rangle = 0.51$ results in a constant value of $\lambda_5 = 3.04452$ for all three models.

As a simple illustration of the differential in *Mathematica* notation, construct the function $f(x) = x^2 + 4x$,

```
f[x_] := Plus[Power[x, 2], Times[4, x]]
```

Then, find the differential $dy \equiv df(x)$ using the built–in function `Dt[ ]`,

```
Apart[Dt[f[x]]]
```

which evaluates to `2 (2 + x) Dt[x]`, the same as $dy = f'(x)\,dx$.

Exercise 37.9.22: How do Calculus texts introduce the concept of the differential?

Solution to Exercise 37.9.22

Set up the exact finite difference between a function $f(x)$ with argument x and that same function after the argument has been increased by some amount Δx,

$$\Delta y = f(x + \Delta x) - f(x)$$

Next, construct this term that you will notice does not mention anything at all about Δx being small, or a limit approaching 0,

$$\frac{f(x + \Delta x) - f(x)}{\Delta x}$$

Subtract from this term the actual *derivative* $f'(x)$, which does invoke the idea of $\Delta x \to 0$, to define an error ϵ,

$$\epsilon = \frac{f(x + \Delta x) - f(x)}{\Delta x} - f'(x)$$

The numerator $f(x + \Delta x) - f(x)$ in the fraction is the exact finite difference Δy as measured on the y–axis as set out in the very first equation above, so that,

$$\epsilon = \frac{\Delta y}{\Delta x} - f'(x)$$

Multiply both sides by Δx and isolate Δy to the left side of the equation,

$$\Delta y = \epsilon \, \Delta x + f'(x) \, \Delta x$$

At this juncture, we see that the exact finite difference Δy is the sum of two terms. But if it is plausible that the error term $\epsilon \, \Delta x$ is relatively small, then the second term $f'(x) \, \Delta x$ alone might serve as good *approximation* to Δy. In fact, this approximation to Δy is labeled as dy and called the *differential*, with,

$$dy = f'(x) \Delta x$$

As a final step in the notation, Δx is also labeled as dx, so that we have,

$$dy = f'(x) \, dx$$

As a *Mathematica* symbolic expression, the differential $dy = f'(x) \, dx$ is,

```
Times[Dt[x], Derivative[1][f][x]]
```

as evaluated by,

```
FullForm[Dt[f[x]]]
```

SOLVED EXERCISES FOR CHAPTER 37

For better or for worse, this is an example of our previous assertion that such *Mathematica* expressions will represent the final actual operational definition of a concept, notwithstanding how else it might have been written in any traditional mathematical notation. The payoff is that whatever lingering mystery might happen to surround a mathematical notation like $dy = f'(x)\, dx$, the expression,

```
Times[Dt[x], Derivative[1][f][x]]
```

always remains computationally, either numerically or symbolically, well–defined.

Exercise 37.9.23: Show a typical example from a Calculus textbook on the application of the differential.

Solution to Exercise 37.9.23

What follows is a typical introductory application of the differential as a linear approximation to some true difference. Consider a hollow metal sphere. The outer thickness of the sphere's shell is 0.05 inches, while its interior radius is 5 inches. Thus, the amount of metal is contained entirely within the volume of the outer shell.

The formula for the volume of a sphere is,

$$\text{Volume} = f(r) = \frac{4}{3}\pi r^3$$

The differential dv is the linear approximation of the volume of a 5.05 inch sphere minus the volume of a 5 inch sphere, in other words, the approximation to the true difference Δv. Thus,

$$\begin{aligned} dv &= f'(r)\, dr \\ &= 4\pi\, r^2 dr \\ &= 4\pi\, (5^2)(0.05) \\ &= 15.7 \text{ cubic inches} \end{aligned}$$

Mathematica calculates this approximation to the true Δv with,

```
Dt[4/3 π Power[r, 3] /. {r → 5, Dt[r] → .05}
```

The exact difference is calculated as,

```
vs[r_] := 4/3 π Power[r, 3]; vs[5.05] - vs[5]
```

15.8656 cubic inches.

Exercise 37.9.24: How does *Mathematica* express the total differential?

Solution to Exercise 37.9.24

Back in Volume II [8, pg. 341] we wrote down the definition of the total differential for a function $f(x, y)$ with two arguments, and then generalized the definition to any number n of arguments x_i with,

$$df = \sum_{i=1}^{n} \frac{\partial f}{\partial x_i} dx_i$$

Suppose we have a function $f(x, y, z)$ with $n = 3$. The total differential is then,

$$df = \frac{\partial f}{\partial x} dx + \frac{\partial f}{\partial y} dy + \frac{\partial f}{\partial z} dz$$

The *Mathematica* expression for the total differential with **FullForm[ ]**,

FullForm[Dt[f[x, y, z]]]

returns for *df* the expression,

Plus[Times[Dt[z], Derivative[0, 0, 1][f][x, y, z]],
    Times[Dt[y], Derivative[0, 1, 0][f][x, y, z]],
    Times[Dt[x], Derivative[1, 0, 0][f][x, y, z]]]

After taking account of the curious reversal of the arguments, it is easy to pattern match with the mathematical expression. Note that, for example, the standard mathematical partial differentiation notation,

$$\frac{\partial f}{\partial y}$$

is equivalent to the *Mathematica* expression,

Derivative[0, 1, 0][f][x, y, z]

As a consistency check, evaluating the *Mathematica* expression for $\frac{\partial f}{\partial y}$,

FullForm[D[f[x, y, z], y]]

also returns,

Derivative[0, 1, 0][f][x, y, z]

Exercise 37.9.25: Write out the generic system of equations that is the solution to a constrained optimization problem relying upon the method of Lagrange multipliers.

Solution to Exercise 37.9.25

In Volume II, the MEP was seen to be a typical example of solving a constrained optimization problem by the method of Lagrange multipliers. The abstract template for the pattern of equations for such a solution involves the total differential.

Consider a generic objective function f that is to be maximized. Suppose that it has four arguments w, x, y, and z. However, the freedom to find this maximum of $f(w, x, y, z)$ is limited by having to conform to, say, two additional functions, $g(w, x, y, z) = 0$ and $h(w, x, y, z) = 0$. The total differential for f, together with these two additional constraint functions, must equal 0, $d(f + \lambda g + \mu h) = 0$. The Lagrange multipliers are λ and μ.

If we were looking at just the total differential df, it would be,

$$df = \frac{\partial f}{\partial w}dw + \frac{\partial f}{\partial x}dx + \frac{\partial f}{\partial y}dy + \frac{\partial f}{\partial z}dz$$

Likewise, the total differential dg is,

$$dg = \frac{\partial g}{\partial w}dw + \frac{\partial g}{\partial x}dx + \frac{\partial g}{\partial y}dy + \frac{\partial g}{\partial z}dz$$

and similarly for the total differential dh. Collecting like terms together,

$$d(f + \lambda g + \mu h) = 0$$

$$\left(\frac{\partial f}{\partial w} + \lambda \frac{\partial g}{\partial w} + \mu \frac{\partial h}{\partial w}\right) dw + \cdots + \left(\frac{\partial f}{\partial z} + \lambda \frac{\partial g}{\partial z} + \mu \frac{\partial h}{\partial z}\right) dz = 0$$

Each term in parentheses must equal 0, and as a consequence the following system of equations develops,

$$\frac{\partial f}{\partial w} + \lambda \frac{\partial g}{\partial w} + \mu \frac{\partial h}{\partial w} = 0$$

$$\frac{\partial f}{\partial x} + \lambda \frac{\partial g}{\partial x} + \mu \frac{\partial h}{\partial x} = 0$$

$$\frac{\partial f}{\partial y} + \lambda \frac{\partial g}{\partial y} + \mu \frac{\partial h}{\partial y} = 0$$

$$\frac{\partial f}{\partial z} + \lambda \frac{\partial g}{\partial z} + \mu \frac{\partial h}{\partial z} = 0$$

When the two constraint equations are added to the mix, we see that we have a total of six equations to solve for the six unknowns w, x, y, z, λ, and μ.

Exercise 37.9.26: What similar role did the total differential play in the development of the MEP formula?

Solution to Exercise 37.9.26

The method of Lagrange multipliers was used to find the maximum value of the objective function, Shannon's information entropy function H,

$$H = -\sum_{i=1}^{n} q_i \ln q_i$$

subject to the constraints of inserting information from some number of constraint function averages, $\sum_{i=1}^{n} F_j(X = x_i)\, q_i$.

The total differential of the information entropy, together with all m constraint functions F_j, was set equal to 0,

$$d\left(H + \sum_{j=1}^{m} \lambda_j F_j\right) = 0$$

The total differential dH is,

$$dH = \sum_{i=1}^{n} \frac{\partial H}{\partial q_i} dq_i$$

The total differentials dF_j are,

$$dF_j = \sum_{i=1}^{n} \frac{\partial F_j}{\partial q_i} dq_i$$

leading to,

$$d\left(H + \sum_{j=1}^{m} \lambda_j F_j\right) = 0$$

$$\sum_{i=1}^{n} \frac{\partial H}{\partial q_i} dq_i + \sum_{i=1}^{n}\sum_{j=1}^{m} \lambda_j \frac{\partial F_j}{\partial q_i} dq_i = 0$$

$$\sum_{i=1}^{n} \left(\frac{\partial H}{\partial q_i} + \sum_{j=1}^{m} \lambda_j \frac{\partial F_j}{\partial q_i}\right) dq_i = 0$$

Suppose that $n = 4$ and $m = 2$. All of the coefficients of the dq_i must equal

SOLVED EXERCISES FOR CHAPTER 37

zero. Then we have the system of equations,

$$\frac{\partial H}{\partial q_1} + \lambda_1 \frac{\partial F_1}{\partial q_1} + \lambda_2 \frac{\partial F_2}{\partial q_1} = 0$$

$$\frac{\partial H}{\partial q_2} + \lambda_1 \frac{\partial F_1}{\partial q_2} + \lambda_2 \frac{\partial F_2}{\partial q_2} = 0$$

$$\frac{\partial H}{\partial q_3} + \lambda_1 \frac{\partial F_1}{\partial q_3} + \lambda_2 \frac{\partial F_2}{\partial q_3} = 0$$

$$\frac{\partial H}{\partial q_4} + \lambda_1 \frac{\partial F_1}{\partial q_4} + \lambda_2 \frac{\partial F_2}{\partial q_4} = 0$$

If we recognize that $F_1(X = x_i)$ is the universal constraint, $F_2(X = x_i)$ will become $F(X = x_i)$. Substituting the partial derivatives yields,

$$-(\ln q_1 + 1) + \lambda_1 + \lambda_2 F(X = x_1) = 0$$

$$-(\ln q_2 + 1) + \lambda_1 + \lambda_2 F(X = x_2) = 0$$

$$-(\ln q_3 + 1) + \lambda_1 + \lambda_2 F(X = x_3) = 0$$

$$-(\ln q_4 + 1) + \lambda_1 + \lambda_2 F(X = x_4) = 0$$

which turn into,

$$Q_1 = \exp\left[(\lambda_1 - 1) + \lambda_2 F(X = x_1)\right]$$
$$= \exp\left[(\lambda_1 - 1)\right] \exp\left[\lambda_2 F(X = x_1)\right]$$

$$Q_2 = \exp\left[(\lambda_1 - 1) + \lambda_2 F(X = x_2)\right]$$
$$= \exp\left[(\lambda_1 - 1)\right] \exp\left[\lambda_2 F(X = x_2)\right]$$

$$Q_3 = \exp\left[(\lambda_1 - 1) + \lambda_2 F(X = x_3)\right]$$
$$= \exp\left[(\lambda_1 - 1)\right] \exp\left[\lambda_2 F(X = x_3)\right]$$

$$Q_4 = \exp\left[(\lambda_1 - 1) + \lambda_2 F(X = x_4)\right]$$
$$= \exp\left[(\lambda_1 - 1)\right] \exp\left[\lambda_2 F(X = x_4)\right]$$

With $\exp\left[(\lambda_1 - 1)\right] = 1/Z$, and $\lambda_2 \equiv \lambda$, the MEP formula for the Q_i emerges as,

$$Q_i = \frac{\exp\left[\lambda F(X = x_i)\right]}{Z}$$

Exercise 37.9.27: What does Wolfram's elementary cellular automaton following Rule 139 look like?

Solution to Exercise 37.9.27

Run the cellular automaton for 20 time steps starting with a single black cell. This *Mathematica* expression,

`ArrayPlot[CellularAutomaton[139, {{1}, 0}, 20], Mesh → True]`

will produce Figure 37.10 showing the pattern of black and white squares for the evolution of Rule 139. You may verify that each cell is colored black or white according to Rule 139 as was illustrated in Figure 37.8 in Exercise 37.9.7.

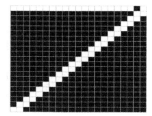

Figure 37.10: *Wolfram's elementary cellular automaton following Rule 139.*

Chapter 38

Relative Entropy, Fisher Information, and the Metric Tensor

38.1 Introduction

The fundamental concepts of relative entropy, squared distance, Fisher information, and the metric tensor are all bound up in some sort of complicated mixture. We certainly have been tossing these ideas about with abandon with only a vague sense of how they are connected. After a while, if only for sanity's sake, it becomes imperative to disentangle the mutual relationships that exist amongst these core concepts that crop up constantly in Information Geometry.

Amari usually starts out any discussion by asserting the relationship between squared distance and relative entropy, but he doesn't provide any derivation of this, saying something to the effect that "it is known to be true." Kullback, however, did provide a rather lengthy and involved proof of the relationship between his information measure and Fisher information. It is Kullback's argument that we are going to deconstruct in this Chapter.

When trying to explain someone else's long argument, I prefer to work in a top–down mode. That is, I want to sketch out, first and foremost, the skeletal outline of the proof, highlighting what seem to be the main points as well as what the main objective of the exercise is all about. Then, I peel off successive layers of the onion by drilling down into the "sub–routines" that were essential components in the outline.

As you might observe, this approach is similar to my appeal for the creators of *Mathematica* programs to rely on a hierarchical outline when explaining their finished (in other words, their totally inaccessible) code.

38.2 Deconstructing Kullback's Argument

38.2.1 Main goal

The overall objective is to show that the relative entropy can be approximated by a quadratic function involving the metric tensor. Kullback however always refers to the quadratic function in the context of the Fisher information matrix.

Kullback expressed the relative entropy as $I(1:2)$, which is equivalent to our current notation of $KL(p,q)$. $I(1:2)$ is a measure of how close two "near by" probability distributions $p \equiv f_1(x)$ and $q \equiv f_2(x)$ are to each other (or, alternatively, as Kullback phrased it, "their discriminability").

Kullback's equation for relative entropy looked like this,

$$I(1:2) = \int f_1(x) \log \frac{f_1(x)}{f_2(x)} d\lambda(x)$$

and the question is whether it can be approximated by a quadratic function, while still retaining Kullback's notation, that looks like this,

$$2I(\theta : \theta + \Delta\theta) \approx \sum_{\alpha=1}^{k} \sum_{\beta=1}^{k} g_{\alpha\beta} \Delta\theta_\alpha \Delta\theta_\beta$$

Kullback says that the two probability distributions $f_i(x)$ appearing in $I(1:2)$ will be expressed in a parametric form as,

$$f_1(x) \equiv f(x, \theta)$$
$$f_2(x) \equiv f(x, \theta + \Delta\theta)$$

where θ is the vector of coordinates for the first distribution, and $\theta + \Delta\theta$ is the vector of coordinates for the second, near by, distribution. There are k parameters in total for both distributions.

The expression for the relative entropy now takes on the form of,

$$I(\theta : \theta + \Delta\theta) = \int f(x, \theta) \log \frac{f(x, \theta)}{f(x, \theta + \Delta\theta)} d\lambda(x)$$

This is equivalent to our more transparent format for the relative entropy,

$$KL(p, q) = \sum_{i=1}^{n} p_i \ln \left(\frac{p_i}{q_i} \right)$$

where distribution p_i has the model parameters $\lambda_j^p \equiv \theta$ and distribution q_i has the model parameters $\lambda_j^p + (\lambda_j^q - \lambda_j^p) \equiv \theta + \Delta\theta \equiv \lambda_j^q$.

38.2.2 Outline of proof

In the proof outline, it is helpful to switch back and forth between Kullback's notation,

$$I(\theta : \theta + \Delta\theta) = \int f(x,\theta) \log \frac{f(x,\theta)}{f(x,\theta+\Delta\theta)} \, d\lambda(x)$$

and the simpler notation,

$$KL(p,q) = \sum_{i=1}^{n} p_i \ln\left(\frac{p_i}{q_i}\right)$$

$$= E_p[\ln p - \ln q]$$

Step 1. Express $\ln q$ in a Taylor series expansion. Expand $\ln q \equiv \log f(x, \theta + \Delta\theta)$ about the point θ.

Step 2. $\ln q \approx$ by some number of terms in the Taylor series. $\log f(x, \theta + \Delta\theta)$ will be approximated by the first three terms in the series. Remaining terms in the series are ignored.

Step 3. $\ln q \approx \ln p$ (first term in the series) plus two more terms in the Taylor series expansion. $\log f(x, \theta + \Delta\theta)$ will be approximated by $\log f(x, \theta)$ plus two more terms in the series.

Step 4. $\ln q - \ln p \approx$ two remaining terms. $\log f(x, \theta + \Delta\theta) - \log f(x, \theta) = \log \frac{f(x,\theta+\Delta\theta)}{f(x,\theta)}$ approximated by two remaining terms.

Step 5. $\ln p - \ln q \approx -$ (two remaining terms). $\log \frac{f(x,\theta)}{f(x,\theta+\Delta\theta)}$ approximated by $-$ (two remaining terms).

Step 6. $E_p[\ln p - \ln q] = KL(p,q)$. $I(\theta : \theta + \Delta\theta) = \int f(x,\theta) \log \frac{f(x,\theta)}{f(x,\theta+\Delta\theta)} \, d\lambda(x)$.

Step 7. $I(\theta : \theta + \Delta\theta) \approx E_p[$ first term on rhs $] = 0 + E_p[-$ second term on rhs $]$.

Step 8. $\int f(x,\theta)(\text{second term}) \, d\lambda(x) = 1/2 \, \boldsymbol{\Delta\theta} \cdot \mathbf{G} \cdot \boldsymbol{\Delta\theta}^{\mathbf{T}}$.

At **Step 8**, $\mathbf{G}$ is the Riemannian metric tensor at point p, with $\mathbf{G} \equiv g_{\alpha\beta}$, the Fisher information matrix. The condensed vector–matrix–transposed vector expression is exactly the same as Kullback's summation expression,

$$1/2 \, \boldsymbol{\Delta\theta} \cdot \mathbf{G} \cdot \boldsymbol{\Delta\theta}^{\mathbf{T}} \equiv 1/2 \sum_{\alpha=1}^{k} \sum_{\beta=1}^{k} g_{\alpha\beta} \Delta\theta_\alpha \Delta\theta_\beta$$

Kullback specifies a total of k coordinates in the vector $\boldsymbol{\theta} = (\theta_1, \cdots, \theta_k)$ for both distributions with the index running from $\alpha = 1$ and $\beta = 1$ to k as compared to our notation with m coordinates with the index running from $r = 1$ and $c = 1$ to m.

Also, we see that the distance measure is now related to the relative entropy through,

$$ds = \sqrt{\boldsymbol{\Delta\theta} \cdot \mathbf{G} \cdot \boldsymbol{\Delta\theta}^{\mathbf{T}}}$$

$$= \sqrt{2 \, KL(p,q)}$$

38.2.3 Step 1 subroutine

What we want to do is to represent the function $\ln q$ around θ in terms of a power series. The generic template for expanding a univariate function $f(x)$ around a by the Taylor series expansion is the following,

$$f(x) \approx f(a) + f'(a)(x-a) + \frac{1}{2!}f''(a)(x-a)^2 + \cdots + \frac{1}{n!}f^{(n')}(a)(x-a)^n + R_n$$

After a pattern match to the above Taylor series template, it appears that we have the following translation equivalencies,

$$x \equiv \theta_q,\ a \equiv \theta_p,\ \theta + \Delta\theta \equiv \theta_p + (\theta_q - \theta_p)$$

Thus, the Taylor series expansion for $\ln q$ is beginning to look like this,

$$f(\theta_q) \approx f(\theta_p) + f'(\theta_p)(\theta_q - \theta_p) + \frac{1}{2}f''(\theta_p)(\theta_q - \theta_p)^2 + \cdots$$

Unfortunately, $f(x, \theta + \Delta\theta)$ is not a univariate function, but the univariate pattern is instructive for what comes next in the multivariate Taylor series expansion.

38.2.4 Step 2 subroutine

We require a formula which tells us the Taylor series expansion for a function with more than one variable. For our current problem, since $\ln q$ is a function of k parameters $\theta = (\theta_1, \cdots, \theta_k)$ and,

$$\ln q \equiv \log f(x, \theta + \Delta\theta)$$

the multivariate Taylor series formula can be written out to the second order as,

$$f(\boldsymbol{\theta_q}) \approx f(\boldsymbol{\theta_p}) + \boldsymbol{\Delta\theta}\boldsymbol{\nabla} f(\boldsymbol{\theta_p}) + \frac{1}{2}\boldsymbol{\Delta\theta}\,\boldsymbol{\nabla}^2 f(\boldsymbol{\theta_p})\,\boldsymbol{\Delta\theta}^T + \cdots$$

where the introduction of the **bold** notation is an indicator that vectors are now being used. Thus,

$$\boldsymbol{\Delta\theta} \equiv (\boldsymbol{\theta_q} - \boldsymbol{\theta_p})$$

The use of the "nabla" symbol $\boldsymbol{\nabla}$ indicates first, the *gradient* function, $\boldsymbol{\nabla} f(\boldsymbol{x})$, and, secondly the *hessian* function, $\boldsymbol{\nabla}^2 f(\boldsymbol{x})$. This is enough progress to lead us directly to **Step 3**.

RELATED CONCEPTS IN IG

38.2.5 Step 3 subroutine

At **Step 3**, we identify the left hand side and the three terms remaining on the right hand side as,

$$\ln q \equiv \log f(\boldsymbol{x}, \boldsymbol{\theta} + \boldsymbol{\Delta\theta})$$

$$\log f(\boldsymbol{\theta_q}) \approx \log f(\boldsymbol{\theta_p}) + \text{ two remaining terms}$$

$$\ln q \approx \ln p + \text{ two remaining terms}$$

38.2.6 Step 4 subroutine

At **Step 4**, we bring the first term on the right hand side over to the left hand side. We are manipulating the expression on the left hand side to eventually get it into the correct format for the relative entropy.

$$\log f(x, \theta + \Delta\theta) - \log f(x, \theta) \approx \text{two remaining terms}$$

$$\log\left[\frac{f(x, \theta + \Delta\theta)}{f(x, \theta)}\right] \approx \boldsymbol{\Delta\theta} \boldsymbol{\nabla} f(\boldsymbol{\theta_p}) + \frac{1}{2} \boldsymbol{\Delta\theta} \boldsymbol{\nabla^2} f(\boldsymbol{\theta_p}) \boldsymbol{\Delta\theta^T}$$

$$\ln\left(\frac{q}{p}\right) \approx \text{two remaining terms}$$

38.2.7 Step 5 subroutine

At **Step 5**, we continue to manipulate the left hand side to get it into the proper format for the relative entropy. The information measure,

$$I(\theta : \theta + \Delta\theta) = \int \log f(x, \theta) \log\left[\frac{f(x, \theta)}{f(x, \theta + \Delta\theta)}\right] d\lambda(x)$$

contains this term,

$$\log\left[\frac{f(x, \theta)}{f(x, \theta + \Delta\theta)}\right]$$

not what we were left with at the end of **Step 4**,

$$\log\left[\frac{f(x, \theta + \Delta\theta)}{f(x, \theta)}\right]$$

However, since $\ln p - \ln q = -(\ln q - \ln p)$

$$-[\log f(x, \theta + \Delta\theta) - \log f(x, \theta)] = \log \frac{f(x, \theta)}{f(x, \theta + \Delta\theta)} \approx -(\text{two remaining terms})$$

38.2.8 Step 6 subroutine

We just remarked on the expression in **Step 5** above,

$$\begin{aligned} I(\theta:\theta+\Delta\theta) &= \int \log f(x,\theta) \log\left[\frac{f(x,\theta)}{f(x,\theta+\Delta\theta)}\right] d\lambda(x) \\ &= E_p[\ln p - \ln q] \end{aligned}$$

So,

$$\int f(x,\theta) \log \frac{f(x,\theta)}{f(x,\theta+\Delta\theta)} d\lambda(x) \approx E_p[-(\text{two remaining terms})]$$

The left hand side is finished at this point. Notice the negative sign in front of the two remaining terms on the right hand side.

38.2.9 Step 7 subroutine

Step 7 focuses exclusively on the right hand side of the expression as it existed at the completion of **Step 6**.

$$\int f(x,\theta) \log \frac{f(x,\theta)}{f(x,\theta+\Delta\theta)} d\lambda(x) \approx E_p[-\boldsymbol{\Delta\theta}\boldsymbol{\nabla} f(\boldsymbol{\theta_p}) - \frac{1}{2} \boldsymbol{\Delta\theta} \boldsymbol{\nabla}^2 f(\boldsymbol{\theta_p}) \boldsymbol{\Delta\theta}^T]$$

Select the first of the two remaining terms in the Taylor series for treatment,

$$E_p[-(\boldsymbol{\Delta\theta}\boldsymbol{\nabla} f(\boldsymbol{\theta_p})) + \text{remaining term})]$$

This involves a vector–vector multiplication with $\boldsymbol{\Delta\theta}$ a row vector consisting of k components, and, the gradient,

$$\boldsymbol{\nabla} f(\boldsymbol{x}) \equiv \frac{\partial \log f(x,\theta)}{\partial \theta_\alpha}$$

a column vector also with k components. Set up the vector multiplication as an explicit multiplication and sum as Kullback shows it,

$$\boldsymbol{\Delta\theta} = (\theta_1^q - \theta_1^p, \theta_2^q - \theta_2^p, \cdots, \theta_k^q - \theta_k^p)$$

with the gradient explicitly,

$$\boldsymbol{\nabla} f(\boldsymbol{\theta_p}) = \begin{pmatrix} \frac{\partial \log f(x,\theta)}{\partial \theta_1} \\ \frac{\partial \log f(x,\theta)}{\partial \theta_2} \\ \cdots \\ \frac{\partial \log f(x,\theta)}{\partial \theta_k} \end{pmatrix}$$

RELATED CONCEPTS IN IG 205

So we have in summation format,

$$\Delta\boldsymbol{\theta}\nabla f(\boldsymbol{\theta_p}) = \sum_{\alpha=1}^{k}(\theta_\alpha^q - \theta_\alpha^p) \times \frac{\partial \log f(x,\theta)}{\partial \theta_\alpha}$$

$$= \sum_{\alpha=1}^{k}\Delta\theta_\alpha \times \frac{\partial \log f(x,\theta)}{\partial \theta_\alpha}$$

What's next is to take the expectation with respect to distribution p of minus this expression.

$$E_p\left[-(\Delta\boldsymbol{\theta}\nabla f(\boldsymbol{\theta_p}))\right] = -E_p\left[\sum_{\alpha=1}^{k}\Delta\theta_\alpha \times \frac{\partial \log f(x,\theta)}{\partial \theta_\alpha}\right]$$

$$= -\int f(x,\theta)\left(\sum_{\alpha=1}^{k}\Delta\theta_\alpha \times \frac{\partial \log f(x,\theta)}{\partial \theta_\alpha}\right)d\lambda(x)$$

$$= -\left(\sum_{\alpha=1}^{k}\Delta\theta_\alpha \int f(x,\theta) \times \frac{\partial \log f(x,\theta)}{\partial \theta_\alpha}d\lambda(x)\right)$$

$$= 0$$

The simplifying resolution that the final expression does, in fact, equal zero is most easily explained by reverting back to our preferred notation for the MEP formalism. The details are shown in Exercise 38.7.6.

Thus, here at the end of **Step 7**, we are concerned only about the expectation of the final remaining term in the Taylor series expansion,

$$\int \log f(x,\theta) \log \frac{f(x,\theta)}{f(x,\theta+\Delta\theta)}d\lambda(x) \approx E_p\left[-\frac{1}{2}\Delta\boldsymbol{\theta}\,\nabla^2 f(\boldsymbol{\theta_p})\,\Delta\boldsymbol{\theta}^T\right]$$

We are faced with the same kind of vector–matrix–vector multiplication we have dealt with before in previous Chapters and Appendix A.

38.2.10 Step 8 subroutine

A k element row vector $\Delta\boldsymbol{\theta}$ multiplies the $k \times k$ hessian matrix,

$$\nabla^2 f(\boldsymbol{\theta_p}) \equiv \frac{\partial^2 \log f(x,\theta)}{\partial \theta_\alpha\,\partial \theta_\beta}$$

resulting in a k element row vector. This k element row vector multiplies a k element column vector $\Delta\boldsymbol{\theta}^T$ resulting in a scalar. The information measure $I(\theta : \theta + \Delta\theta)$ on the left hand side of the approximation must be a scalar.

Kullback decomposed $\nabla^2 f(\boldsymbol{\theta_p})$ into,

$$\frac{\partial^2 \log f(x,\theta)}{\partial \theta_\alpha \, \partial \theta_\beta} = \left[\frac{1}{f(x,\theta)} \frac{\partial^2 f(x,\theta)}{\partial \theta_\alpha \, \partial \theta_\beta} \right] - \left[\frac{1}{f(x,\theta)^2} \frac{\partial f(x,\theta)}{\partial \theta_\alpha} \frac{\partial f(x,\theta)}{\partial \theta_\beta} \right]$$

This kind of decomposition of a second partial derivative of the log of a function is something we have seen before. Jaynes used it for his Equation (11.73) which we explored in depth in Chapter Thirty Six. (See Exercise 36.9.15).

If you were wondering whether the minus sign that was lurking in the background might cause a problem because the information measure could never be negative, it turns out, fortunately for us, that the first term above in brackets is equal to 0. All that's left is the second term above which is conveniently negative, so the whole expression in curly braces below becomes positive.

$$\int \log f(x,\theta) \log \frac{f(x,\theta)}{f(x,\theta+\Delta\theta)} \, d\lambda(x) \approx E_p \left\{ -\frac{1}{2} \boldsymbol{\Delta\theta} \left[-\frac{1}{f(x,\theta)^2} \frac{\partial f(x,\theta)}{\partial \theta_\alpha} \frac{\partial f(x,\theta)}{\partial \theta_\beta} \right] \boldsymbol{\Delta\theta}^T \right\}$$

The right hand side of the Taylor series approximation to Kullback's information measure has finally been reduced down to its final form. When all of the remaining tedious details are worked out in the Exercises, the derivation is complete. Here is Kullback's Equation (6.4) [28, pg.28]

$$I(\theta : \theta + \Delta\theta) = \frac{1}{2} \sum_{\alpha=1}^{k} \sum_{\beta=1}^{k} g_{\alpha\beta} \, \Delta\theta_\alpha \, \Delta\theta_\beta$$

38.3 Amplifying Remarks

As is evident from the above deconstruction, Kullback presents a rather difficult to follow derivation linking his information measure with Fisher's information matrix. In this section, I will make some further amplifying remarks in addition to the step by step process just completed. Basically, these are notational equivalences with my preferred MEP notation. My aim is to achieve, hopefully, even more transparency in Kullback's derivation.

Typical of these kinds of mathematical expositions, Kullback launches directly into the expression for the expansion without any kind of amplifying remarks to soothe the transition for the unprepared reader. My task is to try to fill in some of these missing details that others have dismissed as inconsequential. Therefore, let us engage in my usual discursive re–examination of what the notation is actually referring to.

RELATED CONCEPTS IN IG

The translation between Kullback's notation and my preferred MEP notation begins with,

$$\log f(x,\theta) \equiv \ln P(X = x_i \,|\, \mathcal{M}_p)$$

$$\equiv f(\boldsymbol{\theta_p})$$

$$\log f(x, \theta + \Delta\theta) \equiv \ln P(X = x_i \,|\, \mathcal{M}_q)$$

$$\equiv f(\boldsymbol{\theta_q})$$

$$\boldsymbol{\Delta\theta} \equiv (\boldsymbol{\theta_q} - \boldsymbol{\theta_p})$$

$$\boldsymbol{\theta} + \boldsymbol{\Delta\theta} \equiv \boldsymbol{\theta_p} + (\boldsymbol{\theta_q} - \boldsymbol{\theta_p})$$

Continuing on with further notational equivalencies that hopefully will assist in making the transition back and forth a bit easier,

$$\ln P(X = x_i \,|\, \mathcal{M}_p) \equiv \ln Q_i^p$$

$$\ln P(X = x_i \,|\, \mathcal{M}_q) \equiv \ln Q_i^q$$

$$KL(p,q) \equiv \sum_{i=1}^n Q_i^p \ln\left(\frac{Q_i^p}{Q_i^q}\right)$$

$$\equiv E_p\left[\ln p - \ln q\right]$$

Kullback specified k model parameters, $(\theta_1, \cdots, \theta_\alpha, \cdots, \theta_\beta, \cdots, \theta_k)$, whereas in our MEP notation there are m model parameters, $(\lambda_1, \cdots, \lambda_r, \cdots, \lambda_c, \cdots, \lambda_m)$. Now we can set up the following,

$$\frac{\partial \log f(x,\theta)}{\partial \theta_\alpha} \equiv \frac{\partial \ln Q_i^p}{\partial \lambda_r}$$

$$\frac{\partial \log f(x,\theta)}{\partial \theta_\beta} \equiv \frac{\partial \ln Q_i^p}{\partial \lambda_c}$$

$$\frac{\partial \log f(x,\theta)}{\partial \theta_\alpha}\frac{\partial \log f(x,\theta)}{\partial \theta_\beta} \equiv \frac{\partial \ln Q_i^p}{\partial \lambda_r}\frac{\partial \ln Q_i^p}{\partial \lambda_c}$$

$$\int f(x,\theta) \frac{\partial \log f(x,\theta)}{\partial \theta_\alpha}\frac{\partial \log f(x,\theta)}{\partial \theta_\beta} d\lambda(x) \equiv \sum_{i=1}^n \left[\frac{\partial \ln Q_i^p}{\partial \lambda_r} \times \frac{\partial \ln Q_i^p}{\partial \lambda_c}\right] Q_i^p$$

$$\equiv \sum_{i=1}^n (F_r(X = x_i) - \langle F_r \rangle \times F_c(X = x_i) - \langle F_c \rangle) Q_i^p$$

$$g_{\alpha\beta} \equiv g_{rc}(p)$$

38.4 Length of Curves in a Manifold

The most important conceptual relationship revealed by the derivation at the end of section 38.2.2 is that Kullback's information measure, the relative entropy between two near by probability distributions, can interpreted geometrically as the squared distance between two near by points p and q in the tangent space of a Riemannian manifold at point p,

$$ds^2 \equiv 2\,KL(p,q) \equiv \sum_{r=1}^{m}\sum_{c=1}^{m} g_{rc}(p)(\theta_r^q - \theta_c^p)(\theta_r^q - \theta_c^p)^T$$

Or, alternatively, we could express this as the squared length of a vector with its beginning at $\mathbf{v}$ and its end at $\mathbf{w}$,

$$\|\mathbf{v}-\mathbf{w}\|^2 = (\mathbf{v}-\mathbf{w})\cdot\mathbf{G}\cdot(\mathbf{v}-\mathbf{w})^\mathbf{T}$$

and the norm of this vector equated with the small vector connecting p and q in the tangent space,

$$\|\mathbf{v}-\mathbf{w}\| = \sqrt{(\mathbf{v}-\mathbf{w})\cdot\mathbf{G}\cdot(\mathbf{v}-\mathbf{w})^\mathbf{T}}$$

$$\|\mathbf{p}-\mathbf{q}\| = \sqrt{\mathbf{\Delta\theta}\cdot\mathbf{G}\cdot\mathbf{\Delta\theta^T}}$$

So, the small "element of arc," ds, the distance between p and q, is written as,

$$ds = \sqrt{\mathbf{\Delta\theta}\cdot\mathbf{G}\cdot\mathbf{\Delta\theta^T}}$$

This leads to the classic definition for the length of a curve as,

Definition 2 (Curve length) *Let* $\gamma\colon [a,b] \to S$ *be a curve in the Riemannian manifold* S. *Its length* $s(t)$ *is defined as,*

$$s(t) = \int_a^b \sqrt{\frac{d\boldsymbol{\theta}}{dt}\cdot\mathbf{G}(\boldsymbol{\theta}(t))\cdot\frac{d\boldsymbol{\theta}^\mathbf{T}}{dt}}\,dt$$

The general outline of the argument proceeds something like this. We just went through a long derivation to find the distance between any two near by points, $\|\mathbf{p}-\mathbf{q}\|$. If we wanted to find the distance from point p to some point further away, say the point s, then we would add up the intervening distances to all of the points lying on the path to s as in,

$$ds \approx \|\mathbf{p}-\mathbf{q}\| + \|\mathbf{q}-\mathbf{r}\| + \cdots + \|\mathbf{r}-\mathbf{s}\|$$

Imagine that there are actually many points between r and s, and the above is a compact way of representing this notion. Each such point represents a change in the parameters of a model λ_j. What is done is to suppose that there is a functional relationship between the parameters and another argument traditionally called t, so that it would be acceptable to think that changes in the parameters as t changes would be expressed by $\frac{d\lambda_r}{dt}$.

So now we can express the sum above as,

$$ds = \sum_{i=0}^{n} \frac{\| p(t_i) - p(t_{i+1}) \|}{\Delta t_i} \Delta t_i \qquad (38.1)$$

where $p(t_0) = \mathbf{p}$, $p(t_1) = \mathbf{q}$, $\cdots$, and $p(t_n) = \mathbf{s}$. In the usual way that calculus explains the transition from summation to integration, imagine that Δt_i is getting smaller and smaller. The separation between near by probability distributions is getting smaller and smaller because the difference in the parameters $\lambda_j^{t_i}$ and $\lambda_j^{t_{i+1}}$ is getting smaller and smaller.

Thus, we are led to believe in the plausibility of,

$$s(t) = \int_p^s \sqrt{\sum_{r=1}^{m} \sum_{c=1}^{m} g_{rc}(t) \frac{d\lambda_r}{dt} \frac{d\lambda_c}{dt}} \, dt \qquad (38.2)$$

as the length of the curve between the two probability distributions p and s.

Notice that the metric tensor, Fisher's information matrix, depends upon a changing point with its own unique tangent space. The upshot computationally is that a new metric tensor has to be calculated all along the path from p to s.

The chain rule is used in the above derivation because not only is $\ln p$ changing with λ_r, but λ_r has a functional relationship with t and is changing as t changes.

$$\frac{d \ln p}{dt} = \sum_{r=1}^{m} \left(\frac{\partial \ln p_r}{\partial \lambda_r} \times \frac{d\lambda_r}{dt} \right)$$

In forming the inner product,

$$\left\langle \frac{d \ln p_r}{dt}, \frac{d \ln p_c}{dt} \right\rangle$$

the,

$$\sum_{r=1}^{m} \sum_{c=1}^{m} \left(\frac{\partial \ln p}{\partial \lambda_r} \times \frac{\partial \ln p}{\partial \lambda_c} \right)$$

go to make up $g_{rc}(t)$ while the,

$$\sum_{r=1}^{m} \sum_{c=1}^{m} \left(\frac{d\lambda_r}{dt} \times \frac{d\lambda_c}{dt} \right)$$

also appears in Equation (38.2).

38.5 Related Concepts in Information Geometry

It is time to make at least a preliminary effort to document, together in one place, some of these seemingly related concepts discussed in the past few Chapters. It is worth the effort to try and discover some common thread running through the log likelihood ratio, the maximum likelihood estimator, sufficient statistics, relative entropy, Fisher information, and the Maximum Entropy Principle.

First, review the log likelihood ratio as it appears from our Bayesian and MEP perspective. In Volume II [8, pg. 362], we arrived at the following formula for the log likelihood ratio, that is, the ratio of probability for all of the data when conditioned on two different models, $\mathcal{M}_A$ and $\mathcal{M}_B$,

$$\ln\left[\frac{P(\mathcal{D}\mid\mathcal{M}_A)}{P(\mathcal{D}\mid\mathcal{M}_B)}\right] = N\left[\sum_{j=1}^{m}(\lambda_j^A - \lambda_j^B)\overline{F}_j(X=x_i) + \ln\left(\frac{Z_B}{Z_A}\right)\right] \quad (38.3)$$

Note that the data, the N_i and $\sum N_i = N$, appear on the right hand side with N as the overall multiplier of the term in brackets, and in the *sample* average of the jth constraint function,

$$\overline{F}_j(X=x_i) = \sum_{i=1}^{n}\frac{N_i}{N}F_j(X=x_i)$$

Compare the expression in Equation (38.3) above to the relative entropy of distributions p and q [8, pg. 381],

$$KL(p,q) = \sum_{j=1}^{m}(\lambda_j^p - \lambda_j^q)\langle F_j\rangle_p + \ln\left(\frac{Z_q}{Z_p}\right) \quad (38.4)$$

N does not appear because the relative entropy has nothing to do with any data, and the mathematical expectation of the jth constraint function with respect to distribution p appears where the sample average appeared in the log likelihood ratio. Otherwise, the Lagrange multipliers and the partition functions for the two models arising from the MEP formalism are the same.

However, if the constraint function averages for model $\mathcal{M}_A$,

$$\sum_{i=1}^{n}F_j(X=x_i)Q_i = \langle F_j\rangle$$

in the relative entropy expression are replaced by the actual sample averages from the data,

$$\sum_{i=1}^{n}F_j(X=x_i)\frac{N_i}{N} = \overline{F}_j$$

then the Q_i, in effect, are "estimated" by the normed frequencies. Furthermore, that model has the highest probability over any other model. The sample average is a function of the data that qualifies as a sufficient statistic. It is also the maximum likelihood estimator (MLE).

Finally, with the results of this Chapter, we have the relationship between the relative entropy and the Fisher information matrix that came about from strictly geometric concerns in calculating the length of the curve between points p and s,

$$\sqrt{2KL(p,s)} \approx \int_p^s \sqrt{\sum_{r=1}^m \sum_{c=1}^m g_{rc}(t) \frac{d\lambda_r}{dt} \frac{d\lambda_c}{dt}} \, dt \qquad (38.5)$$

38.6 Connections to the Literature

Kullback's manipulations deriving the link between his information measure and Fisher's information matrix appear in [28, pg. 26–28]. Recall that the time frame for this effort was in the mid 1950s. It is worthwhile, I think, to present his equations verbatim so that the motivated reader can cross–check Kullback's more condensed argument and his notation with my more elaborate reconstruction as outlined in this Chapter and worked out in detail in the Exercises.

> The information measures that we have been studying are related to Fisher's information measure. Consider the parametric case where [the probability distributions] differ according to the values of the k–dimensional parameter $\theta = (\theta_1, \theta_2, \cdots, \theta_k)$. Suppose that θ and $\theta + \Delta\theta$ are neighboring points in the k–dimensional parameter space ... and $f_1(x) = f(x, \theta), f_2(x) = f(x, \theta + \Delta\theta)$. We shall show ... that $I(\theta : \theta + \Delta\theta)$ ... can be expressed as quadratic forms with coefficients defined by Fisher's information matrix.
>
> ...
>
> We may now use the Taylor expansion about θ and obtain
>
> (6.1) $\log f(x, \theta + \Delta\theta) - \log f(x, \theta) =$
>
> $$\sum_{\alpha=1}^k \Delta\theta_\alpha \frac{\partial \log f}{\partial \theta_\alpha} + \frac{1}{2!} \sum_{\alpha=1}^k \sum_{\beta=1}^k \Delta\theta_\alpha \Delta\theta_\beta \frac{\partial^2 \log f}{\partial \theta_\alpha \partial \theta_\beta}$$
>
> $$+ \frac{1}{3!} \sum_{\alpha=1}^k \sum_{\beta=1}^k \sum_{\gamma=1}^k \Delta\theta_\alpha \Delta\theta_\beta \Delta\theta_\gamma \left(\frac{\partial^3 \log f}{\partial \theta_\alpha \partial \theta_\beta \partial \theta_\gamma} \right)_{\theta + t\Delta\theta}$$
>
> ...
>
> We also have
>
> (6.2) $\quad \dfrac{\partial \log f}{\partial \theta_\alpha} = \dfrac{1}{f} \dfrac{\partial f}{\partial \theta_\alpha}; \dfrac{\partial^2 \log f}{\partial \theta_\alpha \partial \theta_\beta} = \dfrac{1}{f} \dfrac{\partial^2 f}{\partial \theta_\alpha \partial \theta_\beta} - \dfrac{1}{f^2} \dfrac{\partial f}{\partial \theta_\alpha} \dfrac{\partial f}{\partial \theta_\beta}$

We may therefore write

$$
(6.3) \quad I(\theta : \theta + \Delta\theta) = \int f(x,\theta) \log \frac{f(x,\theta)}{f(x,\theta + \Delta\theta)} \, d\lambda(x)
$$

$$
= - \int \left(\sum_{\alpha=1}^{k} \Delta\theta_\alpha f \cdot \frac{\partial \log f}{\partial \theta_\alpha} \right) d\lambda(x)
$$

$$
- \frac{1}{2!} \int \left(\sum_{\alpha=1}^{k} \sum_{\beta=1}^{k} \Delta\theta_\alpha \Delta\theta_\beta f \cdot \frac{\partial^2 \log f}{\partial \theta_\alpha \, \partial \theta_\beta} \right) d\lambda(x)
$$

$$
- \frac{1}{3!} \int \left[\sum_{\alpha=1}^{k} \sum_{\beta=1}^{k} \sum_{\gamma=1}^{k} \Delta\theta_\alpha \Delta\theta_\beta \Delta\theta_\gamma f \cdot \left(\frac{\partial^3 \log f}{\partial \theta_\alpha \, \partial \theta_\beta \, \partial \theta_\gamma} \right) \right] d\lambda(x)
$$

$$
= - \sum_{\alpha=1}^{k} \Delta\theta_\alpha \int \frac{\partial f}{\partial \theta_\alpha} \, d\lambda(x)
$$

$$
- \frac{1}{2!} \sum_{\alpha=1}^{k} \sum_{\beta=1}^{k} \Delta\theta_\alpha \Delta\theta_\beta \int \left(\frac{\partial^2 f}{\partial \theta_\alpha \, \partial \theta_\beta} - \frac{1}{f} \frac{\partial f}{\partial \theta_\alpha} \frac{\partial f}{\partial \theta_\beta} \right) d\lambda(x)
$$

$$
- \frac{1}{3!} \sum_{\alpha=1}^{k} \sum_{\beta=1}^{k} \sum_{\gamma=1}^{k} \Delta\theta_\alpha \Delta\theta_\beta \Delta\theta_\gamma \int f \cdot \left(\frac{\partial^3 \log f}{\partial \theta_\alpha \, \partial \theta_\beta \, \partial \theta_\gamma} \right) d\lambda(x)
$$

Accordingly, because of the regularity conditions, to within second order terms, we have

$$
(6.4) \quad I(\theta : \theta + \Delta\theta) = \frac{1}{2} \sum_{\alpha=1}^{k} \sum_{\beta=1}^{k} g_{\alpha\beta} \Delta\theta_\alpha \Delta\theta_\beta
$$

with

$$
g_{\alpha\beta} = \int f(x,\theta) \left(\frac{1}{f(x,\theta)} \frac{\partial f(x,\theta)}{\partial \theta_\alpha} \right) \left(\frac{1}{f(x,\theta)} \frac{\partial f(x,\theta)}{\partial \theta_\beta} \right) d\lambda(x)
$$

and $\mathbf{G} = (g_{\alpha\beta})$ the positive definite Fisher information matrix ...

This is just another example where an author's close familiarity with some piece of subject matter might predispose him to present very condensed "macros." It never seems to occur to such people that those of us who are not steeped in the intricacies of a particular problem, require a much more expanded, more detailed exposition of the justification behind every single step the author is taking. Otherwise, in our naïveté, the argument appears more abstruse than necessary.

Unfortunately, a lot of this obfuscation can be attributed to the psychological need, as Jaynes said of Feller's presentation, to engage in "gamesmanship," or, "one–up–manship." Other writers, and for me this type is personified by Sir Roger Penrose's works for the lay public, just don't realize how much smarter they are

RELATED CONCEPTS IN IG

than the rest of us. In the process of explaining things to us, Penrose seems to assume that the high level functioning of his brain maps over to a similarly high level in my brain, and more the pity for me because that just happens to be a false assumption.

On the brighter side, it is somewhat rewarding to realize that, in the quote below [3, pg. 1702], we can now begin to comprehend larger chunks of Amari's way of writing. This also serves as a good example of his summary assertions about squared distance, Kullback–Leibler divergence, and the metric tensor, albeit absent of any derivations.

> Let us consider a parameterized family of probability distributions $S = \{p(x, \boldsymbol{\xi})\}$, where x is a random variable and $\boldsymbol{\xi} = (\xi_1, \ldots, \xi_n)$ is a real vector parameter to specify a distribution. The family S is regarded as an n–dimensional manifold having $\boldsymbol{\xi}$ as a coordinate system. When the Fisher information matrix $G = (g_{ij})$
>
> $$g_{ij}(\boldsymbol{\xi}) = E\left[\frac{\partial \log p(x, \boldsymbol{\xi})}{\partial \xi_i} \frac{\partial \log p(x, \boldsymbol{\xi})}{\partial \xi_j}\right]$$
>
> where E denotes expectation with respect to $p(x, \boldsymbol{\xi})$, is non–degenerate, S is a Riemannian manifold, and $G(\boldsymbol{\xi})$ plays the role of a Riemannian metric tensor.
>
> The squared distance ds^2 between two nearby distributions $p(x, \boldsymbol{\xi})$ and $p(x, \boldsymbol{\xi} + d\boldsymbol{\xi})$ is given by the quadratic form of $d\boldsymbol{\xi}$
>
> $$ds^2 = \sum g_{ij}(\boldsymbol{\xi}) d\xi^i d\xi^j$$
>
> It is known that this is twice the Kullback–Leibler divergence
>
> $$ds^2 = 2KL[p(x, \boldsymbol{\xi}) \colon p(x, \boldsymbol{\xi} + d\boldsymbol{\xi})]$$
>
> where
>
> $$KL[p \colon q] = \int p(x) \log \frac{p(x)}{q(x)} dx$$

I generally agree with all of this, and the language is consistent with how we have been using it over the past few Chapters. The main point of disagreement is in calling x a "random variable." The correct notation and language is $(X = x_i)$ for some *joint statement* in the n–dimensional state space. Other quibbles over consistency include: Our dual coordinate systems have always been labeled as λ_j and $\langle F_j \rangle$, or, even in Amari's own notation, as θ^i and η_i, not as $\boldsymbol{\xi}$.

It is interesting that Amari interprets Kullback's information measure,

$$I(1 \colon 2) = \int f_1(x) \log \frac{f_1(x)}{f_2(x)} d\lambda(x)$$

as a divergence measure D_α with $\alpha = -1$,

$$D_{(-1)}[p \colon q] = \int p(x) \log \frac{p(x)}{q(x)} dx$$

eschewing the measure theory terminology $d\lambda(x)$ and, apparently, interpreting $f_1(x)$ and $f_2(x)$ in the same way as I do,

$$f_1(x) \equiv \{Q_i^p = P(X = x_i \,|\, \mathcal{M}_k) \to p(x, \boldsymbol{\xi})\}$$

$$f_2(x) \equiv \{Q_i^q = P(X = x_i \,|\, \mathcal{M}_k) \to q(x, \boldsymbol{\xi})\}$$

as probability distributions depending on information as provided within some model, and not as distributions dependent on any data.

Amari always writes his probability expressions along the form of $p(x, \boldsymbol{\xi})$ for a probability distribution containing information inserted by the parameter vector $\boldsymbol{\xi}$. I think that this is less clear than my $P(X = x_i \,|\, \mathcal{M}_k)$. It is important when writing out expressions involving probability that the distinction between joint statements and conditioned upon statements be maintained by the solidus symbol.

I prefer to use summation over the dimension of the state space, in other words, I prefer to sum from $i = 1$ to n instead of integrating over dx, mainly because it seems less mysterious. Besides, it is clear when using the summation notation that it is the statements in the state space that are being summed over. Is it always clear what dx means? It also is somewhat irritating when no region of integration appears in either Kullback's or Amari's notation.

Thus, our formulas for relative entropy look like either of,

$$KL(p, q) = \sum_{i=1}^{n} Q_i^p \ln\left(\frac{Q_i^p}{Q_i^q}\right)$$

$$KL(p, q) = \sum_{i=1}^{n} p_i \ln\left(\frac{p_i}{q_i}\right)$$

where there are no mysteries over the region of summation.

One of the great difficulties in this area of Information Geometry is that different people with different agendas will all rely upon the same mathematics to develop their conceptual arguments. The underlying math is never at fault, but it is sometimes pressed into service for antithetical goals. Distressingly, one often finds that someone who espouses directly opposite conceptual goals to the ones you happen to hold, explains the common underlying mathematics in a particularly pleasing way. So you end up praising the skillful hand servant for her expertise, while denouncing the master that she so faithfully serves.

Just as a most preliminary example of this phenomenon, as it has played out within probability and statistics, is that great historical conceptual divide between what Jaynes labeled the "orthodox statistics" of Fisher, Neyman, Pearson, and all of their erstwhile followers, and the equally ill–defined "Bayesian" camp. Kass and Vos [25] have written a book that, on the face of it, would seem to be quite relevant to our current concerns in Information Geometry. And it goes without saying that their mathematical development of interesting topics within Information Geometry is quite impressive.

But it is very difficult to fold their exposition into the conceptual plan I have been developing because their ultimate agenda is so very different from mine. In a nutshell, they use the mathematics of Differential Geometry to advance the cause of "orthodox statistics" using language with concepts like "sufficient statistics," "parameter estimation," "asymptotic inference, " and on, and on. Their goal is antithetical to developing probability as a generalization of logic.

But this doesn't mean that we must end up throwing the baby out with the bathwater. They often present compelling alternative explanations of fundamental mathematical ideas that enhance our understanding, and, furthermore, pinpoint exactly where disagreements occur between the two approaches. Here is a quote from Kass and Vos discussing the Kullback–Leibler divergence [25, pg. 51],

> The Kullback–Leibler divergence is a measure of discrepancy between two distributions. Defined as the expected log ratio of densities, it plays a fundamental role in determining the limiting behavior of certain log ratios of densities computed from the data (such as the log likelihood ratio statistics). Here we provide the definition and show how the Kullback–Leibler divergence is related to the log likelihood function and Fisher information.
>
> **Definition 2.5.2** The Kullback–Leibler divergence $K(\cdot,\cdot)\colon \mathcal{S} \times \mathcal{S} \to \mathbb{R}$ is
>
> $$K(P_\eta, P_{\eta^\star}) = E_\eta \left\{ \log\left(\frac{p(Y\mid\eta)}{p(Y\mid\eta^\star)}\right) \right\}$$
>
> $$= \int (\log p(y\mid\eta) - (\log p(y\mid\eta^\star)) p(y\mid\eta)\,dy$$
>
> [In terms of the dominating measure ... we would replace dy by $\nu(dy)$.]
>
> ...
>
> ... we immediately obtain the computational form of the Kullback–Leibler divergence
>
> $$K(\eta, \eta^\star) = (\eta - \eta^\star)^T \mu - (\psi(\eta) - \psi(\eta^\star)) \qquad (2.5.16)$$
>
> This provides the basic relationship between the Kullback–Leibler divergence and the log likelihood function, given in the next theorem. The theorem also shows that the MLE in a curved exponential family is obtained by minimizing the Kullback–Leibler divergence between the data to the model.

I wish that those people who are so enamored of this "dominating measure" thing would go out of their way, just once, to enlighten those poor benighted souls like myself who have no idea of what they are referring to. Jaynes's dry remark that he would use measure theory, if he had ever once come across an inferential problem where it was needed, rings in our ears.

At the abstract mathematical level, Definition 2.5.2 causes no problem. We also define a function $KL(p,q)$ which takes two arguments, each from the Riemannian manifold $\mathcal{S}$. However, that first argument is a conditional probability p, conditioned on the information in model $\mathcal{M}_p$, taken from $\mathcal{S}$, and the second argument is another conditional probability q, conditioned on the information in model $\mathcal{M}_q$ also from $\mathcal{S}$. This function $KL(p,q)$ must return a real number scalar, a "squared distance," in $\mathbb{R}$.

$$KL(p,q) = \sum_{i=1}^{n} p_i \ln\left(\frac{p_i}{q_i}\right) \equiv E_p\left[\ln p - \ln q\right]$$

where,

$$p \equiv P(X = x_i \,|\, \mathcal{M}_p) \text{ and } q \equiv P(X = x_i \,|\, \mathcal{M}_q)$$

and a model $\mathcal{M}_k$ inserts information via the m parameters $\{\lambda_1, \lambda_2, \cdots, \lambda_j, \cdots, \lambda_m\}$.

But upon closer examination, is the Kass and Vos definition of relative entropy really the same? They explicitly state that the relative entropy is determined *from the data as a log likelihood ratio*. By using the word *likelihood*, they mean the probability of all of the *data* given the parameters η. The probability distributions appearing as the arguments to Kullback–Leibler divergence function are NOT,

$$p \equiv P(X = x_i \,|\, \mathcal{M}_p) \text{ and } q \equiv P(X = x_i \,|\, \mathcal{M}_q)$$

but instead,

$$P(Y\,|\,\eta) = \prod_{t=1}^{N} P(X_t = x_i \,|\, (\lambda_1, \cdots, \lambda_j, \cdots, \lambda_m)) \equiv P(\mathcal{D}\,|\,\mathcal{M}_k)$$

This interpretation is reinforced with their constant insistence on conflating the concept of the maximum likelihood estimator (MLE), some function of the data, with the relative entropy. Conceptually, the relative entropy is defined prior to, and independently of any data. This proscription doesn't preclude looking at the distance between two distributions, when, *after* the data have been gathered, the information from one distribution is forced to match up with sample averages derived from the data.

In Volume II, Chapter Twenty Seven, [8, pp. 333–334, pp. 360–363], we derived the relationship between the maximum likelihood estimator and relative entropy based on Jaynes's suggestion. The log likelihood ratio took the form of,

$$\ln\left[\frac{P(\mathcal{D}\,|\,\mathcal{M}_A)}{P(\mathcal{D}\,|\,\mathcal{M}_B)}\right] = N\left[\sum_{j=1}^{m}(\lambda_j^A - \lambda_j^B)\overline{F}_j(X = x_i) + \ln\left(\frac{Z_B}{Z_A}\right)\right]$$

which has a visual appearance similar to, but must be conceptually different from, the expression for the relative entropy,

$$KL(p,q) = \sum_{i=1}^{n} p_i \ln\left(\frac{p_i}{q_i}\right) = \sum_{j=1}^{m}(\lambda_j^p - \lambda_j^q)\langle F_j \rangle_p + \ln\left(\frac{Z_q}{Z_p}\right)$$

The log likelihood ratio expression must reflect the impact from the data. That explains the presence of N and the *sample* averages $\overline{F}_j(X = x_i)$.

On the other hand, the relative entropy expression must NOT contain any terms relating to the data, and it does not. There is no term involving N or the N_i, and $\langle F_j \rangle_p$ is *information* inserted into the probability distribution that is independent of any data.

If one attempts to identify an overlap in our arguments through the Kass and Vos Equation (2.5.16),

$$K(\eta, \eta^\star) = (\eta - \eta^\star)^T \mu - (\psi(\eta) - \psi(\eta^\star))$$

which they label as the computational form of the Kullback–Leibler divergence, and our,

$$KL(p,q) = \sum_{i=1}^{n} p_i \ln\left(\frac{p_i}{q_i}\right) = \sum_{j=1}^{m} (\lambda_j^p - \lambda_j^q)\langle F_j \rangle_p + \ln\left(\frac{Z_q}{Z_p}\right)$$

the match up is straightforward with the equivalencies in the model parameters,

$$\eta \equiv \lambda_j^p \text{ and } \eta^\star \equiv \lambda_j^q$$

the constraint function averages, $\mu \equiv \langle F_j \rangle$, and finally the partition functions,

$$\psi(\eta) \equiv Z(\lambda_j^p) \text{ and } \psi(\eta^\star) \equiv Z(\lambda_j^q)$$

Unfortunately, if one does take this tack, it is immediately clear that $K(\eta, \eta^\star)$ is NOT a log likelihood ratio as they demand, but rather a relationship between two conditional probabilities *prior to any data!*

So, to be consistent, they identify μ as the maximum likelihood estimate of the sufficient statistic $F_j(X = x_i)$. Then, the probability distribution p and its parameters λ_j^p must be identified with that one particular model with maximum likelihood. But then the freedom to find the distance between any two arbitrary distributions p and q is lost if one adopts this position.

38.7 Solved Exercises for Chapter Thirty Eight

Exercise 38.7.1: Write out the univariate Taylor series expansion for a function $f(x)$, first in conventional notation, and then followed by the *Mathematica* representation.

Solution to Exercise 38.7.1

The univariate Taylor series expansion around the point a for a generic function $f(x)$ with one argument can be written as,

$$f(x) = f(a) + (x-a)f'(a) + \frac{1}{2!}(x-a)^2 f''(a) + \cdots + \frac{1}{n!}(x-a)^n f^{(n)}(a)$$

Mathematica uses the `Series[ ]` function for Taylor series expansions,

$$\texttt{Series[f[x], \{x, a, 3\}]}$$

which returns out to third order,

`f[a] + f'[a] (x - a) + 1/2 f''[a] (x - a)^2 + 1/6 f^(3) (x - a)^3 + O[x - a]^4`

The multivariate analog to the generic function $f(x)$ out to just the second order, where any usage of the **bold** notation indicates that vectors are involved, can be written with these new symbols,

$$f(\boldsymbol{x}) = f(\boldsymbol{\delta}) + \boldsymbol{\delta}\,\boldsymbol{\nabla} f(\boldsymbol{\delta}) + \frac{1}{2}\boldsymbol{\delta}\,\boldsymbol{\nabla}^2 f(\boldsymbol{\delta})\,\boldsymbol{\delta}^T + \cdots$$

Exercise 38.7.2: How does this begin to get translated for a Taylor series expansion of the relative entropy information measure?

Solution to Exercise 38.7.2

To orient ourselves, investigate how the above *Mathematica* `Series[ ]` function for a Taylor series expansion must be refashioned in order to involve the relative entropy expression. If we construct this as,

$$\texttt{Series[f[}\theta_q\texttt{], \{}\theta_q\texttt{, }\theta_p\texttt{, 3\}] // TraditionalForm}$$

then we are privileged to view the expression,

$$f(\theta_p) + (\theta_q - \theta_p)f'(\theta_p) + \frac{1}{2}(\theta_q - \theta_p)^2 f''(\theta_p) + \frac{1}{6}(\theta_q - \theta_p)^3 f^{(3)}(\theta_p) + O((\theta_q - \theta_p)^4)$$

SOLVED EXERCISES FOR CHAPTER 38

Exercise 38.7.3: Start to pattern match with Kullback's Taylor series expansion.

Solution to Exercise 38.7.3

We are expanding $\ln q$, $\log f(\boldsymbol{x}, \boldsymbol{\theta} + \boldsymbol{\Delta\theta})$ around the point p, $f(\boldsymbol{x}, \boldsymbol{\theta})$.

$$\log f(\boldsymbol{x}, \boldsymbol{\theta} + \boldsymbol{\Delta\theta}) = \log f(\boldsymbol{x}, \boldsymbol{\theta}) + \boldsymbol{\Delta\theta} \frac{\partial \log f(\boldsymbol{x}, \boldsymbol{\theta})}{\partial \boldsymbol{\theta}} + \frac{1}{2} \boldsymbol{\Delta\theta} \frac{\partial^2 \log f(\boldsymbol{x}, \boldsymbol{\theta})}{\partial \boldsymbol{\theta} \partial \boldsymbol{\theta}} \boldsymbol{\Delta\theta}^T$$

$$\log f(\boldsymbol{x}, \boldsymbol{\theta} + \boldsymbol{\Delta\theta}) - \log f(\boldsymbol{x}, \boldsymbol{\theta}) = \boldsymbol{\Delta\theta} \frac{\partial \log f(\boldsymbol{x}, \boldsymbol{\theta})}{\partial \boldsymbol{\theta}} + \frac{1}{2} \boldsymbol{\Delta\theta} \frac{\partial^2 \log f(\boldsymbol{x}, \boldsymbol{\theta})}{\partial \boldsymbol{\theta} \partial \boldsymbol{\theta}} \boldsymbol{\Delta\theta}^T$$

$$\log f(\boldsymbol{x}, \boldsymbol{\theta}) - \log f(\boldsymbol{x}, \boldsymbol{\theta} + \boldsymbol{\Delta\theta}) = -\boldsymbol{\Delta\theta} \frac{\partial \log f(\boldsymbol{x}, \boldsymbol{\theta})}{\partial \boldsymbol{\theta}} - \frac{1}{2} \boldsymbol{\Delta\theta} \frac{\partial^2 \log f(\boldsymbol{x}, \boldsymbol{\theta})}{\partial \boldsymbol{\theta} \partial \boldsymbol{\theta}} \boldsymbol{\Delta\theta}^T$$

$$\log \left[\frac{f(\boldsymbol{x}, \boldsymbol{\theta})}{f(\boldsymbol{x}, \boldsymbol{\theta} + \boldsymbol{\Delta\theta})} \right] = -\boldsymbol{\Delta\theta} \frac{\partial \log f(\boldsymbol{x}, \boldsymbol{\theta})}{\partial \boldsymbol{\theta}} - \frac{1}{2} \boldsymbol{\Delta\theta} \frac{\partial^2 \log f(\boldsymbol{x}, \boldsymbol{\theta})}{\partial \boldsymbol{\theta} \partial \boldsymbol{\theta}} \boldsymbol{\Delta\theta}^T$$

$$\int f(\boldsymbol{x}, \boldsymbol{\theta}) \log \left[\frac{f(\boldsymbol{x}, \boldsymbol{\theta} + \boldsymbol{\Delta\theta})}{f(\boldsymbol{x}, \boldsymbol{\theta})} \right] d\lambda(x) = 0 - \frac{1}{2} \boldsymbol{\Delta\theta} \boldsymbol{\Delta\theta}^T \int f(\boldsymbol{x}, \boldsymbol{\theta}) \frac{\partial^2 \log f(\boldsymbol{x}, \boldsymbol{\theta})}{\partial \boldsymbol{\theta} \partial \boldsymbol{\theta}} d\lambda(x)$$

Exercise 38.7.4: What does the integral on the right hand side of the final step work out to?

Solution to Exercise 38.7.4

This problem has already been addressed in Exercises 36.9.12 and 36.9.15. Here, we employ a stripped–down notation for the integration in order to obtain a more uncluttered view,

$$\frac{\partial^2 \log f}{\partial \theta_\alpha \partial \theta_\beta} = \frac{1}{f} \frac{\partial^2 f}{\partial \theta_\alpha \partial \theta_\beta} - \frac{1}{f} \frac{\partial f}{\partial \theta_\alpha} \frac{1}{f} \frac{\partial f}{\partial \theta_\beta}$$

$$f \left(\frac{1}{f} \frac{\partial^2 f}{\partial \theta_\alpha \partial \theta_\beta} - \frac{1}{f} \frac{\partial f}{\partial \theta_\alpha} \frac{1}{f} \frac{\partial f}{\partial \theta_\beta} \right) = \frac{\partial^2 f}{\partial \theta_\alpha \partial \theta_\beta} - \frac{1}{f} \frac{\partial f}{\partial \theta_\alpha} \frac{\partial f}{\partial \theta_\beta}$$

$$\int \left[\frac{\partial^2 f}{\partial \theta_\alpha \partial \theta_\beta} - \frac{1}{f} \frac{\partial f}{\partial \theta_\alpha} \frac{\partial f}{\partial \theta_\beta} \right] = \int \frac{\partial^2 f}{\partial \theta_\alpha \partial \theta_\beta} - \int \frac{1}{f} \frac{\partial f}{\partial \theta_\alpha} \frac{\partial f}{\partial \theta_\beta}$$

$$= -\int \frac{1}{f} \frac{\partial f}{\partial \theta_\alpha} \frac{\partial f}{\partial \theta_\beta}$$

$$= -\int f \left(\frac{1}{f} \frac{\partial f}{\partial \theta_\alpha} \frac{1}{f} \frac{\partial f}{\partial \theta_\beta} \right)$$

Exercise 38.7.5: Show how to proceed from the resulting expression in the last step above to Kullback's $g_{\alpha\beta}$.

Solution to Exercise 38.7.5

At the last step above, and at the end of the derivation relating his information measure and Fisher's information matrix, Kullback has this version for the metric tensor,

$$g_{\alpha\beta} = \int f(x,\theta) \left(\frac{1}{f(x,\theta)} \frac{\partial f(x,\theta)}{\partial \theta_\alpha} \right) \left(\frac{1}{f(x,\theta)} \frac{\partial f(x,\theta)}{\partial \theta_\beta} \right) d\lambda(x)$$

Proceed from this equation as the starting point, and "reverse engineer" all the way back to the inner product of two vectors.

First, use the oft cited relationship concerning the partial differentiation of the log transform of any function with respect to one of its arguments. Kullback wrote this relationship in the form of,

$$\frac{\partial \log f}{\partial \theta_\alpha} = \frac{1}{f} \frac{\partial f}{\partial \theta_\alpha}$$

This allows us to convert the above beginning expression to,

$$g_{\alpha\beta} = \int f(x,\theta) \left(\frac{\partial \log f(x,\theta)}{\partial \theta_\alpha} \right) \left(\frac{\partial \log f(x,\theta)}{\partial \theta_\beta} \right) d\lambda(x)$$

The integration with respect to the probability distribution $f(x,\theta)$ and the measure $d\lambda(x)$ can be represented in an abbreviated notation as the expectation operator $E[\cdots]$,

$$g_{\alpha\beta} = E\left[\left(\frac{\partial \log f(x,\theta)}{\partial \theta_\alpha} \right) \times \left(\frac{\partial \log f(x,\theta)}{\partial \theta_\beta} \right) \right]$$

Kullback's notation for a probability density function $f(x,\theta)$ is the point p, which appears in the relative entropy equation as,

$$KL(p,q) = \sum_{i=1}^{n} p_i \ln\left(\frac{p_i}{q_i} \right)$$

which, in turn, is equivalent to the Q_i as assigned via the MEP under the information from some model,

$$P(X = x_i \mid \mathcal{M}_k) \equiv Q_i$$

$$Q_i = \frac{e^{\sum_{j=1}^{m} \lambda_j F_j(X=x_i)}}{Z}$$

$$Z = \sum_{i=1}^{n} e^{\sum_{j=1}^{m} \lambda_j F_j(X=x_i)}$$

SOLVED EXERCISES FOR CHAPTER 38

The MEP coordinates λ_j as parameters of the model are equivalent to Kullback's parameters θ_α. With these equivalencies in place,

$$\frac{\partial \log f(x, \theta)}{\partial \theta_\alpha} \equiv \frac{\partial \ln p}{\partial \lambda_r}$$

Kullback's version of the metric tensor can be reformulated to look like,

$$g_{rc}(p) = E_p \left[\left(\frac{\partial \ln p}{\partial \lambda_r} \right) \times \left(\frac{\partial \ln p}{\partial \lambda_c} \right) \right]$$

Working now on the log transform of the MEP version of point p, we have,

$$\frac{\partial \ln p}{\partial \lambda_r} = \frac{\partial \ln \{ \exp [\sum_{j=1}^{m} \lambda_j F_j(X = x_i)]/Z \}}{\partial \lambda_r}$$

$$= \frac{\partial \left(\sum_{j=1}^{m} \lambda_j F_j(X = x_i) - \ln Z \right)}{\partial \lambda_r}$$

$$= F_r(X = x_i) - \frac{\partial \ln Z}{\partial \lambda_r}$$

$$= F_r(X = x_i) - \langle F_r \rangle$$

With this result, we now have,

$$g_{rc}(p) = E_p \left[(F_r(X = x_i) - \langle F_r \rangle) \times (F_c(X = x_i) - \langle F_c \rangle) \right]$$

followed quickly by,

$$g_{rc}(p) = \sum_{i=1}^{n} \left[(F_r(X = x_i) - \langle F_r \rangle) \times (F_c(X = x_i) - \langle F_c \rangle) \right] Q_i$$

The important net result after all of this is that we can definitely assert the equivalency between our metric tensor $g_{rc}(p)$ and Kullback's Fisher information matrix $g_{\alpha\beta}$.

Exercise 38.7.6: Show how the second term in the expansion eventually became 0 in the derivation provided by Exercise 38.7.3.

Solution to Exercise 38.7.6

In the derivation, the second term in the expansion is set to 0 with no justification provided. After the left hand side was transformed into,

$$\int f(\boldsymbol{x}, \boldsymbol{\theta}) \log \left[\frac{f(\boldsymbol{x}, \boldsymbol{\theta} + \boldsymbol{\Delta \theta})}{f(\boldsymbol{x}, \boldsymbol{\theta})} \right] d\lambda(x)$$

the first term on the right hand side would have looked like,
$$-\Delta\theta \int f(x,\boldsymbol{\theta}) \frac{\partial \log f(x,\boldsymbol{\theta})}{\partial \theta} \, d\lambda(x)$$
However, the simplifying resolution that this term does, in fact, equal zero is most easily explained by reverting back to our preferred notation for the MEP formalism.

$$\frac{\partial \log f(x,\theta)}{\partial \theta_\alpha} \equiv \frac{\partial \ln p}{\partial \lambda_j}$$

$$= F_j(X = x_i) - \langle F_j \rangle$$

$$\int f(x,\theta) \frac{\partial \log f(x,\theta)}{\partial \theta_\alpha} \, d\lambda(x) \equiv \sum_{i=1}^{n} \left[Q_i \times \left(\frac{\partial \ln p}{\partial \lambda_j} \right) \right]$$

$$= \sum_{i=1}^{n} Q_i F_j(X = x_i) - \sum_{i=1}^{n} Q_i \langle F_j \rangle$$

$$= \langle F_j \rangle - \left[\langle F_j \rangle \sum_{i=1}^{n} Q_i \right]$$

$$= 0$$

Exercise 38.7.7: Provide an example from *Mathematica* that verifies a step in Exercise 38.7.4.

Solution to Exercise 38.7.7

In the derivation of Exercise 38.7.4, the term,
$$\int \frac{\partial^2 f}{\partial \theta_\alpha \partial \theta_\beta} \, d\lambda(x)$$
was set equal to 0 without any justification. Kullback just asserts it as a fact.

Here is an example using the MEP characterization verifying this. Suppose that $n = 3$ with two constraint functions defined as,
$$F_1(X = x_i) = (1, 0, 0) \text{ and } F_2(X = x_i) = (0, 1, 0)$$
Thus, the numerical assignments are,
$$Q_1 = \frac{e^{\lambda_1}}{e^{\lambda_1} + e^{\lambda_2} + 1}$$

$$Q_2 = \frac{e^{\lambda_2}}{e^{\lambda_1} + e^{\lambda_2} + 1}$$

$$Q_3 = \frac{1}{e^{\lambda_1} + e^{\lambda_2} + 1}$$

Using *Mathematica* to explicitly calculate the sum,

$$\sum_{i=1}^{3} \frac{\partial^2 Q_i}{\partial \lambda_1 \partial \lambda_2}$$

through the partial differentiation function template `D[f, x, y]`,

```
D[Exp[λ₁] / (Exp[λ₁] + Exp[λ₂] + 1), λ₁, λ₂] +
D[Exp[λ₂] / (Exp[λ₁] + Exp[λ₂] + 1), λ₁, λ₂] +
D[1 / (Exp[λ₁] + Exp[λ₂] + 1), λ₁, λ₂]
                                            // FullSimplify
```

returns 0.

Exercise 38.7.8: Put all of the pieces back together to arrive at Kullback's Equation (6.4).

Solution to Exercise 38.7.8

Start off with the last line of Exercise 38.7.3,

$$\int f(\boldsymbol{x}, \boldsymbol{\theta}) \log \left[\frac{f(\boldsymbol{x}, \boldsymbol{\theta} + \Delta \boldsymbol{\theta})}{f(\boldsymbol{x}, \boldsymbol{\theta})}\right] d\lambda(x) = -\frac{1}{2} \Delta \boldsymbol{\theta} \Delta \boldsymbol{\theta}^T \int f(\boldsymbol{x}, \boldsymbol{\theta}) \frac{\partial^2 \log f(\boldsymbol{x}, \boldsymbol{\theta})}{\partial \boldsymbol{\theta} \partial \boldsymbol{\theta}} d\lambda(x)$$

Next, take the results from Exercise 38.7.4,

$$\int f(\boldsymbol{x}, \boldsymbol{\theta}) \log \left[\frac{f(\boldsymbol{x}, \boldsymbol{\theta} + \Delta \boldsymbol{\theta})}{f(\boldsymbol{x}, \boldsymbol{\theta})}\right] d\lambda(x) = \left[-\frac{1}{2} \Delta \boldsymbol{\theta} \Delta \boldsymbol{\theta}^T\right] \times \left[-\int f \left(\frac{1}{f} \frac{\partial f}{\partial \theta_\alpha} \frac{1}{f} \frac{\partial f}{\partial \theta_\beta}\right)\right]$$

After absorbing the lessons from Exercise 38.7.5, we are led to,

$$\int f(\boldsymbol{x}, \boldsymbol{\theta}) \log \left[\frac{f(\boldsymbol{x}, \boldsymbol{\theta} + \Delta \boldsymbol{\theta})}{f(\boldsymbol{x}, \boldsymbol{\theta})}\right] d\lambda(x) = \frac{1}{2} g_{\alpha\beta} \Delta \boldsymbol{\theta} \Delta \boldsymbol{\theta}^T$$

Kullback's Equation (6.4) was written with the matrix–vector multiplications on the right hand side expressed in summation format as,

$$I(\theta : \theta + \Delta\theta) = \frac{1}{2} \sum_{\alpha=1}^{k} \sum_{\beta=1}^{k} g_{\alpha\beta} \Delta\theta_\alpha \Delta\theta_\beta$$

It is helpful to look at the explicit expansion of the summation as,

$$I(\theta : \theta + \Delta\theta) = 1/2 \, [\, g_{11}\Delta\theta_1\Delta\theta_1 + g_{12}\Delta\theta_1\Delta\theta_2 + \cdots + g_{1k}\Delta\theta_1\Delta\theta_k +$$
$$g_{21}\Delta\theta_2\Delta\theta_1 + g_{22}\Delta\theta_2\Delta\theta_2 + \cdots + g_{2k}\Delta\theta_2\Delta\theta_k +$$
$$\cdots$$
$$g_{k1}\Delta\theta_k\Delta\theta_1 + g_{k2}\Delta\theta_k\Delta\theta_2 + \cdots + g_{kk}\Delta\theta_k\Delta\theta_k \,]$$

Finally, substitute the actual differences in coordinates $\Delta\theta_\alpha$ and $\Delta\theta_\beta$ for the two distributions,

$$I(\theta : \theta + \Delta\theta) = 1/2 \, [\, g_{11}(\theta_1^q - \theta_1^p)(\theta_1^q - \theta_1^p) + g_{12}(\theta_1^q - \theta_1^p)(\theta_2^q - \theta_2^p) + \cdots + g_{1k}(\theta_1^q - \theta_1^p)(\theta_k^q - \theta_k^p) +$$
$$g_{21}(\theta_2^q - \theta_2^p)(\theta_1^q - \theta_1^p) + g_{22}(\theta_2^q - \theta_2^p)(\theta_2^q - \theta_2^p) + \cdots + g_{2k}(\theta_2^q - \theta_2^p)(\theta_k^q - \theta_k^p) +$$
$$\cdots$$
$$g_{k1}(\theta_k^q - \theta_k^p)(\theta_1^q - \theta_1^p) + g_{k2}(\theta_k^q - \theta_k^p)(\theta_2^q - \theta_2^p) + \cdots + g_{kk}(\theta_k^q - \theta_k^p)(\theta_k^q - \theta_k^p) \,]$$

Chapter 39

Life on Mars

39.1 Introduction

Eventually one day, as you might have come to expect, your interest is piqued by a seriously complicated model purporting to explain some aspect of the real world. Unfortunately, it turns out that it is impossible to perform any inferential calculations with your complicated model. You then contemplate whether it might be feasible to salvage something from this fiasco by performing the computations not in the vast space of your originally conceived model, but rather in some simpler, more tractable model space.

This is a common problem. The thorny issue of computational complexity was first broached when trying to predict the long term behavior of cellular automata. Wolfram flatly states that there is no way of predicting the exact detail of, say, the Universal Turing Machine, Rule 110, far into the future. And he is right.

So we have to settle for something less. But that something less may still satisfy all of our needs. Statistical mechanics settles for something less when it freely admits that following the consequences of every collision of every atom in a gas is impossible. But physicists can still predict a host of interesting physical phenomena about the gas.

Untold myriad details remain about the gas which statistical mechanics cannot explain, just as we cannot predict the future fine grain distribution of black and white cells of the CA following Rule 110. Perhaps we can live with not knowing about details that are going to change every single time we make a measurement.

An IP cannot function unless there is some stability in its world. We care only about events that will repeat when the causal conditions are under our control. For example, the MEP algorithm provides us with a way of predicting some macroscopic events about a gas when the temperature, pressure, and volume are known or controlled. Jaynes has told us that all of the changing, but uninteresting (to us), details in the physics of the gas cancel out in the end anyway.

Amari has addressed similarly complicated problems by invoking notions from Information Geometry. He has investigated, among other things, whether Ising model spin phenomena, Boltzmann machines, error code correction, and mean field approximation can be simplified. His arguments take advantage of some of the somewhat less difficult concepts in Information Geometry which we now introduce in this Chapter. We attempt to explain these ideas by once again choosing a numerical example that is easy to follow.

39.2 Overview

The problem is treated in Information Geometry by setting up a smooth Riemannian manifold endowed with a metric tensor at each point in the manifold. Suppose further that we call point p the complicated model that lives in a very large space where m is close to n. Call point q the simpler model that lives in a more tractable space where m is much, much less than n. For notational purposes, we'll label the large intractable space $\mathcal{S}^n$ and the more tractable smaller space $\mathcal{S}^m$.

Moreover, we demand that q be, in some sense, a good approximation to p. It is impossible to perform any computations with p, so we'll have to make do with the approximations afforded by q. Amari's solution is to "α–project p" to a point q that lives in a simpler sub–manifold $\mathcal{S}^m$. How this all plays out is the subject of this, and subsequent Chapters.

Up to now, we have labeled the Riemannian manifolds possessing the metric tensors as the structures $(\mathcal{S}, g)$. However, it is quite disconcerting when one arrives at the realization that such an object is really so very abstract that the points in the manifold are *disconnected*!

There is, at this juncture, no *connection* between points in $\mathcal{S}$. A new tangent space with its metric tensor has to be constructed at each and every point. As a consequence, Information Geometry has to augment the current structure, $(\mathcal{S}, g)$, so that one can connect the points and eventually measure distances between them.

Amari discusses his α-connections with the following notation which we shall appropriate and use without any explanation at the present time. As I said before, sometimes you just have to mouth the words before you understand them. So the geometric structure is now labeled as $(\mathcal{S}, g, \nabla^\alpha, \nabla^{-\alpha})$.

For our present purposes, we will restrict ourselves to the two α values[1] of $\alpha = 1$, and $\alpha = -1$. The ∇^α are the mathematical constructs which do permit us to connect the tangent spaces and the points in the manifold. They are the α connections. When $\alpha = 1$, Amari calls $\nabla^{(1)}$ the **e**–connection, and when $\alpha = -1$, he calls $\nabla^{(-1)}$ the **m**–connection.

---

[1] The use of α here has nothing to do with the α parameters of the Dirichlet distribution.

Thankfully, we can now begin to tie in these new concepts with a familiar one, that is, the relative entropy, or Kullback–Leibler information measure. When $\alpha = 1$, we write,

$$D^{(1)}(p \parallel q) \equiv KL(q,p) = \sum_{i=1}^{n} q_i \ln\left(\frac{q_i}{p_i}\right) \tag{39.1}$$

Amari calls this an **e**–projection.

Similarly, for $\alpha = -1$, we write,

$$D^{(-1)}(p \parallel q) \equiv KL(p,q) = \sum_{i=1}^{n} p_i \ln\left(\frac{p_i}{q_i}\right) \tag{39.2}$$

which is what Amari calls the **m**–projection. The general term, not unexpectedly, is called an α–projection. The general α–divergence where $\alpha \neq \pm 1$ has a different appearance than Equations (39.1) and (39.2).

A nice symmetry exists between these two projections when we switch from $\alpha = 1$ to $\alpha = -1$ and switch the positions of q and p,

$$D^{(1)}(p \parallel q) \equiv D^{(-1)}(q \parallel p) = \sum_{i=1}^{n} q_i \ln\left(\frac{q_i}{p_i}\right) \tag{39.3}$$

$$D^{(-1)}(p \parallel q) \equiv D^{(1)}(q \parallel p) = \sum_{i=1}^{n} p_i \ln\left(\frac{p_i}{q_i}\right) \tag{39.4}$$

Thus, the **e**–projection of p to q is the same as the **m**–projection of q to p. Likewise, the **m**–projection of p to q is the same as the **e**–projection of q to p. The Kullback–Leibler relative entropy between q and p is the same as the **e**–projection of p to q. The Kullback–Leibler relative entropy between p and q is the same as the **m**–projection of p to q.

The bottom line is that point p in $\mathcal{S}^n$ can be α–projected to points q, r, s, and so on, in $\mathcal{S}^m$. This is accomplished by finding the minimum distance, or minimum $D^{(\alpha)}$, from p to $\mathcal{S}^m$ when different points in $\mathcal{S}^m$ are chosen.

39.3 An Inferential Example about Life on Mars

We have used dice and kangaroos as motivating examples so far. Let's turn to a made–up scenario involving life on Mars. Suppose that interest centers on some class of models hypothesizing a link between life on Mars and the presence of water. The presence of water happens to be signaled by the presence of two chemical compounds.

The initial vague thoughts behind these models is as follows. The presence of life would imply liquid water. In turn, liquid water would imply the presence of two chemical compounds called Compound C and Compound D. The essential nature of this model is captured with a logic expression such as,

$$\mathcal{M}_k \equiv (L \to W) \wedge ((W \to C) \wedge (W \to D)) \tag{39.5}$$

Suppose that a debate has arisen as to whether it is worthwhile to conduct an experiment with a Mars Rover. An experiment can be performed by the Rover to detect the presence of Compounds C or D in a soil sample. However, it is impossible in this experiment to conduct any measurements for water. What if the experiment finds either Compound C or Compound D, or both, present in the soil sample?

There is no direct deductive chain that would lead to a definitive statement about life on Mars from the results of the experiment. There is, however, the generalization afforded by probability theory that gets us around an invalid use of *modus ponens*. Thus, we appeal to an inferential chain of reasoning to provide not deductive certainty, but rather a quantitative updating to a degree of belief.

The presence of Compounds C and D strengthens the proposition that liquid water exists on Mars. Furthermore, the presence of liquid water strengthens the proposition that life exists on Mars. Would such an inference lead to a significant increase in our certainty about life on Mars? Is there "circumstantial evidence" for life on Mars? Is there some merit in conducting that experiment?

Label the proposition of life on Mars as L, the proposition that liquid water exists on Mars as W, the proposition that compound C is present as C, and the proposition that Compound D is present as D. In other words, the IP asking for the updated state of knowledge about L given C, D, and the class of models capturing the implications discussed above. Would $P(L\,|\,C,D,\mathcal{M}_k)$ increase significantly from the initial base assessment of $P(L\,|\,\mathcal{M}_k)$?

The IP will propose to help settle the debate with a quantitative calculation using Bayes's Theorem.

$$P(L\,|\,C,D,\mathcal{M}_k) = \frac{P(LCD\,|\,\mathcal{M}_k)}{P(CD\,|\,\mathcal{M}_k)} \tag{39.6}$$

Expand the numerator and denominator by the **Sum Rule** to yield,

$P(L\,|\,C,D,\mathcal{M}_k) =$

$$\frac{P(LWCD\,|\,\mathcal{M}_k) + P(L\overline{W}CD\,|\,\mathcal{M}_k)}{P(LWCD\,|\,\mathcal{M}_k) + P(L\overline{W}CD\,|\,\mathcal{M}_k) + P(\overline{L}WCD\,|\,\mathcal{M}_k) + P(\overline{L}\,\overline{W}CD\,|\,\mathcal{M}_k)} \tag{39.7}$$

Assume that each of the four propositions is a binary proposition so that a joint probability table with $n = 16$ cells must be filled in with numerical assignments before Bayes's Theorem can be used for a quantitative solution. The MEP algorithm will be used to make these numerical assignments with specification of the model $\mathcal{M}_k$ synonymous with finding the θ^i and η_i coordinates. This is Amari's notation using the index i, but please don't confuse it with the MEP index i for the ith statement in the state space.

LIFE ON MARS

Just as when the IP reasoned about the kangaroos, the implications in the logic expression of Equation (39.5) may serve as an inspiration for the numerical values that should appear in the 16 cell joint probability table. The logic expression would demand that 0s appear in those cells which represent the DNF terms where a F appears. However, the IP does desire a certain amount of flexibility in its choice of models that frees it from the straight–jacket of strict adherence to the logic expression.

For example, the IP would not want to completely rule out a joint statement like $L\overline{W}\,\overline{C}\,\overline{D}$ where a 0 would be placed under the model for the logic expression, claiming that it is impossible for life to exist if there is no water and no chemical compounds. But if the logic expression is to serve as an inspiration, this cell *should* contain a small numerical assignment for the probability of this joint statement.

The development of such models is standard procedure by now. We leave all of the numerical details of the joint probability table, the coordinates, the constraint functions, and a discussion of the reasonableness of the assignments in light of the logic expression to the Exercises. See Exercises 39.9.11 through 39.9.13 for all of the details about the complicated model behind point p.

39.4 α–Project the Complicated Model

Amari was, of course, thinking about much more complicated problems than our little Mars scenario. Nonetheless, it is still a good surrogate for an initial exposure to α–projection. Like any simulation, it has the advantage that we are able to calculate all the answers for both the complicated and simple models in order to compare them.

Let's take as our complicated model the point p living in $\mathcal{S}^{16}$. One resolution for the MEP algorithm working on p is to choose as the constraint functions the main effects, together with all the possible interactions for the four propositions L, W, C, and D. This leads to a full decomposition of $\mathcal{S}^{16}$ because there are four main effects together with six double interactions, four triple interactions, and one quadruple interaction. The implicit universal constraint provides the final constraint function.

The complicated model giving rise to the numerical assignments for p means that p lives in the Riemannian space where $m = n - 1 = 15$. Fifteen constraint functions and fifteen constraint averages would have to be specified as the information to be inserted into the numerical assignment for p.

These interaction constraint functions are interesting in and of themselves. They capture all the possible degrees of association between the propositions. Thus, for example, if a model wants to include some speculation about an association between life on Mars, water, and Compound D, it includes the triple interaction LWD as part of the model. So p is indeed quite a complicated model because it includes every conceivable mode of association through the main effects, all of the double interactions, all of the triple interactions, and the one quadruple interaction.

Let the various points q, r, s, and so on, representing the less complicated models live, on the other hand, in the lower dimensional sub–manifold $\mathcal{S}^m$ where we'll use $m = 4$ and $m = 6$ as examples. With $m = 4$, the constraint functions and constraint averages represent the information from the main effects only. No interactions of any kind live in this simple space. By definition, the four coordinates η_1 to η_4 can be varied to produce different marginal probabilities for L, W, C, and D, and, as a consequence, to locate different points in $\mathcal{S}^4$.

When this sub–manifold $\mathcal{S}^4$ shows itself to be too impoverished to provide any reasonable approximation to the full complicated model, the IP will move up to $\mathcal{S}^6$ by adding two important interactions. Even though these points residing in this sub–manifold are much simpler than p, they might be an attractive tractable alternative to p.

We have to α–project the point p from $\mathcal{S}^{16}$ into $\mathcal{S}^4$ or $\mathcal{S}^6$ in order to find out which of the points q, r, and so on, has the minimum $D^{(\alpha)}(p \parallel q)$. We say that we **e**–project if we use $\alpha = 1$ and **m**–project if we use $\alpha = -1$. These two projections will produce two distinct points in $\mathcal{S}^m$.

39.4.1 m–projection of p to $\mathcal{S}^4$

First, look at the **m**–projection of p to some point q in the space of models that use just the four main effects. The $\alpha = -1$ relative entropy formula is,

$$D^{(-1)}(p \parallel q) = \sum_{i=1}^{16} p_i \ln\left(\frac{p_i}{q_i}\right)$$

This is the same as the **e**–projection of q back to point p,

$$D^{(1)}(q \parallel p) = \sum_{i=1}^{16} p_i \ln\left(\frac{p_i}{q_i}\right)$$

We want to find the minimum value of this distance–like measure as various points q, r, s, ... are considered in $\mathcal{S}^4$. At this juncture, let's employ a crude, brute force approach to finding the minimum. Choose various points in $\mathcal{S}^4$, calculate the relative entropy between each of these selected points with point p, and locate where the minimum relative entropy occurs.

Refer to Table 39.1 on the next page. This table provides an outline of some computations aiming to directly locate the **m**–projection of point p in $\mathcal{S}^4$. Recall that point q is a simpler model with only four model parameters λ_1 through λ_4, or equivalently $\langle F_1 \rangle$ through $\langle F_4 \rangle$.

Table 39.1: *Locating the minimum value of a divergence function in order to find the α–projection of a point p representing a complicated model to a point q representing a simpler model.*

$\mathcal{M}_k$	$\lambda_1, \lambda_2, \lambda_3, \lambda_4$	$D^{(-1)}(p \parallel q)$	$\langle F_1 \rangle, \langle F_2 \rangle, \langle F_3 \rangle, \langle F_4 \rangle$
1	-1.00 $+1.00$ -1.00 $+1.00$	0.758911 *arbitrary initial* *starting point for q*	0.2689 0.7311 0.2689 0.7311
2	$+0.00$ $+0.00$ $+0.00$ $+0.00$	0.448453 *divergence begins* *to decrease*	0.5000 0.5000 0.5000 0.5000
3	-1.00 $+0.00$ -1.00 $+0.00$	0.348682 *divergence continues* *to decrease*	0.2689 0.5000 0.2689 0.5000
4	-1.00 -1.00 -0.50 $+0.00$	0.259612 *divergence continues* *to decrease*	0.2689 0.2689 0.3775 0.5000
5*	-1.386294 -0.575364 -0.160343 -0.120144	0.210975 *minimum occurs* *at these coordinates*	$(0.20, 0.36, 0.46, 0.47)$ *constraint function averages* *are the same as point p*
6	-1.38 -0.57 -0.16 -0.12	0.210981 *close by coordinates* *near minimum*	0.2010 0.3612 0.4601 0.4700
7	-1.39 -0.58 -0.15 -0.13	0.211004 *close by coordinates* *near minimum*	0.1994 0.3589 0.4626 0.4675

The first column lists seven models with their four coordinates in the second column. These are the various points q, r, s, and so on, in $\mathcal{S}^4$ which are being compared to point p in $\mathcal{S}^{16}$. The third column is the relative entropy measure between the selected points q, r, s, and so on, with point p. We are trying to find the coordinates of the model in the sub–manifold which produce the minimum value of the relative entropy. The coordinates for model $\mathcal{M}_5$ result in the minimum relative entropy.

Therefore, we will take this point as the point q which is the **m**–projection of point p. The fourth and final column lists the corresponding constraint function averages for the first four constraint functions, that is, for the four main effects of L, W, C, and D. At point q, where the minimum relative entropy was located, it will eventually be made clear that the first four constraint function averages $\langle F_1 \rangle_q$ through $\langle F_4 \rangle_q$ match these same dual coordinates at point p.

However, if we were to examine all of the numerical assignments for point q, we would discover that they are not very similar to the assignments for p even though the first four constraint function averages are the same. For example, look at the assignment to the first cell of the joint probability table under model $\mathcal{M}_5$ and point q,

$$Q_1 \equiv P(X = x_1 \,|\, \mathcal{M}_5) \equiv P(LWCD \,|\, q \in \mathcal{S}^4) = 0.016$$

This assignment is not particularly close to the assignment in cell 1 for point p,

$$P(LWCD \,|\, p \in \mathcal{S}^{16}) = 0.10$$

See Exercise 39.9.11 and Figure 39.3 for the joint probability table of p.

Even though we have done the very best we could by finding the point q as the minimum **m**–projection of p, the lack of any interactions in $\mathcal{S}^4$ dooms this initial effort to failure. Nonetheless, we might fix things up by considering slightly more complicated models that do include some relevant interactions.

39.4.2 How do we know we have found the minimum?

We are attempting to find the minimum relative entropy, the shortest distance in the manifold from a fixed point p to another point q residing somewhere in the sub–manifold. All of the points in this sub–manifold are described by information from four constraint function averages. We accomplish this objective by minimizing,

$$\min\{D^{(-1)}(p \,\|\, q)\} = \min_{\{\lambda_j\}} \left\{ \sum_{i=1}^{16} p_i \ln\left(\frac{p_i}{q_i}\right) \right\} \tag{39.8}$$

As usual, the Kullback–Leibler relative entropy can be reworked into,

$$\sum_{i=1}^{16} p_i \ln\left(\frac{p_i}{q_i}\right) = \sum_{i=1}^{16} p_i \ln p_i - \sum_{i=1}^{16} p_i \ln q_i \tag{39.9}$$

LIFE ON MARS 233

Since we are trying to *minimize* the left hand side of the above equation, and $\sum p_i \ln p_i$ is a fixed negative value for the fixed point p, it suffices to *maximize* the second term on the right hand side,

$$\sum_{i=1}^{16} p_i \ln q_i = \sum_{i=1}^{16} p_i \left(\sum_{j=1}^{4} \lambda_j^q F_j(X = x_i) - \ln Z_q \right)$$

$$= \sum_{j=1}^{4} \lambda_j^q \langle F_j \rangle_p - \ln Z_q$$

The expectation is taken with respect to the probability distribution p; thus, the constraint function averages $\langle F_j \rangle_p$ are those specified by the model for p. These four constraint function averages are,

$$\langle F_j \rangle_p = (0.20, 0.36.0.46, 0.47)$$

But if we are maximizing,

$$\sum_{j=1}^{4} \lambda_j^q \langle F_j \rangle_p - \ln Z_q$$

this is the same as minimizing,

$$\ln Z_q - \sum_{j=1}^{4} \lambda_j^q \langle F_j \rangle_p$$

by varying the λ_j^q with fixed $\langle F_j \rangle_p$. Therefore, we are back to the standard MEP algorithm for finding the numerical assignments to q when based on the Legendre transformation. The λ_j^q coordinates shown in Table 39.1 for model $\mathcal{M}_5$ where the minimum $D^{(-1)}(p \parallel q)$ occurred were found in exactly this way.

39.4.3 m–projection of p to $\mathcal{S}^6$

The model behind p, by its very nature, does depend critically on a number of increasingly complicated interactions. These interactions mimic the effect of the logical implications, albeit not in a very transparent fashion. We wouldn't be calling p complicated if it were otherwise.

We would like to begin a search for models that are slightly more involved than the ones we have looked at in $\mathcal{S}^4$. Those simple models excluded all of the interactions. However, we might reasonably hope that models incorporating some interactions, but which still are not as complicated as the ones behind p, might be more successful in finding numerical assignments closer to what we want.

Following this line of thought, suppose the IP considers models that include the interactions LW and LCD in addition to the four main effects. The IP reasons that including this double and triple interaction should capture something of the original logical implication. Liquid water and life are associated in the LW interaction, while life and the two chemical compounds are associated in the LCD interaction.

Carry out the same procedure as before. That is, **m**–project point p into $\mathcal{S}^6$ to find that new unique point q that minimizes the distance–like measure from p. Since now $m = 6$, we will be minimizing the expression,

$$\min_{\lambda_j^q} \left\{ \ln Z_q - \sum_{j=1}^{6} \lambda_j^q \langle F_j \rangle_p \right\}$$

by varying all six λ_j^q while holding all six constraint function averages $\langle F_j \rangle_p$ fixed.

In addition to the first four constraint function averages with respect to p, we also require $\langle F_5 \rangle_p$ as the information for the LW double interaction and $\langle F_{13} \rangle_p$ for the LCD triple interaction.

$$\langle F_5 \rangle_p = 0.16$$

$$\langle F_{13} \rangle_p = 0.11$$

The MEP algorithm finds the minimum distance $D^{(-1)}(p \parallel q)$ at the value of 0.04476361, where the model parameters for point q in $\mathcal{S}^6$ are,

$$\lambda_1^q = -3.32158$$

$$\lambda_2^q = -1.09861$$

$$\lambda_3^q = -0.38271$$

$$\lambda_4^q = -0.33723$$

$$\lambda_5^q = +2.48491$$

$$\lambda_{13}^q = +1.79430$$

Table 39.2 on the next page shows the model for $q_{\min}$ and a couple of close by models q_* and q_{**} whose λ_j^q coordinates are just slightly different than those given above for $q_{\min}$. The distance–like measure for these models from p must be slightly greater.

The proof of the pudding is in the new numerical assignments to the sixteen propositions in the state space. Of course, they cannot be exactly the same as p, but they are much more to our liking. For example, whereas previously,

$$P(LWCD \,|\, q \in \mathcal{S}^4) = 0.016$$

as compared to,

$$P(LWCD \,|\, p) = 0.10$$

now,
$$P(LWCD\,|\,q \in \mathcal{S}^6) = 0.088$$

Table 39.2: *The minimum value of the divergence function when point p representing a complicated model is* **m**-*projected to a point q representing a simpler model.*

$\mathcal{M}_k$	$D^{(-1)}(p \parallel q)$	$P(LWCD \mid q \in \mathcal{S}^6)$
$q_{\min}$	0.04476361	0.0880004
q_*	0.04477635	0.0883621
q_{**}	0.04479207	0.0885553

If you think about it, point q must also reside in the higher dimensional manifold the same as p. Table 39.3 lists all fifteen dual coordinates for point q in terms of Amari's θ^i and η_i.

Table 39.3: *All fifteen dual coordinates for the point q* **m**-*projected from point p to a sub-manifold* $\mathcal{S}^6$. *Amari's index i is used instead of the MEP equivalent j.*

i	Effect	θ^i	η_i
1	L	-3.32158	0.20
2	W	-1.09861	0.36
3	C	-0.38271	0.46
4	D	-0.33723	0.47
5	LW	$+2.48491$	0.16
6	LC	$+0.00000$	0.135621
7	LD	$+0.00000$	0.136813
8	WC	$+0.00000$	0.189591
9	WD	$+0.00000$	0.192747
10	CD	$+0.00000$	0.245099
11	LWC	$+0.00000$	0.108947
12	LWD	$+0.00000$	0.109450
13	LCD	$+1.79430$	0.11
14	WCD	$+0.00000$	0.121775
15	$LWCD$	$+0.00000$	0.088000

The θ^i coordinates are once again rather opaque values. They must, however, take on a value of 0 where an interaction does not exist in a model. Six of the η_i coordinates are what we would surmise based on the examples studied so far. The η_1, η_2, η_3, η_4, η_5, and η_{13} coordinates exactly match the η_i for point p. The four marginal probabilities of L, W, C, and D are reproduced as before, but now we also duplicate the marginal probability of 0.16 for the LW interaction as well as the marginal probability of 0.11 for the LCD interaction.

39.5 Pythagorean Relationships

One notion in Information Geometry that truly lives up to its classical name is the Pythagorean relationship. Expressed in terms of our problem here, consider two other points r and s that also live in $\mathcal{S}^6$, the same sub-manifold as point q which is the **m**-projection of point p living in the higher dimensional manifold S^{16}.

We will numerically examine various triangular relationships from the standpoint of Information Geometry. In particular, we will rely on a formula given by Amari to measure the departure from being a right triangle. For example, triangles $\triangle pqs$ and $\triangle pqr$ should be right triangles, while triangle $\triangle psr$ should not. Figure 39.1 presents a rough sketch of these four points and the triangles formed.

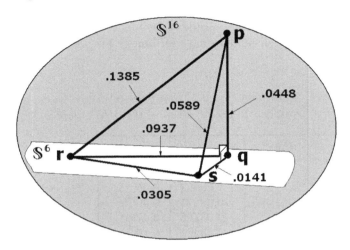

Figure 39.1: *A sketch illustrating the Pythagorean relationships for four points in a Riemannian manifold. This sketch is not to scale.*

Here is a numerical example exploring the triangular relationships. The relative entropy from p to q when q was the **m**-projection of p onto $\mathcal{S}^6$ has already been calculated as,

$$D^{(-1)}(p \parallel q) = 0.0447636$$

Therefore, the squared distances from p to r and p to s must be larger than this value because they must be larger for any point other than q by definition. Remember that we found q by minimizing the distance from p to $\mathcal{S}^6$. The squared distance from p to r is calculated as,

$$D^{(-1)}(p \parallel r) = 0.13848062$$

while the squared distance from p to s is,

$$D^{(-1)}(p \parallel s) = 0.05888912$$

verifying the larger squared distances compared to $D^{(-1)}(p \parallel q)$.

But we are most interested in triangle $\triangle psr$ to confirm whether the formula that measures the departure from being a right triangle,

$$D^{(-1)}(p \parallel s) + D^{(-1)}(s \parallel r) - D^{(-1)}(p \parallel r) = \sum_{i=1}^{16}(p_i - s_i)\ln\left(\frac{r_i}{s_i}\right) \quad (39.10)$$

is upheld. This formula, Amari's Equation (3.61) [2, pg. 67], will be explored more fully in the next Chapter.

The numerical assignments for these four points p, q, s, and r forming the various triangles as shown in the sketch above need to be found. Two assignments, that for p and q, have already been found in the discussion of the previous sections.

Points s and r are in the same sub–manifold as point q, so any of the coordinates λ_1, λ_2, λ_3, λ_4, λ_5, and λ_{13} might be varied to produce other models in $\mathcal{S}^6$. Let point s lie on the coordinate curve with point q so that, say, only λ_5 is changed from q's value of 2.41869 to another point s where $\lambda_5 = 2$ and the other five coordinates remain the same. The MEP algorithm finds the new numerical assignments s_i for point s under this changed model.

In like manner, suppose that r changes both λ_5 and λ_{13}. Let $\lambda_5 = 1$ and $\lambda_{13} = 2$ with the MEP algorithm once again finding the numerical assignments r_i under this new model. When Equation (39.10) is used to find if there is any discrepancy from a right triangle, both $\triangle pqr$ and $\triangle pqs$ come out with a value of 0. They are both right triangles.

$$D^{(-1)}(p \parallel q) + D^{(-1)}(q \parallel r) - D^{(-1)}(p \parallel r) = \sum_{i=1}^{16}(p_i - q_i)\ln\left(\frac{r_i}{q_i}\right)$$
$$= 0$$

$$D^{(-1)}(p \parallel q) + D^{(-1)}(q \parallel s) - D^{(-1)}(p \parallel s) = \sum_{i=1}^{16}(p_i - q_i)\ln\left(\frac{s_i}{q_i}\right)$$
$$= 0$$

However, $\triangle psr$ is not a right triangle, as the calculation below indicates. The squared length $D^{(-1)}(s \parallel r)$ is 0.0305.

$$D^{(-1)}(p \parallel s) + D^{(-1)}(s \parallel r) - D^{(-1)}(p \parallel r) = \sum_{i=1}^{16}(p_i - s_i)\ln\left(\frac{r_i}{s_i}\right)$$
$$= -0.0490962$$
$$0.0589 + D^{(-1)}(s \parallel r) - 0.1385 = -0.0491$$
$$D^{(-1)}(s \parallel r) = 0.0305$$

39.6 Updating the IP's State of Knowledge

The IP will use Bayes's Theorem as presented back in Equation (39.7) to calculate its updated state of knowledge about life on Mars after it has received positive confirmation of the two chemical compounds C and D in the Martian soil from the Rover experiment. However, the numerical assignments in Bayes's Theorem can come from any model. We have examined three models in particular for the purposes of this example.

First, look at the answer we would like to compute if only we could solve for the complicated model behind point p. (Remember that the simpler model q is a surrogate for some model p that is beyond our current computational capabilities. As a fanciful example, suppose the correct theoretical model is known from String Theory, but the resulting numerical computation is intractable.)

Refer to Exercise 39.9.11 for the numerical assignments under p. For the sake of the example, if we unrealistically assume that the numerical assignments can be calculated from the model for p, then,

$$P(L\,|\,C,D,p\in\mathcal{S}^{16}) = \frac{0.10 + 0.01}{0.10 + 0.01 + 0.06 + 0.08} = 0.44$$

For the purposes of the simulation, we know that the correct updating of the IP's degree of belief in life on Mars goes up from 20% to 44% given the encouraging results from the experiment.

Suppose the IP first tries to find some tractable probability distribution by α-projecting point p to a point q in the smaller sub–space where no interactions exist. Then,

$$P(L\,|\,C,D,q\in\mathcal{S}^{4}) = \frac{0.0156 + 0.0277}{0.0156 + 0.0277 + 0.0623 + 0.1107} = 0.2002$$

The IP's degree of belief in life on Mars hardly budges at all from the marginal probability for life before the experiment was conducted.

But the IP rejects this projection as too simplistic because it is strongly believed from other scientific knowledge that the interactions are important. That was why the IP set up a third model, still tractable but slightly more complex than the one above, by adding one double and one triple interaction, namely, the LW interaction and the LCD interaction. Now,

$$P(L\,|\,C,D,q\in\mathcal{S}^{6}) = \frac{0.0880 + 0.0220}{0.0880 + 0.0220 + 0.0338 + 0.1013} = 0.4488$$

The IP's degree of belief in life on Mars based on this more tractable model is almost the same as the original complicated model. Of course, the IP doesn't know how good this revised model is. It remains a judgment call on the IP's part whether to investigate further models, or to take this updated state of knowledge as the new baseline for further experiments.

39.7 Inference and Logic

It is important to understand that this story about life on Mars has been treated from the standpoint of probability theory as an extension of logic. That is why we spent so much time discussing the implicational logic expression in Equation (39.5). It is interesting to see how it might have arisen from vague musings about the problem, and how it inspires some actual values appearing in a joint probability table.

Contrast this approach with the more familiar data analytical approach. In this Chapter, there was no mention of mass quantities of data, parameter estimation, likelihood equations, hypothesis testing, and the like. There was one known fact; the experiment detected the presence of Compounds C and D. From this observation, the IP updated its state of knowledge about the existence of life on Mars.

The IP could not *deduce* anything in this story. Life was not established as a fact, therefore water could not be deduced. Neither was water established as a fact, therefore Compounds C and D could not be deduced. Only some of the consequences from the presence of water were established as facts, namely the existence of the two compounds. Deductively, we were stymied at the very start.

It is invalid in logic to reason backwards from consequences to the premises, or from the effects to the causes. Reasoning by inference, however, does permit the IP to reason backwards from effects to causes at the cost of assuming that certain information was inserted into a model guiding that inference. The title of Laplace's 1774 paper which started off this whole enterprise of "inverse probability" was provocatively titled,

The probability of causes when the events are known

Of course, not relying on vast amounts of data meant that somehow or other the numerical assignment to q was judged to be driven by a very good model. If you know that a die is fairly constructed from its physical characteristics, do you need to toss it 20,000 times? Possessing a model the IP believes in to the exclusion of all others permits it to make strong inferences from just a single observation!

No *data* entered into the MEP algorithm used to assign numerical values to any of the probability distributions p, q, r, or s. On the other hand, *information* was inserted into a probability distribution by the IP in the form of constraint functions and their averages.

Information Geometry provides a deeper rationale for why the MEP algorithm is a good algorithm. The constraint functions and averages are explained in terms of the dual coordinate system. And we can think of a manifold with a dual coordinate system abstractly from the standpoint of differential calculus to study how changes in one coordinate system affect its dual coordinate system.

The entropy of a probability distribution is seen to be associated with a distance-like measure between points in a manifold. Maximizing or minimizing entropy is then seen to be analogous to finding the shortest squared distance between points.

Many critics of MEP find no comfortable intuitive justification for the procedure of entropy maximization. They want to know why is it so important to maximize this rather subjective and admittedly epistemological entity?

Maybe finding the shortest distance between two points is a more motivating notion for these critics. An example from the earliest days of modern physics posed the question of finding the shortest path a bead would take when sliding along a wire and subjected to gravitational forces. So variational calculus has proved to be a very powerful device. Entropy maximization is simply another example.

Another key insight that Information Geometry provides us is an antidote to the ever popular description of entropy as some form of disorder or chaos. Perhaps it would be better to visualize entropy as some distance–like measure that wants to be minimized or maximized in order to achieve the goal of maximizing missing information rather than some silly metaphor like messy bedrooms or rusting cars.

Entropy maximization, when all is said and done, is really *secondary* to the primary goal of satisfying all the constraints provided as the information. Satisfying the constraints dictates that the tractable points live on the same sub–manifold.

But just because they live in the same space does not mean that they are all equally good candidates. We simply want that particular point in this sub–manifold that is the closest to the complicated model. Entropy maximization is that insurance policy we take out to eliminate any risk that we might in fact be including some information that we didn't intend to include!

39.8 Connections to the Literature

Let me begin by reiterating my firmly held conviction that Amari does the best job of anyone that I have read at directly translating the mathematical abstractions inherent in Differential Geometry over to the more concrete probabilistic and inferential concerns of Information Geometry. We have relied heavily upon his extensive mathematical insights and will continue to do so.

However, we must part ways over many of his conceptual interpretations. A glaring example of conceptual discordance occurs right where you would hope that all of the mathematical groundwork will begin to pay off in a deeper understanding of inference.

The beginning section of Chapter 4 of Amari and Nagaoka's book [2],

Methods of Information Geometry

entitled "Statistical inference and differential geometry" has been selected for my critical comments. Just as I did in Chapter Thirty Two, Volume II, with Jaynes's derivation of Laplace's *Rule of Succession*, I will provide a detailed, sentence by sentence, critical annotation of Amari's conceptual view of probability and inference.

LIFE ON MARS

These bedrock ideas, which may be adequately expressed verbally for the most part, and require a minimum of mathematical symbolism, must hang together in a coherent fashion. No amount of mathematical finessing can paper over jarring discordances. Perhaps there are some readers who will remember what I said in the *Apologia* to Volume I.

My disagreement with the "Establishment" does not center on whether some piece of mathematics is correct or not. I always assume that the mathematics has been done correctly. What does demand a critical deconstruction is whether a consistent and coherent argument has been laid down concerning the fundamental core concepts, definitions, and motivating ideas behind inferencing.

Basically, what becomes very clear after laboring over Amari, Murray and Rice, Kass and Vos, and many others, is that, in trying to integrate the contribution from Information Geometry to inference, they simply did not ever fully accept the Bayesian viewpoint.

By a Bayesian viewpoint, I mean that an IP should conceptually divide an inference into two main stages: 1) the formal manipulation rules of probability theory, and 2) the question of how to assign numerical values to probabilities. In the end, although much lip service is given over to mentioning the Bayesian perspective, they were all ultimately wedded to an orthodox interpretation as promulgated by Fisher and his followers, and to all of the subsequent mutilations post Fisher.

The following quote from Amari and Nagaoka [2, pg. 81] appears verbatim and in small type face. My critical comments are interspersed within the flow of their text at appropriate points and appear in **boldface type**.

Chapter 4
Statistical inference and
differential geometry

Suppose that we are given data generated according to some unknown probability distribution.

Once again, data are NOT generated by probability distributions. This is a flagrant example of Jaynes's *Mind Projection Fallacy*.

Probability distributions reflect a state of knowledge as held by an Information Processor about joint statements in some state space. The data, that is, any number of actual occurrences of those joint statements, are generated by some physical causal process in the real world, and not by a state of knowledge, or degree of belief, as it exists in the mind of a conscious entity.

Of course, I may have misunderstood and the subject matter of discourse is not inference, but rather psychokinesis!

Statistical inference is the process of extracting information concerning the underlying probability distribution from the data.

Absolutely not! Inference is a generalization of deduction that permits some quantitative conclusion to be reached about propositions even when logic would rule such an attempt "invalid."

These models under consideration by the IP may not allow for any logical conclusions. However, they do allow the IP to change its quantitative degrees of belief. This generalizes deduction. There is no "underlying" probability distribution, just as there is no "true" probability distribution. There are only probability distributions that are consistent with the information inserted under some model.

There is no "process of extracting information concerning the underlying probability distribution from the data." Information is inserted by each and every model before, and prior to, the observation of any data. The role of the data is to change an IP's state of knowledge about all of the models that it had established prior to any data. In other words, the role of the data is to make that revision from $P(\mathcal{M}_k) \to P(\mathcal{M}_k \mid \mathcal{D})$.

If we have prior knowledge concerning the underlying mechanism generating the data, in other words, if we know the shape of the unknown distribution, then the possible candidates may be constrained to a parameterized family of distributions. We call such a family a **statistical model**.

Slippage into the *Mind Projection Fallacy* once again. Prior knowledge concerning some underlying mechanism causing the data would be reflected by differing probabilities over model space, $P(\mathcal{M}_k \mid \mathcal{I})$, as opposed to, say, a uniform distribution over model space when no such prior knowledge exists.

In addition, all of our numerical assignments, in fact, MUST BE a parameterized family of distributions because they are all conditioned on some model, $Q_i \equiv P(X = x_i \mid \mathcal{M}_k)$. Unfortunately, the above verbiage is typical of the rampant confusion over which space we are talking about.

A statistical model is just one $\mathcal{M}_k$, a *statement* asserting that some numerical assignment to the probabilities in the state space is TRUE. It happens to be synonymous with the specific information that has been inserted into the probability distribution by either of the dual parameters. The entire family consists of an infinite number of such models.

4.1 Estimation based on independent observations

Consider a family of probability distributions $S = \{p(x; \xi)\}$ parameterized by $\xi = [\xi^i]$ for $i = 1, \cdots, n$. Under appropriate regularity conditions, S may be viewed as an n–dimensional manifold for which ξ is a (local) coordinate system.

This is perfectly fine, and we have accepted to use Amari's different notation. We prefer $Q_i \equiv P(X = x_i \mid \mathcal{M}_k)$ to $p(x; \xi)$, and writing

λ_j and $\langle F_j \rangle$ from the MEP formula as the dual coordinates, or parameters of the model, as opposed to $\xi = [\xi^i]$. Amari is constantly switching notation between $\xi = [\xi^i]$ and θ^i and η_i to designate the parameters.

Now let $x_1, \cdots, x_N$ be independent observations of the random variable x distributed according to $p(x; \xi)$. Letting $x^N = (x_1, \cdots, x_N)$, the task of statistical inference is to infer the probability distribution $p(x; \xi)$ given the N data points x^N.

I maintain that nobody has ever observed a "random variable x." What is observed is that one of the joint statements from the state space has happened. When N of these events have taken place there are N data points.

The task of statistical inference is NOT to infer the probability distribution $p(x; \xi)$ given the N data points x^N. One of the tasks of statistical inference is to calculate $P(\mathcal{M}_k \,|\, \mathcal{D})$, the reorientation of the degree of belief in the truth of a particular model when conditioned on all of the data points.

$p(x; \xi) \equiv P(X = x_i \,|\, \mathcal{M}_k)$ is not inferred at all, and certainly, as the notation clearly indicates, has nothing whatsoever to do with any data $\mathcal{D}$. $p(x; \xi) \equiv P(X = x_i \,|\, \mathcal{M}_k)$ is the repository of some partial information emanating from some model that dictates the state of knowledge about any statement in the state space.

For example, **estimation** is a kind of inference task, where the goal is to find an estimate $\hat{\xi}$ of ξ; alternatively, one might consider the task of **testing** where the goal is to decide if the hypothesis $H_0: \xi = \xi_0$ is accepted against the alternative hypothesis $H_1: \xi \neq \xi_0$, or is rejected.

Here we experience first hand Amari's unequivocal commitment to the orthodox statistical view. The correct Bayesian perspective banishes words like *estimation* and *testing*, along with many others, from the inferential vocabulary.

Parameters of models are never estimated. Parameters are either clearly specified together with the model, or they never appear at all if the model has never been introduced. The Bayesian approach does not concern itself with any kind of estimators!

The classical orthodox notion of hypothesis testing has also been superseded by the modern Bayesian approach. Basically, the all or none idea of accepting or rejecting an hypothesis has been replaced by reorienting model space conditioned on the data. Even this task is subsidiary to the main task of predicting a future observation where the future observation in question must be averaged over *all* of the updated models.

Here is a quote from Bernardo & Smith's book *Bayesian Theory* [6, pg. 424] that illustrates the difference in approach between committed Bayesians and the old school.

A final general comment. Frequentist procedures centre their attention on producing inference statements about *unobservable* parameters. As we shall see ... such an approach typically fails to produce a sensible solution to the more fundamental problem of *predicting future observations*. [emphasis in the original.]

We now return after this brief interruption to quoting from Amari.

We have already seen how a Riemannian metric based on the Fisher information matrix and one–parameter family of affine connections called the α–connections may be introduced on a manifold S representing a statistical model. Since the probability distribution governing x^N can be written using the distribution of a single data point as

$$p_N(x^N;\xi) = \prod_{t=1}^{N} p(x_t;\xi)$$

we also have

$$\log p_N(x^N;\xi) = \sum_{t=1}^{N} \log p(x_t;\xi)$$

It is a curious fact that the probability distribution for the data always seems to be prefaced by some remark about "independent identically distributed random variables" as if this phrase were a major requirement that the individual data points must satisfy if inference is to proceed in a correct fashion. The data ARE NOT aware of anything that is forcing them to be either "independent," or "identically distributed."

Only an IP burdened down with epistemological constraints creates a probability distribution over the data. The data, ontological in nature, are not bound by any such epistemological concerns such as possessing the characteristics of being independent and identically distributed. It is, in fact, the IP's use of the formal manipulation rules of probability theory that give rise to what we label as the independent identically distributed nature of the data.

For example, the probability for the joint distribution of all the data points and a model $\mathcal{M}_k$ is, through application of the *Product Rule*,

$$\begin{aligned}
P(X_1, X_2, \cdots, X_N, \mathcal{M}_k) &= P(X_N, \cdots, X_2, X_1, \mathcal{M}_k) \\
&= P(X_N \mid \cdots X_2, X_1, \mathcal{M}_k) \times \cdots \times \\
&\quad P(X_2 \mid X_1, \mathcal{M}_k) \times P(X_1 \mid \mathcal{M}_k) \times P(\mathcal{M}_k)
\end{aligned}$$

Because the probability of a joint statement at any trial t is, by definition, independent of any previous data, and thus dependent only upon the model, this last result reduces to,

$$P(X_N \mid \cdots X_2, X_1, \mathcal{M}_k) \times \cdots \times P(X_1 \mid \mathcal{M}_k) \times P(\mathcal{M}_k) =$$

$$P(X_N \mid \mathcal{M}_k) \times \cdots \times P(X_2 \mid \mathcal{M}_k) \times P(X_1 \mid \mathcal{M}_k) \times P(\mathcal{M}_k)$$

Therefore, we have through the *Sum Rule*,

$$P(X_1, X_2, \cdots, X_N, \mathcal{M}_k) = \prod_{t=1}^{N} P(X_t \mid \mathcal{M}_k) \times P(\mathcal{M}_k)$$

$$P(X_1, X_2, \cdots, X_N) = \sum_{k=1}^{M} \left[\prod_{t=1}^{N} P(X_t \mid \mathcal{M}_k) \times P(\mathcal{M}_k) \right]$$

leading to,

$$\ln P(\mathcal{D}) = \ln \left\{ \sum_{k=1}^{M} \exp \left[\sum_{t=1}^{N} \ln P(X_t \mid \mathcal{M}_k) + \ln P(\mathcal{M}_k) \right] \right\}$$

By viewing x^N as a random variable, we find $S_N = \{p(x^N; \xi)\}$, to be, like $S = S_1$, a manifold with ξ as a coordinate system. ... Hence distinguishing between the geometries of S^N [sic] (should be S_N) and S serves no purpose, and it suffices to simply consider the geometry of S.

$x^N \equiv (X_1, X_2, \cdots, X_N)$ is the data; it is not a random variable. The only thing which might be considered a random variable is not any statement, nor any statement which has actually occurred (a data point), but rather $F(X = x_i)$, the constraint function. But we think it better not to confuse the issue with unnecessary terms (and thereby choose to invoke Ockham's razor). So we banish the term "random variable" from our vocabulary as well.

Also, S is not the same as what Amari labels as S_1. S happens to be the manifold containing all the points p, q, r, and so on, and these are, by definition, the $P(X = x_i \mid \mathcal{M}_k)$. S_1 where $N = 1$ is a completely different space. For just one data point,

$$P(\mathcal{D}) \equiv P(X_1 = x_i) = \sum_{k=1}^{M} P(X_1 = x_i \mid \mathcal{M}_k) P(\mathcal{M}_k)$$

This is enough to give you the general flavor of the conceptual errors present at the outset. They are enough to block any further hope that a recovery might be imminent further down the road. Indeed, Amari goes on to discuss estimators, the asymptotic properties of estimators, consistency, efficiency, asymptotic Cramér–Rao inequality, in other words, all of the orthodox baggage that has weighed down inference for so many years.

Since I have gone to some length to critically assess Amari's faulty conceptual notions, let me rebalance things by quoting other instances where we essentially agree. This involves Amari's explanation of the Normal distribution as a member of the exponential family. It also gives me an excuse to repeat myself on some aspects of notation.

Shortly following the above passage we have [2, pg. 85]

4.1 Exponential families and observed points

In the definition of an exponential family:

$$p(x;\theta) = \exp\{C(x) + \theta^i F_i(x) - \psi(\theta)\}, \qquad (4.14)$$

the n functions $F_1(x), \ldots, F_n(x)$ are random variables. Hence let us rename them as the n random variables

$$x_i = F_i(x) \qquad (i = 1, \ldots, n),$$

and let $x = [x_1, \ldots, x_n]$. Suppose we also define the probability density functions on the n–dimensional random variable $x = [x_i]$ with respect to the dominating measure

$$\mathrm{d}\mu(x) = \exp\{C(x)\}\,\mathrm{d}x$$

As mentioned, I believe we are all better served by eliminating the concept of *random variable*. However, one can see here that at least Amari labels my constraint functions $F_j(X = x_i)$ as random variables as opposed to the actual statements $(X = x_i)$. There are only ever m constraint functions, not n, with the allowable maximum of $m = n - 1$. Also, one must be careful to remember that Amari's index i is my index j. This is also the closest anyone gets to disambiguating the "dominant measure" notion. Since $C(x) = 0$, $\mathrm{d}\mu(x) = \mathrm{d}x$, or more accurately, $\mathrm{d}\mu(x) = \mathrm{d}F_j(x)$.

Then Equation (4.14) may be rewritten without loss of generality as

$$p(x;\theta) = \exp\{\theta^i x_i - \psi(\theta)\} \qquad (4.15)$$

We use this representation in the discussion below. As was already seen, the exponential family $S = \{p(x;\theta)\}$ is a dually flat space, with its e–affine coordinate system given by the natural parameters θ, its m–affine coordinate system given by the expectation parameters

$$\eta_i = E_\theta[x_i]$$

We also have

$$E_\theta[(x_i - \eta_i)(x_j - \eta_j)] = g_{ij}(\theta)$$

where g_{ij} is the Fisher information matrix with respect to the natural parameters.

In my notation, the Fisher information matrix is written as,

$$g_{rc}(p) = \sum_{i=1}^{n}(F_r(x_i) - \langle F_r \rangle)(F_c(x_i) - \langle F_c \rangle)\,Q_i$$

Amari then presents the Normal distribution as an example of an exponential family.

Example 4.1 (Normal distribution: see Examples 2.1 and 2.5) *Consider a family of probability distributions parameterized by $[\mu, \sigma]$ of the form*

$$p(x;\mu,\sigma) = \frac{1}{\sqrt{2\pi}\sigma}\exp\left\{-\frac{1}{2\sigma^2}(x-\mu)^2\right\},$$

This is a 2-dimensional space formed by normal distributions, and it may be rewritten as

$$p(x; \mu, \sigma) = \exp\left\{ \frac{\mu}{\sigma^2} x - \frac{1}{2\sigma^2} x^2 - \frac{\mu^2}{2\sigma^2} \log(\sqrt{2\pi}\sigma) \right\}$$

Now introduce the coordinate system $\theta = [\theta^1, \theta^2]$ defined by

$$\theta^1 = \frac{\mu}{\sigma^2} \qquad \theta^2 = -\frac{1}{2\sigma^2}$$

and the random variable $x = [x_1, x_2]$ defined by

$$x_1 = F_1(x) = x \text{ and } x_2 = F_2(x) = x^2$$

We then see that we have the exponential family given by

$$p(x; \theta) = \exp\{\theta^i x_i - \psi(\theta)\} \text{ and}$$

$$\psi(\theta) = \frac{\mu^2}{2\sigma^2} + \log(\sqrt{2\pi}\sigma)$$

The expectation parameters $\eta = [\eta_1, \eta_2]$ are given by

$$\eta_1 = E[x_1] = \mu \text{ and}$$

$$\eta_2 = E[x_2] = E[x^2] = \mu^2 + \sigma^2$$

This characterization of the Normal distribution is exactly the same as given by the MEP. See Chapter Thirty in Volume II [8]. However, I would like to point out the distinction that the dimension of the state space is the real line between $-\infty$ and $+\infty$ and $n \to \infty$. This dimension never changes.

Thus, the dimension of the state space is not drastically reduced at all by considering just two parameters. The model space has been drastically reduced by considering information consisting of just two parameters. This is why the Normal distribution, as ubiquitous as it is, is really just a very, very special case for all the possibilities that might exist over the real line. Anyone care to talk about Black Swans?

39.9 Solved Exercises for Chapter Thirty Nine

Exercise 39.9.1: Derive a shorter logic expression equivalent to Equation (39.5). The goal of this exercise is to highlight those cells in the joint probability table where 0s would be expected.

Solution to Exercise 39.9.1

The original logic expression $\mathcal{M}_k \equiv (L \to W) \wedge ((W \to C) \wedge (W \to D))$ proposed as inspiration for various models in the Life on Mars scenario can be shortened to $\overline{L}\,\overline{W} \vee WCD$, "Either no Life and no Water, or Water and both Compounds." which is interesting in and of itself.

Expanding each of these two terms via the **Sum Rule** shows the six cells of the joint probability table which should have non–zero assignments. The first four cells are derived from $\overline{L}\,\overline{W}$,

(1) $\overline{L}\,\overline{W}CD$, (2) $\overline{L}\,\overline{W}\,\overline{C}D$, (3) $\overline{L}\,\overline{W}C\overline{D}$, and (4) $\overline{L}\,\overline{W}\,\overline{C}\,\overline{D}$

and the final two cells are derived from WCD,

(5) $LWCD$ and (6) $\overline{L}WCD$

The six cells identified by this procedure are cells 13, 14, 15, 16, 1, and 9 in the joint probability table. The remaining ten cells would be expected to have 0s from the logic expression.

As can be seen by looking ahead to the joint probability table in Exercise 39.9.11, these six cells do generally have higher numerical values, while the remaining ten cells have lower numerical assignments. As we said, the logic expression was merely a rough guide for how some complicated model with many interaction terms might assign numerical values.

Exercise 39.9.2: Select a couple of the joint statements that would be ruled out under the above logic expression.

Solution to Exercise 39.9.2

If a cell in the joint probability table has an assignment of 0, then the joint statement indexed by that cell can not happen under this model. We know from the above exercise that ten joint statements are ruled out as impossible under this model. For example, the joint statement indexed by cell 2, $LW\overline{C}D$, has a 0 assigned, therefore, the possibility that Life exists, Water exists, Compound D exists, but Compound C does not exist is ruled out. As another example consider cell 12, $\overline{L}W\overline{C}\,\overline{D}$, where a 0 would be placed. This model rules out the fact that Water exists, but Life, Compound C, and Compound D do not.

Exercise 39.9.3: Volume I explained the generic Boolean notation for a functional assignment. List all of the functional assignments to the sixteen settings of the four binary variables in the logic expression.

Solution to Exercise 39.9.3

A quadruple from $\mathbf{B} \times \mathbf{B} \times \mathbf{B} \times \mathbf{B}$ is mapped to $\mathbf{B}$ by a functional assignment where $\mathbf{B}$ is the carrier set $\{T, F\}$. Notationally, $f \colon \mathbf{B}^4 \to \mathbf{B}$. For example, $f(TFTF) = F$. The functional assignments for all 16 possible settings of the four variables is shown below in Table 39.4.

Table 39.4: *The functional assignments for Equation (39.5).*

$TTTT$	$TTTF$	$TTFT$	$TTFF$	$TFTT$	$TFTF$	$TFFT$	$TFFF$
T	F	F	F	F	F	F	F

$FTTT$	$FTTF$	$FTFT$	$FTFF$	$FFTT$	$FFTF$	$FFFT$	$FFFF$
T	F	F	F	T	T	T	T

Exercise 39.9.4: What is the DNF for this logic expression?

Solution to Exercise 39.9.4

We have essentially answered this question in the previous exercise. The DNF for the logic expression in Equation (39.5) consists of the six terms where the functional assignment is T. Thus, the DNF expansion of the logic expression of the motivating model consists of the following six terms,

$$LWCD \vee \overline{L}WCD \vee \overline{L}\,\overline{W}CD \vee \overline{L}\,\overline{W}C\overline{D} \vee \overline{L}\,\overline{W}\,\overline{C}D \vee \overline{L}\,\overline{W}\,\overline{C}\,\overline{D}$$

the same as found in Exercise 39.9.1.

Exercise 39.9.5: Surely all those terms in the DNF shown in the previous Exercise can't reflect the logic expression? For example, the third term, $\overline{L}\,\overline{W}CD$ is a joint statement saying that both life and water don't exist, yet Compounds C and D do. Isn't this a direct contradiction of what the model is trying to say?

Solution to Exercise 39.9.5

The DNF terms are correct. Reflect back on the somewhat strange definition of implication for two variables in the logic expression $A \to B$. The functional assignment is F only when A is T and B is F. In the other three cases, the functional assignment is T. The DNF for $A \to B$ was $AB \vee \overline{A}B \vee \overline{A}\,\overline{B}$. The strange consequence for the implication logic function is that, for example, A can be FALSE and B can be TRUE. This situation is not ruled out as impossible. The only situation that is ruled out as impossible for the implication function (by having a probability assignment of 0 and by not appearing in the DNF) is if A is TRUE and B is FALSE.

Turn your attention to any of the six terms in the current DNF for Equation (39.5): Is L true and W false? No. Is W true and C false? No. Is W true and D false? No. Just as $\overline{A}B$ in the two variable case gets a functional assignment of T by the implication logic function, so too does $\overline{L}\,\overline{W}CD$ receive a functional assignment of T.

Exercise 39.9.6: Use *Mathematica* to establish the tautology existing between the full DNF as found in Exercise 39.9.4 and the shorter form mentioned in Exercise 39.9.1.

Solution to Exercise 39.9.6

The *Mathematica* built–in function **TautologyQ[ ]** may have one argument *boolean function* and will return **True** or **False** depending on whether a tautology truly exists. The *boolean function* appearing as the argument will be **Equivalent[ ]** with two arguments. The first argument is the logic expression for the long form of the DNF, while the second argument is the shorter form.

```
logicExpression1 =
    Or[And[L, W, C, D], And[Not[L], W, C, D],
        And[Not[L], Not[W], C, D],
        And[Not[L], Not[W], C, Not[D]],
        And[Not[L], Not[W], Not[C], D],
        And[Not[L], Not[W], Not[C], Not[D]]]
logicExpression2 =
    Or[And[Not[L], Not[W]], And[W, C, D]]
```

Checking for the tautology,

TautologyQ[Equivalent[logicExpression1, logicExpression2]]

returns the expected answer of **True**. You may wish at this point to review the discussion of logical tautologies in section 2.5 of Volume I.

The functional assignments of T and F shown in Table 39.4 were found from an application of **BooleanTable[ ]** to the original logic expression,

BooleanTable[And[Implies[L, W],
    And[Implies[W, C], Implies[W, D]]], {L, W, C, D}]

which returns the list,

{True, False, False, False, False, False, False, False,
 True, False, False, False, True, True, True, True}

SOLVED EXERCISES FOR CHAPTER 39

Exercise 39.9.7: Mull over the intuitive plausibility for the probability assignments of 0 to the ten joint statements under the model reflected in the logic expression.

Solution to Exercise 39.9.7

It is fun to simply translate any joint statement into its verbal equivalent and then assess the implications of impossibility for this statement. The DNF revealed in Table 39.5 indicates that all of the joint statements beginning with L, except for the first, have a 0 assigned under this model, and are therefore impossible.

For example, the joint statement in cell 2, $LWC\overline{D}$, is impossible. If Life does exist, then Water must exist as well. But if Water exists, Compound D must be found. This statement in cell 2 denies the existence of Compound D, so it is assigned a probability of 0.

And, in a more extreme case, examine the joint statement in cell 8, $L\overline{W}\,\overline{C}\,\overline{D}$. Life exists, but Water and neither Compound exists, so the information the model must insert into the probability distribution assigns a 0 to this joint statement.

There are three additional joint statements involving a situation where Life does not exist. Examine cell 10 for the joint statement $\overline{L}WC\overline{D}$ which is assigned a probability of 0 and therefore impossible under this model. Life does not exist, but Water does. Therefore, Compound D must be found, but this is denied in the statement. A similar examination of the other two statements where Life does not exist, but a 0 is assigned, show that Water is present, but either or both of Compounds C and D are absent. Once again, all of these 0s reflect the information inserted into the probability distribution under this model.

Exercise 39.9.8: Revisit the implication logic function $A \to B$ in order to discuss the conditions under which an updated state of knowledge about A approaches various values.

Solution to Exercise 39.9.8

This discussion "primes the pump" for the next exercise which asks for a similar discursive discussion about the whole Life on Mars scenario as based on the more complex implicational logic expression of Equation (39.5). Under what conditions does $P(A \mid B)$ increase and when does it approach 1, 0, and $P(A)$?

Asking about A knowing only B is an "invalid" application of the classical logic syllogism called *modus ponens* and the certain deductions ensuing from its application. Of course, probability theory as a generalization of logic will correctly return $P(B \mid A) = 1$ in the correct "forward" application of the syllogism.

However, as we have learned, probability theory will also permit us to make an inference in the "backwards" direction, reasoning from the truth of B to an updated belief about the truth of A, captured by $P(A \mid B)$.

So, repeating the standard approach using Bayes's Theorem, we have,

$$P(A \mid B) = \frac{P(AB)}{P(B)} = \frac{P(AB)}{P(AB) + P(\overline{A}B)}$$

or, in terms of the cells of the joint probability table,

$$P(A \mid B, \mathcal{M}_k) = \frac{\text{Cell 1}}{\text{Cell 1} + \text{Cell 2}}$$

From the three terms where T appears in the DNF expansion of $A \to B$, we may insert different non–zero numerical assignments into cells 1, 2, and 4 under various models. A 0 for F would always be inserted into cell 3. This reflects the essential definition of the implication logic function that F is assigned when A is TRUE and B is FALSE.

Now consider various models as inspired by $A \to B$ shown in Figure 39.2. Panel (a) shows the Boolean functional assignments for $A \to B$. The remaining five panels, (b) through (f), show numerical assignments to the four cells under different models together with the corresponding marginal probabilities.

In panels (b), (c), and (d) the marginal probability for A is $P(A \mid \mathcal{M}_k) = 0.20$. In panel (b), $P(A \mid B, \mathcal{M}_1)$ increases from 20% to 40%. Panel (c) shows $P(A \mid B, \mathcal{M}_2)$ increasing from 20% to 95.24%. In this case, the model assigns values where $P(A \mid B, \mathcal{M}_2)$ is approaching 1. In panel (d), $P(A \mid B, \mathcal{M}_3)$ hardly increases at all from the marginal probability of 20%. In panel (e), $P(A \mid B, \mathcal{M}_4)$ is approaching a probability of 0. The final panel (f) shows that we hit a natural barrier, here around 50% for $P(A \mid B, \mathcal{M}_5)$, when the probability for all of the non–zero statements are about the same.

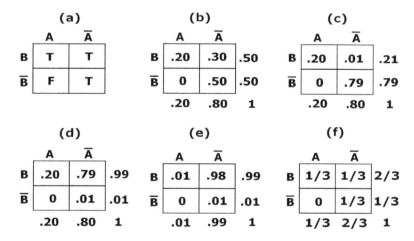

Figure 39.2: *Six joint probability tables reflecting various models for the implication logic function $A \to B$.*

Exercise 39.9.9: Assimilate the lessons learned in the last few exercises by discussing the general characteristics of the joint probability table for the Life on Mars scenario.

Solution to Exercise 39.9.9

Start from the initial premise that any numerical assignment reflected in a joint probability table should generally follow the dictates laid down by the implicational logic expression of Equation (39.5). Six of the sixteen cells in the joint probability table will have non–zero values assigned because of the DNF expansion of the logic function. These are cells 1, 9, 13, 14, 15, and 16. The remaining ten cells in the table will have values close to 0 assigned based on the same DNF rationale. These are cells 2, 3, 4, 5, 6, 7, 8, 10, 11, and 12.

The complicated model p with its main effects and interaction terms behind its numerical assignment doesn't follow this pattern exactly. But the six cells 1, 9, 13, 14, 15, and 16 do have relatively larger numerical assignments, while the ten cells 2, 3, 4, 5, 6, 7, 8, 10, 11, and 12 have smaller numerical assignments. This is why I called the logic expression an "inspiration" for the actual model underlying p.

Repeat the same argument as in the last exercise. Write down Bayes's Theorem for finding the updated state of knowledge about life on Mars given the presence of the two chemical compounds,

$$P(L\,|\,C, D, \mathcal{M}_k) = $$

$$\frac{P(LWCD\,|\,\mathcal{M}_k) + P(L\overline{W}CD\,|\,\mathcal{M}_k)}{P(LWCD\,|\,\mathcal{M}_k) + P(L\overline{W}CD\,|\,\mathcal{M}_k) + P(\overline{L}WCD\,|\,\mathcal{M}_k) + P(\overline{L}\,\overline{W}CD\,|\,\mathcal{M}_k)}$$

or expressed in terms of the cells involved,

$$P(L\,|\,C, D, \mathcal{M}_k) = \frac{\text{Cell 1} + \text{Cell 5}}{\text{Cell 1} + \text{Cell 5} + \text{Cell 9} + \text{Cell 13}}$$

With this in hand, it becomes easy to think about the direction in which the probability will go under different conditions. If cell 1 is large, cell 5 close to 0, and cells 9 and 13 as small as possible, then $P(L\,|\,C, D, \mathcal{M}_k)$ will approach 1. But the fact that the marginal probability for L starts off relatively low at 0.20 means that the cells in $\overline{L}$ (cells 9 through 16) will have to be relatively larger. The only way to make $P(L\,|\,C, D, \mathcal{M}_k)$ approach 1 is to make the marginal probabilities for C and D as small as possible. This will make the relevant cells 9 and 13 smaller, and thus help achieve this overall goal.

This is just another illustration of the lessons to be learned from an inferential approach to problem solving. If Compounds C and D are relatively rare, then discovering their presence on the surface of Mars just makes the IP's degree of belief that much stronger in liquid water and life. In the actual example studied in the text, Compounds C and D were not that rare, so updating the degree of belief in life on Mars is going to hit a natural barrier around, say, 50% as we observed.

Exercise 39.9.10: How did Jaynes make this point?

Solution to Exercise 39.9.10

As Jaynes dryly informed us, finding a man with a mask and a bagful of jewelry outside a burglarized shop raises our degree of belief that he is the burglar because this circumstance is otherwise such a rare and unusual event that is not normally experienced. If we saw *innocent* men with masks and bags of jewelry walking around burgled shops late at night as a routine matter of course, then we wouldn't raise our degree of belief of guilt for this particular man!

As Jaynes tells us in this little fanciful story, there is always some explanation for these events that can't strictly be ruled out. Maybe, just maybe, the man is coming home from a costume party when he passes his jewelry business where a passing car has flung some stones up and broken the shop's window. He is merely collecting his jewelry for safekeeping. The travesty that passes for our judicial system refuses to reason inferentially based on "circumstantial evidence," but demands some sort of absolute deductive certainty that eliminates bizarre and unlikely possibilities like the one above, which, of course, is never attainable.

Exercise 39.9.11: Show a joint probability table for a point p that follows the complicated model with fifteen constraint functions.

Solution to Exercise 39.9.11

Figure 39.3 shows a joint probability table at the top of the next page with $n = 16$ cells for the four binary propositions L, W, C, and D. A few marginal probabilities are also provided. For example, the marginal probability for L is 0.20, while the marginal probability for LW is 0.16.

This is a numerical assignment under a model that tries to closely follow the strictures of the logical expression,

$$\mathcal{M}_k \equiv (L \to W) \land ((W \to C) \land (W \to D))$$

without being bound by the restriction of placing exact 0s in the ten prescribed cells of the joint probability table.

This joint probability table reflects the numerical assignment that simulates a complicated model that we are pretending is too difficult for any computations. It has been labeled as the point p living in $\mathcal{S}^{16}$. It will be α–projected to other points $q, r, s, \ldots$ that live in lower dimensional spaces. But as upcoming exercises will demonstrate, we will need to compute with p in order to observe the implications of the various formulas that Amari provides.

SOLVED EXERCISES FOR CHAPTER 39

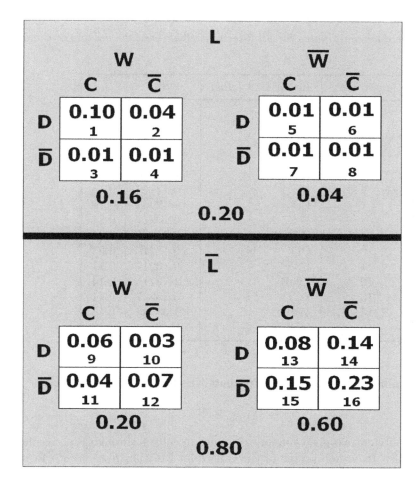

Figure 39.3: *A joint probability table with numerical assignments inspired by the logic expression for the life on Mars scenario.*

Exercise 39.9.12: List the dual coordinates for the complicated model underlying the numerical assignment to p.

Solution to Exercise 39.9.12

The 15 θ^i and η_i coordinates are listed in Table 39.5 at the top of the next page. Remember that Amari's index i is the same as the index j as we use it in the MEP representation. The η_i coordinates, together with the constraint functions, were specified as the information in the model. The θ^i coordinates were found by the MEP algorithm. Note that there are four main effects, six double interactions, four triple interactions, and one quadruple interaction.

Table 39.5: *The dual coordinates for all fifteen constraints of the complicated model behind point p.*

Constraint i	Interaction Label	θ^i	η_i
1	L	−3.13549	0.20
2	W	−1.18958	0.36
3	C	−0.42744	0.46
4	D	−0.49644	0.47
5	LW	1.18958	0.16
6	LC	0.42744	0.13
7	LD	0.49644	0.16
8	WC	−0.13217	0.21
9	WD	−0.35086	0.23
10	CD	−0.13217	0.25
11	LWC	0.13217	0.11
12	LWD	1.73716	0.14
13	LCD	0.13217	0.11
14	WCD	1.38943	0.16
15	LWCD	−0.46864	0.10

The η_i coordinates are the marginal probabilities for the ith effect. Thus,

$$\eta_1 \equiv \langle F_1 \rangle = 0.20$$

is the sum of all eight assigned numerical values in cells 1 through 8 of the joint probability table. It is the average of the constraint function $F_1(X = x_i)$ for the main effect of L.

In like manner, all of the η_i coordinates can be matched up with the main effects, double interactions, triple interactions, and the quadruple interaction. For example, take the LWD triple interaction. This is a marginal probability that sums over the two cells,

$$P(LW\overline{C}D \mid p \in \mathcal{S}^{16}) + P(LW\overline{C}\,\overline{D} \mid p \in \mathcal{S}^{16})$$

From Figure 39.3, the numerical assignments in cell 1 and cell 2 sum to,

$$0.10 + 0.04 = 0.14$$

Thus, part of the information in the complicated model for p is the specification that $\eta_{12} \equiv \langle F_{12} \rangle = 0.14$.

The notation $\eta_{15} \equiv \langle F_{15} \rangle$ would seem to indicate, as above, that a sum of probabilities has been taken over certain cells. But this is the quadruple interaction $LWCD$. The quadruple interaction $LWCD$ is just one cell, cell 1, and represents the specification of a particular probability assignment for Q_1, where here part of the information for point p is that $Q_1 = 0.10$.

SOLVED EXERCISES FOR CHAPTER 39

Exercise 39.9.13: List the values of all fifteen constraint functions. Then provide the associated constraint function averages for the complicated model underlying point p.

Solution to Exercise 39.9.13

The fifteen constraint functions are split across two tables. Table 39.6 shows the first ten constraints, that is, the four main effects L, W, C, and D, and the six double interactions, LW, LC, LD, WC, WD, and CD. It helps to refer back to Figure 39.3 to verify that each constraint function is picking out the appropriate cells of the joint probability table.

Table 39.6: *The detailed listing of the mapping from statements to numbers as represented by the first ten constraint functions. The constraint function averages are shown in the last row.*

i	L	W	C	D	LW	LC	LD	WC	WD	CD
1	1	1	1	1	1	1	1	1	1	1
2	1	1	0	1	1	0	1	0	1	0
3	1	1	1	0	1	1	0	1	0	0
4	1	1	0	0	1	0	0	0	0	0
5	1	0	1	1	0	1	1	0	0	1
6	1	0	0	1	0	0	1	0	0	0
7	1	0	1	0	0	1	0	0	0	0
8	1	0	0	0	0	0	0	0	0	0
9	0	1	1	1	0	0	0	1	1	1
10	0	1	0	1	0	0	0	0	1	0
11	0	1	1	0	0	0	0	1	0	0
12	0	1	0	0	0	0	0	0	0	0
13	0	0	1	1	0	0	0	0	0	1
14	0	0	0	1	0	0	0	0	0	0
15	0	0	1	0	0	0	0	0	0	0
16	0	0	0	0	0	0	0	0	0	0
$\langle F_j \rangle_p$	0.20	0.36	0.46	0.47	0.16	0.13	0.16	0.21	0.23	0.25

Table 39.7 shows the second half of the table with the remaining five constraints consisting of the four triple interactions, LWC, LWD, LCD, and WCD, together with the one quadruple interaction, $LWCD$.

The constraint function averages $\langle F_j \rangle_p$ are shown in the bottom row of the table. The $(X = x_i)$ index the 16 joint statements where,

$$(X = x_1) \equiv LWCD, (X = x_2) \equiv LW\overline{C}D, \cdots, (X = x_{16}) \equiv \overline{L}\,\overline{W}\,\overline{C}\,\overline{D}$$

Table 39.7: *The detailed listing of the mapping from statements to numbers as represented by the final five constraint functions. The constraint function averages are shown in the last row.*

i	LWC	LWD	LCD	WCD	LWCD
1	1	1	1	1	1
2	0	1	0	0	0
3	1	0	0	0	0
4	0	0	0	0	0
5	0	0	1	0	0
6	0	0	0	0	0
7	0	0	0	0	0
8	0	0	0	0	0
9	0	0	0	1	0
10	0	0	0	0	0
11	0	0	0	0	0
12	0	0	0	0	0
13	0	0	0	0	0
14	0	0	0	0	0
15	0	0	0	0	0
16	0	0	0	0	0
$\langle F_j \rangle_p$	0.11	0.14	0.11	0.16	0.10

Exercise 39.9.14: Show the joint probability table for the m–projection of point $p \in S^{16}$ to a point $q \in S^4$.

Solution to Exercise 39.9.14

With the numerical assignments q_i for point q as shown in Figure 39.4 at the top of the next page, we can utilize this alternative formula for the minimum distance,

$$D^{(-1)}(p \parallel q) = \sum_{i=1}^{16} p_i \ln p_i - \sum_{i=1}^{16} p_i \ln q_i$$

$$= -H(p) - \sum_{i=1}^{16} p_i \ln q_i$$

to verify the minimum squared distance of the **m**–projection of point p as it was shown in Table 39.1.

The numerical assignments p_i under the model for point p were listed in Exercise 39.9.11 in Figure 39.3. The negative of the information entropy at point p is then,

SOLVED EXERCISES FOR CHAPTER 39

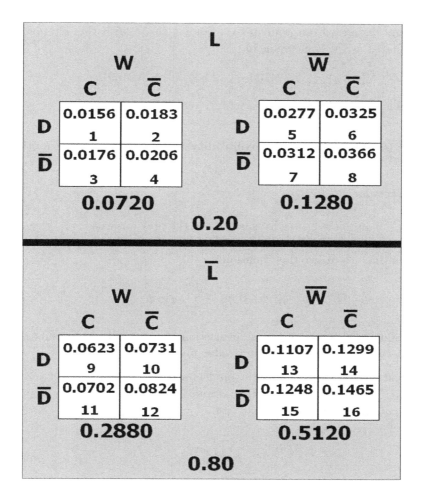

Figure 39.4: *A joint probability table containing the numerical assignments when the complicated model of point p is α-projected down into a point q in $\mathcal{S}^4$.*

$$-H(p) = \sum_{i=1}^{16} p_i \ln p_i$$

$$= 0.10 \ln 0.10 + \cdots + 0.23 \ln 0.23$$

$$= -2.324136$$

The second term is calculated similarly,

$$\sum_{i=1}^{16} p_i \ln q_i = 0.10 \ln 0.0156 + \cdots + 0.23 \ln 0.1465$$

$$= -2.535110$$

So, the minimum distance from point p to a point q in $\mathcal{S}^4$ with this alternative formula is calculated as the same value of,

$$D^{(-1)}(p \parallel q) = -2.324136 - (-2.535110) = 0.210975$$

as shown in Table 39.1 under model $\mathcal{M}_5$.

Exercise 39.9.15: Show the joint probability table for the m–projection of point $p \in \mathcal{S}^{16}$ to a point $q \in \mathcal{S}^6$.

Solution to Exercise 39.9.15

The solution to this problem proceeds exactly as in the previous exercise. With the numerical assignments q_i for point q as shown in Figure 39.5 at the top of the next page, we will continue to utilize this equation,

$$D^{(-1)}(p \parallel q) = -H(p) - \sum_{i=1}^{16} p_i \ln q_i$$

The minimum squared distance of the **m**–projection of point p down to point q in the new sub–space of $\mathcal{S}^6$ was the minimum value presented in Table 39.2.

The numerical assignments p_i under the model for point p were listed in Exercise 39.9.11 in Figure 39.3. The negative of the information entropy at point p remains the same as calculated in the previous exercise,

$$\begin{aligned} -H(p) &= \sum_{i=1}^{16} p_i \ln p_i \\ &= -2.324136 \end{aligned}$$

The second term will be different since the numerical assignments q_i are different,

$$\begin{aligned} \sum_{i=1}^{16} p_i \ln q_i &= 0.10 \ln 0.088 + \cdots + 0.23 \ln 0.2081 \\ &= -2.368900 \end{aligned}$$

So, the minimum distance from point p to a point q in $\mathcal{S}^6$ with this alternative formula is calculated as the same value appearing in Table 39.2,

$$D^{(-1)}(p \parallel q) = -2.324136 - (-2.368900) = 0.044764$$

Since $-H(p)$ will remain the same for all projections, it suffices to maximize $\sum p_i \ln q_i$ in order to minimize the divergence function. We just saw that the new point q in $\mathcal{S}^6$ led to a larger $\sum p_i \ln q_i$ and therefore a smaller divergence function value.

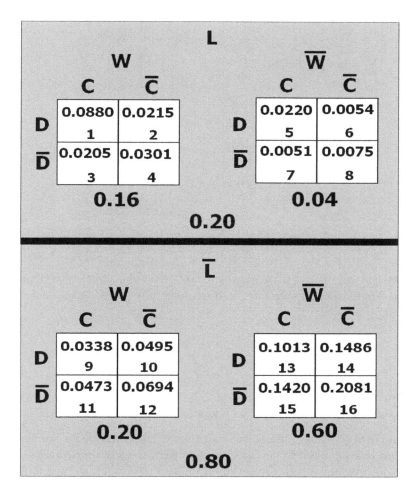

Figure 39.5: *A joint probability table containing the numerical assignments when the complicated model of point p is **m**–projected down into a point q in $\mathcal{S}^6$.*

In the limit, if the **m**–projection results in a point q that has its q_i equal to p_i, then the divergence function reaches its absolute minimum at 0. But maximizing $\sum p_i \ln q_i$ is the same as minimizing the Legendre transformation as shown on the next page. Repeating the beginning steps,

$$D^{(-1)}(p \parallel q) = \sum_{i=1}^{n} p_i \ln\left(\frac{p_i}{q_i}\right)$$

$$= \sum_{i=1}^{n} p_i \ln p_i - \sum_{i=1}^{n} p_i \ln q_i$$

the negative value of the first term will be offset by subtracting the negative value of the second term resulting in a positive quantity greater than or equal to zero.

$$\sum_{i=1}^{16} p_i \ln q_i \equiv E_p\left[\ln q\right]$$

$$E_p\left[\ln q\right] = E_p\left[\sum_{j=1}^{m} \lambda_j^q F_j(X = x_i) - \ln Z_q\right]$$

$$= \sum_{j=1}^{m} \lambda_j^q \langle F_j \rangle_p - \ln Z_q$$

$$H_{\max}(q) = \min_{\lambda_j^q} \{\ln Z_q - \sum_{j=1}^{m} \lambda_j^q \langle F_j \rangle_p \}$$

This value at the last step is the maximized entropy of the distribution q. It must always be a value equal to or greater than 0. If we make it negative, it becomes the second term, and,

$$\sum p_i \ln q_i \equiv -\left[\min_{\lambda_j^q} \{\ln Z_q - \sum_{j=1}^{m} \lambda_j^q \langle F_j \rangle_p\}\right]$$

The only subtlety is that the actual information for the model underlying q is the appropriate m constraint function averages taken from the model underlying p.

It makes sense to verify from Figure 39.5 that the new information for point q in $\mathcal{S}^6$ is, in fact, what it is supposed to be. Two new interactions were added to the four main effects already present in the projection of point p to $\mathcal{S}^4$. These interactions were introduced in order to provide a model that influenced the probability for Life on Mars if water and/or the two chemical compounds were found.

By summing over the appropriate cells of the joint probability table in Figure 39.5, we see that the LW double interaction is correctly found as $\langle F_5 \rangle_p = 0.16$. The LCD triple interaction is correctly found as $\langle F_{13} \rangle_p = 0.11$.

The LW double interaction is $Q_1 + Q_2 + Q_3 + Q_4 = 0.16$, and the LCD triple interaction is $Q_1 + Q_5 = 0.11$. To double-check, refer back to Tables 39.6 and 39.7 to verify the constraint function values under the columns for LW and LCD.

Chapter 40

Triangular Relationships

40.1 Introduction

It seems that one of the most familiar geometrical relationships has an analogy in Information Geometry. These are the "triangular relationships" that we learn about in Euclidean geometry and trigonometry.

As Amari has emphasized, these triangular relationships play an analogous role amongst three points representing probability distributions. These distributions now must be thought of as residing in an n–dimensional Riemannian manifold instead of a Euclidean space. Amari highlights various theorems involving these triangular relationships as an important part of Information Geometry.

For example, in trigonometry one is exposed to the *Law of Cosines* relating the length of sides and angles in any triangle. Of course, this all takes place in ordinary Euclidean space. But something very similar to the Law of Cosines carries over into the generalization of Euclidean space, the space of Riemannian manifolds where our family of probability distributions lives.

40.2 Law of Cosines

40.2.1 Trigonometry

Here is a definition for the *Law of Cosines* which any book on trigonometry will give you.

Definition 3 (Law of Cosines) *The square of any side of a triangle is equal to the sum of the squares of the other two sides minus two times the length of the other sides and the cosine of their included angle.*

Figure 40.1 shows $\triangle ABC$ with the vertices of the triangle labeled as A, B, and C. The length of the side opposite $\angle A$ is a, the length of the side opposite $\angle B$ is b, and the length of the side opposite $\angle C$ is c. By the *Law of Cosines*, the squared length of, say, side b is the squared length of side a plus the squared length of side c minus two times the length of side a times the length of side c times the cosine of $\angle B$,

$$b^2 = a^2 + c^2 - 2\,ac\,\cos(\angle B) \tag{40.1}$$

Exercise 40.7.1 has a numerical example.

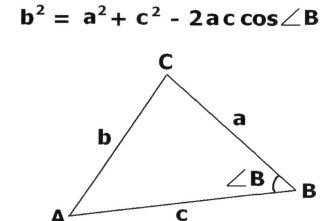

Figure 40.1: *The Law of Cosines for a triangle.*

40.2.2 Linear Algebra

Moving up one level of mathematical generalization from trigonometry to Linear Algebra, the *Law of Cosines* is expressed as,

$$\|A - B\|^2 = \|A\|^2 + \|B\|^2 - 2\,\|A\|\|B\|\,\cos\theta \tag{40.2}$$

where A and B are vectors, $\|A-B\|^2$ is the squared length of the side of the triangle made up from the other two sides, that is, the vectors A and B, with θ the included angle between these two vectors. Figure 40.2 has a sketch of these vectors forming their triangular relationship.

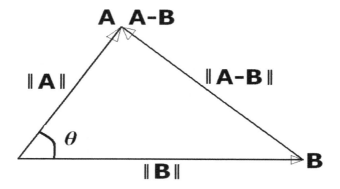

Figure 40.2: *The Law of Cosines for vectors.*

Using the definition of the squared length of a vector with the "central dot" notation, it is relatively easy to demonstrate the Law of Cosines for vectors. The central dot notation is the same as that implemented in the *Mathematica* operator of two arguments,

$$\text{Dot}[\,arg1,\ arg2\,]$$

which for arguments of two vectors a_i and b_i produces a scalar product of vectors `Dot[a, b]` evaluated as $\sum_{i=1}^{n} a_i b_i$. In the Einstein summation convention this is $a_i b^i$ where a_i is a covariant vector and b_i is a contravariant vector.

$$\begin{aligned}
\|A - B\|^2 &= (A - B) \cdot (A - B) \\
&= (A \cdot A) - (A \cdot B) - (B \cdot A) + (B \cdot B) \\
&= (A \cdot A) - 2(A \cdot B) + (B \cdot B) \\
&= \|A\|^2 + \|B\|^2 - 2(A \cdot B) \\
A \cdot B &= \|A\|\|B\| \cos\theta \\
\|A - B\|^2 &= \|A\|^2 + \|B\|^2 - 2\|A\|\|B\| \cos\theta \quad (40.3)
\end{aligned}$$

There is a clear analogy between this expression for vectors in an n–dimensional vector space and a triangle in a two dimensional Euclidean space.

At this juncture, we do, however, have a sub–problem to solve. At the next–to–the–last line in the above derivation, it is not obvious that $A \cdot B = \|A\|\|B\| \cos\theta$. How do we formulate $\cos\theta$ in the context of Linear Algebra so that we can substitute it into the above expression?

Figure 40.3 is a very busy diagram of the vectors involved in the derivation. The first thing to do is to locate vector A and vector B in the diagram. The angle between vectors A and B is θ. Vector P is the *projection* of A onto B. Vector P is seen to be some fraction of vector B represented by $P = cB$. Now locate vector $A + (-P) = A - P = A - cB$ which is the addition of vector $(-P)$ to vector A. We would like vector $(A - cB)$ to be orthogonal to vector B.

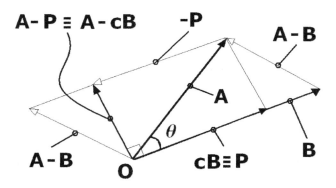

Figure 40.3: *Finding $\cos\theta$ in the vector representation.*

Thus,
$$(A - cB) \cdot B = 0$$
$$A \cdot B - c(B \cdot B) = 0$$
$$A \cdot B = c(B \cdot B)$$
$$c = \frac{A \cdot B}{B \cdot B}$$

Now focus on angle θ in the diagram. By definition,
$$\cos\theta = \frac{\text{adjacent}}{\text{hypotenuse}}$$
$$\cos\theta = c\frac{\|B\|}{\|A\|}$$
$$\cos\theta = \frac{A \cdot B}{B \cdot B}\frac{\|B\|}{\|A\|}$$
$$B \cdot B = \|B\|^2$$
$$\cos\theta = \frac{A \cdot B}{\|A\|\|B\|}$$

40.3 Triangular Relationships in IG

With the Law of Cosines for triangles in two dimensional Euclidean space, and its analog for vectors in an n–dimensional vector space as motivating background, let's see how the Law of Cosines carries over into IG. Amari provides this equation for the triangular relationship among the three points p, q, and r in a Riemannian manifold with divergence measure $D(*\|*)$.

$$D(p\|q) + D(q\|r) - D(p\|r) = \int p(x) - q(x)\{\log r(x) - \log q(x)\}\,dx \qquad (40.4)$$

The $D(*\|*)$, as they appear in the above equation, are to be interpreted as the **m**–projection $D^{(-1)}(*\|*)$, or as the Kullback–Leibler relative entropy $KL(*,*)$.

From the very beginning of these books, I have set out as one of my goals an attempt to "unwind some of the macros" that appear in the literature. In other words, I want to provide the reader with a more expanded, more detailed look at the bald, relatively unmotivated, equations that are plopped into the reader's lap. Let's work through the twists and turns of the above triangular relation equation with this goal in mind.

Amari's divergence measure $D(*\|*)$, as it eventually turns out, is the Kullback–Leibler relative entropy. My unwinding will then begin with $KL(*,*)$. Also, let's proceed with a summation definition of relative entropy over the n–dimensional state space, so that there is no mystery whatsoever about what x might be referring to in an integration expression.

Start off then by listing the three Kullback–Leibler relative entropy equations for distributions p, q, and r,

$$KL(p, q) = \sum_{i=1}^{n} p_i \ln\left(\frac{p_i}{q_i}\right)$$

$$KL(q, r) = \sum_{i=1}^{n} q_i \ln\left(\frac{q_i}{r_i}\right)$$

$$KL(p, r) = \sum_{i=1}^{n} p_i \ln\left(\frac{p_i}{r_i}\right)$$

The second procedural manipulation will be to substitute,

$$\ln x - \ln y \text{ for } \ln\left(\frac{x}{y}\right)$$

in any relative entropy expression. After this much has been accomplished, multiply through by the appropriate distribution and distribute the summation sign.

So we now have,

$$KL(p,q) = \sum_{i=1}^{n} p_i (\ln p_i - \ln q_i)$$

$$= \sum p_i \ln p_i - \sum p_i \ln q_i$$

$$KL(q,r) = \sum_{i=1}^{n} q_i (\ln q_i - \ln r_i)$$

$$= \sum q_i \ln q_i - \sum q_i \ln r_i$$

$$KL(p,r) = \sum_{i=1}^{n} p_i (\ln p_i - \ln r_i)$$

$$= \sum p_i \ln p_i - \sum p_i \ln r_i$$

Third, imagine that these three points p, q, and r form a triangle. For a concrete example, refer back to the points p, q, and r forming the triangle in Figure 39.1 constructed to illustrate the Life on Mars scenario. If the Pythagorean theorem were to hold, then the squared length of $\overline{pq}$ and $\overline{qr}$ is equal to the squared length of $\overline{pr}$. So, in some sense, $KL(p,q) + KL(q,r) - KL(p,r)$ is measuring the degree of departure from a right triangle.

Substitute these transformed relative entropy expressions just derived, cancel the $\sum p_i \ln p_i$ terms, and group the remaining four terms appropriately,

$$KL(p,q) + KL(q,r) - KL(p,r) = \left(\sum p_i \ln p_i - \sum p_i \ln q_i\right) + \left(\sum q_i \ln q_i - \sum q_i \ln r_i\right)$$

$$- \left(\sum p_i \ln p_i - \sum p_i \ln r_i\right)$$

$$= -\sum p_i \ln q_i + \sum q_i \ln q_i - \sum q_i \ln r_i + \sum p_i \ln r_i$$

$$= \sum p_i \ln r_i - \sum p_i \ln q_i - \sum q_i \ln r_i + \sum q_i \ln q_i$$

Now, factor out $\sum p_i$ and $\sum q_i$,

$$KL(p,q) + KL(q,r) - KL(p,r) = \sum p_i \ln r_i - \sum p_i \ln q_i - \sum q_i \ln r_i + \sum q_i \ln q_i$$

$$= \sum p_i (\ln r_i - \ln q_i) - \sum q_i (\ln r_i - \ln q_i)$$

$$= \sum_{i=1}^{n} (p_i - q_i)(\ln r_i - \ln q_i) \qquad (40.5)$$

With Equation (40.5), we have finally arrived at the discrete analog to Amari's integral expression. We might take one more step and turn the right hand side into something that looks very much like a typical relative entropy expression,

$$KL(p,q) + KL(q,r) - KL(p,r) = \sum_{i=1}^{n}(p_i - q_i)\ln\left(\frac{r_i}{q_i}\right) \qquad (40.6)$$

In loose geometric terms, this final expression might be phrased as "an information measure of the inability of points p, q, and r to form a right triangle."

40.4 Amari's Theorem 3.7

In addition to the formula of Equation (40.4), Amari presented another version of this triangular relationship. Here is my unwinding, if you will, of Amari's less than transparent macro–level theorem. At first, I present Amari's original IG flavored notation. In the next section, I translate this over into my preferred MEP notation.

40.4.1 Amari's IG notation

Theorem 3.7 gives the "triangular" relationship between three points in terms of the dual coordinates θ^i and η_i,

$$D(p \parallel q) + D(q \parallel r) - D(p \parallel r) = \{\theta^i(p) - \theta^i(q)\}\{\eta_i(r) - \eta_i(q)\} \qquad (40.7)$$

Instead of the Kullback–Leibler relative entropy expressions, divergence functions are employed on the left hand side of Equation (40.7). First, express each divergence function term on the left hand side by Amari's Equation (3.44) [2, pg. 61],

$$D(p \parallel q) = \psi(p) + \varphi(q) - \theta^i(p)\,\eta_i(q)$$

$$D(q \parallel r) = \psi(q) + \varphi(r) - \theta^i(q)\,\eta_i(r)$$

$$D(p \parallel r) = \psi(p) + \varphi(r) - \theta^i(p)\,\eta_i(r)$$

The right hand sides are just the Legendre transformation in another guise. What will eventually be the right hand side of Theorem 3.7 now looks like this,

$$D(p \parallel q) + D(q \parallel r) - D(p \parallel r) = [\psi(p) + \varphi(q) - \theta^i(p)\,\eta_i(q)] + [\psi(q) + \varphi(r) - \theta^i(q)\,\eta_i(r)]$$
$$- [\psi(p) + \varphi(r) - \theta^i(p)\,\eta_i(r)]$$

Subtracting off the third term results in,

$$D(p \parallel q) + D(q \parallel r) - D(p \parallel r) = \varphi(q) - \theta^i(p)\,\eta_i(q) + \psi(q) - \theta^i(q)\,\eta_i(r) + \theta^i(p)\,\eta_i(r)$$

Rearranging terms results in,

$$D(p \| q) + D(q \| r) - D(p \| r) = \varphi(q) + \psi(q) + \theta^i(p)\,\eta_i(r) - \theta^i(q)\,\eta_i(r) - \theta^i(p)\,\eta_i(q)$$

Apply Amari's Equation (3.36) on the first two terms [2, pg. 59],

$$\varphi = \theta^i\,\eta_i - \psi$$

$$\varphi(q) + \psi(q) = \theta^i(q)\,\eta_i(q)$$

Substitute this result into the current state of the derivation,

$$\begin{aligned} D(p \| q) + D(q \| r) - D(p \| r) &= \varphi(q) + \psi(q) + \theta^i(p)\,\eta_i(r) - \theta^i(q)\,\eta_i(r) - \theta^i(p)\,\eta_i(q) \\ &= \theta^i(q)\,\eta_i(q) + \theta^i(p)\,\eta_i(r) - \theta^i(q)\,\eta_i(r) - \theta^i(p)\,\eta_i(q) \\ &= \theta^i(p)\,\eta_i(r) - \theta^i(q)\,\eta_i(r) - \theta^i(p)\,\eta_i(q) + \theta^i(q)\,\eta_i(q) \end{aligned}$$

Factor the right hand side of the last expression to finally arrive at Theorem 3.7,

$$\theta^i(p)\,\eta_i(r) - \theta^i(q)\,\eta_i(r) - \theta^i(p)\,\eta_i(q) + \theta^i(q)\,\eta_i(q) = \{\,\theta^i(p) - \theta^i(q)\,\}\{\,\eta_i(r) - \eta_i(q)\,\}$$

$$D^{(1)}(p \| q) + D^{(1)}(q \| r) - D^{(1)}(p \| r) = \{\,\theta^i(p) - \theta^i(q)\,\}\{\,\eta_i(r) - \eta_i(q)\,\} \quad (40.8)$$

Notice that the divergence function expressions have been disambiguated in the final line by explicitly specifying that $\alpha = 1$. Amari's Theorem 3.7 as restated above in my Equation (40.8) *is different* than the triangular relationship spelled out previously in Equation (40.6). The exercises go into all of the detail of why they are different and the care that must be taken in disambiguating the **e**–projection from the **m**–projection, and *vice versa*.

For example,

$$D^{(-1)}(p \| q) + D^{(-1)}(q \| r) - D^{(-1)}(p \| r) = \{\,\theta^i(q) - \theta^i(r)\,\}\{\,\eta_i(q) - \eta_i(p)\,\}$$

is the correct Theorem 3.7 result for the $\alpha = -1$ projection.

In addition, working through Amari's derivations, but this time with the MEP notation, helps to identify potential missteps in whether the **e**–projection or the **m**–projection was used. The MEP perspective also keeps us properly aligned with the Kullback–Leibler definitions for relative entropy.

40.4.2 My MEP notation and Amari's notation

It should be enlightening then to translate Amari's notation back into my notation which is based on the MEP. Amari's Equation (3.36),

$$\varphi = \theta^i \eta_i - \psi$$

shows the negative information entropy, $\varphi \equiv -H(Q_i)$, appearing on the left hand side of the equation. Therefore, in translating over into MEP notation,

$$\begin{aligned}
\varphi(p) &\equiv E_p[\ln p_i] \\
&= E_p\left[\sum_{j=1}^{m} \lambda_j^p F_j(X = x_i) - \ln Z_p\right] \\
&= \sum_{j=1}^{m} \lambda_j^p \langle F_j \rangle_p - \ln Z_p \\
&= \theta^i(p)\, \eta_i(p) - \psi(p)
\end{aligned}$$

where the translation from $\theta^i(p)$ to λ_j^p, $\eta_i(p)$ to $\langle F_j \rangle_p$, and $\psi(p)$ to $\ln Z_p$ is evident. I explicitly show the summation over the m parameters, whereas Amari uses the Einstein summation convention in writing expressions like $\theta^i \eta_i$.

The MEP algorithm utilizes the Legendre transformation in order to calculate the maximum information entropy $H_{\max}(Q_i)$. It does this by varying the Lagrange multipliers λ_j^p while keeping constraint function averages $\langle F_j \rangle_p$ fixed in finding the minimum,

$$H_{\max}(p) = \min_{\lambda_j^p} \left\{ \ln Z_p - \sum_{j=1}^{m} \lambda_j^p \langle F_j \rangle_p \right\}$$

Amari's Equation (3.44),

$$D(p \parallel q) = \psi(p) + \varphi(q) - \theta^i(p)\, \eta_i(q)$$

is the **e**–projection, not the **m**–projection, of p to q. Therefore, it is equivalent to the relative entropy between distributions q and p, $KL(q, p)$.

Careful attention must always be paid to the order of the points p and q appearing in Amari's divergence notation as well as whether it is $D^{(1)}(*\|*)$ or $D^{(-1)}(*\|*)$. The next section goes into great detail on this issue because it is so very confusing.

With this clarification, it is easy to translate the dual of the Kullback–Leibler relative entropy notation back into Amari's notation,

$$\begin{aligned}
KL(q,p) &= E_q\left[\ln\left(\frac{q_i}{p_i}\right)\right] \\
&= E_q[\ln q_i] - E_q[\ln p_i] \\
&= \sum_{i=1}^{n} q_i \ln q_i - E_q[\ln p_i] \\
&= -H(q) - \left[\sum_{j=1}^{m} \lambda_j^p \langle F_j\rangle_q - \ln Z_p\right] \\
&= \varphi(q) + \psi(p) - \sum_{j=1}^{m} \lambda_j^p \langle F_j\rangle_q \\
&= \varphi(q) + \psi(p) - \theta^i(p)\,\eta_i(q) \\
D(p\,\|\,q) &= \psi(p) + \varphi(q) - \theta^i(p)\,\eta_i(q)
\end{aligned}$$

So we will want to document these kinds of clarifications in the dizzying array of notation for the divergence function whenever we get the opportunity.

$$KL(q,p) \equiv D^{(-1)}(q\,\|\,p) \equiv D^{(1)}(p\,\|\,q) \equiv D(p\,\|\,q)$$

40.5 Divergence Notation

It has most likely not escaped the reader's attention that aspects of notation *qua* notation, not even considering any inherent mathematical complexity, is definitely a contributing factor hindering an easy grasp of Information Geometry. It owes much of this affliction to what it has borrowed from Differential Geometry.

A prominent example bedevils us in this Chapter with Amari's varying notation for the divergence function as it used to capture his triangular relationships. In the last section, we had to pay careful attention to the kind of projection implied by the divergence notation in order to arrive at the correct match up with Amari's equation.

I intend to provide the reader with some help in this regard by going through, and hopefully clarifying, some of this confusing notation for the divergence function. I proceed by following the rough chronological order of Amari's work. I would like to point out that a common standardized notation for the divergence function was never once adopted over the entire time span when this function was discussed.

Because explicating this history of changing notation in any detail is so very tedious, I have relegated it to the Exercises. When you have mustered up enough courage to face it, please proceed to Exercise 40.7.8.

40.6 Connections to the Literature

Section 40.4, which developed the triangular relation among the points p, q, and r, is my expansion of Amari's Theorem 3.7 [2, pg. 62].

> **Theorem 3.7** Let $\{[\theta^i], [\eta_i]\}$ be mutually dual affine coordinate systems of a dually flat space (S, g, ∇, ∇^*), and let D be a divergence on S. Then a necessary and sufficient condition for D to be the (g, ∇)-divergence is that for all $p, q, r \in S$ the following **triangular relation** holds:
>
> $$D(p \parallel q) + D(q \parallel r) - D(p \parallel r) = \{\theta^i(p) - \theta^i(q)\}\{\eta_i(r) - \eta_i(q)\}$$

Later on, [2, pg. 67] Amari presents his Equation (3.61). However, I have placed my development of his equation in section 40.3 and in my Equation (40.5) prior to explaining Theorem 3.7. I did this because Amari's preliminary remarks provided the clue for my disentangling Theorem 3.7 by introducing the relative entropy expression $KL(p, q)$ as discussed in section 40.3.

> Let us proceed to investigate the canonical divergence. Substituting Equations (3.56) and (3.57) into Equation (3.44) we see that the $(g, \nabla^{(1)})$-divergence on the exponential family $S = \{p_\theta\}$ is given by
>
> $$D^{(1)}(p_\theta \parallel p_{\theta'}) = E_{\theta'}[\log p_{\theta'} - \log p_\theta],$$
>
> which is the 1–divergence defined by Equation (3.26), or in other words, the dual of the Kullback divergence, and consequently the $(g, \nabla^{(-1)})$-divergence is the Kullback divergence $D^{(-1)}$. The triangular relation (3.49) in this case is essentially equivalent to the following relation for the Kullback divergence $D = D^{(-1)}$:
>
> $$D(p \parallel q) + D(q \parallel r) - D(p \parallel r) = \int \{p(x) - q(x)\}\{\log r(x) - \log q(x)\}\, dx \quad (3.61)$$
>
> which is elementary but often useful in applications.

The notation, the surrounding language, and general motivation given in Kass and Vos's book [25] are all somewhat at odds with our presentation. Even though, from time to time, one can still discern, albeit with a certain amount of wishful thinking and hoping, similar underlying patterns.

I have attempted here in this Chapter to highlight some fundamental IG concepts involving familiar triangular relationships. What follows are two passages that seem to bridge seemingly evident similarities between Kass and Vos's treatment and my treatment presented in this Chapter. It is for others to judge whether the significantly more complicated mathematical context provided by Kass and Vos is worth the effort.

Here is how they begin to lay out the background material necessary to present the triangular relationships we have been discussing in this Chapter [25, pg. 248],

Lemma 9.3.2 Suppose $D(\cdot, \cdot)$ is a divergence with divergence parameter $\xi \in \Xi$ and metric matrix $g_{rs}(\xi)$. Then there exists a dual divergence D^* having dual parameters $\xi^* \in \Xi^*$ and a pair of potential functions $\psi \colon \Xi \to R$ and $\phi \colon \Xi^* \to R$ such that

$$D(\xi_1, \xi_2) = \psi(\xi_1) + \phi(\xi_2^*) - \xi_1^r \xi_{2r*}^* \qquad (9.3.10)$$

$$D^*(\xi_1^*, \xi_2^*) = \phi(\xi_1^*) + \psi(\xi_2) - \xi_2^r \xi_{1r}^* \qquad (9.3.11)$$

where

$$\frac{\partial \psi}{\partial \xi^r} = \xi_r^* \qquad \frac{\partial^2 \psi}{\partial \xi^s \partial \xi^r} = \frac{\partial \xi_r^*}{\partial \xi^s} = g_{rs} \qquad (9.3.12)$$

$$\frac{\partial \phi}{\partial \xi_r^*} = \xi^r \qquad \frac{\partial^2 \phi}{\partial \xi_s^* \partial \xi_r^*} = \frac{\partial \xi^r}{\partial \xi_s^*} = g^{rs} \qquad (9.3.13)$$

and g^{rs} are the components of the inverse of the metric matrix. Obviously, D and D^* are related by

$$D(p_1, p_2) = D^*(p_2, p_1)$$

This lemma leads to their main theorem on the Law of Cosines as it applies to divergence functions [25, pg. 255]. It is a concise abstract summary of the essential geometric notion of the Law of Cosines in trigonometry, the Pythagorean theorem, and orthogonality, all generalized to probability distributions residing in Riemannian manifolds,

Theorem 9.3.5 For any three points $p_1, p_2, p_3 \in \mathcal{S}$,

$$D(p_1, p_3) = D(p_1, p_2) + D(p_2, p_3) - \langle v(p_2, p_1), v^*(p_2, p_3) \rangle \qquad (9.3.35)$$

Proof. From (9.3.10) we have

$$D(\xi_1, \xi_2) = \psi(\xi_1) + \phi(\xi_2^*) - \xi_1^r \xi_{2r}^*$$

so that

$$D(\xi_1, \xi_2) + D(\xi_2, \xi_3) = D(\xi_1, \xi_3) + (\xi_1^r - \xi_2^r)(\xi_{3r}^* - \xi_{2r}^*) \qquad (9.3.36)$$

Equation (9.3.35) follows from rewriting (9.3.36) in parameter–free form and using the duality of ∂^r and ∂_s (i.e., $\langle \partial^r, \partial_s \rangle = \delta_s^r$) and equations (9.3.32) and (9.3.34) to write

$$\langle v(p_2, p_1), v^*(p_2, p_3) \rangle = \langle (\xi_1^r - \xi_2^r) \partial_r, (\xi_{3s}^* - \xi_{2s}^*) \partial^s \rangle = (\xi_1^r - \xi_2^r)(\xi_{3s}^* - \xi_{2s}^*) \delta_r^s$$

∎

Equation (9.3.35) is related to the following identity for squared distances

$$\|v(p_1, p_3)\|^2 = \|v(p_1, p_2)\|^2 + \|v(p_2, p_3)\|^2 - 2\langle v(p_2, p_1), v(p_2, p_3) \rangle \qquad (9.3.37)$$

which shows that the divergence behaves like one–half times a squared distance. When $v(p_2, p_1)$ and $v^*(p_2, p_3)$ are orthogonal, (9.3.35) is called the Pythagorean relationship for divergences.

For the final attribution, I will show Amari's Theorem 3.20 [2, pp. 77–78]. For me, it is only by passing back and forth between these very difficult theorems and things that are well understood that I have any glimmer of hope of making any progress.

Unfortunately, this is a negative example because I have NOT figured out this theorem. It also serves as a warning example of what you might expect as common fare when you dip your toe into these murky waters.

Theorem 3.20 *Let ∇ and $\nabla^\star$ be symmetric connections, $g = \langle \ , \ \rangle$ a Riemannian metric and D a divergence on S. Then the following conditions (i) and (ii) are equivalent.*

(i) The triple $(g, \nabla, \nabla^\star)$ is a dualistic structure induced from D in the sense described in §3.2.

(ii) The following approximation is valid at every point q, as other points p and r approach q:

$$D(p \parallel q) + D(q \parallel r) - D(p \parallel r) = \langle \mathcal{E}_q^{-1}(p), \mathcal{E}_q^{\star -1}(r) \rangle_q + o(\Delta^3)$$

where $\mathcal{E}_q$ and $\mathcal{E}_q^\star$ are respectively the exponential maps for ∇ and $\nabla^\star$ at q, and

$$\Delta \stackrel{\text{def}}{=} \max \{ \parallel \xi(p) - \xi(q) \parallel, \parallel \xi(r) - \xi(q) \parallel \}$$

for an arbitrary coordinate system $\xi = [\xi^i]$.

Understandably, it is difficult psychologically for an author to confess up to his inadequacies. But I think that we would all better appreciate someone who tells us what he knows when he also tells us honestly what he does not know. Personally, I would love to read technical expositions where the author says something to the effect, "You know, I never did quite grasp the meaning of Concept X by Nobel prize winner Professor Y, and the reason I could not understand it was because ..."

40.7 Solved Exercises for Chapter Forty

Exercise 40.7.1: Work out an example of the Law of Cosines for a triangle in Euclidean space.

Solution to Exercise 40.7.1

Figure 40.4 shows an arbitrary triangle $\triangle ABC$ with side a of length 5 and side c of length 7. The included angle between these two sides is $\angle B = 30°$.

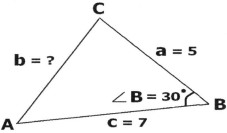

Figure 40.4: *A numerical example of the Law of Cosines to find the length of side b in triangle ABC. The drawing is merely suggestive; it is not accurate.*

$$\begin{aligned} b^2 &= a^2 + c^2 - 2ac\cos\theta \\ &= 25 + 49 - 2(5)(7)\cos(30°) \\ b &= \sqrt{74 - 70(0.866025)} \\ &= 3.66 \end{aligned}$$

Exercise 40.7.2: Work out an example of the Law of Cosines as it applies to vectors.

Solution to Exercise 40.7.2

Find the length of the vector $\| A - B \|$ as it is portrayed in Figure 40.2. The Law of Cosines formula for vectors in a triangular relationship was developed as,

$$\| A - B \|^2 = \| A \|^2 + \| B \|^2 - 2 \| A \| \| B \| \cos\theta$$

Suppose that vector B lies along the x–axis with a length of 6. $B = (6, 0)$. Suppose that vector A has both x and y–coordinates at 2.5, so $A = (2.5, 2.5)$. The angle θ

between the vectors A and B is equal to $\theta = 45°$. The length of the vector $\| A - B \|$ can be calculated with the following *Mathematica* function,

```
LawOfCosines[a_?VectorQ, b_?VectorQ] :=
            Sqrt[Power[Norm[a], 2] + Power[Norm[b], 2] -
            2 Norm[a] Norm[b] Cos[VectorAngle[a, b]]]
```

A call to,

```
LawOfCosines[{2.5, 2.5}, {6, 0}]
```

returns a value of 4.30116 for $\| A - B \|$, the length of the vector completing the triangle. We can find that angle $\theta = 45°$ with,

```
N[(180 VectorAngle[{2.5, 2.5}, {6, 0}]) / π]
```

Exercise 40.7.3: Find the same answer for the angle between the vectors from first principles.

Solution to Exercise 40.7.3

From the development in section 40.2.2, we know that,

$$\cos \theta = \frac{A \cdot B}{\| A \| \| B \|}$$

Without relying upon the *Mathematica* built-in function **VectorAngle[ ]**, we could implement the above formula to find that the angle $\theta = 45°$ with,

```
N[180 ArcCos[(Dot[{2.5, 2.5}, {6, 0}]) /
            (Norm[{2.5, 2.5}] Norm[{6, 0}])] / π]
```

Exercise 40.7.4: Give the m–projection version of Equation (40.7).

Solution to Exercise 40.7.4

Equation (40.7) was written as,

$$D^{(1)}(p \| q) + D^{(1)}(q \| r) - D^{(1)}(p \| r) = \{\theta^i(p) - \theta^i(q)\}\{\eta_i(r) - \eta_i(q)\}$$

The **m**–projection version is,

$$D^{(-1)}(p \| q) + D^{(-1)}(q \| r) - D^{(-1)}(p \| r) = \{\theta^i(q) - \theta^i(r)\}\{\eta_i(q) - \eta_i(p)\}$$

Following the same strategy as was employed in section 40.4.1, translate all three divergence functions into Amari's notation,

$$D^{(-1)}(p \parallel q) = \psi(q) + \varphi(p) - \theta^i(q)\,\eta_i(p)$$

$$D^{(-1)}(q \parallel r) = \psi(r) + \varphi(q) - \theta^i(r)\,\eta_i(q)$$

$$D^{(-1)}(p \parallel r) = \psi(r) + \varphi(p) - \theta^i(r)\,\eta_i(p)$$

Substitute these results into the **m**–projection version of Equation (40.7),

$$\begin{aligned}D^{(-1)}(p \parallel q) + D^{(-1)}(q \parallel r) - D^{(-1)}(p \parallel r) &= [\psi(q) + \varphi(p) - \theta^i(q)\,\eta_i(p)] + \\ &\quad [\psi(r) + \varphi(q) - \theta^i(r)\,\eta_i(q)] - \\ &\quad [\psi(r) + \varphi(p) - \theta^i(r)\,\eta_i(p)]\end{aligned}$$

Subtracting off the third term in brackets results in,

$$\begin{aligned}D^{(-1)}(p \parallel q) + D^{(-1)}(q \parallel r) - D^{(-1)}(p \parallel r) &= \psi(q) - \theta^i(q)\,\eta_i(p) + \varphi(q) - \theta^i(r)\,\eta_i(q) + \theta^i(r)\,\eta_i(p) \\ &= \psi(q) + \varphi(q) - \theta^i(q)\,\eta_i(p) - \theta^i(r)\,\eta_i(q) + \theta^i(r)\,\eta_i(p)\end{aligned}$$

Apply Amari's Equation (3.36), the Legendre transformation, to the first two terms,

$$\psi(q) + \varphi(q) = \theta^i(q)\,\eta_i(q)$$

and rearranging terms results in,

$$\begin{aligned}D^{(-1)}(p \parallel q) + D^{(-1)}(q \parallel r) - D^{(-1)}(p \parallel r) &= \psi(q) + \varphi(q) - \theta^i(q)\,\eta_i(p) - \theta^i(r)\,\eta_i(q) + \theta^i(r)\,\eta_i(p) \\ &= \theta^i(q)\,\eta_i(q) - \theta^i(q)\,\eta_i(p) - \theta^i(r)\,\eta_i(q) + \theta^i(r)\,\eta_i(p) \\ &= \theta^i(q)\,\{\eta_i(q) - \eta_i(p)\} - \theta^i(r)\,\{\eta_i(q) - \eta_i(p)\} \\ &= \{\theta^i(q) - \theta^i(r)\}\{\eta_i(q) - \eta_i(p)\}\end{aligned}$$

Exercise 40.7.5: Use the alternative m–projection version of Theorem 3.7 just proved in the last exercise to show that both $\triangle pqr$ and $\triangle pqs$ of Figure 39.1 are right triangles.

Solution to Exercise 40.7.5

In section 39.5, Equation (40.6) was already used to show that both triangles $\triangle pqr$ and $\triangle pqs$ were right triangles. Now use,

SOLVED EXERCISES FOR CHAPTER 40

$$D^{(-1)}(p \parallel q) + D^{(-1)}(q \parallel r) - D^{(-1)}(p \parallel r) = \{\theta^i(q) - \theta^i(r)\}\{\eta_i(q) - \eta_i(p)\}$$

to show the same thing.

Here is an easy *Mathematica* implementation of the above result. Set up the user–defined function **Thm3pt7[ ]** with four arguments all consisting of lists, the two θ^i coordinates and the two η_i coordinates.

```
Thm3pt7[theta1_List, theta2_List, eta1_List, eta2_List] :=
                    Total[(theta1 - theta2) (eta1 - eta2)]
```

In order to find out if $\triangle pqr$ is a right triangle, we require the list of $\theta^i(q)$ and $\theta^i(r)$ coordinates together with the list of $\eta_i(q)$ and $\eta_i(p)$ coordinates. These have all been computed as part of the MEP algorithm at some point in previous exercises, but they need to be explicitly evaluated once again.

```
thetaq = {-3.32158, -1.09861, -0.382707, -0.337225, 2.48491, 0,
          0, 0, 0, 0, 0, 0, 1.7943, 0, 0}
```

This reminds us that point q was the $\alpha = -1$–projection, the **m**–projection, of point p into the sub–space $\mathcal{S}^6$. The first four $\theta^i(q)$ coordinates reflect the separate marginal probabilities for Life, Water, Compound C, and Compound D, (L, W, C, and D), while $\lambda_5 = 2.48491 \equiv \theta^5(q)$ reflects the double interaction of Life and Water (LW) and $\lambda_{13} = 1.79430 \equiv \theta^{13}(q)$ reflects the triple interaction of Life, Compound C, and Compound D (LCD).

Continuing on, the list for $\theta^i(r)$ is,

```
thetar = {-3.32158, -1.09861, -0.382707, -0.337225, 1, 0, 0, 0,
          0, 0, 0, 0, 2, 0, 0}
```

Point r lives in the same sub–space $\mathcal{S}^6$ as point q, but it is not the **m**–projection of point p. Point r has been moved somewhat farther away from point q by changing $\lambda_5^q = 2.48491$ to $\lambda_5^r = 1$ and $\lambda_{13}^q = 1.79430$ to $\lambda_{13}^r = 2$. The model behind the probability distribution for r still retains information from the four marginal probabilities and the two interactions.

To finish up, we need the lists for $\eta_i(q)$ and $\eta_i(p)$. We will show $\eta_i(p)$ first since it contains all of the information behind the complicated model for p. These were all of the $\langle F_j \rangle_p$ shown in Tables 39.6 and 39.7.

```
etap = {.20, .36, .46, .47, .16, .13, .16, .21, .23, .25, .11, .14,
        .11, .16, .10}
```

Since point q was the **m**–projection of point p, six constraint function averages, $\langle F_1 \rangle_q$ through $\langle F_5 \rangle_q$ and $\langle F_{13} \rangle_q$ had to match those of point p. Thus,

```
etaq = {.20, .36, .46, .47, .16, .135621, .136813, .189591,
        .192747, .245099, .108947, .10945, .11, .121775, .088}
```

Now, it is a simple matter to evaluate,

$$\texttt{Thm3pt7[thetaq, thetar, etaq, etap]}$$

which returns the hoped for value of 0. Doing the same thing[1] for triangle $\triangle pqs$ with,

$$\texttt{Thm3pt7[thetaq, thetas, etaq, etap]}$$

also returns the hoped for value of 0. The coordinates $\theta^i(s)$ reflect the change in the model of point s to $\lambda_5^s = 2$ and back to the same coordinate for the LCD triple interaction $\lambda_{13}^q = \lambda_{13}^s = 1.7943$.

```
thetas = {-3.32158, -1.09861, -0.382707, -0.337225, 2, 0, 0, 0,
          0, 0, 0, 0, 1.7943, 0, 0}
```

Thus, we can assert that the Pythagorean theorem holds for these distributions,

$$D^{(-1)}(p \parallel q) + D^{(-1)}(q \parallel r) = D^{(-1)}(p \parallel r)$$

$$D^{(-1)}(p \parallel q) + D^{(-1)}(q \parallel s) = D^{(-1)}(p \parallel s)$$

Exercise 40.7.6: Derive the dual version of Equation (40.6).

Solution to Exercise 40.7.6

Review the steps taken in section 40.3 to derive Equation (40.6) as,

$$KL(p,q) + KL(q,r) - KL(p,r) = \sum_{i=1}^{n}(p_i - q_i)\ln\left(\frac{r_i}{q_i}\right)$$

to find the dual version,

$$KL(q,p) + KL(r,q) - KL(r,p) = \sum_{i=1}^{n}(q_i - r_i)\ln\left(\frac{q_i}{p_i}\right)$$

So, *mutatis mutandis* and condensing the notation,

$$\begin{aligned}
KL(q,p) + KL(r,q) - KL(r,p) &= \sum q \ln\frac{q}{p} + \sum r \ln\frac{r}{q} - \sum r \ln\frac{r}{p} \\
&= \sum q \ln q - \sum q \ln p + \sum r \ln r - \sum r \ln q \\
&\quad - \left(\sum r \ln r - \sum r \ln p\right) \\
&= \sum q \ln q - \sum q \ln p - \sum r \ln q + \sum r \ln p
\end{aligned}$$

---

[1] Does anybody remember my tip of the hat to Douglas Hofstadter for his deep analysis of what "doing the same thing" really means?

Now factor,

$$\begin{aligned}
KL(q,p) + KL(r,q) - KL(r,p) &= \sum q \ln q - \sum q \ln p - \sum r \ln q + \sum r \ln p \\
&= \sum q (\ln q - \ln p) - \sum r (\ln q - \ln p) \\
&= \sum (q - r)(\ln q - \ln p) \\
&= \sum_{i=1}^{n} (q_i - r_i) \ln\left(\frac{q_i}{p_i}\right)
\end{aligned}$$

Exercise 40.7.7: Does this dual version also result in saying that $\triangle pqr$ is a right triangle?

Solution to Exercise 40.7.7

No, it does not result in the same thing. Implementing the above dual version of Equation (40.6) in *Mathematica*,

```
dualVersion[p1_List, p2_List, p3_List] :=
                     Total[(p1 - p2) Log[p1 / p3]]
```

and evaluating `dualVersion[q, r, p]` results in a non–zero answer of,

$$KL(q,p) + KL(r,q) - KL(r,p) = 0.0102225$$

Thus, the **e**–projection of point p must result in a different point $q^{(e)}$ than our previous **m**–projection of point p to $q^{(m)} \equiv q$.

Exercise 40.7.8: Attempt to document the changing notation for Amari's divergence function.

Solution to Exercise 40.7.8

To start in [1, pg. 35], we have an α–divergence defined as,

$$D_\alpha(\theta, \theta') = \psi(\theta) + \phi(\eta') - \theta \cdot \eta'$$

Then, the $-\alpha$–divergence is defined as $D_{-\alpha}(\theta, \theta') = D_\alpha(\theta', \theta)$. For the specific case of $\alpha = 1$, Amari tells us that,

$$D_1(\theta', \theta) \equiv D_{-1}(\theta, \theta') = I[p(x, \theta') : p(x, \theta)] = \int p(x, \theta) \log \frac{p(x, \theta)}{p(x, \theta')} dP$$

With this much established, we move on to [2, pg. 61] where,

$$D(p \parallel q) = \psi(p) + \varphi(q) - \theta^i(p) \eta_i(q)$$

which is Equation (3.44), the expression used previously. But we have already determined that this $D(p \parallel q)$ had to be the same as $KL(q,p)$. So $KL(q,p) = D_1(\theta', \theta) = D_{-1}(\theta, \theta')$, with $\theta' \equiv p$ and $\theta \equiv q$. Thus,

$$D_1(p,q) \equiv D_{-1}(q,p) = \int p(x,\theta) \log \frac{p(x,\theta)}{p(x,\theta')} \, \mathrm{d}P \equiv \sum_{i=1}^{n} q_i \ln\left(\frac{q_i}{p_i}\right) \equiv KL(q,p)$$

But Amari then uses this notation in his Equation (3.46) [2, pg. 61],

$$D^*(p \parallel q) = D(q \parallel p)$$

which means that,

$$D(q \parallel p) = KL(p,q)$$

$$= \sum_{i=1}^{n} p_i \ln\left(\frac{p_i}{q_i}\right)$$

$$D^*(p \parallel q) = \sum_{i=1}^{n} p_i \ln\left(\frac{p_i}{q_i}\right)$$

Next up in this frazzled attempt to keep pace with an ever changing divergence notation is [3, pg. 1407] where the notation changes again to a divergence $K(P,Q)$ written as,

$$K(\theta_P, \eta_Q) = \psi(\theta_P) + \varphi(\eta_Q) - \theta_{P_i}\eta_{Q_i}$$

where the analogy to Equation (3.44) is evident. However, in the very next equation he goes on to write,

$$K(P \parallel Q) = E_{\theta_P}\left[\log \frac{p(r; \theta_P)}{p(r; \theta_Q)}\right]$$

We must therefore have accumulated these equivalencies,

$$K(P \parallel Q) \equiv KL(p,q) \equiv \sum_{i=1}^{n} p_i \ln\left(\frac{p_i}{q_i}\right) \equiv D^*(p \parallel q) \equiv D(q \parallel p)$$

We are not through yet. In [4, pg. 1703], Amari presents another divergence formula in Equation (22),

$$D[p \colon p'] = \psi(\boldsymbol{\theta}) + \varphi(\boldsymbol{\eta}') - \boldsymbol{\theta} \cdot \boldsymbol{\eta}'$$

Thus, pattern–matching with Equation (3.44) we must have,

$$D[p \colon p'] \equiv D(p \parallel q) \equiv KL(q,p)$$

But in the next equation, Equation (23), it is stated that,

$$D[p \colon p'] = E_\theta\left[\log \frac{p(x,\boldsymbol{\theta})}{p(x,\boldsymbol{\theta}')}\right]$$

SOLVED EXERCISES FOR CHAPTER 40

This is not consistent since,

$$D[p\colon p'] \equiv KL(q,p) = E_{\theta'}\left[\log \frac{p(x,\boldsymbol{\theta}')}{p(x,\boldsymbol{\theta})}\right]$$

Finally, in [5] we hope to find a consistent way out of this labyrinth into which we have so innocently wandered. Once again, a different notation is dredged up for the divergence expressions. The first Amari labels as the Kullback–Leibler divergence,

$$D_{-1}[q\colon p] = \sum q \log \frac{q}{p}$$

and the second as its reverse,

$$D_1[q\colon p] = \sum p \log \frac{p}{q}$$

The thread that leads us back out of the labyrinth is provided by,

$$D_\alpha[q\colon p] = D_{-\alpha}[p\colon q]$$

There is only one hitch in all of this and it centers on the fact that, in this particular case, the roles of p and q have been reversed from the way we have been thinking about them up till now. That is, here q is the complicated distribution and p is the simpler distribution that results when q is α–projected down into a smaller dimensional sub–manifold. So, if we just reverse the roles of p and q as given above, we have in one final frenzy of equivalencies,

$$\begin{aligned}
D_{-1}[p\colon q] &= \sum p \log \frac{p}{q} \\
D_1[p\colon q] &= \sum q \log \frac{q}{p} \\
D_\alpha[p\colon q] &= D_{-\alpha}[q\colon p] \\
D_1[p\colon q] &= D_{-1}[q\colon p] \\
D_1[p\colon q] &= D[p \parallel q] \\
D_{-1}[p\colon q] &= D^*[p \parallel q]
\end{aligned}$$

The **m**–projection of a complicated distribution p down to a simpler distribution q has the distance–like measure of $KL(p,q)$, while the **e**–projection of p to q is $KL(q,p)$.

To conclude all of this, we must provide some kind of resolution to this notation dilemma. We will henceforth standardize on the notation,

$$D^{(-1)}(p \parallel q) \equiv KL(p,q) = \sum_{i=1}^{n} p_i \ln\left(\frac{p_i}{q_i}\right)$$

and,
$$D^{(1)}(p \parallel q) \equiv KL(q,p) = \sum_{i=1}^{n} q_i \ln\left(\frac{q_i}{p_i}\right)$$

This maintains the consistency with Amari's equation (3.44),
$$D^{(1)}(p \parallel q) \equiv D(p \parallel q) = \psi(p) + \varphi(q) - \theta^i(p)\,\eta_i(q)$$

and our notation through,
$$KL(q,p) = \sum_{i=1}^{n} q_i \ln\left(\frac{q_i}{p_i}\right)$$
$$= \sum_{i=1}^{n} q_i \ln q_i - \sum_{i=1}^{n} q_i \ln p_i$$
$$= -H(q_i) - \left(\sum_{j=1}^{m} \lambda_j^p \langle F_j \rangle_q - \ln Z_p\right)$$
$$= \ln Z_p - H(q_i) - \sum_{j=1}^{m} \lambda_j^p \langle F_j \rangle_q$$
$$= \psi(p) + \varphi(q) - \theta^i(p)\,\eta_i(q)$$

Chapter 41

Orthogonal Decompositions and Foliations

41.1 Introduction

It has been my admittedly limited experience that addressing the intricacies of Information Geometry is made somewhat easier by first turning to the MEP for an initial orientation. I have tried to promote the strategic viewpoint that one should always approach the difficulties inherent in IG by focusing on the main goal of solving an inferential problem. A primary component for accomplishing this objective is to use the MEP for numerical assignments to probabilities.

Or, to say it another way, I recommend the perspective that thinks of models as inserting all of the *active* information into probability distributions, while at the same time letting entropy take care of all of the *missing* information.

However, I don't think there would be much of an argument over the fact that the opposite tack is taken in much of the current literature. There, it seems that the MEP, if mentioned at all, is relegated to a minor corollary, so to speak, of any main developments achieved through the efforts of Differential Geometry.

Amari is quite typical in regarding the MEP as a minor offshoot from his primary geometric goals. Although, one has to be happy that at least he makes the effort to acknowledge the MEP and fits it into his ongoing development.

In this Chapter, I would like to play on the theme of the tension that exists concerning the relative importance of the MEP within IG. I am going to expand on Amari's treatment of *mutually dual foliations* as a typically abstract topic from IG's perspective. However, I will always try to turn the discussion around to a comment that these concepts are quite familiar, and perhaps even mundane when looked at from the perspective of the maximum entropy principle.

And, as mentioned above, it is interesting to dwell, at least briefly, on the difference that Amari places on the minor importance of the MEP in the grand scheme of things *vis-à-vis* my elevation of the MEP to a quintessential role. As a curious footnote, Amari rarely credits Jaynes for any of his seminal MEP contributions. In some cases, Amari references others for IG concepts that Jaynes had mentioned years earlier.

41.2 Mixed Coordinate System

From the MEP perspective, we have the dual coordinate system consisting of the Lagrange multipliers λ_j and the constraint function averages $\langle F_j \rangle$. They enable us to represent probability distributions as repositories of information under some model.

In Amari's IG notation, this MEP notation gets translated over into the e–affine coordinates θ^i and the m–affine coordinates η_i. The conceptual basis changes to thinking of points in a Riemannian manifold $\mathcal{S}^n$. In all of our previous discussions, a probability distribution Q_i, or a point p, could be expressed in either one of these coordinate systems, irrespective of the particular MEP or IG choice, but we did not think to mix the two coordinate systems.

In order to discuss the concepts of orthogonality and foliations, Amari introduced the idea of a **mixed coordinate system** that uses both the θ^i and η_i coordinates, appropriately partitioned to represent a point in a "flat" space $(S, g, \nabla, \nabla^\star)$.

Complementary to Amari's more general discussion, I will always fall back on concrete numerical examples when discussing these topics centered on orthogonal decompositions and foliations. For an easy beginning example, I will return to the $n = 4$ dimensional state space of the original kangaroo scenario extensively discussed in Chapter Twenty One of Volume II [8, pg. 125]. The emphasis there was on the information contained in progressively more complicated models leading up to models defining correlations.

Only $m = n - 1 = 3$ Lagrange multipliers were needed for the information to define a model for a probability distribution Q_i that contained correlations. Alternatively, only three constraint function averages were required as the dual coordinate system.

The following three constraint function averages were chosen as one example for a "complicated" model that implemented a correlation between beer and hand preference,

$$\langle F_1 \rangle = 0.75$$

$$\langle F_2 \rangle = 0.75$$

$$\langle F_3 \rangle = 0.70$$

FOLIATIONS

The Lagrange multipliers for this model were calculated by my MEP algorithm as,

$$\lambda_1 = -1.38629$$

$$\lambda_2 = -1.38629$$

$$\lambda_3 = +4.02535$$

With this context in hand as a refresher, we can follow Amari's prescription for how to set up a mixed coordinate system. The total range of $m = 3$ coordinates is divided into two sections, called section I and section II. The η_i coordinates are partitioned as (η_I, η_{II}), while the θ^i coordinates are partitioned as (θ^I, θ^{II}).

The coordinates can be partitioned in any acceptable way. For this example, suppose that section I consists of the first two coordinates, and section II consists of the final third coordinate. The interesting geometric feature is that any point can be thought of as the intersection of two sub–manifolds as defined by the coordinates in each section.

For example, point p above has the mixed coordinate system of $\xi = (\eta_I, \theta^{II})$. This translates into $\xi = (0.75, 0.75, 4.02535)$. Point p occurs at the intersection of a sub–manifold defined by $\eta_I(p)$ and a sub–manifold defined by $\theta^{II}(p)$. Amari labels these intersecting sub–manifolds as $M(\eta_I)$ and $E(\theta^{II})$, presumably standing for m–affine coordinates and e–affine coordinates respectively.

The picture in Figure 41.1 on the next page, adapted from Amari's own sketch, aids immeasurably in the intuition for this so-called *mutually dual foliation* of the Riemannian space $\mathcal{S}^n$. As the picture indicates, we will continue the numerical example by finding the mixed coordinates for three additional points in the manifold labeled as q, r, and s.

We have just found the mixed coordinates for point p as the intersection of the two sub–manifolds $\{M[\eta_I(p)], E[\theta^{II}(p)]\}$. Doing the same thing for point q, we must find $\{M[\eta_I(p)], E[\theta^{II}(r)]\}$. This is the intersection of the same sub–manifold $M[\eta_I(p)]$ that p lives in, but q must be in a different $E[\theta^{II}(r)]$. It is defined to live in the same E sub–manifold as point r. Thus, $\xi = (0.75, 0.75, 3.5)$ for point q where θ^3 has a fixed value of 3.5 for all points living in this sub–manifold.

Continuing on in the same vein, we see from the sketch that point r exists at the intersection of $\{M[\eta_I(r)], E[\theta^{II}(r)]\}$. The M sub–manifold that r lives in must be different than points p and q. Therefore, its first two mixed coordinates $\eta_I(r)$ must be different than $\eta_I(p)$ or $\eta_I(q)$. Suppose that $\eta_1(r) \equiv \langle F_1 \rangle_r = 0.72$ and $\eta_2(r) \equiv \langle F_2 \rangle_r = 0.78$. However, since r lives in $E[\theta^{II}(r)]$ its third coordinate must be 3.5. Thus, $\xi = (0.72, 0.78, 3.5)$ for point r.

Finally, consider point s. Orienting ourselves by looking at the sketch, we see that this point exists at the intersection of $\{M[\eta_I(r)], E[\theta^{II}(p)]\}$. The M sub–manifold that s lives in must be different than points p and q, but is the same as

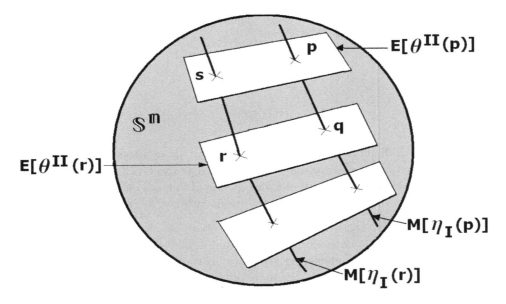

Figure 41.1: *A sketch of how the intersecting sub–manifolds in a foliation determine the mixed coordinate system for any point. The four points, p, q, r, and s shown in the figure represent four different numerical assignments to the probability of the statements in the state space.*

point r. Point s lives, however, in same E sub–manifold as point p. Therefore, the mixed coordinate system is $\xi = (0.72, 0.78, 4.02535)$ for point s.

A table summarizing the m–affine, the e–affine, and the recently found mixed coordinates would be helpful at this juncture. Table 41.1, at the top of the next page, lists all four points, p, q, r, and s in the first column. This is followed by a representation in the three coordinate systems in the succeeding three columns. The final column provides the numerical assignment to the four statements in the state space under the information defined by specifying the model's parameters. The information must be the same for any of the three coordinate systems.

Suppose that point q is the α–projection of point p onto the sub–manifold $E[\theta^{II}(r)]$. The squared distance–like measure represented by the concept of the canonical divergence function $D^{(-1)}(p \parallel \star)$, where $\star$ could be any point in $E[\theta^{II}(r)]$, takes on its minimum value at q.

The intersecting manifolds $E[\theta^{II}(r)]$ and $M[\eta_I(p)]$ are orthogonal at q. Thus, $\triangle pqr$ is a right triangle and the IG version of the Pythagorean theorem says that,

$$D^{(-1)}(p \parallel r) = D^{(-1)}(p \parallel q) + D^{(-1)}(q \parallel r)$$

Table 41.1: *The m–affine, e–affine, and mixed coordinate systems for four points in a Riemannian manifold. The invariant probability distributions under any of these three coordinate systems are shown in the final column.*

Point	η_i	θ^i	η_I, θ^{II}	$P(X = x_i \mid \mathcal{M}_k)$
p	0.75 0.75 0.70	−1.38629 −1.38629 +4.02535	0.75 0.75 4.02535	(0.70, 0.05, 0.05, 0.20)
q	0.75 0.75 0.68759	−1.10053 −1.10053 +3.50000	0.75 0.75 3.50	(0.687590, 0.062410, 0.062410, 0.187590)
r	0.72 0.78 0.681851	−1.56168 −0.61669 +3.50000	0.72 0.78 3.50	(0.681851, 0.038149, 0.098149, 0.181851)
s	0.72 0.78 0.692697	−1.95414 −0.79174 +4.02535	0.72 0.78 4.02535	(0.692697, 0.027303, 0.087303, 0.192697)

Calculating each one of these three divergence functions directly according to its definition as a relative entropy confirms this triangular relationship.

$$D^{(-1)}(p \parallel r) = D^{(-1)}(p \parallel q) + D^{(-1)}(q \parallel r)$$

$$D^{(-1)}(p \parallel r) = \sum_{i=1}^{4} p_i \ln\left(\frac{p_i}{r_i}\right)$$

$$= 0.0172168$$

$$D^{(-1)}(p \parallel q) = \sum_{i=1}^{4} p_i \ln\left(\frac{p_i}{q_i}\right)$$

$$= 0.0031628$$

$$D^{(-1)}(q \parallel r) = \sum_{i=1}^{4} q_i \ln\left(\frac{q_i}{r_i}\right)$$

$$= 0.0140540$$

$$0.0172168 = 0.0031628 + 0.0140540$$

The dual form of this triangular relationship exists for $\triangle psr$. The $\alpha = 1$ form of the divergence function illustrates that point s is the α–projection of point p along a different geodesic. See Figure 41.2 on the next page for a sketch of these

Pythagorean relationships among the four probability distributions p, q, r, and s. Clearly, the squared distance–like measure along the same hypotenuse of triangles $\triangle pqr$ and $\triangle psr$, that is, $\overline{pr}^2$, will depend upon the α connection.

$$D^{(1)}(p \parallel r) = D^{(1)}(p \parallel s) + D^{(1)}(s \parallel r)$$

$$D^{(1)}(p \parallel r) = \sum_{i=1}^{4} r_i \ln\left(\frac{r_i}{p_i}\right)$$

$$= 0.0206674$$

$$D^{(1)}(p \parallel s) = \sum_{i=1}^{4} s_i \ln\left(\frac{s_i}{p_i}\right)$$

$$= 0.0177068$$

$$D^{(1)}(s \parallel r) = \sum_{i=1}^{4} r_i \ln\left(\frac{r_i}{s_i}\right)$$

$$= 0.0029606$$

$$0.0206674 = 0.0177068 + 0.0029606$$

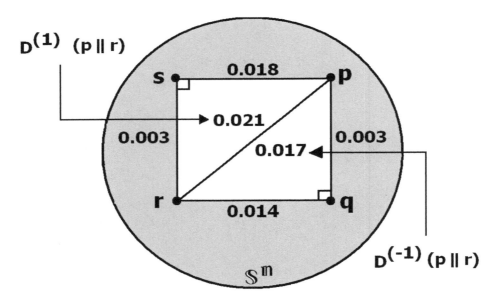

Figure 41.2: *A sketch of the Pythagorean relationships among the four probability distributions p, q, r, and s (not to scale). The $\alpha = -1$ and $\alpha = 1$ projections are shown.*

41.3 Model Interactions and Foliations

The above material served to prime the pump for the real objective, as viewed from an information processing standpoint, of how to decompose the effects due to any level of interactions in complicated models. In the introductory kangaroo scenario, there are the two marginal probabilities for beer and hand preference available for independence models. There is only the one double interaction available to model correlations.

Therefore, we want to examine those mutually dual foliations which shed light on independence models and models with correlations. Fortunately, only one small conceptual change is required from what we have accomplished in the previous section. If models with no correlations are desired, then the Lagrange multiplier λ_3 will equal 0. Any information from the constraint function $F_3(X = x_i) = (1, 0, 0, 0)$ and its average $\langle F_3 \rangle = Q_1$ will not enter into any MEP assignment with $\lambda_3 = 0$.

Thus, $E[\theta^{II}]$ for the mixed coordinates of points q and r will now be fixed at the constant value of $\theta^3 = 0$. We want these points to live in the sub-manifold where there are no correlations, or in other words, no double interaction. These are different points than before, but with the same labels of $q, r, s \ldots$

Point p, as an example of a complicated model containing information from a double interaction, will be exactly the same as before. As already mentioned, point p exists at the intersection of the two sub-manifolds $\{M[\eta_I(p)], E[\theta^{II}(p)]\}$.

The mixed coordinates for p are the same as before with $\xi = (0.75, 0.75, 4.02535)$. The e-affine coordinates also remain at $(\theta^1, \theta^2, \theta^3) = (-1.38629, -1.38629, 4.02535)$, and the m-affine coordinates at $(\eta_1, \eta_2, \eta_3) = (0.75, 0.75, 0.70)$. The assignment under this model is the by now familiar,

$$Q_i \equiv P(X = x_i \mid \mathcal{M}_p) = (0.70, 0.05, 0.05, 0.20)$$

Since $\lambda_3 = 4.02535$, this model is definitely incorporating information about the interaction between hand and beer preference.

However, when we turn to point q we implement the changes for a model with no interactions. Point q will be the α-projection of point p into the sub-manifold where there is no double interaction. Point q exists at the intersection of $\{M[\eta_I(p)], E[\theta^{II}(r)]\}$. Its new mixed coordinates are $\xi = (0.75, 0.75, 0)$. Its e-affine coordinates are $(\theta^1, \theta^2, \theta^3) = (-1.09861, -1.09861, 0)$, while its m-affine coordinates are $(\eta_1, \eta_2, \eta_3) = (0.75, 0.75, 0.5625)$. Its numerical assignment under this independence model is recognized as,

$$Q_i \equiv P(X = x_i \mid \mathcal{M}_q) = (0.5625, 0.1875, 0.1875, 0.0625)$$

Point r must be another independence model since it also has a model parameter of $\lambda_3 = 0$. It lives in the same E sub-manifold as point q, but at the intersection of a different M sub-manifold than q. The consequence of this is that (η_1, η_2) will change from $(0.75, 0.75)$. Say it changes to $(0.70, 0.80)$.

Point r exists then at the intersection of $\{M[\eta_I(r)], E[\theta^{II}(r)]\}$. Perhaps it would help to remember that section I consists of (η_1, η_2), and section II consists of θ^3. The new mixed coordinates are $\xi = (0.70, 0.80, 0)$. The e–affine coordinates are found to be $(\theta^1, \theta^2, \theta^3) = (0.847298, 1.38629, 0)$, while the m–affine coordinates are $(\eta_1, \eta_2, \eta_3) = (0.70, 0.80, 0.56)$. The numerical assignment to the point r, or better, to the probability distribution Q_i under this independence model, is calculated as,

$$Q_i \equiv P(X = x_i \,|\, \mathcal{M}_r) = (0.56, 0.14, 0.24, 0.06)$$

Finally, we come to point s. Since this distribution exists at the intersection of $\{M[\eta_I(r)], E[\theta^{II}(p)]\}$, it also, like p, must be a model that incorporates the information from the double interaction. The new mixed coordinates, with the same λ_3 as p, are $\xi = (0.70, 0.80, 4.02535)$. The e–affine coordinates are found to be $(\theta^1, \theta^2, \theta^3) = (-2.27704, -0.42483, 4.025535)$, while the m–affine coordinates are $(\eta_1, \eta_2, \eta_3) = (0.70, 0.80, 0.681392)$. The numerical assignment to the point s, or better, to the probability distribution Q_i under this correlated model is calculated as,

$$Q_i \equiv P(X = x_i \,|\, \mathcal{M}_s) = (0.681392, 0.018608, 0.118608, 0.181392)$$

Table 41.2 performs the same service as the previous table by gathering in one place the different coordinate systems together with the numerical assignment for these four points. Points q, r, and s are different than the points in the last section with same label.

Table 41.2: *The m–affine, e–affine, and mixed coordinate systems, together with the probability assignment for four points in a Riemannian manifold. Both points q and r live in the sub–manifold where there is no interaction.*

Point	η_i	θ^i	η_I, θ^{II}	$P(X = x_i \,|\, \mathcal{M}_k)$
p	0.7500 0.7500 0.7000	-1.38629 -1.38629 $+4.02535$	0.75000 0.75000 4.02535	$(0.70, 0.05, 0.05, 0.20)$
q	0.7500 0.7500 0.5625	-1.09861 -1.09861 0	0.75000 0.75000 0	$(0.5625, 0.1875,$ $0.1875, 0.0625)$
r	0.7000 0.8000 0.5600	$+0.84730$ $+1.38629$ 0	0.70000 0.80000 0	$(0.56, 0.14, 0.24, 0.06)$
s	0.7000 0.8000 0.6814	-2.27704 -0.42483 $+4.02535$	0.70000 0.80000 4.02535	$(0.681392, 0.018608,$ $0.118608, 0.181392)$

41.4 Information Decomposition

Intuition will be helped for what is to come by drawing another picture of the intersecting sub–manifolds representing the mutually dual foliation of the entire Riemannian space. To that objective, see Figure 41.3 below.

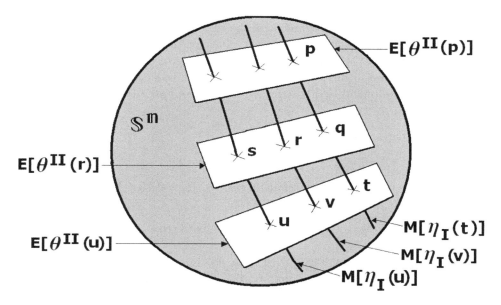

Figure 41.3: *Mutually dual foliation of the Riemannian space designed to illustrate information decomposition.*

For a change of pace, change the definitions of Section I and Section II. Let Section I now consist of just the first coordinate, while section II consists of the last two coordinates. Since point p occurs at the intersection of the two sub–manifolds $\{M[\eta_I(t)], E[\theta^{II}(p)]\}$, the mixed coordinates ξ for point p are now,

$$\xi = (\eta_I(t), \theta^{II}(p)) = (0.75, -1.38629, 4.02535)$$

Thus, it must be that the one fixed value for $\eta_I(t) = 0.75$, while the two fixed values for $\theta^{II}(p) = (-1.38629, 4.02535)$. The numerical assignment, of course must remain the same at,

$$Q_i \equiv P(X = x_i \mid \mathcal{M}_p) = (0.70, 0.05, 0.05, 0.20)$$

Since the emphasis has shifted to the decomposition of information measures related to models containing varying levels of interactions, we let the θ^i take on more and more 0 values. During this initial exploration, things remain very simple because for all of the "complicated" models, of which p is an example, there is only one double interaction. The model parameter for p has $\lambda_3 \equiv \theta^3 = 4.02535$ to enforce this interaction constraint function.

All of the points intersecting $E[\,\theta^{II}(p)\,]$ will enforce the double interaction to some extent. They will vary only to the extent that $M[\,\eta_I\,]$ varies. For example, the point (not shown) occurring at the intersection of $\{M[\,\eta_I(u)\,], E[\,\theta^{II}(p)\,]\}$ will be a different "complicated" model enforcing the double interaction, but with a different numerical assignment. None of these points living in the sub–manifold $E[\,\theta^{II}(p)\,]$ will be independence models.

Sub–manifold $E[\,\theta^{II}(r)\,]$ starts to implement the plan to make the contributions from a constraint function ineffective. The fixed values for the section II θ^i values are $\theta^2 = 1.09861$ and $\theta^3 = 0$. Thus, with $\theta^3 = 0$ all the points intersecting this sub–manifold will lack any kind of interaction. They will all be independence models, but with θ^2 not 0 these probability distributions will still contain information from both marginal probabilities.

The mixed coordinates ξ for point r are then,

$$\xi = (\eta_I(v), \theta^{II}(r)) = (0.60, 1.09861, 0)$$

where it is now obvious that $\eta_I(v) = 0.60$ changing from $\eta_I(t) = 0.75$. The numerical assignment for the probability distribution at r is,

$$Q_i \equiv P(X = x_i \,|\, \mathcal{M}_r) = (0.45, 0.15, 0.30, 0.10)$$

The $\alpha = -1$ projection of point p down to the sub–manifold $E[\,\theta^{II}(r)\,]$ where there are no interactions, by definition, must be that point $q^\star$ in the sub–manifold that attains the minimum relative entropy from point p. That minimum occurs at point q where the first two constraint function averages $\langle F_1 \rangle \equiv \eta_1$ and $\langle F_2 \rangle \equiv \eta_2$ for q must match those for p. We see that this is true because at the intersection of $\{M[\,\eta_I(t)\,], E[\,\theta^{II}(r)\,]\}$ the mixed coordinates are $\xi = (0.75, 1.09861, 0)$ leading to the numerical assignment at q of,

$$Q_i \equiv P(X = x_i \,|\, \mathcal{M}_q) = (0.5625, 0.1875, 0.1875, 0.0625)$$

Before we finish by providing a convenient summary table for all of the points shown in the mutually dual foliation of the Riemannian space for the introductory kangaroo scenario, consider the case of point u. The last sub–manifold $E[\,\theta^{II}(u)\,]$ was constructed so that not only was the interaction rendered ineffective as above in $E[\,\theta^{II}(r)\,]$, but, in addition, so was the second marginal probability. Both θ^2 and θ^3 assume the fixed value of 0 in this sub–manifold. However, the dual manifold $M[\,\eta_I(u)\,]$ has a fixed value of 0.50.

This all leads to the mixed coordinates $\xi = (0.50, 0, 0)$ for point u and the numerical assignment under this model of,

$$Q_i \equiv P(X = x_i \,|\, \mathcal{M}_u) = (0.25, 0.25, 0.25, 0.25)$$

Thus, point u is the very special probability distribution that we have been calling the "fair model," or the probability distribution ensuing from the information in a model favoring no joint statement.

FOLIATIONS

Amari makes a theorem out of showing that the canonical divergence function involving the complicated model at point p and the fair model at point u can be decomposed by the various α–projections starting at point p. And it turns out to be a very compelling intuitive result that closely follows the Pythagorean relationships we have studied in detail. For example, in Figure 41.3,

$$D^{(-1)}(p \parallel u) = D^{(-1)}(p \parallel q) + \cdots + D^{(-1)}(v \parallel u)$$

Table 41.3 summarizes the coordinate situation for all seven points that were shown in Figure 41.3.

Table 41.3: *The m–affine, e–affine, and mixed coordinate systems, together with the probability assignment for seven points in a Riemannian manifold. Points q, r, and s live in the sub–manifold where there is no interaction. Points t and v live in the sub–manifold where there is no effect from the interaction or the second marginal probability. Point u contains the information from the fair model where there are no effects from the marginal probabilities or the interaction.*

Point	η_i	θ^i	η_I, θ^{II}	$P(X = x_i \mid \mathcal{M}_k)$
p	0.7500 0.7500 0.7000	−1.38629 −1.38629 +4.02535	0.75000 −1.38629 4.02535	(0.70, 0.05, 0.05, 0.20)
q	0.7500 0.7500 0.5625	+1.09861 +1.09861 0	0.75000 1.09861 0	(0.5625, 0.1875, 0.1875, 0.0625)
r	0.6000 0.7500 0.4500	+0.40547 +1.09861 0	0.60000 1.09861 0	(0.45, 0.15, 0.30, 0.10)
s	0.5000 0.7500 0.3750	0 +1.09861 0	0.50000 1.09861 0	(0.375, 0.125, 0.375, 0.125)
t	0.7500 0.5000 0.3750	+1.09861 0 0	0.75 0 0	(0.375, 0.375, 0.125, 0.125)
v	0.6000 0.5000 0.3000	+0.40547 0 0	0.60 0 0	(0.30, 0.30, 0.20, 0.20)
u	0.5000 0.5000 0.2500	0 0 0	0.50 0 0	(0.25, 0.25, 0.25, 0.25)

Let's verify from first principles that the Pythagorean theorem holds for the triangle $\triangle ptu$.

$$D^{(-1)}(p \parallel u) = D^{(-1)}(p \parallel t) + D^{(-1)}(t \parallel u)$$

$$\sum_{i=1}^{4} p_i \ln\left(\frac{p_i}{u_i}\right) = \sum_{i=1}^{4} p_i \ln\left(\frac{p_i}{t_i}\right) + \sum_{i=1}^{4} t_i \ln\left(\frac{t_i}{u_i}\right)$$

$$0.515161 = 0.384349 + 0.130812$$

What is an alternative formula for the left hand side since point u has the numerical assignment of $Q_i = 1/n$ under the fair model?

$$\sum_{i=1}^{n} p_i \ln\left(\frac{p_i}{u_i}\right) = -H_{\max}(p) - (-\ln n)$$

$$\ln n - H_{\max}(p) = \ln 4 + \sum_{i=1}^{4} p_i \ln p_i$$

$$= 1.3863 + (0.70 \ln 0.70 + \cdots + 0.20 \ln 0.20)$$

$$= 0.515161$$

41.5 Decomposition for a Larger State Space

Amari presents an example of his ideas about information decomposition involving a triple interaction. Since we have at our disposal the familiar enhanced kangaroo scenario with a state space of dimension $n = 8$, we can check to see if our thoughts are in synch with his on these issues.

The kangaroo scenario that added a third binary variable of fur color to the already existing binary variables of hand and beer preference has been studied extensively, beginning in Chapter Twenty Two of Volume II [8, pg. 149]. It will serve then as our new running numerical example for the ideas introduced in this Chapter as applied to a larger state space.

The most complicated models in the enhanced kangaroo scenario happen to possess $m = n - 1 = 7$ constraint functions in order to include information on all three marginal probabilities, all three double interactions, and, finally, the one triple interaction. Our generic point p, standing in for an example of such a complicated model, has the numerical assignment from the MEP formula of,

$$Q_i \equiv P(X = x_i \mid \mathcal{M}_p) = \frac{\exp\left[\sum_{j=1}^{m} \lambda_j F_j(X = x_i)\right]}{Z(\lambda_1, \cdots, \lambda_7)}$$

$$= (0.40, 0.10, 0.18, 0.07, 0.05, 0.05, 0.12, 0.03)$$

I would like to strongly emphasize that p must be considered, first and foremost, as just an ordinary, run–of–the–mill MEP assignment from the class of complicated models. The model behind p imparts certain information concerning all of the interactions. Therefore, it has nothing whatsoever to do with any data.

Nonetheless, as we are aware from our previous study of this scenario, this particular MEP assignment *is* special in that it is the maximum likelihood model. The seven constraint function averages $\langle F_j \rangle$ defining the information for p exactly match the sample averages $\overline{F}_j$ from the $N = 100$ observations of the kangaroos.

It is conceptually important to note that the IP becomes aware of this fact only *after* the data have been collected, and certainly not *before* any of the observations on the kangaroos. Then, p was just one of the possible MEP assignments to the probabilities for the statements in the state space.

All seven Lagrange multipliers for the probability distribution p have non–zero values, indicating, as we expect, that all seven constraint functions are involved in the assignment. However, the IP could set $\lambda_7 = 0$ in order to negate the effect of the triple interaction under a new model. We would then have a slightly less complicated model incorporating the information from all three marginal probabilities, and all three double interactions. Then, m drops from 7 to 6.

In like manner, we could successively set $\lambda_6 = 0$, $\lambda_5 = 0$, and, finally, $\lambda_4 = 0$ to eliminate in succession all of the double interactions. With $m = 3$, all that remains are the class of independence models.

Amari sets up a different foliation of the entire Riemannian space than the type we have been examining up till now. He labels this new foliation as an "hierarchical e–structure" with the notation of $\boldsymbol{E}_k(0)$. The 0 argument indicates that some of the θ^i coordinates have been set to 0. For each $\boldsymbol{E}_k(0)$, the 0s start with θ^{k+1}. These sub–manifolds are nested within each other such that $\boldsymbol{E}_k(0)$ is a sub–manifold of $\boldsymbol{E}_{k+1}(0)$.

So, for example,

$$\boldsymbol{E}_0(0) \equiv (\theta^1 = 0, \theta^2 = 0, \cdots, \theta^7 = 0)$$

$$\boldsymbol{E}_1(0) \equiv (\theta^1, \theta^2 = 0, \theta^3 = 0, \cdots, \theta^7 = 0)$$

$$\vdots \equiv \vdots$$

$$\boldsymbol{E}_6(0) \equiv (\theta^1, \theta^2, \theta^3, \theta^4, \theta^5, \theta^6, \theta^7 = 0)$$

If a θ^i has not been set to 0 by the $\boldsymbol{E}_k(0)$ it can take on any value. Notice that the lowest sub–manifold $\boldsymbol{E}_0(0)$ in the hierarchy consists of just a single point u with the numerical assignment of $Q_i = 1/8$.

Amari retains the idea of a mixed coordinate system defined as,

$$\xi_k = (\eta_1, \eta_2, \cdots, \eta_k, \theta^{k+1}, \cdots, \theta^{n-1})$$

For example, the point with the mixed coordinate system of,

$$\xi_3 = (\eta_1, \eta_2, \eta_3, \theta^4 = 0, \theta^5 = 0, \theta^6 = 0, \theta^7 = 0)$$

belongs to the sub–manifold $\boldsymbol{E}_3(0)$. And so, Amari uses almost the same language as we do when he says that, "The submanifold $\boldsymbol{E}_k$ includes only the probability distributions that do not have effects higher than k."

As usual, the **m**–projection of the point p (representing the class of complicated models inserting information from all three double interactions and the one triple interaction) down to some sub–manifold $\boldsymbol{E}_k(0)$ is that point residing in $\boldsymbol{E}_k(0)$ whose relative entropy with p is the smallest. For all of the points $q \in \boldsymbol{E}_k(0)$, there is one special point $q^\star$ that has the minimum relative entropy with p. By the Pythagorean theorem,

$$D^{(-1)}(p \parallel q) = D^{(-1)}(p \parallel q^\star) + D^{(-1)}(q^\star \parallel q)$$

The curve connecting p and $q^\star$ is called the m–geodesic and is orthogonal to $\boldsymbol{E}_k(0)$. The curve connecting q and $q^\star$ is called the e–geodesic and is in $\boldsymbol{E}_k(0)$, so it is orthogonal to the m–geodesic that connects p and $q^\star$.

Look at the **m**–projection of p down to $\boldsymbol{E}_6(0)$. We are trying to find the point called $q^\star$ in $\boldsymbol{E}_6(0)$ closest to p. In other words, what is the best model with only double interactions that comes closest to the more complicated model that includes all of these double interactions, but also includes the one triple interaction?

By definition, all points q in $\boldsymbol{E}_6(0)$ have $\theta^7 = 0$ in order to make sure that the constraint function $F_7(X = x_i)$ does not have any influence. Thus, every single model q in $\boldsymbol{E}_6(0)$ has all three double interactions, but only one of them, namely $q^\star$, is special in the sense that its relative entropy is the absolute minimum with regard to p.

In the mixed system of coordinates, all of the $q \in \boldsymbol{E}_6(0)$ have the coordinates $\xi_6 = (\eta_1, \cdots, \eta_6, 0)$ where η_1 through η_6 are free to vary. Select some arbitrary q, with, say, the mixed coordinates $\xi_6 = (0.70, 0.80, 0.55, 0.60, 0.52, 0.44, 0)$.

This arbitrary point q is a completely legitimate probability distribution under a model with information from all three double interactions. But it lacks information from any triple interaction. The numerical assignment for this distribution is,

$$Q_i \equiv P(X = x_i \mid \mathcal{M}_q) = (0.429, 0.091, 0.171, 0.009, 0.011, 0.019, 0.189, 0.081)$$

Select a candidate $q^\star$ without the benefit of knowing what criteria it must meet. This point, like q, must also live in the sub–manifold $\boldsymbol{E}_6(0)$. Suppose that $q^\star$ has mixed coordinates $\xi_6 = (0.69, 0.81, 0.54, 0.61, 0.51, 0.45, 0)$.

Once again, this assignment is a completely legitimate probability distribution under a different model with information from all three double interactions, but missing the information from the triple interaction. The numerical assignment for

FOLIATIONS

our first candidate $q^\star$ is,

$$Q_i \equiv P(X = x_i \,|\, \mathcal{M}_{q^\star})$$

$$= (0.438475, 0.0715247, 0.171525, 0.00847533,$$

$$0.0115247, 0.0184753, 0.188475, 0.0915247)$$

Suppose further that this candidate $q^\star$ is indeed better than the arbitrary q just introduced. However, we don't know yet whether it is that special $q^\star$ which is at the minimum distance from p.

We already know that the mixed coordinates for p are,

$$\xi_6 = (\eta_1^p, \eta_2^p, \cdots, \theta_p^7) = (0.75, 0.75, 0.60, 0.58, 0.50, 0.45, 1.8281271)$$

The final Lagrange multiplier λ_7^p for p does not equal 0 because it must enforce the triple interaction. We now have all of the ingredients needed to compute the triangular relationships among the points.

However,

$$D^{(-1)}(p \,\|\, q) - [\, D^{(-1)}(p \,\|\, q^\star) + D^{(-1)}(q^\star \,\|\, q) \,] \neq 0$$

The Pythagorean theorem does not hold for this choice of $q^\star$. This $q^\star$ is thus not the **m**–projection of p to $\boldsymbol{E}_6(0)$. There must exist another $q^\star$ such that its m–geodesic *is* orthogonal to $\boldsymbol{E}_6(0)$. If, in fact, we choose $q^\star$ such that its first six η_i coordinates match p's first six η_i coordinates, then we find that this $q^\star$ is the one we have been looking for.

The dual version of Amari's Theorem 3.7, studied in the last Chapter, says that this formula,

$$D^{(-1)}(p \,\|\, q^\star) + D^{(-1)}(q^\star \,\|\, q) - D^{(-1)}(p \,\|\, q) = \{\, \theta^i(q^\star) - \theta^i(q) \,\}\{\, \eta_i(q^\star) - \eta_i(p) \,\}$$

will assess the departure from the IG version of the Pythagorean theorem.

In this case, the non–zero value of,

$$\{\, \theta^i(q^\star) - \theta^i(q) \,\}\{\, \eta_i(q^\star) - \eta_i(p) \,\} = -0.0177829$$

tells us that we have to search for another $q^\star$ that, when inserted into the above formula, yields 0.

It is satisfying to see, from a strictly numerical assessment, why the minimum divergence function $D^{(-1)}(p \,\|\, q^\star)$ must have the first six η_i for $q^\star$ matching up with the corresponding η_i for p. In the implied summation of the expression,

$$\{\, \theta^i(q^\star) - \theta^i(q) \,\}\{\, \eta_i(q^\star) - \eta_i(p) \,\}$$

the first six terms in $\{\, \eta_i(q^\star) - \eta_i(p) \,\}$ will all equal 0, thus annihilating the actual differences in the first six terms in $\{\, \theta^i(q^\star) - \theta^i(q) \,\}$. Then, by definition,

$$\theta^7(q^\star) = \theta^7(q) = 0$$

so that the final actual difference between $\eta_7(q^\star)$ and $\eta_7(p)$ is also annihilated. The result must be 0, and the m–geodesic from p to q is orthogonal to the e–geodesic from q to $q^\star$. All of the computational details to verify these assertions will be forthcoming while exploring Exercises 41.8.11 through 41.8.16.

41.6 IG's Reaction to the MEP

One detects a difference in tone with regard to the relative importance of the MEP that is strongly dependent on an author's core philosophical approach to probability theory. Does he or she approach probability from an information–centric philosophy, or from an abstract geometric philosophy? Take as an example the contrasting emphasis placed on the MEP by my exposition highlighting MEP models as carriers of information versus Amari's characterization of the MEP as a minor corollary ensuing from the Pythagorean relationship.

For me, the MEP is intimately involved in the second of the two fundamental pillars that underlie inferencing. The formal rules for manipulating probability expressions have no need for the MEP. But when it comes time to actually assign numerical values to the degrees of belief, the MEP blossoms into an indispensable tool.

The MEP appears right at the outset of an inferencing problem, at least in principle. After the state space and constraint functions have been defined, the MEP becomes the operational algorithm for incorporating information into a probability distribution through some model. Then, any joint probability,

$$Q_i \equiv P(X = x_i \,|\, \mathcal{M}_k)$$

as might be required later by the formal manipulation rules can be given a numerical assignment.

At this early stage of the inferencing problem, there can be no relative entropy calculations before the MEP has made its assignments to points p, q, r, and so on. The MEP is the necessary precursor before any geometric notion of distance–like measures can be invoked.

Moreover, the rationale for the MEP, as I happen to view it, stands on its own, independent of the need for any buttressing from geometric arguments. Jaynes's argument illustrated that a generic variational technique can be leveraged to find a maximum value of all the missing information subject to the active information inserted by the IP in the form of constraint function averages. From then on, it becomes a procedure requiring numerical optimization. No further support for this rationale is required from any geometric considerations.

Compare this information–centric attitude, as I and many others have adopted, with Amari's geometric characterization of the MEP as a barely mentionable by–product of projecting points in a manifold down to lower sub–manifolds. The MEP is mentioned only after points, coordinate systems, divergence functions, α–connections, and so on, are assumed to exist in the Riemannian space. This supporting scaffolding required for the geometric approach is a lot more complex than that erected under the information–centric approach.

As introduced in this Chapter, the MEP is viewed off–handedly as a rather curious by–product of projecting points to sub–manifolds. As Amari tells us, the **m**–projection "is closely related to the maximal [sic] entropy principle ..." The IG version of the Pythagorean Theorem is expressed using divergence functions,

$$D^{(-1)}(p \| u) = D^{(-1)}(p \| q) + D^{(-1)}(q \| u)$$

where p is any complicated model, q is an **m**–projection of p, and u is the fair model where the $Q_i = 1/n$.

Repeating for the umpteenth time the derivation of the relative entropy between any complicated model p and the fair model u because that is how I do things, we find that,

$$KL(p,u) = \sum_{i=1}^{n} p_i \ln\left(\frac{p_i}{u_i}\right)$$

$$= \sum_{i=1}^{n} p_i \ln p_i - \sum_{i=1}^{n} p_i \ln u_i$$

$$= \sum_{i=1}^{n} p_i \ln p_i - \ln u_i \sum_{i=1}^{n} p_i \quad \text{(because the } u_i \text{ are constants)}$$

$$= -H(p) - \left(\sum_{j=1}^{m} \lambda_j^u F_j(X = x_i) - \ln Z_u\right)$$

Since all $\lambda_j^u = 0$

$$KL(p,u) = -H(p) + \ln Z_u$$

$$= \ln n - H(p)$$

Do the same thing for the two terms on the right hand side of the Pythagorean theorem. The first term on the right hand side, the relative entropy between p and q is,

$$KL(p,q) = \sum_{i=1}^{n} p_i \ln\left(\frac{p_i}{q_i}\right)$$

$$= \sum_{i=1}^{n} p_i \ln p_i - \sum_{i=1}^{n} p_i \ln q_i$$

$$= \sum_{i=1}^{n} p_i \ln p_i - E_p[\ln q_i]$$

$$= -H(p) - E_p\left(\sum_{j=1}^{m} \lambda_j^q F_j(x_i) - \ln Z_q\right)$$

$$= -H(p) - \sum_{j=1}^{m} \lambda_j^q \langle F_j \rangle_p + \ln Z_q$$

The second term on the right hand side, the relative entropy between q and u, is,

$$KL(q,u) = \sum_{i=1}^{n} q_i \ln\left(\frac{q_i}{u_i}\right)$$

$$= \sum_{i=1}^{n} q_i \ln q_i - \sum_{i=1}^{n} q_i \ln u_i$$

$$= \ln n - H(q)$$

Putting this all back together again, we have,

$$D^{(-1)}(p \parallel u) = D^{(-1)}(p \parallel q) + D^{(-1)}(q \parallel u)$$

$$KL(p,u) = KL(p,q) + KL(q,u)$$

$$\ln n - H(p) = -H(p) - \sum_{j=1}^{m} \lambda_j^q \langle F_j \rangle_p + \ln Z_q + (\ln n - H(q))$$

$$0 = -H(q) + \ln Z_q - \sum_{j=1}^{m} \lambda_j^q \langle F_j \rangle_p$$

$$H(q) = \ln Z_q - \sum_{j=1}^{m} \lambda_j^q \langle F_j \rangle_p$$

Thus, the Pythagorean relationship,

$$D^{(-1)}(p \parallel u) = D^{(-1)}(p \parallel q) + D^{(-1)}(q \parallel u)$$

is true if the MEP algorithm finds the assignments to q based on the information in $\langle F_j \rangle_p$,

$$H_{\max}(q) = \ln Z_q - \sum_{j=1}^{m} \lambda_j^q \langle F_j \rangle_p$$

This relationship is seen to be the standard Legendre transformation technique we have always employed as an algorithm for finding the MEP assignment to q. Here, the information will consist of some number of constraint function averages from p where λ_j^q does not equal 0. In the examples we have been using from the $n = 8$ kangaroo scenario, only $\lambda_7^q = 0$. Thus, the information from p's three marginal probabilities, and its three double interactions, but not the information about the triple interaction, were used to find q.

Amari encapsulates all of this geometric development about the MEP into a theorem that states:

The m–projection of p to q in E_k maximizes the entropy among the $q \in M_k$ having the same k–marginals as p.

While true, to my way of thinking this doesn't have quite the intuitive punch of the information–centric derivation.

Finding the MEP assignment to q involves the shortest squared distance between point u and point q, namely, the e–geodesic between u and q because q has to be as close as possible to the point encapsulating no information. Furthermore, point p must be **m**–projected to point q, in other words, the m–geodesic between p and q must be found because q must possess those constraint function averages wherever λ_j^q does not equal 0. For example, if q happens to live in $E_6(0)$, then q will take on the information from p's first six constraint function averages.

Points p, q, and u all lie at the intersection of certain sub–manifolds E_k and M_k where the specification of the mixed coordinate system makes everything work out properly. These statements are clarified with a numerical example in Exercise 41.8.20 accompanied by a sketch of the points and their intersecting sub–manifolds.

Doesn't all of this strike you as a rather convoluted way of arriving at the MEP? As I look at it, it seems especially so when compared to the much more direct approach that involves the concepts of active information in the form of constraint function averages, and its counterpart, the missing information as quantified by information entropy.

But, in the end, this is what is so fascinating. It's not about the underlying mathematics; it is about the core concepts that drive the human mind to construct a personally aesthetically pleasing and coherent structure.

41.7 Connections to the Literature

Most of the material in this Chapter was motivated by Amari's paper entitled [4],

Information Geometry on Hierarchy of Probability Distributions

As a convenient summary, I will quote extensively from this section of his paper,

> IV. SIMPLE EXAMPLE: TRIPLEWISE INTERACTIONS.

Interspersed along the way will be the correspondence between the numerical results from my $n = 8$ enhanced kangaroo scenario when bounced off of Amari's more abstract example as he presents it in the above mentioned section of his paper.

The goal of this exercise is to highlight the contrasts between an information–centric approach and the geometric approach. The information–centric approach leans heavily on viewing the MEP as the vehicle through which an IP introduces information into a probability distribution at the very beginning of the inferential problem. The latter approach concentrates on calculating geometric quantities like squared distances between points, and does not see fit to introduce the concept of information until the very end.

Let's begin with the familiar notions of a joint probability distribution together with Amari's unique way of describing it. Fortunately, his choice of an example with a discrete state space where $n = 8$, and an emphasis on interactions, provides a nice overlap with my extensive discussion of these same issues within the enhanced kangaroo scenario. It's a profitable exercise to identify in the deconstruction of an inferential problem the place where two people take two different paths.

> The general results are explained by a simple example of the set of joint probability distributions $S_3 = \{p(\boldsymbol{x})\}$, $\boldsymbol{x} = (x_1, x_2, x_3)$, of binary random variables X_1, X_2 and X_3, where $x_i = 0$ or 1. We can expand $\log p(\boldsymbol{x})$
>
> $$\log p(\boldsymbol{x}) = \sum \theta_i x_i + \sum \theta_{ij} x_i\, x_j + \theta_{123} x_1 x_2 x_3 - \psi$$
>
> obtaining the log–linear model ... This shows that S_3 is an exponential family. The canonical or e–affine coordinates are $\boldsymbol{\theta} = (\theta_1, \theta_2, \theta_3; \theta_{12}, \theta_{23}, \theta_{31}; \theta_{123})$ which are partitioned as
>
> $$\boldsymbol{\theta} = (\boldsymbol{\theta}_1, \boldsymbol{\theta}_2, \boldsymbol{\theta}_3)$$
>
> $$\boldsymbol{\theta}_1 = (\theta_1, \theta_2, \theta_3)$$
>
> $$\boldsymbol{\theta}_2 = (\theta_{12}, \theta_{23}, \theta_{31})$$
>
> $$\boldsymbol{\theta}_3 = (\theta_{123})$$
>
> This defines a hierarchical e–structure in S_3 where $\boldsymbol{\theta}_2$ represents pairwise interactions and $\boldsymbol{\theta}_3$ represents the triple interaction, although they are not

orthogonal. The corresponding m-affine coordinates are partitioned as
$$\boldsymbol{\eta} = (\boldsymbol{\eta}_1, \boldsymbol{\eta}_2, \boldsymbol{\eta}_3)$$
where $\boldsymbol{\eta}_1 = (\eta_1, \eta_2, \eta_3)$ $\boldsymbol{\eta}_2 = (\eta_{12}, \eta_{23}, \eta_{13})$, and $\boldsymbol{\eta}_3 = (\eta_{123})$, with
$$\eta_i = E[x_i] = \text{Prob}\{x_i = 1\}$$
$$\eta_{ij} = E[x_i x_j] = \text{Prob}\{x_i = x_j = 1\}$$
$$\eta_{123} = E[x_1 x_2 x_3] = \text{Prob}\{x_1 = x_2 = x_3 = 1\}$$

Despite the differences in basic concepts as reflected in language that talks about random variables, or fails to mention anything about the information in models, Amari sets up an inferential problem that is completely analogous to the $n = 8$ kangaroo scenario.

We also consider the set of joint probability distributions $Q_i \equiv P(X = x_i \,|\, \mathcal{M}_k)$ for $i = 1$ to 8. The $(X = x_i)$, however, are *statements*, and the mapping from statements to real numbers, the constraint functions $F_j(X = x_i)$, are the only things which might be thought of as "random variables."

From the outset, the $F_j(X = x_i)$ are defined to capture the information in the three marginal probabilities, the three double interactions, and the one triple interaction through the respective constraint function averages. This is much more transparent than binary random variables taking on the values of 0 or 1.

And, certainly our MEP characterization for the Q_i as,
$$P(X = x_i \,|\, \mathcal{M}_k) = \frac{\exp\left[\sum_{j=1}^{m} \lambda_j F_j(X = x_i)\right]}{Z(\lambda_1, \lambda_2, \cdots, \lambda_m)}$$
is an exponential family whose logarithmic expansion is,
$$\ln Q_i = \sum_{j=1}^{m} \lambda_j F_j(X = x_i) - \ln Z$$

Our Lagrange multipliers λ_j are the same as Amari's e-affine coordinates θ_i. In most of his articles, Amari uses the notation of θ^i for these canonical coordinates, so we will adhere to that convention. Our constraint function averages $\langle F_j \rangle$ are the same as Amari's m-affine coordinates η_i.

Although not a big issue, I prefer to stick to the simplicity of the full breakdown represented by $\sum_{j=1}^{m} \lambda_j F_j(X = x_i)$ rather than the partitioning scheme that Amari lays down. In my scheme, we can step down by one at every level to assess, for example, when the information from only two of the double interactions is effective, or the information from only one double interaction is effective, and so on.

And, I think it is clearer when defining the η_i to always go back to the MEP in order to understand that,
$$\eta_i \equiv \langle F_j \rangle = \sum_{i=1}^{n} F_j(X = x_i) \, Q_i$$

For example, take the second double interaction as represented by the symbols BH in the kangaroo scenario. Using BH as a short form will allow us to specify the information in the beer and hand preference correlation. Since the fourth constraint function $F_4(X = x_i) = (1, 0, 1, 0, 0, 0, 0, 0)$ picks out the first and third cell in the joint probability table, $\langle F_4 \rangle$ specifies the marginal probability,

$$\eta_4 \equiv \langle F_4 \rangle = Q_1 + Q_3 = P(BHF) + P(BH\overline{F}) = P(BH)$$

During any examination of the effects of the BH interaction, it is clear that the Lagrange multiplier λ_4 will be involved in Q_1 and Q_3.

It is also clearer to avoid the double and triple subscripting that Amari adopts for his coordinates. I opt to increment the coordinates by one as we move through the interactions. So, for example, Amari's notation for the triple interaction,

$$\boldsymbol{\eta}_3 = (\eta_{123}) = E[x_1 x_2 x_3] = \text{Prob}\{x_1 = x_2 = x_3 = 1\}$$

is for me simply,

$$\eta_7 \equiv \langle F_7 \rangle = Q_1 = P(X = x_1 \,|\, \mathcal{M}_k)$$

For me then, the subscripts 1, 2, and 3 refer to the marginal probabilities $P(B)$, $P(H)$, and $P(F)$ for the main effects, the subscripts 4, 5, and 6 to the marginal probabilities $P(BH)$, $P(BF)$, and $P(HF)$ of the double interactions, with the final subscript 7 referring to the marginal probability (here just a single joint probability) $P(BHF)$ for the triple interaction.

Pick up the thread where Amari defines the hierarchical e-structure for this problem.

We have a hierarchical e-structure

$$\boldsymbol{E}_0 \subset \boldsymbol{E}_1 \subset \boldsymbol{E}_2 \subset \boldsymbol{E}_3 = \boldsymbol{S}_3$$

where $\boldsymbol{E}_0$ is a singleton $p_0(\boldsymbol{x}) = 1/8$ with $\boldsymbol{\theta} = 0$. $\boldsymbol{E}_1$ is defined by $\boldsymbol{\theta}_{1+} = (\boldsymbol{\theta}_2, \boldsymbol{\theta}_3) = 0$, $\boldsymbol{E}_2$ is defined by $\boldsymbol{\theta}_{2+} = \boldsymbol{\theta}_3 = 0$. On the other hand, $\boldsymbol{\eta}_{1-} = \boldsymbol{\eta}_1 = (\eta_1, \eta_2, \eta_3)$ gives the marginal distributions of X_1, X_2 and X_3, and $\boldsymbol{\eta}_{2-} = (\boldsymbol{\eta}_1, \boldsymbol{\eta}_2)$ gives the all the [sic] pairwise marginal distributions.

There are no issues here except that one must pay close attention to the notation. $\boldsymbol{E}_0(0)$ was defined earlier in this Chapter as $(\theta^1 = 0, \theta^2 = 0, \cdots, \theta^7 = 0)$. So, for us, $\boldsymbol{E}_0(0)$ is simply the one distribution under the fair model with $Q_i = 1/8$. $\boldsymbol{E}_1(0)$ is the sub-manifold where the triple interaction and all three double interactions have been nullified. $\boldsymbol{E}_2(0)$ is the sub-manifold where only the triple interaction has been nullified. $\boldsymbol{E}_3 \equiv \mathcal{S}^8$ is the full manifold where every point resides. We place point p and any other point enforcing a triple interaction (referred to as "complicated models") in $\boldsymbol{E}_3$.

A point q with an assignment of,

$$Q_i \equiv P(X = x_i \,|\, \mathcal{M}_q) = \frac{\exp\left[\sum_{j=1}^{7} \lambda_j^q F_j(X = x_i)\right]}{Z(\lambda_1, \cdots, \lambda_7)}$$

lives in sub-manifold $\boldsymbol{E}_2(0)$ because $\lambda_7^q = 0$.

FOLIATIONS

And we are in agreement as well that, for example,

$$\boldsymbol{\eta}_1 = (\eta_1, \eta_2, \eta_3) \equiv (\langle F_1 \rangle, \langle F_2 \rangle, \langle F_3 \rangle)$$

contains, as we would express it, the information on the three marginal probabilities, the probability for beer preference $P(B)$, the probability for hand preference $P(H)$, and the probability for fur color $P(F)$.

Return now to Amari's description of the mixed coordinate system.

> Consider the two mixed cut coordinates: $\boldsymbol{\xi}_1 = (\boldsymbol{\eta}_{1-}; \boldsymbol{\theta}_{1+})$ and $\boldsymbol{\xi}_2 = (\boldsymbol{\eta}_{2-}; \boldsymbol{\theta}_{2+})$. Since $\boldsymbol{\theta}_{1+}$ are orthogonal to the coordinates that specify the marginal distributions $\boldsymbol{\eta}_{1-}$, $\boldsymbol{\theta}_{1+} = (\boldsymbol{\theta}_2, \boldsymbol{\theta}_3)$ represents the effect of mutual interactions of X_1, X_2, and X_3, independently of their marginals. Similarly, E_2 defined by $\theta_{123} = 0$ is composed of all the distributions which have no intrinsic triplewise interactions but pairwise correlations. The two distributions given by $\boldsymbol{\xi}_2 = (\boldsymbol{\eta}_{2-}; \boldsymbol{\theta}_3)$ and $\boldsymbol{\xi}_2' = (\boldsymbol{\eta}_{2-}; \boldsymbol{\theta}_3')$ have the same pairwise marginal distributions but differ only in the pure triplewise interactions. Since θ_{123} is orthogonal to $\boldsymbol{\eta}_{2-} = (\boldsymbol{\eta}_1, \boldsymbol{\eta}_2)$, it represents purely triplewise interactions as is well known in the log–linear model ...

This is all very confusing. You have to spend a lot of time carefully making sure that everything does, in fact, match up. The first mixed system of coordinates mentioned translates into,

$$\boldsymbol{\xi}_1 = (\boldsymbol{\eta}_{1-}; \boldsymbol{\theta}_{1+})$$
$$= (\boldsymbol{\eta}_1, \boldsymbol{\theta}_2, \boldsymbol{\theta}_3)$$
$$= (\eta_1, \eta_2, \eta_3, \theta_{12}, \theta_{23}, \theta_{31}, \theta_{123})$$
$$= (\langle F_1 \rangle, \langle F_2 \rangle, \langle F_3 \rangle, \lambda_4, \lambda_5, \lambda_6, \lambda_7)$$

and the second mixed system of coordinates mentioned translates into,

$$\boldsymbol{\xi}_2 = (\boldsymbol{\eta}_{2-}; \boldsymbol{\theta}_{2+})$$
$$= (\boldsymbol{\eta}_1, \boldsymbol{\eta}_2, \boldsymbol{\theta}_3)$$
$$= (\eta_1, \eta_2, \eta_3, \eta_{12}, \eta_{23}, \eta_{13}, \theta_{123})$$
$$= (\langle F_1 \rangle, \langle F_2 \rangle, \langle F_3 \rangle, \langle F_4 \rangle, \langle F_5 \rangle, \langle F_6, \rangle, \lambda_7)$$

Thus, the two distributions given by the mixed cut coordinates $\boldsymbol{\xi}_2 = (\boldsymbol{\eta}_{2-}; \boldsymbol{\theta}_3)$ and $\boldsymbol{\xi}_2' = (\boldsymbol{\eta}_{2-}; \boldsymbol{\theta}_3')$ are translated more transparently into numerical assignments for probability distributions where the information is specified by a mixture of dual parameters in the models,

$P(X = x_i \mid \mathcal{M}_k)$ defined by information $(\langle F_1 \rangle, \langle F_2 \rangle, \langle F_3 \rangle, \langle F_4 \rangle, \langle F_5 \rangle, \langle F_6, \rangle, \lambda_7)$

$P(X = x_i \mid \mathcal{M}_{k'})$ defined by information $(\langle F_1 \rangle, \langle F_2 \rangle, \langle F_3 \rangle, \langle F_4 \rangle, \langle F_5 \rangle, \langle F_6, \rangle, \lambda_7')$

where it is now much easier to see, as Amari asserts, that these are two models that incorporate the same information for the three marginal probabilities and the three double interactions, but differ in the information about the triple interaction through λ_7 and λ_7'.

We now reach the part of Amari's argument where he defines the "orthogonal decompositions" for his example. His important objective in doing so is to propose an alternative to the standard definition of "mutual entropy," a concept that is not present in the MEP, but arose from Information Theory.

The partitioned coordinates $\boldsymbol{\theta} = (\boldsymbol{\theta}_1, \boldsymbol{\theta}_2, \boldsymbol{\theta}_3)$ are not orthogonal, so that we cannot say that $\boldsymbol{\theta}_2$ summarizes all the pure pairwise correlations, except for the special case $\boldsymbol{\theta}_3 = 0$. Given $p(\boldsymbol{x}, \boldsymbol{\theta})$, we need to separate pairwise correlations and triplewise interactions invariantly and obtain the "orthogonal" quantitative decomposition of these effects.

We project p to $\boldsymbol{E}_1$ and $\boldsymbol{E}_2$, giving $p^{(1)}$ and $p^{(2)}$ respectively. Then, $p^{(1)}$ is the independent distribution having the same marginals as p, and $p^{(2)}$ is the distribution having the same pairwise marginals as p but no triplewise interaction. By putting

$$D_2 = D[p : p^{(2)}]$$

$$D_1 = D[p^{(2)} : p^{(1)}]$$

$$D_0 = D[p^{(1)} : p_0]$$

we have the decomposition

$$D[p : p_0] = D_2 + D_1 + D_0$$

Here, D_2 represents the amount of purely triplewise interaction, D_1 the amount of purely pairwise interaction, and D_0 the degree of deviations of the marginals of p from the uniform distribution p_0. We have thus the "orthogonal" quantitative decomposition of interactions.

It will be helpful for my deconstruction of these remarks to refer to Figure 41.4 appearing on the next page. There are five points p, q, r, s, and u located in the dually flat space of $\mathcal{S}^n$. Point p is the same for both Amari and me, while I relabel his $p^{(2)}$ as q, his $p^{(1)}$ as s, and his p_0 as u. Point p, as always, refers to any "complicated" model with $m = 7$ constraint function averages.

Point q is the $\alpha = -1$ projection, that is, the m–projection of p onto $\boldsymbol{E}_2$. The probability distribution q contains the same information as the probability distribution p concerning the marginal probabilities and the double interactions,

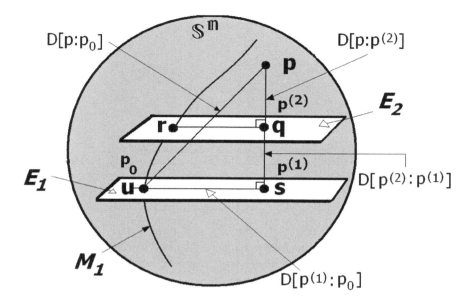

Figure 41.4: *Another mutually dual foliation of the Riemannian space designed to illustrate information decomposition.*

but eliminates the information from the triple interaction. $q \equiv p^{(2)}$ lives in the sub–manifold $\boldsymbol{E}_2$ where $\boldsymbol{\theta}_3 = \theta_{123} = \lambda_7^q = 0$.

Point r is simply another distribution, different than q, but still living in $\boldsymbol{E}_2$, and therefore lacking any triple interaction information. Like q, it does contain, however, information about the three marginal probabilities, and three double interactions.

Point s is the further **m**–projection of p onto $\boldsymbol{E}_1$ where models living there lack all three double interactions as well as the triple interaction. By definition, $(\boldsymbol{\theta}_2 = 0, \boldsymbol{\theta}_3 = 0)$. Nevertheless, these models still retain information about the marginal probabilities.

The final point $u \equiv p_0$ represents the uniform distribution where $Q_i = 1/8$. It is the "fair model." I show it living in $\boldsymbol{E}_1$ along with s, where its $\boldsymbol{\theta}_1$ coordinates are all equal to 0, but it obviously lives in $\boldsymbol{E}_0$ as well.

With this correspondence established, we can rely on the MEP algorithm to calculate those q_i whose first six $\langle F_j \rangle$ match those of p, and which additionally have the maximum entropy of any q that lives in $\boldsymbol{E}_2$. The p_i of the probability distribution p have been discussed many times already.

The q_i are found via the MEP algorithm by incorporating the above information that excludes any triple interaction, but otherwise matches all of the remaining information in p about marginal probabilities and double interactions. Thus, D_2 can be calculated first. From its definition,

$$D_2 = D[p : p^{(2)}]$$

$$= D^{(-1)}(p \parallel q)$$

$$= \sum_{i=1}^{8} p_i \ln\left(\frac{p_i}{q_i}\right)$$

$$= 0.40 \ln\left(\frac{0.40}{0.383456}\right) + 0.10 \ln\left(\frac{0.10}{0.116544}\right) + \cdots + 0.03 \ln\left(\frac{0.03}{0.0465444}\right)$$

$$= 0.015057$$

When point p is **m**–projected to point $s \equiv p^{(1)}$ living in $\boldsymbol{E}_1$, the probability distribution s contains the same information on the three marginal probabilities as p, but has eliminated all information about the double and triple interactions. $s \equiv p^{(1)}$ lives in the sub–manifold $\boldsymbol{E}_1$ where,

$$(\boldsymbol{\theta}_2, \boldsymbol{\theta}_3) = (\theta_{12}, \theta_{23}, \theta_{31}, \theta_{123}) = (\lambda_4^s = \lambda_5^s = \lambda_6^s = \lambda_7^s = 0)$$

Now, D_1 can be calculated as,

$$D_1 = D[p^{(2)} : p^{(1)}]$$

$$= D^{(-1)}(q \parallel s)$$

$$= \sum_{i=1}^{8} q_i \ln\left(\frac{q_i}{s_i}\right)$$

$$= 0.383456 \ln\left(\frac{0.383456}{0.3375}\right) + 0.116544 \ln\left(\frac{0.116544}{0.1125}\right) + \cdots + 0.0465444 \ln\left(\frac{0.0465444}{0.025}\right)$$

$$= 0.0318369$$

Point u is p_0. It also lives in $\boldsymbol{E}_1$ where,

$$(\boldsymbol{\theta}_2, \boldsymbol{\theta}_3) = (\theta_{12}, \theta_{23}, \theta_{31}, \theta_{123}) = (\lambda_4^u = \lambda_5^u = \lambda_6^u = \lambda_7^u = 0)$$

but at a different location than point s. It occurs at the intersection of $\boldsymbol{E}_1$ and the sub–manifold $\boldsymbol{M}_1$,

$$\boldsymbol{M}_1 = \{\boldsymbol{\eta}_1, \boldsymbol{\eta}_2, \boldsymbol{\eta}_3\} = \{\eta_1 = 1/2, \eta_2 = 1/2, \eta_3 = 1/2, \boldsymbol{\eta}_2, \boldsymbol{\eta}_3\}$$

This is the dual sub–manifold where the points that live there have fixed values as shown above for the first three constraint function averages, but allow the last four constraint function averages to vary. Thus, u has the mixed coordinates of,

$$\boldsymbol{\xi}_1 = \{\eta_1 = 1/2, \eta_2 = 1/2, \eta_3 = 1/2, \theta_{12} = \theta_{23} = \theta_{31} = \theta_{123} = 0\}$$

and obviously $u_i = 1/8$. The third term D_0 can be calculated as,

$$D_0 = D[p^{(1)} : p_0]$$

$$= D^{(-1)}(s \parallel u)$$

$$= \sum_{i=1}^{8} s_i \ln\left(\frac{s_i}{u_i}\right)$$

$$= \ln 8 - H_{\max}(s)$$

$$= 0.281760$$

With these computational details taken care of, we can check to see whether the decomposition according to the triple interaction, then all of the double interactions, and finally, the marginal distributions for the $n = 8$ enhanced kangaroo scenario adheres to,

$$D[p : p_0] = D_2 + D_1 + D_0$$

First of all, the left hand side $D[p : p_0]$, the divergence measure from the complicated model in p to the fair model in u is easily calculated as,

$$D[p : p_0] = D^{(-1)}(p \parallel u) = \ln 8 - H_{\max}(p) = 0.328653$$

Thus, the overall decomposition is verified with,

$$D_2 + D_1 + D_0 = 0.015057 + 0.031837 + 0.281760 = 0.328653 = D[p : p_0]$$

On the basis of this example, we might be willing to conjecture that Amari was right to construct his *Theorem 8* [4, pg. 1705], stating that,

$$D[p : p_0] = \sum_{k=1}^{n} D[p^{(k)} : p^{(k-1)}]$$

Amari's objective was to demonstrate that this information geometric definition of an orthogonal decomposition was at variance with Information Theory's standard definition for "mutual information." This allows us another opportunity to try to identify the correspondence with Amari's scenario and our enhanced kangaroo scenario [4, pg. 1706].

It is interesting to show a new type of decomposition of entropy or information from this result. We have

$$D[p : p_0] = \sum p(\boldsymbol{x}) \log \frac{p(\boldsymbol{x})}{p_0(\boldsymbol{x})}$$

$$= -H(X_1, X_2, X_3) + 3$$

$$D[p^{(1)} : p_0] = -\sum_{i=1}^{3} H(X_i) + 3$$

where $H(X_1, X_2, X_3)$ and $H(X_i)$ are the total and marginal entropies of $\boldsymbol{X} = (X_1, X_2, X_3)$. Hence,

$$H[X_1, X_2, X_3] = \sum_{i=1}^{3} H(X_i) - D[p : p^{(2)}] - D[p^{(2)} : p^{(1)}]$$

One can define mutual information among X_1, X_2, and X_3 by

$$I[X_1, X_2, X_3] = \sum p(x_1, x_2, x_3) \log \frac{p(x_1, x_2, x_3)}{p(x_1) p(x_2) p(x_3)}$$

$$= \sum H(X_i) - H[X_1, X_2, X_3]$$

Then, [Eq. (73) $D[p : p_0] = D_2 + D_1 + D_0$] gives a new invariant positive decomposition of $I[X_1, X_2, X_3]$

$$I[X_1, X_2, X_3] = D[p : p^{(2)}] + D[p^{(2)} : p^{(1)}]$$

In information theory, the mutual information among three variables is sometimes defined by

$$I_3(X_1 : X_2 : X_3) = I(X_1 : X_2) - I(X_1, X_2 \mid X_3)$$

Unfortunately, this quantity is not necessarily non–negative ... Hence, our decomposition is new and is completely different from the conventional one ...

$$I(X_1, X_2, X_3) = I(X_1 : X_2) + I(X_2 : X_3) + I(X_1 : X_3) - I(X_1 : X_2 : X_3)$$

Continuing on with our task of identifying the correspondence between Amari's explanation and our own example, we easily recognize that Amari has reverted to the traditional communications convention of expressing information in bits and taking logs with respect to base 2 in his,

$$D[p : p_0] = -H(X_1, X_2, X_3) + 3$$

We will, for consistency, retain $\ln n$ in all of our expressions. Thus, $H(X_1, X_2, X_3)$ is the information entropy for any full joint probability over all eight statements in the state space under a complicated model p,

$$D[p : p_0] = D^{(-1)}(p \parallel u)$$

$$= \sum_{i=1}^{8} p_i \ln \left(\frac{p_i}{u_i} \right)$$

$$= -H(p) + \ln 8$$

$$= \ln 8 - H(X_1, X_2, X_3)$$

Similarly, his $D[\,p^{(1)}:p_0\,] = -\sum_{i=1}^{3} H(X_i) + 3$ is deconstructed as,

$$D[\,p^{(1)}:p_0\,] = D^{(-1)}(s \parallel u)$$

$$= \sum_{i=1}^{8} s_i \ln\left(\frac{s_i}{u_i}\right)$$

$$= -H_{\max}(s) + \ln 8$$

$$= -[\,H(X_1) + H(X_2) + H(X_3)\,] + \ln 8$$

$$= \ln 8 - \sum_{i=1}^{3} H(X_i)$$

Point s is the **m**–projection of point q down into the sub–manifold $\boldsymbol{E}_1$. This sub–manifold holds all of the independent distributions where,

$$(\boldsymbol{\theta}_2, \boldsymbol{\theta}_3) = (\theta_{12}, \theta_{23}, \theta_{31}, \theta_{123}) = (\lambda_4^s = \lambda_5^s = \lambda_6^s = \lambda_7^s = 0)$$

but $\boldsymbol{\theta}_1 = (\theta_1, \theta_2, \theta_3) = (\lambda_1, \lambda_2, \lambda_3)$ can assume values to represent various marginal probabilities of $P(B)$, $P(H)$, and $P(F)$. Since these probability distributions are over only two statements, the information entropy of these marginal probabilities in our example works out to,

$$H(X_1) \equiv H[P(B)]$$

$$= -(0.75 \ln 0.75 + 0.25 \ln 0.25)$$

$$= 0.56234$$

$$H(X_2) \equiv H[P(H)]$$

$$= -(0.75 \ln 0.75 + 0.25 \ln 0.25)$$

$$= 0.56234$$

$$H(X_3) \equiv H[P(F)]$$

$$= -(0.60 \ln 0.60 + 0.40 \ln 0.40)$$

$$= 0.67301$$

$$H(X_1) + H(X_2) + H(X_3) = 1.79768$$

$$H_{\max}(s) = 1.79768$$

$$D[p^{(1)} : p_0] = \ln 8 - 1.79768$$

$$= 0.28176$$

We are now in a position to verify that the entropy of the full joint probability distribution for the complicated model can be decomposed as follows:

$$H[X_1, X_2, X_3] = \sum_{i=1}^{3} H(X_i) - D[p : p^{(2)}] - D[p^{(2)} : p^{(1)}]$$

For this purpose, let's adopt the more consistent notation in Amari's *Theorem 8* in order to write down the decomposition as,

$$D[p : p_0] = \sum_{k=1}^{n} D[p^{(k)} : p^{(k-1)}]$$

$$D[p^{(3)} : p^{(0)}] = D[p^{(1)} : p^{(0)}] + D[p^{(2)} : p^{(1)}] + D[p^{(3)} : p^{(2)}]$$

Now, let's set up the equivalencies to translate the notation from Amari's quote above into our notation,

$$D[p : p_0] \equiv D[p^{(3)} : p^{(0)}]$$

$$D[p^{(3)} : p^{(0)}] \equiv D^{(-1)}(p \parallel u)$$

$$D[p^{(1)} : p^{(0)}] \equiv D^{(-1)}(s \parallel u)$$

$$D[p^{(2)} : p^{(1)}] \equiv D^{(-1)}(q \parallel s)$$

$$D[p^{(3)} : p^{(2)}] \equiv D^{(-1)}(p \parallel q)$$

$$D^{(-1)}(p \parallel u) = D^{(-1)}(s \parallel u) + D^{(-1)}(q \parallel s) + D^{(-1)}(p \parallel q)$$

Substitute what we have just found out for the left hand side of the equation and the first term on the right hand side to show the decomposition of the entropy for

FOLIATIONS

the full joint probability distribution $H(p) \equiv H[X_1, X_2, X3]$,

$$D^{(-1)}(p \parallel u) = \ln 8 - H_{\max}(p)$$

$$D^{(-1)}(s \parallel u) = \ln 8 - H_{\max}(s)$$

$$H(p) = H(s) - D[p^{(2)} : p^{(1)}] - D[p^{(3)} : p^{(2)}]$$

$$H[X_1, X_2, X_3] = \sum_{i=1}^{3} H(X_i) - D[p^{(3)} : p^{(2)}] - D[p^{(2)} : p^{(1)}]$$

$$= \sum_{i=1}^{3} H(X_i) - D^{(-1)}(p \parallel q) - D^{(-1)}(q \parallel s)$$

This expression is the intermediate building–block (the lemma) that we need for Amari's definition of the mutual information given that the standard definition of mutual information within the discipline of Information Theory is defined as,

$$I[X_1, X_2, X_3] = \sum H(X_i) - H[X_1, X_2, X_3]$$

Then,

$$I[X_1, X_2, X_3] = D[p^{(3)} : p^{(2)}] + D[p^{(2)} : p^{(1)}]$$

$$= D^{(-1)}(p \parallel q) + D^{(-1)}(q \parallel s)$$

For our example, Amari's definition of the mutual information among the variables of beer preference, hand preference, and fur color works out to,

$$I[X_1, X_2, X_3] = D^{(-1)}(p \parallel q) + D^{(-1)}(q \parallel s)$$

$$= 0.015057 + 0.031837$$

$$= 0.046894$$

The final few lines of this quoted section of Amari's paper are revealing in that Amari has to develop complicated formulas to actually find these projections. If the MEP algorithm is relied on, then everything remains simple. For me, this is merely another confirmation of the utility of the MEP in solving inferential problems.

> It is easy to obtain $p^{(1)}$ from given p. The coordinates of $p^{(2)}$ are obtained by solving [the set of equations that express the constraint function averages as the partial differentiation of the partition function with respect to the Lagrange multipliers, and that express the Lagrange multipliers as the partial differentiation of the negative entropy with respect to the constraint function averages] In the present case, the mixed–cut of $p^{(2)}$ is $\boldsymbol{\xi}_2 = (\boldsymbol{\eta}_1, \boldsymbol{\eta}_2; 0)$. Hence,

the $\boldsymbol{\eta}$–coordinates of $p^{(2)}$ are $(\boldsymbol{\eta}_1, \boldsymbol{\eta}_2; \overline{\eta}_{123})$ where the marginals η_i and η_{ij} are the same as p and $\overline{\eta}_{123}$ is determined such that θ_{123} becomes 0. Since we have

$$\theta_{123} = \log \frac{p_{111} p_{100} p_{010} p_{001}}{p_{110} p_{101} p_{011} p_{000}}$$

by putting $\theta_{123} = 0$, the $\overline{\eta}_{123}$ of $p^{(2)}$ is given by solving (82) shown at the bottom of the page.

Equation (82) at the bottom of the page happens to be an extremely complicated equation for finding $\overline{\eta}_{123}$ of $p^{(2)}$. This may all be true, but it is so much easier to grasp with the MEP at your disposal. A strictly utilitarian *Mathematica* MEP program constructed for this Chapter, and presented in the Exercises, will illustrate how one can avoid Amari's overly complicated prescription.

Finally, we can observe from the following quote from the same paper how Amari and Information Geometry in general are forced to treat the MEP. Rather than approaching it directly as a standard problem in variational calculus (as in Jaynes's seminal exposition which I elaborated on in Volume II), the MEP is viewed as a geometric relationship among the distances separating three points living in a Riemannian manifold.

A point q, which is the object of focus as an MEP distribution, is interpreted as a point that has to be as close as possible to two other points p and u. q has to be as close as possible to a point p which lives in a higher dimension than q, but possesses the information that q also wants. At the same time, q has to be as close as possible to another point u which might live in the same sub–manifold as q, but which possesses the maximum amount of missing information. Finding the q that satisfies these constraints is the MEP assignment for q.

Theorem 7. The projection $p^{(k)}$ of p to $\boldsymbol{E}_k$ is the maximizer of entropy among $q \in \boldsymbol{M}_k$ having the same k–marginals as p

$$p^{(k)} = \arg \max_{q \in \boldsymbol{M}_k} H[q]$$

This relationship is useful for calculating $p^{(k)}$.

Needless to say, we concur with the massive understatement that occurs at the end of the theorem since the sentiment expressed in the theorem forms the basis for the MEP algorithm. But the foregoing serves as a good example of how the MEP was curiously relegated to a minor role amidst all of the surrounding exotica of IG.

SOLVED EXERCISES FOR CHAPTER 41   317

41.8 Solved Exercises for Chapter Forty One

Exercise 41.8.1: Write a *Mathematica* program to perform various MEP calculations for the simple kangaroo scenario.

Solution to Exercise 41.8.1

The program that follows utilizes the Legendre transformation template seen many times before in my books. This approach is one way to implement an MEP algorithm that finds both the numerical assignments and resulting maximum entropy value for any discrete probability distribution. For any continuous distribution, we rely on the traditionally known probability density functions because these are derivable by the MEP as well.

The goal is obviously to find that particular numerical assignment reflecting the active information under some designated model. At the same time, it is desirable to eliminate all of the information not in that model.

Computationally, the operational implementation needed to fashion this missing information is to maximize the information entropy through some MEP algorithm. Of course, the MEP algorithm must also retain all of the non–missing information to be included in the distribution under the specific model. This active information appears in the form of m constraint function averages.

This first program is the basis for later refinements needed in the enhanced kangaroo scenario where the state space is bumped up to $n = 8$ from the $n = 4$ state space of the simple scenario. The most important new feature in these later programs is the addition of a constraint on the allowable values of the Lagrange multipliers.

This requirement for the numerical optimization routine comes into play when we are investigating sub–manifolds where some number of Lagrange multipliers will assume the value of 0 because they are excluding information from more complicated models. For example, such models might be ignoring higher level interactions. As we shall see shortly, this constraint on the parameters is easily implemented within the *Mathematica* **NMinimize[ ]** function.

The highest level description begins with a user defined function **MEP40[** *arg* **]** with one argument, a list of three constraint function averages.

MEP40[cfa_List] := Module[
                              (* *MEP algorithm is implemented* *)
                    Panel[Column[
                            (* *various results are displayed* *) **]]**
                                              (* *end Module* *) **]**

This code implements the MEP algorithm as based on the Legendre transformation and has appeared before,

```
MEP40[cfa_List] := Module[
  {Qsub, zsymbolic, znumeric, lambda,
   lnumeric, numerator, solution,
   entropy, qi, checkcfa, m = Length[cfa],
   cm = {{1,1,0,0}, {1,0,1,0}, {1,0,0,0}}},
   lambda = Map[Subscript[λ, #]&, Range[m]];
   Qsub = Map[Subscript[Q, #]&, Range[m+1]];
   zsymbolic = Total[Exp[Dot[lambda, cm]]];
   solution = NMinimize[Log[zsymbolic]-Dot[lambda, cfa], lambda];
   entropy = First[solution];
   znumeric = First[zsymbolic /. Rest[solution]];
   lnumeric = First[lambda /. Rest[solution]];
   numerator = N[Exp[Dot[lnumeric, cm]]];
   qi = numerator / znumeric;
   checkcfa = Dot[cm, qi];
   Panel[Column[ (* various results are displayed *) ]]
                                              (* end Module *) ]
```

Note especially the explicit form of the $F_j(X = x_i)$ as contained in the generic constraint matrix cm = $\{\{\cdots\}, \{\cdots\}, \cdots, \{\cdots\}\}$.

This program can be used to check that the numerical assignment to point p, the complicated model with the information from both marginal probabilities and the one double interaction, is, in fact, $Q_i = (0.70, 0.05, 0.05, 0.20)$, as shown in the first row of Table 41.2.

Evaluate **MEP40[{.75, .75, .70}]** with the argument a list of the three constraint function averages $\langle F_1 \rangle$, $\langle F_2 \rangle$, and $\langle F_3 \rangle$ for the complicated model p. The results appear below in Figure 41.5.

Assigned probabilities

Q_1	Q_2	Q_3	Q_4
0.7	0.05	0.05	0.2

Lagrange multipliers and constraint function averages

λ_1	λ_2	λ_3
−1.3862944	−1.3862944	4.0253517
<F_1>	<F_2>	<F_3>
0.75	0.75	0.7

Relative entropy from assigned probabilities to the fair model
0.515161

Figure 41.5: Mathematica *output showing the assigned probabilities, the Lagrange multipliers, the three constraint function averages, and the relative entropy between the complicated model of point p in the simple kangaroo scenario and the uniform distribution.*

SOLVED EXERCISES FOR CHAPTER 41

Exercise 41.8.2: Give a detailed analysis of the matrix **cm** as it appears in the *Mathematica* program of the previous exercise.

Solution to Exercise 41.8.2

The expression **cm** defined in **Module[ ]** is a matrix,

$$\text{cm} = \{\{1,\ 1,\ 0,\ 0\},\ \{1,\ 0,\ 1,\ 0\},\ \{1,\ 0,\ 0,\ 0\}\}$$

As far as *Mathematica* is concerned, this is a list of lists containing the m constraint functions $F_j(X = x_i)$. The index i on the statement $(X = x_i)$, where i runs from 1 through n, represents the ith observed value for any of the $n = 4$ joint statements of the introductory kangaroo scenario. For example, $(X = x_3)$ is the joint statement "A kangaroo prefers Foster's beer and is left–handed.".

The index j on $F_j(X = x_i)$ is the jth mapping, where j runs from 1 through m, of the joint statements to a real number. For this problem, $m = n - 1 = 3$ is the maximum number of constraint functions. Here, all of the mappings are either 0 or 1. For example, $F_2(X = x_3) = 1$ is the second mapping from the above joint statement to the real number 1. The first mapping to all four joint statements, as in $F_1(X = x_i) = (1, 1, 0, 0)$, is the first list in the matrix **cm**.

Setting $F_1(X = x_i) = (1, 1, 0, 0)$ permits an average like $Q_1 + Q_2 = \langle F_1 \rangle = 3/4$ to be specified, and is the information the IP wishes to insert into a probability distribution about hand preference. Setting $F_2(X = x_i) = (1, 0, 1, 0)$ permits an average like $Q_1 + Q_3 = \langle F_2 \rangle = 3/4$ to be specified, and is the information the IP wishes to insert into a probability distribution about beer preference. The final mapping, $F_3(X = x_i) = (1, 0, 0, 0)$, permits an average like $Q_1 = \langle F_3 \rangle = 0.70$ to be specified, and is the information the IP wishes to insert into a probability distribution about the interaction between beer preference and hand preference.

Finally, observe how the matrix **cm** fits into the computation of the partition function.

zsymbolic = Total[Exp[Dot[lambda, cm]]];

Working our way outwards from within this nested expression, we first have the vector matrix multiplication in **Dot[lambda, cm]** which evaluates to the list,

$$\{\lambda_1 + \lambda_2 + \lambda_3,\ \lambda_1,\ \lambda_2,\ 0\}$$

(If **lambda** is thought of as a (1×3) row vector, then the multiplication shown above in **Dot[lambda, cm]** is a conformable operation, $(1 \times 3) \cdot (3 \times 4)$, leading to a (1×4) row vector.)

Then, exponentiating this list, `Exp[Dot[lambda, cm]]`, evaluates to the list,

$$\{e^{\lambda_1+\lambda_2+\lambda_3}, e^{\lambda_1}, e^{\lambda_2}, 1\}$$

Finally, the partition function `zsymbolic` is calculated as the sum of the elements in the previous list, `Total[Exp[Dot[lambda, cm]]]`,

$$1 + e^{\lambda_1} + e^{\lambda_2} + e^{\lambda_1+\lambda_2+\lambda_3}$$

Exercise 41.8.3: Complete the above *Mathematica* program with code that presents the results in the format as shown in Figure 41.5.

Solution to Exercise 41.8.3

Orienting ourselves at the `Panel[Column[ ]]` function within the highest level template,

```
MEP40[cfa_List] := Module[
                (* MEP algorithm is implemented *)
                Panel[Column[ (* various results are displayed *) ]]
                                            (* end Module *) ]
```

I formatted the results as follows,

```
Panel[Column[
  {Column[{Style["Assigned probabilities",
          {FontFamily → "Verdana", Red, FontSize → 20}],
          Grid[{Qsub, qi}, Frame → All, Spacings → {1, 1},
          BaseStyle → {FontFamily → "ArialBold", FontSize → 20},
          Background → {None, {LightGreen, LightBlue}}]}],
  Spacer[50],
  Column[{Style["Lagrange multipliers and
                  constraint function averages",
          {FontFamily → "Verdana", Red, FontSize → 20}],
          Grid[{lambda, Chop[Map[Function[x, NumberForm[x, {9, 7}]],
          lnumeric]], {"⟨F₁⟩", "⟨F₂⟩", "⟨F₃⟩"}, checkcfa},
          Frame → All, Spacings → {1, 1},
          BaseStyle → {FontFamily → "ArialBold", FontSize → 20},
          Background → {None, {Yellow, LightYellow,
                                Yellow, LightYellow}}]}],
  Spacer[50],
  Column[{Style["Relative entropy from assigned probabilities to
                  fair model.",
          {FontFamily → "Verdana", Red, FontSize → 20}],
          Spacer[20],
          Style[Log[m + 1] - entropy,
          {FontFamily → "ArialBold", Blue, FontSize → 20}]}]}],
          ImageSize →{700, 420}]]
```

Notice that the relative entropy to the uniform distribution shown in the expression **Log[m + 1] - entropy** is the calculation $\ln n - H_{\max}(\star)$ used extensively in this Chapter.

It is quite often the case that the code for the presentation of the results in some pleasant fashion is much more complicated than the actual relevant computation. We see an example of this here. I bring this up merely to reiterate my plea to those who wish to explain their *Mathematica* code to make an extra effort to clearly delineate where such distinctions take place.

Exercise 41.8.4: What minor change do we need to make to the above program in order to constrain the value of the third Lagrange multiplier?

Solution to Exercise 41.8.4

Suppose that we are still in the realm of the simple kangaroo scenario, but now we want to look at models that negate the double interaction. These would be the class of independence models that still retain the information about the marginal probabilities for beer and hand preference. Therefore, we would be investigating models where $\lambda_3 = 0$.

To investigate $E_k(0)$ sub–manifolds where some Lagrange multipliers have been set to 0, incorporate the following changes to the above program. Create a new function called **MEP41[** *arg1, arg2* **]** with two arguments. The first argument is the same as before, a list of all three constraint function averages. The second argument is the value that the third Lagrange multiplier should always assume within the **NMinimize[ ]** numerical optimization routine. The other two Lagrange multipliers will still vary freely in order to find the solution.

Thus, the global template now looks like this,

MEP41[cfa_List, constraint_] := Module[
  (MEP algorithm with constraint on third Lagrange multiplier *)*
    Panel[Column[*(* various results are displayed *)* **]]**
                                                  (end Module *)* **]**

The only other change takes place within **NMinimize[ ]**,

NMinimize[{Log[zsymbolic] - Dot[lambda, cfa], λ_3 == constraint},
       lambda];

Exercise 41.8.5: Verify the numerical assignment to point q in Table 41.3 with the above program.

Solution to Exercise 41.8.5

All that is required is a call to **MEP41[{.75, .75, .70}, 0]** which returns the results shown in Figure 41.6 at the top of the next page.

Assigned probabilities

Q_1	Q_2	Q_3	Q_4
0.5625	0.1875	0.1875	0.0625

Lagrange multipliers and constraint function averages

λ_1	λ_2	λ_3
1.0986123	1.0986123	0.0000000
$\langle F_1 \rangle$	$\langle F_2 \rangle$	$\langle F_3 \rangle$
0.75	0.75	0.5625

Relative entropy from assigned probabilities to the fair model.

0.261624

Figure 41.6: *The* Mathematica *output showing the assigned probabilities, the three Lagrange multipliers, the three constraint function averages, and the relative entropy between the less complicated model behind point q and the uniform distribution in the simple kangaroo scenario.*

This is the independence model that appeared in Jaynes's discussion of the kangaroo scenario. My analysis of this situation can be reviewed in Volume II, sections 21.3.1 and 21.5.

Exercise 41.8.6: Modify the previous program in the same way to verify the numerical assignment to point t in Table 41.3.

Solution to Exercise 41.8.6

The global template is slightly modified to take account of the fact that now two Lagrange multipliers, λ_2 and λ_3, will be set to 0. Point t lives in the sub–manifold where the double interaction on beer hand preference, and the second marginal probability on beer preference, have both been negated.

MEP42[cfa_List, constraint1_, constraint2_] := Module[
(* MEP algorithm with constraints on second and third Lagrange multiplier *)
      Panel[Column[*(* various results are displayed *)* **]]**
                  (* end Module *) **]**

Just as before, make the permitted change within **NMinimize[ ]** allowing any parameter to be fixed at a given value during the numerical optimization routine to find the minimum value of $\ln Z - \sum_{j=1}^{m} \lambda_j \langle F_j \rangle$,

NMinimize[{Log[zsymbolic] - Dot[lambda, cfa],
        λ_2 == constraint1, λ_3 == constraint2}, lambda];

`MEP42[{.75, .75, .70}, 0, 0]` returns the expected numerical assignment,
$$Q_i = (0.375, 0.375, 0.125, 0.125)$$
under the information in the model behind point t.

It is easy to confirm the information in this model as contained in the first constraint function average,
$$\langle F_1 \rangle = Q_1 + Q_2 = 0.375 + 0.375 = 0.75$$
This information concerns the marginal probability for hand preference.

To ensure the maximum amount of missing information under this model, we see that $\langle F_2 \rangle = Q_1 + Q_3 = 0.50$, and $\langle F_3 \rangle = Q_1 = 0.375$. Remember that the model parameters λ_2 and λ_3 have been fixed at 0 during the minimization, thus negating any extra information in $\langle F_2 \rangle$ and $\langle F_3 \rangle$.

Exercise 41.8.7: What do the relative entropy results for points p, q, and t indicate?

Solution to Exercise 41.8.7

Each successive probability distribution is getting closer to the fair model with its numerical assignment of $Q_i = (1/4, 1/4, 1/4, 1/4)$. In other words, each successive model has more and more missing information. This must be the case because these three points are getting closer to point u which has the maximum amount of missing information.

In more detail, the squared distance to point u is getting smaller as we progress from point p through point t,
$$D^{(-1)}(p \parallel u) = \ln 4 - H_{\max}(p) = 0.515161$$
$$D^{(-1)}(q \parallel u) = \ln 4 - H_{\max}(q) = 0.261624$$
$$D^{(-1)}(t \parallel u) = \ln 4 - H_{\max}(t) = 0.130812$$

Exercise 41.8.8: Verify the Pythagorean theorem for $\triangle ptu$ in Figure 41.3 with a *Mathematica* program.

Solution to Exercise 41.8.8

In section 41.4, a calculation from first principles showed that the Pythagorean theorem held for the triangle $\triangle ptu$.
$$D^{(-1)}(p \parallel u) = D^{(-1)}(p \parallel t) + D^{(-1)}(t \parallel u)$$
$$\sum_{i=1}^{4} p_i \ln\left(\frac{p_i}{u_i}\right) = \sum_{i=1}^{4} p_i \ln\left(\frac{p_i}{t_i}\right) + \sum_{i=1}^{4} t_i \ln\left(\frac{t_i}{u_i}\right)$$
$$0.515161 = 0.384349 + 0.130812$$

Write a *Mathematica* function **PythagoreanThm[** *arg1, arg2, arg3* **]**, having three arguments, to verify the above calculation. Each one of these three arguments consists of a list of the numerical assignments under a particular model.

```
PythagoreanThm[p1_List, p2_List, p3_List] :=
   Module[{re1, re2, re3},
     re1 = Total[p1 Log[p1 / p2]];
     re2 = Total[p2 Log[p2 / p3]];
     re3 = Total[p1 Log[p1 / p3]];
     Grid[{{"p to t", "t to u", "p to u", "(p to t) + (t to u)"},
        {re1, re2, re3, re1 + re2}}, Frame → All, Spacings → {1, 1},
        BaseStyle → {FontFamily → "ArialBold", FontSize → 20},
        Background → {None, {LightBlue, LightYellow}}]]
```

A call to this function with each argument containing the numerical assignments under points p, t, and u,

```
PythagoreanThm[{.70,.05,.05,.20}, {.375,.375,.125,.125},
               {.25,.25,.25,.25}]
```

confirms the fact that these three probability distributions satisfy the Pythagorean theorem.

The theorem holds because point t is the **m**–projection of point p to that point which has a minimum relative entropy with p. Point t has the mixed coordinates $(\eta_I, \theta^{II}) = (0.75, 0, 0)$. Point p has been α–projected down to a sub–manifold where any double interaction of beer and hand preference no longer exists, where the marginal probability for beer preference has changed to $P(B) = 1/2$, but where the marginal probability for hand preference is still $P(H) = 3/4$.

The model parameters for the probability distribution t of $\lambda_2 = \lambda_3 = 0$ ensure that $P(B) = 1/2$ and $P(BH) = 0.375$.

Exercise 41.8.9: Verify the information decomposition of the complicated model p via the various α–projections.

Solution to Exercise 41.8.9

The information decomposition of point p as it proceeds down through various projections to point u is expressed as,

$$D^{(-1)}(p \parallel u) = D^{(-1)}(p \parallel q) + D^{(-1)}(q \parallel t) + D^{(-1)}(t \parallel u)$$

The assignments for all of these points have been tabulated in Table 41.3. It's easy enough to just add a fourth list of assignments for another probability distribution in a *Mathematica* function called **infodecomp[ ]**.

```
infodecomp[p1_List, p2_List, p3_List, p4_List] :=
            Module[{re1, re2, re3, re4},
```

```
                    re1 = Total[p1 Log[p1 / p2]];
                    re2 = Total[p2 Log[p2 / p3]];
                    re3 = Total[p3 Log[p3 / p4]];
                    re4 = Total[p1 Log[p1 / p4]];
                    {re4, re1, re2, re3, re1 + re2 + re3}]
```

Evaluating,

`infodecomp[{.70,.05,.05,.20}, {.5625,.1875,.1875,.0625},`
`            {.375,.375,.125,.125}, {.25,.25,.25,.25}]`

shows that,

$$0.515161 = 0.253537 + 0.130812 + 0.130812$$

Exercise 41.8.10: Show the orthogonality in $\triangle pqu$ in two different ways.

Solution to Exercise 41.8.10

Two formulas have been derived to compute whether two curves that meet at a common point are orthogonal. Here, the geodesic joining points p and q intersects the geodesic joining points u and q at the common point q. Are the two curves orthogonal?

The first formula for determining whether orthogonality exists is given by my reworking of Amari's Theorem 3.7 into a form appropriate for the relative entropy. If the curves are orthogonal, then the expression below involving the dual coordinates θ^i and η_i for p, q, and u must equal 0.

$$\{\theta^i(q) - \theta^i(u)\}\{\eta_i(q) - \eta_i(p)\} = 0$$

It is easy to see that this is true because the first two constraint function averages for the MEP distributions p and q are the same. The third term is 0 because the Lagrange multiplier $\lambda_3 = 0$ for both MEP distributions q and u. Thus, the three terms are,

$$(1.09863 - 0)(0.75 - 0.75) + (1.09863 - 0)(0.75 - 0.75) + (0 - 0)(0.5625 - 0.70) = 0$$

The second formula was also derived in Chapter Forty as Equation (40.6),

$$KL(p,q) + KL(q,u) - KL(p,u) = \sum_{i=1}^{4}(p_i - q_i)\ln\left(\frac{u_i}{q_i}\right) = 0$$

This might be viewed as a sterner test since it is not immediately obvious how it will turn out. But the calculation, done easily enough with the *Mathematica* program,

`orthog[p1_List, p2_List, p3_List] := Total[ (p1 - p2) Log[p3 / p2]]`

also returns 0.

We already know from previous exercises that,
$$D^{(-1)}(p \parallel u) = 0.515161$$
and that,
$$D^{(-1)}(q \parallel u) = 0.261624$$
Thus, if the Pythagorean theorem holds for $\triangle pqu$, and now we know that it does because the two curves are orthogonal, $D^{(-1)}(p \parallel q)$ must equal,
$$D^{(-1)}(p \parallel u) - D^{(-1)}(q \parallel u) = 0.515161 - 0.261624 = 0.253537$$
which it does.

All of the previous exercises referred to the basic kangaroo scenario of dimension $n = 4$. The following exercises and named points will now refer to the $n = 8$ enhanced kangaroo scenario.

Exercise 41.8.11: What minor, but essential, change needs to be made in the *Mathematica* program for the $n = 8$ kangaroo scenario?

Solution to Exercise 41.8.11

The constraint matrix **cm** needs to be enlarged to account for the $m = 7$ possible constraint functions $F_j(X = x_i)$. The new constraint matrix for **MEP80[*arg1*]** is,

```
cm = {
       (* Beer *)              {1, 1, 1, 1, 0, 0, 0, 0},
       (* Hand *)              {1, 0, 1, 0, 1, 0, 1, 0},
       (* Fur *)               {1, 1, 0, 0, 1, 1, 0, 0},
       (* BH interaction *)    {1, 0, 1, 0, 0, 0, 0, 0},
       (* BF interaction *)    {1, 1, 0, 0, 0, 0, 0, 0},
       (* HF interaction *)    {1, 0, 0, 0, 1, 0, 0, 0},
       (* BHF interaction *)   {1, 0, 0, 0, 0, 0, 0, 0}
     }
```

The commented code identifies the three marginal probabilities for beer preference, hand preference, and fur color, the three double interactions involving beer–hand interaction, the beer–fur color interaction, and the hand–fur color interaction, and finally the one triple interaction involving the beer–hand–fur color interaction.

For example, the constraint function $F_6(X = x_i)$ captures any information in the *HF* double interaction with $\langle F_6 \rangle = Q_1 + Q_5$. The information in the model for p specifies that $Q_1 + Q_5 = 0.45$.

The rest of the code remains the same. However, the argument to,

$$\text{MEP80[} arg1 \text{]}$$

must now be a list of $m = 7$ constraint function averages.

Exercise 41.8.12: Use MEP80[*arg1*] to verify the numerical assignments dictated by the complicated model behind point p.

Solution to Exercise 41.8.12

Evaluating,

$$\text{MEP80}[\{.75, .75, .60, .58, .50, .45, .40\}]$$

results in Figure 41.7 below.

Assigned probabilities to complicated model behind p

Q_1	Q_2	Q_3	Q_4	Q_5	Q_6	Q_7	Q_8
0.4	0.1	0.18	0.07	0.05	0.05	0.12	0.03

Lagrange multipliers and constraint function averages

λ_1	λ_2	λ_3	λ_4	λ_5	λ_6	λ_7
0.8472979	1.3862944	0.5108256	−0.4418328	−0.1541507	−1.3862944	1.8281271
<F_1>	<F_2>	<F_3>	<F_4>	<F_5>	<F_6>	<F_7>
0.75	0.75	0.6	0.58	0.5	0.45	0.4

Value of divergence function from complicated model p to fair model u
0.328653

Figure 41.7: *The Mathematica output showing the assigned probabilities, the seven Lagrange multipliers, the seven constraint function averages, and finally the relative entropy between the complicated model behind point p and the uniform distribution behind point u for the $n = 8$ kangaroo scenario.*

Exercise 41.8.13: How is any point living in the sub–manifold $E_3(0)$ described?

Solution to Exercise 41.8.13

In the hierarchical e–structure foliation of the Riemannian manifold $\mathcal{S}^8$, $E_3(0)$ is a sub–manifold whose coordinates are,

$$E_3(0) = (\theta^1, \theta^2, \theta^3, \theta^4 = \theta^5 = \theta^6 = \theta^7 = 0)$$

Thus, any probability distribution living in $E_3(0)$ contains no information about any level of interaction among the beer, hand, and fur color variables. Every probability distribution in $E_3(0)$ contains information only about the marginal probabilities, $P(B)$, $P(H)$, and $P(F)$, for beer preference, hand preference, and fur color.

Exercise 41.8.14: What are the mixed coordinates for a point living in the sub–manifold $E_4(0)$?

Solution to Exercise 41.8.14

The mixed coordinate system for such a point is,

$$\xi_4 = (\eta_1, \eta_2, \eta_3, \eta_4, 0, 0, 0)$$

Label some point as point r living in $E_4(0)$. Let the mixed coordinates for r be,

$$\xi_4(r) = (\eta_1 = 0.75, \eta_2 = 0.75, \eta_3 = 0.60, \eta_4 = 0.58, 0, 0, 0)$$

Any point living in $E_4(0)$ includes information about not only the three marginal probabilities, but also includes information about a double interaction between beer preference and hand preference. Specifically, the information in the model behind point r is,

$$P(B) \equiv \langle F_1 \rangle = Q_1 + Q_2 + Q_3 + Q_4 = 0.75$$

$$P(H) \equiv \langle F_2 \rangle = Q_1 + Q_3 + Q_5 + Q_7 = 0.75$$

$$P(F) \equiv \langle F_3 \rangle = Q_1 + Q_2 + Q_5 + Q_6 = 0.60$$

$$P(BH) \equiv \langle F_4 \rangle = Q_1 + Q_3 = 0.58$$

Exercise 41.8.15: Consider all of the points $q \in E_5(0)$. Which point $q^\star$ of all these points is at the end of the m–geodesic that starts at point p?

Solution to Exercise 41.8.15

Point p is that probability distribution designated as the complicated model since it incorporates information from all $m = 7$ constraint function averages. In other words, it inserts information about all three marginal probabilities, all three double interactions, and the one triple interaction.

All of the points $q \in E_5(0)$ are represented by a mixed coordinate system,

$$\xi_5(q) = (\eta_1, \eta_2, \eta_3, \eta_4, \eta_5, 0, 0)$$

In order to begin the computations, arbitrarily select a candidate $q^\star$ from among the $q \in E_5(0)$ as,

$$\xi_5(q^\star) = (\eta_1 = 0.70, \eta_2 = 0.80, \eta_3 = 0.55, \eta_4 = 0.60, \eta_5 = 0.52, 0, 0)$$

The relative entropy between p and this candidate $q^\star$ is,

$$KL(p, q^\star) = \sum_{i=1}^{8} p_i \ln \left(\frac{p_i}{q_i^\star} \right)$$

$$= 0.12896$$

Is this value the minimum value of the relative entropy between p and any point $q \in \boldsymbol{E}_5(0)$? No, it is not. We are able to find a new candidate q^* with mixed coordinates,

$$\xi_5(q^*) = (\eta_1 = 0.74, \eta_2 = 0.76, \eta_3 = 0.59, \eta_4 = 0.59, \eta_5 = 0.50, 0, 0)$$

whose relative entropy $KL(p, q^*) = 0.0189609$ is lower than our first choice. This new point is much closer to point p.

Ultimately, the minimum value of the relative entropy $KL(p, q^*) = 0.0153156$ occurs at a point q^* whose coordinates are,

$$\xi_5(q^*) = (\eta_1 = 0.75, \eta_2 = 0.75, \eta_3 = 0.60, \eta_4 = 0.58, \eta_5 = 0.50, 0, 0)$$

This minimum value must occur where the Pythagorean theorem holds for the three points point p, point q^*, and point q that constitute a right triangle,

$$D^{(-1)}(p \parallel q) = D^{(-1)}(p \parallel q^*) + D^{(-1)}(q^* \parallel q)$$

The **m**–geodesic is the shortest distance from point p to point q^*, and perpendicular to any e–geodesic curve in $\boldsymbol{E}_5(0)$ connecting some q to q^*.

From the above Pythagorean theorem, any $q \in \boldsymbol{E}_5(0)$ that is NOT q^* is obviously going to be at a greater squared distance to p by an amount $D^{(-1)}(q^* \parallel q)$.

Exercise 41.8.16: Use the two formulas from Exercise 41.8.10 to show that the curve from point q to point q^* and the curve from point p to point q^* in $\triangle pq^*q$ are orthogonal.

Solution to Exercise 41.8.16

The first formula says that the two curves are orthogonal if,

$$\{\theta^i(q^*) - \theta^i(q)\}\{\eta_i(q^*) - \eta_i(p)\} = 0$$

or, in our more familiar MEP notation,

$$\sum_{j=1}^{7}[(\lambda_j^{q^*} - \lambda_j^q) \times (\langle F_j \rangle_{q^*} - \langle F_j \rangle_p)] = 0$$

For the first five terms $j = 1$ to 5, $\langle F_j \rangle_{q^*} = \langle F_j \rangle_p$, so these first five terms must equal 0. For the next two terms where $j = 6$ and $j = 7$, $\lambda_j^{q^*} = \lambda_j^q = 0$, so these final two terms must equal 0 as well. The two curves are orthogonal.

The second formula says that the two curves are orthogonal if,

$$\sum_{i=1}^{8}(p_i - q_i^*)\ln\left(\frac{q_i}{q_i^*}\right) = 0$$

Use the *Mathematica* function **orthog[** *arg1, arg2, arg3* **]** from Exercise 41.8.10 with the three arguments the list of numerical assignments for p, q^*, and q to verify that this computation also equals 0.

Exercise 41.8.17: Why is point p called the maximum likelihood model?

Solution to Exercise 41.8.17

Conceptually prior to any data being gathered about the beer preferences, hand preferences, and fur color from some number of kangaroos, models are permitted to incorporate information about marginal probabilities, together with information about all the double and triple interactions among these three variables.

A perfectly legitimate model $\mathcal{M}_{\text{ML}}$, where the model subscript gives the game away, might be one where this information is represented by the following model parameters,

$$\langle F_1 \rangle = 0.75$$

$$\langle F_2 \rangle = 0.75$$

$$\langle F_3 \rangle = 0.60$$

$$\langle F_4 \rangle = 0.58$$

$$\langle F_5 \rangle = 0.50$$

$$\langle F_6 \rangle = 0.45$$

$$\langle F_7 \rangle = 0.40$$

If the MEP algorithm is provided with this information, the ensuing numerical assignment is as listed in Figure 41.7. This assignment is believed, prior to any data, just as much and just as little as any other acceptable assignment.

However, *after* the data have been collected, and say that the data is the same as that shown in the contingency table of Figure 22.2, it is observed that the normed frequency counts from the $N = 100$ kangaroos sampled are exactly the same as the assigned probabilities under the model for point p,

$$\frac{N_1}{N} = \frac{40}{100} = 0.40 = Q_1, \cdots, \frac{N_8}{N} = \frac{3}{100} = 0.03 = Q_8$$

Moreover, all of the *sample* averages, for example, the *sample* average for preferring Foster's beer,

$$\overline{F}_1 = (N_1 + N_2 + N_3 + N_4)/N = 75/100 = 0.75$$

exactly match the information that was inserted by the model behind point p.

The ratio of the posterior probability for model $\mathcal{M}_{\text{ML}}$ when compared to any other model $\mathcal{M}_k$ will always be greater than 1, and for the vast majority of models, much greater than 1.

For an extreme example, in a calculation carried out in Exercise 29.9.25 in Volume II, the ratio of the maximum likelihood model for point p compared to the fair model behind point u, after conditioning on the data, was the very large value of,

$$\frac{P(\mathcal{M}_{\text{ML}} \mid \mathcal{D})}{P(\mathcal{M}_{\text{Fair}} \mid \mathcal{D})} = \exp[\, N \times KL(p, u)\,] = \exp[\, 100 \times 0.32865 \,] \approx 1.88 \times 10^{14}$$

Exercise 41.8.18: Describe the MEP from the viewpoint of Information Geometry, and thus in a slightly different way than as a problem for variational calculus that was adopted in Volume II.

Solution to Exercise 41.8.18

Figure 41.8 below provides the context for the following discussion. The inferential problem remains the same kangaroo scenario involving the three variables of beer preference, hand preference, and fur color. This is a state space of dimension $n = 8$. A complicated model incorporating information from the three marginal probabilities, the three double interactions, and the one triple interaction provides a numerical assignment to point p existing in $\mathcal{S}^8$.

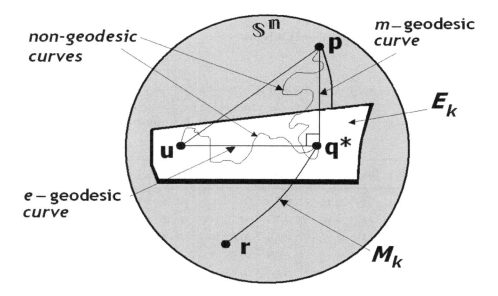

Figure 41.8: *How Information Geometry characterizes the MEP by visualizing points and distances within a Riemannian manifold. Point $q^\star$ is the maximum entropy distribution because it is as close as possible to point u and point p.*

Suppose the IP is interested in how IG characterizes a point $q^\star$ that possesses less information than p because it is neglecting some interactions. Where is that point $q^\star$ located in orthogonal sub–manifolds $\boldsymbol{E}_k$ and $\boldsymbol{M}_k$ with respect to three points p, u, and r? Point r is located on the same sub–manifold $\boldsymbol{M}_k$ as $q^\star$, while $q^\star$ is the $\alpha = -1$ projection of p to $\boldsymbol{E}_k$.

For an example, let's pick the sub–manifolds $\boldsymbol{E}_5$ and $\boldsymbol{M}_5$ because the IP is interested in projecting the complicated model p down to a less complicated model $q^\star$ where there is no information about both the BHF triple interaction among beer preference, hand preference, and fur color, and the final HF double interaction between hand preference and fur color.

All points $q \in \boldsymbol{E}_5$ have coordinates where $\theta^6 = \theta^7 = 0$. Conversely, the first five θ^i coordinates for any point $q \in \boldsymbol{E}_5$ can take on any value. Thus, both u and $q^\star$ live in $\boldsymbol{E}_5$. Of course, we are well aware that *all* of the θ^i coordinates for u equal 0.

Point p does not live in $\boldsymbol{E}_5$, therefore θ^6 and θ^7 may, and do, have non–zero values. Point r does not live in $\boldsymbol{E}_5$ either because its θ^6 and θ^7 coordinates also have non–zero values. And, finally, any point $q \in \boldsymbol{E}_5$ that is not $q^\star$, even though it must have a 0 coordinate value for θ^6 and θ^7, will have at least one of its η_1 through η_5 coordinates different than p.

The three points p, $q^\star$, and r all live on the $\boldsymbol{M}_5(p)$ sub–manifold. By definition, the coordinates for any point living on $\boldsymbol{M}_5(p)$ must share common values for the first five η_i coordinates. But η_6 and η_7 are free to vary for any point in $\boldsymbol{M}_5(p)$. Points p, $q^\star$, and r have η_6 and η_7 coordinates with different values.

This fact is summarized in Table 41.4 below.

Table 41.4: *A comparison of the η_i coordinates for the four points p, $q^\star$, r, and u. The three points p, $q^\star$, and r must share the same value for the first five coordinates. Point u does not share any common η_i values since it does not live in the $\boldsymbol{M}_5(p)$ sub–manifold.*

Effect	Information	p	$q^\star$	r	u
B	$\langle F_1 \rangle \equiv \eta_1$	0.75	0.75	0.75	1/2
H	$\langle F_2 \rangle \equiv \eta_2$	0.75	0.75	0.75	1/2
F	$\langle F_3 \rangle \equiv \eta_3$	0.60	0.60	0.60	1/2
BH	$\langle F_4 \rangle \equiv \eta_4$	0.58	0.58	0.58	1/4
BF	$\langle F_5 \rangle \equiv \eta_5$	0.50	0.50	0.50	1/4
HF	$\langle F_6 \rangle \equiv \eta_6$	0.45	0.45467	0.44	1/4
BHF	$\langle F_7 \rangle \equiv \eta_7$	0.40	0.38667	0.41	1/8

Amari characterized $q^\star$ as the MEP distribution because of the two Pythagorean relationships existing among the four points p, $q^\star$, u, and r. In the first triangle $\triangle pq^\star u$,

$$D^{(-1)}(p \parallel u) = D^{(-1)}(p \parallel q^\star) + D^{(-1)}(q^\star \parallel u)$$

In other words, $q^\star$ is closer to p by the squared distance $D^{(-1)}(q^\star \parallel u)$. This is true because $q^\star$ is the **m**–projection of p down to $\boldsymbol{E}_5$.

In the second triangle $\triangle rq^\star u$,

$$D^{(-1)}(r \parallel u) = D^{(-1)}(r \parallel q^\star) + D^{(-1)}(q^\star \parallel u)$$

In other words, $q^\star$ is closer to u by the squared distance $D^{(-1)}(r \parallel q^\star)$. It is that point in the sub–manifold $\boldsymbol{M}_5(p)$ that is closest to u. p is farther away from u than $q^\star$. r is even farther away from u than p. Any point $r \in \boldsymbol{M}_5(p)$ is farther away from u when compared to $q^\star$.

Point $q^\star$ wants to be as close as it possibly can get to point u. u has the most missing information of any probability distribution in the manifold. Therefore, $q^\star$ also wants to possess the property of having the most missing information.

However, it is constrained in this endeavor by the active information that has been inserted into it by a model. This information takes the form of five constraint function averages detailing the information from three marginal probabilities and two double interactions. Because $q^\star$ lives in $\boldsymbol{E}_5$ and $\lambda_6 = \lambda_7 = 0$, it ignores any information about the BHF triple interaction and the HF double interaction. The information in the model for $q^\star$ matches the information on the three marginal probabilities and two double interactions that are contained as information in p.

So $q^\star$ has to be as close as possible to p and as close as possible to u. It is as close as possible to p when $q^\star$ is the **m**–projection of p to $\boldsymbol{E}_5(q)$. It is also that point in $\boldsymbol{M}_5(p)$ that is as close as possible to u. In other words, $q^\star$ has the maximum entropy of any distribution that contains the same information as p for the first five constraint function averages.

The relative entropy between $q^\star$ and u is minimized when $H(q^\star)$ is maximized through the relationship,

$$D^{(-1)}(q^\star \parallel u) = \ln n - H_{\max}(q^\star)$$

But this expression is the same as maximizing $H(q^\star)$ subject to the constraint that $q^\star$ must have the same first five constraint function averages as p because $q^\star$ had to be as close as possible to p. The MEP algorithm finds this $H_{\max}(q^\star)$ through the usual,

$$H_{\max}(q^\star) = \min_{\lambda_j} \left\{ \ln Z_{q^\star} - \sum_{j=1}^{7} \lambda_j^{q^\star} \langle F_j \rangle_p \right\}$$

Remember that $\lambda_6^{q^\star} = \lambda_7^{q^\star} = 0$ so the final two constraint function averages from p do not enter into the picture.

It helps to recall that, computationally, during the *Mathematica* evaluation of the MEP algorithm, five fixed constraint function averages from p are fed in as arguments and do not vary, while the Lagrange multipliers are free to vary during the numerical optimization routine that converges on to the maximum entropy numerical assignment.

Exercise 41.8.19: Investigate more closely what is taking place for the function being minimized by `NMinimize[ ]` within the MEP algorithm.

Solution to Exercise 41.8.19

The Legendre transformation relating (1) the information entropy, $H_{\max}(q)$, (2) the partition function, Z_q, and (3) the dual coordinates, λ_j^q and $\langle F_j \rangle_p$, for some probability distribution q has usually been written as,

$$H_{\max}(q) = \min_{\lambda_j} \left\{ \ln Z_q - \sum_{j=1}^{m} \lambda_j^q \langle F_j \rangle_p \right\}$$

The function $\ln Z_q - \sum_{j=1}^{m} \lambda_j^q \langle F_j \rangle_p$ is numerically minimized by varying the Lagrange multipliers λ_j^q, while holding the constraint function averages $\langle F_j \rangle_p$ fixed. The numerical routine used by `NMinimize[ ]` returns the minimum value of the function which is the solution for the information entropy $H_{\max}(q)$, and the values of the Lagrange multipliers λ_j^q which brought about this minimum value.

Leaning upon those concepts from Information Geometry just recently explored, the probability distribution $q^\star$, or point $q^\star$, for which a numerical assignment is sought, is the **m**–projection of some other point p possessing m constraint function averages $\langle F_j \rangle_p$. But point $q^\star$ lives in a sub–manifold $E_k(0)$ where some of its mixed coordinates $\theta^i = 0$. Therefore, options to `NMinimize[ ]` are exercised to allow constraints to be placed on some of the Lagrange multipliers.

Here are two detailed examples taken from our two kangaroo scenarios. For the first example taken from the $n = 4$ state space, insert the log of the partition function, $\ln Z_q$, explicitly into `NMinimize[ ]` with the desired constraint on the model parameter that $\lambda_3 = 0$. The goal is to have the MEP algorithm find the numerical assignment to a probability distribution $q^\star$ with the same information as p concerning the marginal probability for beer and hand preference, but lacking any information about a correlation between hand and beer preference.

The point $q^\star$ lives at the orthogonal intersection of the sub–manifolds $E_2(q)$ and $M_2(p)$. It has the mixed coordinates,

$$\xi(q^\star) = (\eta_1 = 0.75, \eta_2 = 0.75, \theta^3 = 0)$$

It is the $\alpha = -1$ projection of point p with the mixed coordinates of,

$$\xi(p) = (\eta_1 = 0.75, \eta_2 = 0.75, \theta^3 = 4.02535)$$

SOLVED EXERCISES FOR CHAPTER 41

After inserting the explicit values for $\ln Z_q$ and $\sum_{j=1}^{m} \lambda_j^q \langle F_j \rangle_p$, have *Mathematica* evaluate,

NMinimize[{Log[Exp[λ_1 + λ_2 + λ_3] + Exp[λ_1] + Exp[λ_2] + 1] -
  (λ_1 0.75 + λ_2 0.75 + λ_3 0.70), λ_3 == 0}, {λ_1, λ_2, λ_3}]

The evaluation of the above expression returns the answer,

$$\{1.12467, \{\lambda_1 \to 1.09861, \lambda_2 \to 1.09861, \lambda_3 \to 0\}\}$$

The information entropy of the probability distribution q^* is $H_{\max}(q^*) = 1.12467$. The relative entropy between q^* and u is then confirmed as,

$$\ln n - H_{\max}(q^*) = \ln 4 - 1.12467 = 1.38629 - 1.12467 = 0.26162$$

For the second example taken from the $n = 8$ state space, do the same thing. Insert the log of the partition function, $\ln Z_q$, explicitly into **NMinimize[ ]** with the desired constraint on the model parameters that $\lambda_6 = \lambda_7 = 0$. The goal is to have the MEP algorithm find the numerical assignment to a probability distribution q^* that has the same information as p, but lacks information about any triple interaction involving beer preference, hand preference, and fur color, as well as any double interaction between hand preference and fur color.

The point q^* lives at the orthogonal intersection of the sub–manifolds $\boldsymbol{E_5(q)}$ and $\boldsymbol{M_5(p)}$. It has the mixed coordinates,

$$\xi(q^*) = (\eta_1 = 0.75, \eta_2 = 0.75, \eta_3 = 0.60, \eta_4 = 0.58, \eta_5 = 0.50, \theta^6 = 0, \theta^7 = 0)$$

Point q^* is also the $\alpha = -1$ projection of point p, (the **m**–projection), which has the mixed coordinates of,

$$\xi(p) = (\eta_1 = 0.75, \eta_2 = 0.75, \eta_3 = 0.60, \eta_4 = 0.58, \eta_5 = 0.50,$$
$$\theta^6 = -1.386294, \theta^7 = 1.8281271)$$

Insert the explicit values for $\ln Z_q$ and $\sum_{j=1}^{m} \lambda_j^q \langle F_j \rangle_p$ for this new situation, and have *Mathematica* evaluate,

NMinimize[{Log[Exp[λ_1 + λ_2 + λ_3 + λ_4 + λ_5 + λ_6 + λ_7] +
  Exp[λ_1 + λ_3 + λ_5] +
  Exp[λ_1 + λ_2 + λ_4] +
  Exp[λ_1] +
  Exp[λ_2 + λ_3 + λ_6] +
  Exp[λ_3] +
  Exp[λ_2] + 1] -
  (λ_1 0.75 + λ_2 0.75 + λ_3 0.60 + λ_4 0.58 + λ_5 0.50 +
  λ_6 0.45 + λ_7 0.40), λ_6 == 0, λ_7 == 0},
  {λ_1, λ_2, λ_3, λ_4, λ_5, λ_6, λ_7}]

The evaluation of the above expression returns the answer,

{1.7661, {$\lambda_1 \to$ 0.165985, $\lambda_2 \to$ 0.753772, $\lambda_3 \to$ -0.405465,
$\lambda_4 \to$ 0.473458, $\lambda_5 \to$ 1.09861, $\lambda_6 \to$ 0, $\lambda_7 \to$ 0}}

The information entropy of the probability distribution $q^\star$ is $H_{\max}(q^\star) = 1.7661$. The relative entropy between $q^\star$ and u is then confirmed as,

$$\ln n - H_{\max}(q^\star) = \ln 8 - 1.7661 = 2.0794 - 1.7661 = 0.3133$$

The next exercise will verify this squared distance between points $q^\star$ and u when they are treated as parts of a right triangle in the $\mathcal{S}^8$ Riemannian manifold.

Exercise 41.8.20: Numerically confirm that the triangular relationships are as discussed in the previous exercises.

Solution to Exercise 41.8.20

Figure 41.9 appearing below is a diagram highlighting the two triangles $\triangle pq^\star u$ and $\triangle rq^\star u$. Points u, $q^\star$, and an arbitrary q all live in $\boldsymbol{E}_5(q)$. Points p, $q^\star$, and r all live in $\boldsymbol{M}_5(p)$. Point $q^\star$ lives at the orthogonal intersection of $\boldsymbol{E}_5(q)$ and $\boldsymbol{M}_5(p)$.

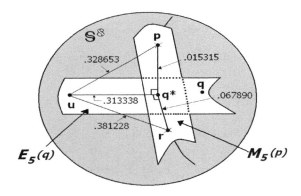

Figure 41.9: *The squared distances involved in the two triangles $\triangle pq^\star u$ and $\triangle rq^\star u$. Both triangles are right triangles so the Pythagorean theorem can be used.*

The relative entropies are filled in along the two sides and the hypotenuse of both triangles. Making the association between the canonical divergence function with $\alpha = -1$ and the Kullback–Leibler relative entropy,

$$D^{(-1)}(p \parallel q) \equiv KL(p, q) = \sum_{i=1}^{8} p_i \ln\left(\frac{p_i}{q_i}\right)$$

we have the following squared distances,

$$KL(p, u) = 0.328653$$

$$KL(p, q^*) = 0.015315$$

$$KL(q^*, u) = 0.313338$$

$$KL(r, u) = 0.381228$$

$$KL(r, q^*) = 0.067890$$

Thus, it is easy to verify the Pythagorean theorem for $\triangle pq^*u$ as,

$$D^{(-1)}(p \parallel u) = D^{(-1)}(p \parallel q^*) + D^{(-1)}(q^* \parallel u)$$

$$0.328653 = 0.015315 + 0.313338$$

and for $\triangle rq^*u$ as,

$$D^{(-1)}(r \parallel u) = D^{(-1)}(r \parallel q^*) + D^{(-1)}(q^* \parallel u)$$

$$0.381228 = 0.067890 + 0.313338$$

No point $q \in \boldsymbol{E}_5(q)$ can get closer to p than q^*. The squared distance $KL(p, q^*)$ is shorter than $KL(p, u)$ by 0.313338. The squared distance $KL(p, q^*)$ would be shorter than any arbitrary q for $KL(p, q)$ by $KL(q, q^*)$.

No point $r \in \boldsymbol{M}_5(p)$ can get closer to u than q^*. Points p and r are both farther away from u at $KL(p, u) = 0.328653$ and $KL(r, u) = 0.3812228$ compared to $KL(q^*, u) = 0.313338$.

Exercise 41.8.21: After the past few exercises, revisit what Amari had to say about projections and the MEP.

Solution to Exercise 41.8.21

It is instructive at this juncture to mull over Amari's characterization of how the MEP fits into his overall scheme of things [4, pg. 1705].

> *D. Maximal Principle*
> The projection $p^{(k)}$ is closely related with the maximal entropy principle. [Here, Amari references Jaynes's 1982 article, *On The Rationale of Maximum–Entropy Methods*]. The projection $p^{(k)}$ belongs to $\boldsymbol{M}_k(p)$ which consists of all the probability distributions having the same k-marginal distributions as p, that is, the same $\boldsymbol{\eta}_{k-}$ coordinates. For any $q \in \boldsymbol{M}_k$, $p^{(k)}$ is its projection to $\boldsymbol{E}_k$. Hence, because of the Pythagoras theorem
> $$D[q \colon p_0] = D[q \colon p^{(k)}] + D[p^{(k)} \colon p_0]$$
> the minimizer of $D[q \colon p_0]$ for $q \in \boldsymbol{M}_k$ is $p^{(k)}$.

Let's match things up with our notation before we proceed any further. Our point $q^\star$ is the projection $p^{(k)}$. And, it is true that $q^\star$ belongs to the same sub-manifold $\boldsymbol{M}_k(p)$ as the complicated model behind point p because all of the points in $\boldsymbol{M}_k(p)$ must possess the same five coordinates of $\langle F_1 \rangle$ through $\langle F_5 \rangle$. What Amari calls p_0, I call point u for the fair model assigning $1/n$ to all statements in the state space. But I labeled as point r, not point q, as an example for any $r \in \boldsymbol{M}_k$. Any point q for me belonged to $\boldsymbol{E}_5(q)$ where the final two λ_j coordinates must always be equal to 0.

So, the expression Amari has written for the Pythagorean theorem,

$$D[q\colon p_0] = D[q\colon p^{(k)}] + D[p^{(k)}\colon p_0]$$

becomes for me,

$$KL(r, u) = KL(r, q^\star) + KL(q^\star, u)$$

If you want to minimize $KL(r, u)$ for any $r \in \boldsymbol{M}_k(p)$, then that r must be, in fact point $q^\star$.

Amari continues on with,

> We have
>
> $$D[q\colon p_0] = \sum q(\boldsymbol{x}) \log q(\boldsymbol{x}) - \sum q(\boldsymbol{x}) \log p_0(\boldsymbol{x})$$
>
> $$= -H[q] - \text{const}$$
>
> because $\sum q(\boldsymbol{x}) \log p_0(\boldsymbol{x})$ depends only on the marginal distributions of q and is constant for all $q \in \boldsymbol{M}_k$. Hence, we have the geometric form of the maximum principle ...
>
> *Theorem 7:* The projection $p^{(k)}$ to $\boldsymbol{E}_k$ is the maximizer of entropy among $q \in \boldsymbol{M}_k$ having the same k–marginals as p
>
> $$p^{(k)} = \arg \max_{q \in \boldsymbol{M}_k} H[q]$$
>
> This relation is useful for calculating $p^{(k)}$.

Amari provides us with an alternative geometric way of conceptualizing the MEP in the above *Theorem 7*. It is complementary to Jaynes's standard variational calculus approach to the MEP. In one manner, it might be even better because of the emphasis on the *active information* resident in point p that point $q^\star$ must possess, and the maximum amount of *missing information* in point u that point $q^\star$ must simultaneously possess. That these information properties are easily seen to be fulfilled with relatively simple geometric arguments like the Pythagorean theorem is quite remarkable.

Amari writes the expression,

$$D[q\colon p_0] = \sum q(\boldsymbol{x}) \log q(\boldsymbol{x}) - \sum q(\boldsymbol{x}) \log p_0(\boldsymbol{x})$$

$$= -H[q] - \text{const}$$

which gets translated into finding the minimum for $KL(r,u)$ which must occur when $r \equiv q^\star$,

$$KL(r,u) = \sum_{i=1}^{n} r_i \ln\left(\frac{r_i}{u_i}\right)$$

$$= \sum_{i=1}^{n} r_i \ln r_i - \sum_{i=1}^{n} r_i \ln(1/n)$$

$$= -H(r) - \sum_{i=1}^{n} r_i(\ln 1 - \ln n)$$

$$= -H(r) - \sum_{i=1}^{n} r_i(0 - \ln n)$$

$$= -H(r) - (-\ln n \sum_{i=1}^{n} r_i)$$

$$= \ln n - H(r)$$

$$\max\left[H(r) \in M_k(p)\right] \equiv H(q^\star) \quad \text{(because } q^\star \text{ is as close as possible to } u\text{)}$$

$$\min\left[KL(r,u)\right] = \ln n - H(q^\star)$$

So, once again, we have the finding that the relative entropy between any r and u is minimized when some $H(q)$ is maximized. In other words, we must find that $q^\star$ that has maximum entropy subject to being the **m**–projection of point p.

Exercise 41.8.22: Use previously derived facts about orthogonality to add plausibility to Amari's Theorem 3.7.

Solution to Exercise 41.8.22

In Euclidean plane geometry, the Pythagorean Theorem applies to a right triangle where two sides meet at a 90° angle. The cosine of 90° is 0 and the term *orthogonal* is used to describe the right angle where the two sides meet. The issues emanating from *orthogonality* are central to Information Geometry.

This kind of geometric intuition about orthogonality is much the same within Information Geometry. The Pythagorean relationship among the three points p, q, and r requires that two geodesics intersecting at point q, say, be orthogonal. The angle ϕ between two tangent vectors, call them **v** and **w** for now, was defined by,

$$\cos(\phi) = \frac{\langle \mathbf{v}, \mathbf{w}\rangle}{\sqrt{\langle \mathbf{v}, \mathbf{v}\rangle}\sqrt{\langle \mathbf{w}, \mathbf{w}\rangle}}$$

So we want $\cos(\phi) = 0$ and this will be the case if the numerator in the above expression is 0. If the two tangent vectors are orthogonal, their inner product is 0, or, $\langle \mathbf{v}, \mathbf{w} \rangle = 0$.

To proceed further, expand these two vectors according to their dual bases. (You might want to review the topics in Chapter Thirty Four to refresh your memory about the following steps). This following creative decision in the proof is important: expand $\mathbf{v}$ with respect to the *first* dual coordinate system, and expand $\mathbf{w}$ with respect to the *second* dual coordinate system. Still expressing things in a generic notation because it makes the exposition easier to follow,

$$\mathbf{v} = v^j \, \mathbf{e}_j$$

$$\mathbf{w} = w_k \, \mathbf{e}^k$$

Now we can re-write the inner product as,

$$\langle \mathbf{v}, \mathbf{w} \rangle = \langle v^j \, \mathbf{e}_j, w_k \, \mathbf{e}^k \rangle$$

When the operation on the right hand side is carried out, the significance of selecting the dual basis vectors is made clear. Their inner product is the identity matrix so that,

$$\langle v^j \, \mathbf{e}_j, w_k \, \mathbf{e}^k \rangle = v^j \, \langle \mathbf{e}_j, \mathbf{e}^k \rangle \, w_k$$

$$\langle \mathbf{e}_j, \mathbf{e}^k \rangle = \mathbf{I}$$

$$\langle v^j \, \mathbf{e}_j, w_k \, \mathbf{e}^k \rangle = v^j \cdot \mathbf{I} \cdot w_k$$

$$= v^j \cdot w_k$$

In the above derivation we are taking advantage of the common convention in differential geometry, called the Einstein summation convention, of not explicitly writing down the summation over indices.

At this juncture, after the global result has been revealed in a relatively transparent fashion, we want to revert back from this generic notation to the notation we have been using all along. Unfortunately, that notation is little more opaque than that presented above. That is why we began in a more abstract vein.

Return to the step where we found the inner product $\langle \mathbf{v}, \mathbf{w} \rangle$. We are going to do the same thing, but now in terms of parametric representation of the geodesic curves. The derivation begins by recalling the chain rule for derivatives,

$$\frac{\mathrm{d}f(y(t))}{\mathrm{d}t} = \sum \left[\frac{\mathrm{d}y^j(t)}{\mathrm{d}t} \times \frac{\partial f}{\partial \theta^j} \right]$$

Process this to yield,

$$\frac{\partial f}{\partial \theta^j} = \mathbf{e}_j$$

$$\mathbf{e}_j = \partial_i$$

$$\sum \left[\frac{dy^j(t)}{dt} \times \frac{\partial f}{\partial \theta^j}\right] = \sum \left[\frac{d\theta^j(t)}{dt} \times \mathbf{e}_j\right]$$

$$\mathbf{v} = \frac{d\theta^j(t)}{dt}\partial_j$$

Exactly the same thing is done with $\mathbf{w}$ but with respect to its dual basis vectors so that,

$$\mathbf{w} = \frac{d\eta_k(t)}{dt}\partial^k$$

Now the inner product between $\mathbf{v}$ and $\mathbf{w}$ looks like,

$$\langle \mathbf{v}, \mathbf{w} \rangle = \left\langle \frac{d\theta^j(t)}{dt}\partial_j, \frac{d\eta_k(t)}{dt}\partial^k \right\rangle$$

This is just different notation for the same operation that resulted in $\langle \mathbf{e}_j, \mathbf{e}^k \rangle = \mathbf{I}$ where now the analogous operation is $\langle \partial_j, \partial^k \rangle = \delta_j^k = \mathbf{I}$ so that we are left with,

$$\langle \mathbf{v}, \mathbf{w} \rangle = \frac{d\theta^j(t)}{dt} \cdot \frac{d\eta_k(t)}{dt} = 0$$

We want to find the tangent vector on the m–geodesic curve at q. The m–geodesic curve from p to q is characterized as a straight line,

$$\theta^j = t\theta^j(p) + (1-t)\theta^j(q)$$

Taking the derivative,

$$\frac{d\theta^j}{dt} = \theta^j(p) - \theta^j(q)$$

In the same way, the tangent vector of the e–geodesic curve connecting r to q is found from the straight line characterization of the e–geodesic,

$$\eta_j = t\eta_k(r) + (1-t)\eta_k(q)$$

by taking the appropriate derivative,

$$\frac{d\eta_k(t)}{dt} = \eta_k(r) - \eta_k(q)$$

Finally, after all of this, we arrive at,

$$\langle \mathbf{v}, \mathbf{w} \rangle = \{\theta^j(p) - \theta^j(q)\} \cdot \{\eta_k(r) - \eta_k(q)\} = 0$$

as the orthogonality condition that makes the Pythagorean relationship valid in the Riemannian manifold.

Exercise 41.8.23: Show how *Mathematica* allows you to monitor the step by step progress of a numerical routine.

Solution to Exercise 41.8.23

Like all programming languages, *Mathematica* has a great variety of debugging tools. One such debugging feature is helpful in monitoring the progress taking place within numerical routines like **NMinimize[ ]**. An option to **NMinimize[ ]** called **StepMonitor** is illustrated in this exercise.

We know that **NMinimize[ ]** will return $H_{\max}(p)$, the maximum value of the information entropy, as well as the Lagrange multipliers λ_j that bring about this maximum for some distribution p. The information inserted by the model for the distribution p is contained in the constraint function averages. For our familiar example, the distribution p for the $n = 4$ kangaroo scenario is found by specifying the three constraint function averages of $\langle F_1 \rangle = 0.75$, $\langle F_2 \rangle = 0.75$, and $\langle F_3 \rangle = 0.70$.

Executing the MEP algorithm,

```
NMinimize[Log[Exp[λ₁ + λ₂ + λ₃] + Exp[λ₁] + Exp[λ₂] + 1]
         - (λ₁ .75 + λ₂ .75 + λ₃ .70 ), {λ₁, λ₂, λ₃}]
```

will return the nested list,

```
{0.871133, {λ₁ → -1.38629, λ₂ → -1.38629, λ₃ → 4.02535}}
```

where we see the results for the information entropy and the model parameters,

$$H_{\max}(p) = 0.871133$$

$$\lambda_1 = -1.38629$$

$$\lambda_2 = -1.38629$$

$$\lambda_3 = +4.02535$$

for the "complicated" model behind point p involving the double interaction.

Construct the following *Mathematica* expression to monitor the step by step progress of the values currently being used for the three Lagrange multipliers, and the resulting value of $H_{\max}(p) = \ln Z_p - \sum_{j=1}^{3} \lambda_j^p \langle F_j \rangle_p$.

```
Block[{count = 1, func = Log[Exp[λ₁ + λ₂ + λ₃] + Exp[λ₁] + Exp[λ₂]
               + 1] - (λ₁ .75 + λ₂ .75 + λ₃ .70)},
      {NMinimize[func, {λ₁, λ₂, λ₃}, Method → "NelderMead",
       StepMonitor :→ Print[{count++, λ₁, λ₂, λ₃, func}]]}]
```

This *Mathematica* code will return what happened at each of the 64 steps that it took **NMinimize[ ]** to reach the end point of the numerical routine.

Chapter 42

Boltzmann Machines

42.1 Introduction

Boltzmann machines provide us with an opportunity to discuss how certain general inferential problems often have unique, traditional, and perhaps, to some extent, accidental histories attached to them. These kind of contingent histories are quite understandable whenever a problem has been studied on its own merits over a long period of time.

The topic of artificial neural networks (ANNs) presents us with one such class of inferential problems which have experienced a unique history. At first glance, ANNs appear to be somewhat detached from the conceptual underpinnings and the jargon that I have tried to associate with inference as a generalization of logic. Swimming against the tide, I would like instead to emphasize the commonalities shared by all inferential problems rather than their differences due to some particular contingent and accidental history.[1]

My task is made easier because much of the traditional history of ANNs explored the notion that the cognitive processes mimicked by ANNs could be solved through an analogy with the equations developed for statistical mechanics. Jaynes showed us how the traditional equations appearing in statistical mechanics could be transferred over to inference through the auspices of the maximum entropy principle. We are halfway home because of this path clearing effort that links statistical mechanics, ANNs, the MEP, and probability theory.

In addition, this elementary introduction to Boltzmann machines extends the discussion on how intractable, complicated models can be simplified to tractable less complicated models through Information Geometry.

---

[1] Should the reader be puzzled by this admittedly strange language talking of "contingent and accidental histories," I can only plead that I have been unduly influenced by Wolfram's use of a similar term in his *A New Kind of Science* to describe how present–day mathematics has evolved.

The suggestive descriptive label, **Boltzmann machines**, that is attached to our novel inferential scenario introduced in this Chapter follows *Stigler's Law of Eponymy*. This law states, only half tongue-in-cheek, that when a name becomes linked to some scientific concept, it shall never be the name of the true originator!

It will come as no surprise that the 19th Century Austrian physicist Ludwig Boltzmann never had anything to do with such a thing as a "Boltzmann machine." If this law is correct, then Stigler would be the first to admit that he was not the originator of *Stigler's Law of Eponymy*!

Credit, and the true origin of the label, seems to belong to some pioneering work done in the closely related areas of parallel distributed processing, machine learning, and neural networks. This effort gained currency around the mid 1980s.

Many researchers recognized the potential in this new approach, inspired as it was by the biology of neural networks, to advance the field of artificial intelligence. Somehow the idea that ANNs were simply offering an interesting new approach to solving an inferential problem got lost in the shuffle. We will see examples of this in some of the ANN related jargon that, when parsed, turns out to be exactly the same concepts that appear in all of our inferential scenarios.

This research helped to stimulate a true renaissance in cognitive information processing that continues to the present day. For our present purposes, these efforts relied heavily upon probability theory, and what was, at the time, a revolutionary acceptance of the Bayesian paradigm. Boltzmann's name got attached to a type of neural network because concepts like an energy function and the canonical equation from statistical mechanics played an integral role. "Learning" by a Boltzmann machine was cast in the then radical garb of updating through Bayesian methods.

Many scientists have made substantial contributions to our understanding of the role played by Boltzmann machines since they first became popular. Typically cited are areas as diverse as combinatorial optimization, classification, content addressable memories, learning through examples, and so on. The ideas behind Boltzmann machines are now seeding hot new research areas in AI like "deep learning."

For our present elementary purposes though, we shall rely heavily on David MacKay's excellent and concise summary of Boltzmann machines appearing in Chapter 43 of his book,

Information Theory, Inference, and Learning Algorithms

Amari has also discussed Boltzmann machines either as an example in their own right as complicated models, or as simplifications of even more complicated models. This latter approach is the one we shall promote in this Chapter.

Surprisingly, Boltzmann machines are relatively easy to explain if we continue to rely on our standard expository preference for simple inferential scenarios. Thus, it seems worthwhile to go ahead and make simple Boltzmann machines the goal for the next set of Information Geometry concepts and associated numerical examples.

BOLTZMANN MACHINES

The effort taken to examine Boltzmann machines has a further pay off because they generalize to other interesting problems. Physicists like to study the magnetic properties of matter through "Ising models" and "infinite range spin glasses." Likewise, cognitive scientists study associative memory through "hierarchical networks of visible and hidden neurons."

For us, though, life is less dramatic. I will examine Boltzmann machines strictly as probabilistic representations. I will attempt to make the gap between these exotic applications and already understood inferential scenarios as small as possible. With this voluntarily enforced discipline to restrict the discussion to our already examined components of inferencing as information processing, I believe that these other interpretations, relying as they must on their accidental and contingent histories, become somewhat more palatable.

42.2 Preliminary Description

A Boltzmann machine consists of a set of interconnected nodes. Each node in the Boltzmann machine can be in an ON state or an OFF state. At other times, the nodes may be characterized as having a "spin state" with UP or DOWN spin values, or sometimes they are simply designated by ± 1. The essential point is that each node, or statement as we will think of them, is a binary variable with only two possible measured states. Things haven't changed in this regard when compared to most of our inferential examples.

We'll take a simple instantiation of a Boltzmann machine with five nodes labeled as A, B, C, D, and E. All five nodes are interconnected in a manner suggested by the sketch in Figure 42.1 at the top of the next page. Ten lines are drawn to indicate a connection between nodes A and B, A and C, and so on, up through a connection between nodes D and E.

The "strength" of the interconnection between any two nodes is indicated by λ_6 between nodes A and B, and so on, up to λ_{15} between nodes D and E. A node is ON if it is colored light gray and OFF if it colored dark gray. Thus, nodes A, C and E are ON, while nodes B and D are OFF.

This kind of simple Boltzmann machine leads us to consider it as analogous to our point $q^\star$ considered as a simplified version of some more "intractable" model behind a point p. Think of this intractable model as possessing enough constraint functions to capture not only all of the information in the main effects, but also the information in all possible interaction terms for five variables.

Consequently, there would be 31 terms consisting of 1) five main effects, 2) ten double interactions, 3) ten triple interactions, 4) five quadruple interactions, and 5) the final quintuple interaction. We will think of point p then as a probability distribution for any such "complicated" model chosen from the class of models consisting of these 31 terms.

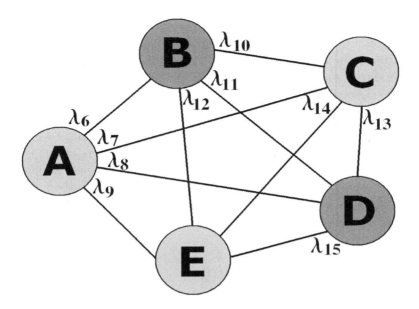

Figure 42.1: *Sketch of a Boltzmann machine with five nodes. The overall state of the Boltzmann machine can be captured by a joint statement asserting which nodes are ON and which are OFF. The interconnection strength between nodes is indicated by a λ_j.*

Although, in these beginning stages, p as just outlined above does remain tractable numerically, the ever increasing complexity involved in treating all of these higher order interactions is still pretty daunting. I don't see very many explanations, even in the "hard sciences," that take advantage of a quintuple interaction.

The state space consists of the $n = 2^5 = 32$ joint statements. A typical joint statement might be, "Node A is ON and node B is OFF and node C is ON and node D is OFF and node E is ON." This joint statement is expressed in the notation,

$$(A = a_1, B = b_2, C = c_1, D = d_2, E = e_1)$$

where $x_1 \equiv$ ON and $x_2 \equiv$ OFF. A numerical value assigned as a legitimate probability to a joint statement under some model $\mathcal{M}_k$ is written as,

$$P(A = a_1, B = b_2, C = c_1, D = d_2, E = e_1 \mid \mathcal{M}_k)$$

This longer version for the probability of a joint statement will often be simplified to an expression like,

$$Q_{11} \equiv P(X = x_{11} \mid \mathcal{M}_k) \equiv P(A\overline{B}C\overline{D}E \mid \mathcal{M}_k)$$

where the probability of some specific joint statement is identified by indexing it to a cell in the joint probability table. For example, Q_{11} occupies the appropriate cell in the joint probability table shown ahead in Figure 42.2 on page 350.

Suppose that the IP is engaging the Boltzmann machine in an attempt to solve some inferential task. The IP is seeking to predict whether node A is ON or OFF on the next trial given the ON–OFF status of the other four nodes and given, as well, the known outcomes from some number N of previous trials.

Another familiar probabilistic expression captures the degree of belief that the following statement: "Node A is OFF at trial $N+1$ given that node B is ON, node C is ON, node D is ON, and node E is OFF at trial $N+1$, and given the known settings of all five nodes at N previous trials." is TRUE, would be,

$$P(A_{N+1} = a_2 \mid B_{N+1} = b_1, C_{N+1} = c_1, D_{N+1} = d_1, E_{N+1} = e_2, \mathcal{D})$$

The notation $\mathcal{D}$ is for the data consisting of the known measured values at each of N trials for each of the five statements,

$$\mathcal{D} \equiv \{(A_1, B_1, C_1, D_1, E_1), (A_2, B_2, C_2, D_2, E_2), \cdots, (A_N, B_N, C_N, D_N, E_N)\}$$

42.3 The Perspective from IG

Every Boltzmann machine is a point in a dually flat Riemannian manifold. But we can say more than that. Every Boltzmann machine is a point $q^\star$ living at the intersection of two orthogonal sub–manifolds $\boldsymbol{E}_k(q)$ and $\boldsymbol{M}_k(p)$.

$\boldsymbol{E}_k(q)$ defines that sub–manifold where all of the points $q \in \boldsymbol{E}_k$ represent simple models which do not possess information about interaction terms higher than double interactions. $\boldsymbol{M}_k(p)$ defines that sub–manifold where all of the points $p, r, s \in \boldsymbol{M}_k$ represent complicated models which do possess information about all of the higher interaction terms.

Point p is a special point in $\boldsymbol{M}_k(p)$ because it happens to match the maximum likelihood model $\mathcal{M}_{\text{ML}}$. When point p is **m**–projected down into $\boldsymbol{E}_k(q)$, we realize the Boltzmann machine $q^\star$.

From the point of view of Information Geometry, the IP seeks to α–project point p to a point $q^\star$ living at the orthogonal intersection of the sub–manifolds $\boldsymbol{E}_k(q)$ and $\boldsymbol{M}_k(p)$. Then $q^\star$ will be based on a model substantially less complicated than p. Every point in the dually flat Riemannian manifold $\mathcal{S}^n$ represents some probability distribution. Imagine that a Boltzmann machine of ultimate interest will consist of a large number of nodes rather than merely the five of this example. Some method must be found for generating tractable models.

For example, if models where the five main effects, together with one double interaction, one triple interaction, and one quadruple interaction are considered, then $m = 8$. Such models represent the α–projection of point p in $\mathcal{S}^{32}$ to a point $q^\star$ at the orthogonal intersection of $\boldsymbol{E}_8(q)$ and $\boldsymbol{M}_8(p)$. The probability distribution $q^\star$ will possess the same information as the probability distribution p for the main effects and the stated interactions, but will possess no other information whatsoever.

In this Chapter, m will take on a value of 10 because we want to duplicate a Boltzmann machine. The IP will be examining models consisting solely of the ten double interactions. No models including any of the main effects, triple, quadruple, or quintuple interactions will be studied. Thus, we will be studying α–projections of a point p in $\mathcal{S}^{32}$ to many points q in a sub–manifold at the intersection of $\boldsymbol{E}_{10}(q)$ and $\boldsymbol{M}_{10}(p)$.

The relative entropy will be used as a distance measure to find the minimum distance between point p and the various points q in the sub–manifold. Where that minimum distance occurs is at point $q^\star$ in $\boldsymbol{M}_{10}(p)$. This unique point $q^\star$ also is at the minimum distance to a point u when both of these points are in $\boldsymbol{E}_{10}(q)$.

Point u is the probability distribution with the absolute maximum amount of missing information. Point $q^\star$ must be as close as possible to point u so that it too has the maximum amount of missing information, while at the same time retaining the active information it inherited from p. This is the triangular relationship among the points p, $q^\star$, and u.

42.4 Boltzmann Machines and Thermodynamics

This choice of inferential models, comprised of only the double interactions, mimics the original Boltzmann machine inspiration that looked at the synaptic connection between two neurons. Then, physicists realized that Boltzmann machines could be generalized to concepts in statistical mechanics like spin glasses and Ising models. We shall appropriate as topics for discussion, the especially lucid introduction to Boltzmann machines as described by MacKay.

He presents probability distributions for the Boltzmann machine which look very similar to the Boltzmann distributions from statistical mechanics.

$$P(\mathbf{x} \mid \mathbf{W}) = \frac{1}{Z(\mathbf{W})} \exp\left[-E(\mathbf{x})\right] \qquad (42.1)$$

Therefore, we are not surprised to see the appearance of an "energy function,"

$$E(\mathbf{x}) = -\frac{1}{2}\mathbf{x}^\mathbf{T}\mathbf{W}\mathbf{x} \qquad (42.2)$$

However, the temperature parameter and Boltzmann constant $k_B T$, an essential feature of the canonical Boltzmann distribution, seem to be missing. They crop up again only when "simulated annealing" is employed to find the best models.

The vector $\mathbf{x}$ in the above probability expression represents the state of the Boltzmann machine when taking account of all of its ON–OFF nodes. $\mathbf{W}$ is a matrix of "weights" connecting any two nodes.

As I said in the **Introduction**, I will convert everything that might be expressed in the jargon of the idiosyncratic contingent history behind Boltzmann machines back to our standard inferential scenario. For starters, MacKay begins with a probability distribution, Equation (42.1), from statistical mechanics.

BOLTZMANN MACHINES

I would rather prefer to begin with MEP assigned probabilities Q_i, which will take this form under the information provided by the given model $\mathcal{M}_k$,

$$Q_i \equiv P(X = x_i \mid \mathcal{M}_k) = \frac{1}{Z(\lambda_1, \lambda_2, \cdots, \lambda_m)} \exp\left[\sum_{j=1}^{m} \lambda_j F_j(X = x_i)\right] \quad (42.3)$$

The summation is now altered to show the indices $j = 6$ through $m^\star = 15$ to reserve the remaining parameters for the main effects, or any of the higher order interactions, should we want to refer to those models.

$$Q_i \equiv P(X = x_i \mid \mathcal{M}_k) = \frac{1}{Z(\lambda_6, \lambda_7, \cdots, \lambda_{15})} \exp\left[\sum_{j=6}^{15} \lambda_j F_j(X = x_i)\right] \quad (42.4)$$

MacKay's energy function, when expanded out with the requisite vector–matrix–vector multiplications, looks like this,

$$-E(\mathbf{x}) = \frac{1}{2}\mathbf{x}^T\mathbf{W}\mathbf{x} \equiv \underbrace{w_{12}x_1x_2 + w_{13}x_1x_3 + \cdots + w_{45}x_4x_5}_{\text{10 terms}} \quad (42.5)$$

In MEP notation, the energy function looks like this,

$$\sum_{j=6}^{15} \lambda_j F_j(X = x_i) = \underbrace{\lambda_6 F_6(X = x_i) + \lambda_7 F_7(X = x_i) + \cdots + \lambda_{15} F_{15}(X = x_i)}_{\text{10 terms}} \quad (42.6)$$

where the $\mathbf{W}$ matrix for the Boltzmann machine is the same as the subset of the Lagrange multipliers, and all ten $x_i x_j$ terms are represented by $F_j(X = x_i)$.

There are no x_i^2 terms in a Boltzmann machine, therefore all w_{ii} elements of the matrix vanish. Also, there is no distinction between the connection strength between, say, nodes 2 and 5 and nodes 5 and 2, so terms like $w_{25} x_2 x_5 + w_{52} x_5 x_2$ are divided by half. From the total of $5^2 = 25$ elements in the $\mathbf{W}$ matrix, only the $(5 \times 4)/2 = 10$ lower left off–diagonal elements w_{12} through w_{45} are needed.

The analog in our preferred MEP representation is that there are only ten double interaction terms in an ANOVA type decomposition. These consist of the AB double interaction, the AC double interaction, and so on, up to the tenth and final DE double interaction. The AB interaction is captured by the constraint function $F_6(X = x_i)$ with its associated Lagrange multiplier λ_6, the AC interaction is captured by the constraint function $F_7(X = x_i)$ with its associated Lagrange multiplier λ_7, and so on, up to where the tenth and final DE interaction is captured by the constraint function $F_{15}(X = x_i)$ with its associated Lagrange multiplier λ_{15}.

The explanation of logistic regression in Chapter Twenty Three of Volume II went into quite some detail on how to set up constraint functions and their averages in order to implement any level of interaction by a model. We will use the same strategy here to assign probabilities for the Boltzmann machine. The exercises will go into all the detail of how to do this.

For right now, let's get a feel for the MEP assignment to some cell in the joint probability table. And, in order to do this, we need to systematically generate a joint probability table for a state space of dimension $n = 32$.

Figure 42.2 below shows an abbreviated joint probability table for clarity. Just the subscript i is shown in each cell instead of Q_i. For example, the first 16 cells constitute $P(E = e_1)$, while the second 16 cells make up $P(E = e_2)$. Even easier, go back to our logic notation with just $P(E)$ and $P(\overline{E})$.

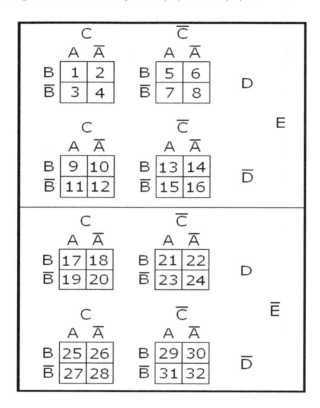

Figure 42.2: *The joint probability table for the Boltzmann machine example with five nodes. Each cell is labeled with the index i for the assignment Q_i under some model.*

As a check, observe that the probability for node A consists of the 16 cells,

$$P(A) = Q_1 + Q_3 + Q_5 + Q_7 + \cdots + Q_{25} + Q_{27} + Q_{29} + Q_{31} \quad (42.7)$$

and the probability for, say, node C consists of the 16 cells,

$$P(C) = Q_1 + Q_2 + Q_3 + Q_4 + \cdots + Q_{25} + Q_{26} + Q_{27} + Q_{28} \quad (42.8)$$

BOLTZMANN MACHINES

The models comprising all of the points $q \in \boldsymbol{E}_k$, that is to say, any Boltzmann machine, will have to specify information about the double interactions. In other words, they will specify marginal probabilities like $P(AB)$, $P(AC)$, and so on. For example, the parameter $\langle F_6 \rangle$, the constraint function average for $F_6(X = x_i)$, will incorporate information about the marginal probability $P(AB \mid \mathcal{M}_k)$ through,

$$P(AB \mid \mathcal{M}_k) = \langle F_6 \rangle = \sum_{i=1}^{32} F_6(X = x_i) \, Q_i \tag{42.9}$$

Examining the joint probability table, we are able to discern that $P(AB \mid \mathcal{M}_k)$ is the marginal probability over eight cells,

$$P(AB \mid \mathcal{M}_k) = Q_1 + Q_5 + Q_9 + Q_{13} + Q_{17} + Q_{21} + Q_{25} + Q_{29} \tag{42.10}$$

Thus, $F_6(X = x_1) = F_6(X = x_5) = \cdots = F_6(X = x_{29}) = 1$ while all the remaining $F_6(X = x_i) = 0$.

The numbering for the constraint functions implementing the available double interactions started at 6 because we wanted to reserve the numbers 1 through 5 for the main effects. The numbers 16 through 32 would be reserved for implementing any triple, quadruple, or quintuple interactions should we want to consider those more complicated models.

Suppose the IP wanted the MEP assignment in the seventh cell of the joint probability table. Here is the shorter expression,

$$Q_7 \equiv P(X = x_7 \mid \mathcal{M}_k) = P(A\overline{B}\,\overline{C}DE \mid \mathcal{M}_k) \tag{42.11}$$

Given the pattern of 1s and 0s for all ten constraint functions $F_j(X = x_7)$, the numerator for Q_7 looks like $\exp[\lambda_8 + \lambda_9 + \lambda_{15}]$. The partition function Z_q will be very complicated because it consists of the sum over all 32 possible numerators.

$$Z_q = \sum_{i=1}^{32} \exp\left[\sum_{j=6}^{15} \lambda_j \, F_j(X = x_i)\right] \tag{42.12}$$

Thus, $\exp[\lambda_8 + \lambda_9 + \lambda_{15}]$ would be just one component in the partition function as contributed by Q_7. There would be 31 other similar terms in Z_q. Thus, it is quite evident that as problems become more complicated, calculation of the partition function becomes a major nuisance.

Don't worry about sorting all of this out right now. The above was just a quick overview of the some of the major components in implementing the MEP assignments to the joint probability table. The exercises will go into all of the details when you are ready.

42.5 The Kangaroos Meet the Boltzmann Machine

I would not leave you adrift at this point without a numerical example. Let's enlist the aid of our friends the kangaroos since they have helped us out so many times in the past.

Suppose an inferential problem presents itself where we would like to update our state of knowledge about a couple of interesting characteristics of the kangaroos that are difficult to discern directly. Let's agree that these less discernible traits are a kangaroo's intelligence and its beer preference.

There are three other more readily observable characteristics consisting of hand preference, fur color, and place of residence. Each of these five characteristics will be measured as binary variables so that we can set up a state space of $n = 32$.

Map the generic notation used for the ON–OFF nature of the nodes in the Boltzmann machine to statements and their two possible measurements as follows:

$A \rightarrow$ Intelligence

$a_1 =$ Above average and $a_2 =$ Below average

$B \rightarrow$ Beer Preference

$b_1 =$ Foster's and $b_2 =$ Corona

$C \rightarrow$ Hand Preference

$c_1 =$ Right and $c_2 =$ Left

$D \rightarrow$ Fur Color

$d_1 =$ Sandy and $d_2 =$ Beige

$E \rightarrow$ Residence

$e_1 =$ Melbourne and $e_2 =$ Sydney

The probability expression,

$$P(A = a_1, B = b_2 \mid C = c_1, D = d_2, E = e_1, \mathcal{M}_k) \equiv P(A, \overline{B} \mid C, \overline{D}, E, \mathcal{M}_k) \qquad (42.13)$$

indicates a probability computed by Bayes's Theorem and based on the information in model $\mathcal{M}_k$ for a kangaroo possessing above average intelligence who prefers to drink Corona given that it is known that the kangaroo is a right handed beige kangaroo who lives in Melbourne.

By Bayes's Theorem, this probability is,

$$P(A, \overline{B} \mid C, \overline{D}, E, \mathcal{M}_k) = \frac{P(A, \overline{B}, C, \overline{D}, E \mid \mathcal{M}_k)}{P(C, \overline{D}, E \mid \mathcal{M}_k)} \qquad (42.14)$$

From the joint probability table, the cell numbers involved are,

$$P(A, \overline{B} \mid C, \overline{D}, E, \mathcal{M}_k) = \frac{Q_{11}}{Q_9 + Q_{10} + Q_{11} + Q_{12}} \qquad (42.15)$$

The probability in the denominator, $P(C, \overline{D}, E \mid \mathcal{M}_k)$, is seen to be the sum over the four joint probabilities,

$$P(A, B, C, \overline{D}, E \mid \mathcal{M}_k) + P(\overline{A}, B, C, \overline{D}, E \mid \mathcal{M}_k) + P(A, \overline{B}, C, \overline{D}, E \mid \mathcal{M}_k) + P(\overline{A}, \overline{B}, C, \overline{D}, E \mid \mathcal{M}_k)$$

Consider one particular model, explored further in Exercise 42.9.3, merely as an illustration of the above calculation by Bayes's Theorem. By the way, this particular model is not the Boltzmann machine model which is taken up in the next section.

The probability for a kangaroo possessing above average intelligence who prefers to drink Corona given that it is known that the kangaroo is a right handed beige kangaroo who lives in Melbourne, is calculated by Bayes's Theorem as,

$$\begin{aligned} P(A, \overline{B} \mid C, \overline{D}, E, \mathcal{M}_k) &= \frac{Q_{11}}{Q_9 + Q_{10} + Q_{11} + Q_{12}} \\ &= \frac{0.04263}{0.04688 + 0.00085 + 0.04263 + 0.01367} \\ &= 0.409739 \end{aligned}$$

The other three possibilities for a kangaroo known to be a right handed beige kangaroo living in Melbourne are,

$$\begin{aligned} P(A, B \mid C, \overline{D}, E, \mathcal{M}_k) &= 0.450630 \\ P(\overline{A}, B \mid C, \overline{D}, E, \mathcal{M}_k) &= 0.008204 \\ P(\overline{A}, \overline{B} \mid C, \overline{D}, E, \mathcal{M}_k) &= 0.131426 \end{aligned}$$

The greatest probability is that the kangaroo is above average intelligence and prefers to drink Foster's given its already observed traits. As an essential check, all four of these probabilities must sum to 1,

$$0.450630 + 0.008204 + 0.409739 + 0.131426 = 1$$

If there were no association between intelligence, beer preference, and these known traits, then the probabilities for all four possibilities would be equal to 1/4. The model $\mathcal{M}_k$ must then be inserting information about some sort of relationship between intelligence, beer preference, and these three traits. Why do we know that? Because we do see marked differences from 1/4 in the conditional probabilities.

42.6 Prediction Using a Boltzmann Machine

But the discussion in the last section is just idle speculation about the consequences from one particular model. Speculation about, say, intelligence and beer preference based on knowing hand preference, fur color, and place of residence based on one model is as good as speculation based on any other model. Without any data, all models remain on the same equal footing.

Our ultimate goal is to make a prediction about any next kangaroo given the data from some number of N kangaroos. Suppose that we have been lucky enough to sample a total of $N = 640$ kangaroos on all five traits. What kind of predictions can the Boltzmann machine provide us after undergoing some "learning" from these data?

42.6.1 The data

The contingency table containing these data appear in Figure 42.3 below. As usual, each cell of the contingency table contains an N_i frequency count.

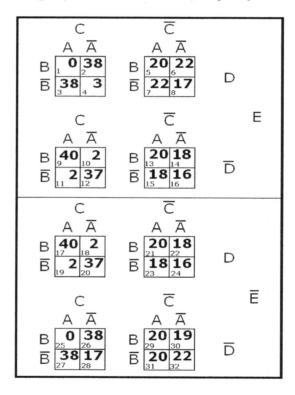

Figure 42.3: *The thirty two cell contingency table containing the observed frequency counts N_i in a sample of 640 kangaroos. The Boltzmann machine will "learn" from these data.*

The sum of all the frequency counts must add up to all of the kangaroos sampled, $\sum_{i=1}^{32} N_i = 640$. For example, there are $N_{15} = 18$ left handed, smart Corona drinking kangaroos who have beige fur color and live in Melbourne in cell 15 of the contingency table.

From this table, we want to find the sample averages that correspond to the constraint function averages. In particular, we need to find the ten sample averages $\overline{F}_6$ through $\overline{F}_{15}$ that correspond to the information about all ten double interactions in $\langle F_6 \rangle$ through $\langle F_{15} \rangle$.

In this way, we can establish our Boltzmann machine and, at the same time, the model behind point q^*. Point p is a "complicated model," that is, the maximum likelihood model, $\mathcal{M}_{\text{ML}}$. The 31 sample averages for point p correspond to all 31 constraint function averages which designate the information in the main effects, the ten double interactions, the ten triple interactions, the five quadruple interactions, and the single quintuple interaction.

These sample averages are easy enough to calculate. They are,

$$\overline{F}_j = \frac{1}{N} \sum_{i=1}^{32} N_i \, F_j(X = x_i)$$

For example, the sample average for the second double interaction AC between intelligence and hand preference is,

$$\overline{F}_7 = \frac{1}{640} \sum_{i=1}^{32} N_i \, F_7(X = x_i)$$

The calculation is as follows,

$$\sum_{i=1}^{32} N_i \, F_7(X = x_i) = (0 \times 1) + (38 \times 0) + (38 \times 1) + (3 \times 0) + \cdots + (40 \times 1) + \cdots + (22 \times 0)$$

$$= N_1 + N_3 + N_9 + N_{11} + N_{17} + N_{19} + N_{25} + N_{27}$$

$$= 160$$

$$\overline{F}_7 = \frac{160}{640}$$

$$= 0.25$$

Carrying out exactly the same calculations for all ten sample averages reveals that they are all equal to $1/4$,

$$\overline{F}_6 = \overline{F}_7 = \cdots = \overline{F}_{15} = 0.25$$

With these calculations of the sample averages for the double interactions, we have an inkling that trouble is afoot.

42.6.2 The Boltzmann machine fails miserably

The point $q^\star$ is the **m**–projection of point p into the sub–manifold $\boldsymbol{E}_k(q)$. All the points that live in this sub–manifold must have their θ^1 through θ^5 as well as their θ^{16} through θ^{31} coordinates all equal to 0. Only the coordinates θ^6 through θ^{15} can vary for different points. Since $q^\star$ is the **m**–projection of point p, its dual coordinates η_6 through η_{15} must match these same coordinates for p. But η_6 through η_{15} for point p are all equal to 0.25 as we just discovered.

When the MEP algorithm is used to find the numerical assignments to point $q^\star$, each Q_i must then equal $1/n = 1/32$. Under this fair model, any probability of intelligence and beer preference conditional on hand preference, fur color, and place of residence must equal $1/4$.

Thus, this Boltzmann machine fails miserably at its task. It cannot do any pattern completion or memory association. No matter what the known traits of the kangaroo happen to be, the Boltzmann machine cannot distinguish between any kangaroo on intelligence and beer preference.

42.6.3 Inferencing still works

However, despite the failure of the Boltzmann machine, the framework for solving any inferencing problem will still provide the optimal predictions achievable from these data. The generally applicable prediction formulas for this inferential scenario were developed in Volumes I and II.

The probability for any future frequency count $\{M_1, M_2, \cdots, M_n\}$ when these counts are conditioned on the known data $\mathcal{D} \equiv \{N_1, N_2, \cdots, N_n\}$ is equal to,

$$P(M_1, M_2, \cdots, M_n \mid \mathcal{D}) = \frac{M!\,(N+n-1)!}{\prod_{i=1}^{n} N_i!\,(M+N+n-1)!} \times \frac{\prod_{i=1}^{n}(M_i+N_i)!}{\prod_{i=1}^{n} M_i!}$$

This formula was derived by averaging over the posterior probability $P(\mathcal{M}_k \mid \mathcal{D})$ of *every single model*. See Exercises 20.9.1 and 20.9.2 on pages 108–109 in Volume II if you would like to review a derivation.

There need be no concern that the answer provided by the above formula is some sort of "approximation" based on accepting a single model, good or bad. Actually, it would be better to say that every single model, good, bad, or indifferent, made its unique contribution to the prediction. The predictions from good models were heavily weighted, the predictions from bad models hardly counted at all.

When interest centers on the "next" occurrence given the data, this formula simplifies to,

$$P(M_1 = 0, \cdots, M_i = 1, \cdots, M_n = 0 \mid \mathcal{D}) = \frac{N_i + 1}{N + n} \qquad (42.16)$$

BOLTZMANN MACHINES

Thus, the joint probability that the next kangaroo is a stupid right handed Corona drinker with sandy fur color who hails from Sydney is,

$$P(M_1 = 0, \cdots, M_{20} = 1, \cdots, M_{32} = 0 \,|\, \mathcal{D}) = \frac{38}{640 + 32}$$

$$P(A_{N+1} = a_2, B_{N+1} = b_2, C_{N+1} = c_1, D_{N+1} = d_1, E_{N+1} = e_2 \,|\, \mathcal{D}) = 0.056548$$

Bayes's Theorem, as a general manipulation rule for probabilities, is applicable no matter what the probabilities are conditioned on, so,

$$P(\overline{A}_{N+1}, \overline{B}_{N+1} \,|\, C_{N+1}, D_{N+1}, \overline{E}_{N+1}, \mathcal{D}) = \frac{P(\overline{A}_{N+1}, \overline{B}_{N+1}, C_{N+1}, D_{N+1}, \overline{E}_{N+1} \,|\, \mathcal{D})}{P(C_{N+1}, D_{N+1}, \overline{E}_{N+1} \,|\, \mathcal{D})}$$

Since the denominator in Bayes's Theorem is the sum over four joint probabilities, we now have,

$$P(\overline{A}_{N+1}, \overline{B}_{N+1} \,|\, C_{N+1}, D_{N+1}, \overline{E}_{N+1}, \mathcal{D}) = \frac{38}{41 + 3 + 3 + 38} = 0.440759$$

Similar calculations for the other three possibilities reveal that, contrary to the Boltzmann machine, general inferencing will "complete the pattern" for a right handed, sandy fur colored kangaroo hailing from Sydney. It will notice that the highest probability is calculated as $41/85 = 0.48$ for a smart Foster's drinker.

Unfortunately, it is almost divided on this conclusion because this probability is not much greater than the probability that the kangaroo is, in fact, a dumb Corona drinker as calculated above as $38/85 = 0.44$.

It is pretty sure, however, that the kangaroo with these three traits is not a dumb Foster's drinker or a smart Corona drinker. Remember that the Boltzmann machine placed equal probability on all four possibilities, and therefore could not "complete the pattern" for any "clamped environmental vector."

42.6.4 A designed system for inferencing

There is a distinction to be made between *supervised* and *unsupervised* learning for Boltzmann machines. What we have just demonstrated might be viewed as an example of unsupervised learning where the inferential consequences are driven exclusively by the data in Figure 42.3.

In contrast, we might consider that what we want, in fact, is a *designed* system. It is not unreasonable to think that this goal is always the impetus behind any AI artifact. Such a system is engineered from the outset so that its output meets some design criteria. Thus, we might want to design a Boltzmann machine whose output criterion is to "complete the pattern" when a number of "environmental variables" are "clamped" to the "input units" of the Boltzmann machine.

Let's continue to rely upon the existing kangaroo scenario in order to illustrate an inferential system *designed* to automatically and correctly classify all kangaroos into one of four categories. These four categories are:

Category 1. (A, B) The smart Foster's drinkers.

Category 2. $(\overline{A}, B)$ The dumb Foster's drinkers.

Category 3. $(A, \overline{B})$ The smart Corona drinkers.

Category 4. $(\overline{A}, \overline{B})$ The dumb Corona drinkers.

The supposed known characteristics of hand preference, fur color, and place of residence actually define a correct categorization. For example, when the system is presented with an incomplete pattern of a left handed beige colored kangaroo from Melbourne, it should correctly complete the partial pattern by outputting with probability 1 that this kangaroo is a dumb Corona drinker. This is part of the design specification of the machine.

Assuming that the definition of the state space has been taken care of, every inference using probability theory always begins with the joint probability table. As we discovered in Volume I when we examined elementary cellular automata from a probabilistic standpoint, one could always set up a joint probability table such that the assignments reproduced the deductions made from a particular Boolean function.

When this technique was applied to how Wolfram defined his elementary cellular automata, Bayes's Theorem could always reproduce the correct black or white cell at time step $N+1$ with probability 1. The conditioning in Bayes's Theorem was on the colors of three cells at time step N.

The same technique can be used here by judiciously assigning the values of $1/8$ and 0 under the information in some model (the designed system) to all 32 cells in the joint probability table. For example, Q_1 is assigned $1/8$, while Q_2, Q_3, and Q_4 are assigned 0. When completing the pattern for a right handed sandy colored kangaroo from Melbourne, Bayes's Theorem instructs us to categorize this kangaroo unequivocally in category 1 as a smart Foster's drinker.

$$P(A, B \mid C, D, E, \mathcal{M}_{\text{DS}}) = \frac{P(A, B, C, D, E \mid \mathcal{M}_{\text{DS}})}{P(C, D, E \mid \mathcal{M}_{\text{DS}})} = \frac{Q_1}{Q_1 + Q_2 + Q_3 + Q_4} = 1$$

There are eight possibilities for the three environmental variables (C, D, E) that could be clamped onto the input units of the Boltzmann machine in order for the correct pattern to be completed. These range from (C, D, E) through $(\overline{C}, \overline{D}, \overline{E})$. Any one of these eight possibilities for the known variables is going to result in certainty for the categorization.

The incomplete pattern of input variables mentioned above of a left handed beige colored kangaroo from Melbourne is categorized by the designed system as a dumb Corona drinker,

$$P(\overline{A}, \overline{B} \mid \overline{C}, \overline{D}, E, \mathcal{M}_{\text{DS}}) = \frac{P(\overline{A}, \overline{B}, \overline{C}, \overline{D}, E \mid \mathcal{M}_{\text{DS}})}{P(\overline{C}, \overline{D}, E \mid \mathcal{M}_{\text{DS}})} = \frac{Q_{16}}{Q_{13} + Q_{14} + Q_{15} + Q_{16}} = 1$$

because the information in the designed system has $1/8$ assigned to Q_{16}, while Q_{13}, Q_{14}, and Q_{15} were assigned 0.

Notice especially that there is no subscript $N+1$ attached to the prediction for intelligence and beer preference. A single model $\mathcal{M}_{\text{DS}}$ with the appropriate type of information was able to implement the engineered system. There was no need to condition the prediction on N pieces of data, or average over all the models with respect to all $P(\mathcal{M}_k \mid \mathcal{D})$.

Perhaps my initial assessment of the Boltzmann machine by highlighting its failure mode was a bit harsh. The Boltzmann machine, like all ANNs and all AI for that matter, are conceptualized, at least implicitly, to be engineered or designed systems. The way I have depicted them, through a manufactured example where the data were lacking in any double interactions, shackles any teleological goals the Boltzmann machine might aspire to.

But how well can a Boltzmann machine emulate this latest "complicated model" that was designed to correctly complete patterns for the kangaroos? The MEP has no problem with assigning 0 and $1/8$ to their proper locations within the joint probability table. It accomplishes this task in the usual way by taking account of the information from all of the interaction types. However, the Boltzmann machine can only avail itself of the information from the ten double interactions.

As you might have come to expect by now, the Boltzmann machine cannot hope to duplicate the performance of the designed system for general inferencing. It can not make predictions that are certain.

Worse than that, however, is that the Boltzmann machine can do no better with its predictions than the fair model. The designed system with its placement of numerical assignments of 0 and $1/8$ into the joint probability table, has all of its model's ten double interactions exactly equal to the fair model's ten double interactions, namely $1/4$.

There can be no hope that the Boltzmann machine, which, from the perspective of Information Geometry, is the α-projection of this designed system from the full Riemannian manifold down into a sub-manifold where the first five coordinates and last sixteen coordinates are all 0, might be able to distinguish the smart and dumb Foster's and Corona drinkers based on their known traits of hand preference, fur color, and place of residence.

As Exercise 42.9.10 will demonstrate in a little more detail, all of the relevant information in the designed system which distinguishes it from the information resident in the Boltzmann machine, is contained in two triple interactions, all five quadruple interactions, and the one quintuple interaction.

42.7 Posterior Probability of Models

Just to be clear, we are temporarily leaving behind any kind of designed system as was broached in the last section. We are now focusing once again on the so–called unsupervised model based on the empirical data of section 42.6.1 and Figure 42.3.

Because MacKay discusses the likelihood ratio for the Boltzmann machine at length, let's review the ratio of posterior probabilities for any two models where the model in the numerator is the maximum likelihood model based on the sample averages from any data. This ratio will provide the relative importance of the prediction made by any model for any future frequency count.

Recall that this maximum likelihood model $\mathcal{M}_{\text{ML}}$ was, in fact, the complicated model behind the point p that was **m**–projected to the point q^*. This latter point lives in a special sub–manifold that contains information about the ten double interactions. It has information about only those interactions, and not any higher order interactions.

From Volume II, we have the generic formula based on the output from the MEP algorithm for calculating the ratio of the posterior probability of any two models,

$$\frac{P(\mathcal{M}_A \mid \mathcal{D})}{P(\mathcal{M}_B \mid \mathcal{D})} = \exp\left\{ N \times \left[\sum_{j=1}^{m} (\lambda_j^A - \lambda_j^B) \overline{F}_j(X = x_i) + \ln\left(\frac{Z_B}{Z_A}\right) \right] \right\} \quad (42.17)$$

The maximum likelihood model will have more say than any other model for a prediction because its relative weight from the above ratio will be greater than any other model. It will be enormously favored over competing models that don't reflect the data as well as it is able to do. Any non–maximum likelihood model that is fairly close to the data will, however, get to participate in the prediction, with the proviso that its prediction will not be weighted quite as heavily as that made by the maximum likelihood model.

For a relevant example, consider the relative weighting of the prediction that the next kangaroo will be a smart right handed sandy colored Foster's drinker from Melbourne. The model for the Boltzmann machine had a probability of 1/32 attached to this prediction. On the other hand, the maximum likelihood model has a probability of 0 attached to this same prediction.

The Boltzmann machine model's probability for this prediction is almost literally not counted. The maximum likelihood model's relative weight compared to the Boltzmann machine model will be calculated by,

$$\frac{P(\mathcal{M}_{\text{ML}} \mid \mathcal{D})}{P(\mathcal{M}_{\text{BM}} \mid \mathcal{D})} = \exp\left\{ 640 \times \left[\sum_{j=1}^{32} (\lambda_j^{\text{ML}} - 0) \overline{F}_j(X = x_i) + \ln\left(\frac{32}{Z_{\text{ML}}}\right) \right] \right\}$$

When the Lagrange multipliers λ_j^{ML} and the partition function Z_{ML} for model $\mathcal{M}_{\text{ML}}$, together with the sample averages from the data $\overline{F}_j(X = x_i)$, are inserted into this formula, the relative weighting in favor of $\mathcal{M}_{\text{ML}}$ is very decisive at,

$$\frac{P(\mathcal{M}_{\text{ML}} \mid \mathcal{D})}{P(\mathcal{M}_{\text{BM}} \mid \mathcal{D})} = 1.84 \times 10^{69}$$

Carefully note, however, that the important final prediction is an average over every model. The Boltzmann machine model's assignment of $Q_1 = 1/32$, together with the maximum likelihood model's assignment of $Q_1 = 0$ are just two of the many assignments in the computation of this final average. The final prediction for the next kangaroo to be a smart, right handed, sandy colored Foster's drinker from Melbourne, when based on all of the data is,

$$P(A_{N+1}, B_{N+1}, C_{N+1}, D_{N+1}, E_{N+1} \mid \mathcal{D}) = \frac{N_1 + 1}{N + n} = \frac{1}{672}$$

This is seen to be close to the maximum likelihood model's prediction of 0 based on the frequency count in cell 1, but just a little higher because the formal manipulation rules of probability theory took account of every single model's prediction.

42.8 Connections to the Literature

One of the essential features of a Boltzmann machine resides in the issue of how to make it learn to do what you want it to do, whether that learning is supervised or unsupervised. MacKay [29, pp. 522–523] identifies a version of Bayes's Theorem for the posterior probability of the matrix $\mathbf{W}$ as essential to this concept of *Boltzmann machine learning*. Learning in a Boltzmann machine centers on the updating of the connections strengths in the matrix $\mathbf{W}$ based on the known data.

MacKay provides this version of Bayes's Theorem,

$$P(\mathbf{W} \mid \{x^{(n)}\}_1^N) = \frac{\left[\prod_{n=1}^N P(x^{(n)} \mid \mathbf{W})\right] P(\mathbf{W})}{P(\{x^{(n)}\}_1^N)}$$

I find this notation and the resulting interpretation as applied to learning in a Boltzmann machine to be somewhat confusing and misleading. First of all, I think it is better to stick to the fundamental probabilistic relationships where equations like this are seen to be nothing more than a particular application of the representation for the posterior probability of a model,

$$P(\mathcal{M}_k \mid \mathcal{D}) = \frac{P(\mathcal{D} \mid \mathcal{M}_k) P(\mathcal{M}_k)}{P(\mathcal{D})}$$

In addition, in this latter standard representation, we avoid the blunder of thinking we are assigning a probability distribution over parameters rather than a probability distribution over models.

In any case, his subsequent development of the log likelihood is fascinating for us because it follows exactly the quintessential relationship between maximum likelihood and the MEP that Jaynes had explicated earlier on. But, once again, the interpretation placed on these findings is different for different people depending on how one views the big picture.

In MacKay's notation, the log likelihood for the Boltzmann machine is expressed as,

$$\ln \left[\prod_{n=1}^{N} P(x^{(n)} \mid \mathbf{W}) \right] = \sum_{n=1}^{N} \left[\frac{1}{2} \mathbf{x}^{(n)T} \mathbf{W} \mathbf{x}^{(n)} - \ln Z(\mathbf{W}) \right]$$

In the standard generic notation we have been developing in order to maintain consistency with the MEP, this log likelihood would be expressed as,

$$\ln [P(\mathcal{D} \mid \mathcal{M}_k)] = \ln W(N) + \sum_{i=1}^{n} N_i \ln Q_i$$

One of the nice features of the MEP is that the expansion of $\ln Q_i$ results in an expression that allows one to more easily see various relationships like the one MacKay proposed. The slow sequence of steps involving the expansion of $\ln Q_i$ is shown next. The constant term for a fixed set of data, the log of the multiplicity factor, $\ln W(N)$, is dropped at the outset because it will always cancel when the ratio of the posterior probability of models is formed.

$$\ln [P(\mathcal{D} \mid \mathcal{M}_k)] = \sum_{i=1}^{n} N_i \left(\sum_{j=1}^{m} \lambda_j F_j(X = x_i) - \ln Z \right)$$

$$= \sum_{i=1}^{n} N_i \left(\sum_{j=1}^{m} \lambda_j F_j(X = x_i) \right) - \sum_{i=1}^{n} N_i \ln Z$$

$$= \sum_{i=1}^{n} N_i \left(\sum_{j=1}^{m} \lambda_j F_j(X = x_i) \right) - \ln Z \sum_{i=1}^{n} N_i$$

$$= \sum_{i=1}^{n} N_i \left(\sum_{j=1}^{m} \lambda_j F_j(X = x_i) \right) - N \ln Z$$

$$= N \left[\sum_{i=1}^{n} \frac{N_i}{N} \left(\sum_{j=1}^{m} \lambda_j F_j(X = x_i) \right) - \ln Z \right]$$

$$= N \left(\sum_{j=1}^{m} \lambda_j \overline{F}_j - \ln Z \right)$$

In conventional Fisherian statistics, one finds the maximum likelihood estimate by differentiating the log likelihood with respect to its parameters, followed by equating this result to 0. In marching towards this objective, MacKay isolates the second term in his log likelihood expression, and differentiates with respect to all of the parameters in the matrix $\mathbf{W}$,

$$\frac{\partial}{\partial w_{ij}} \ln Z(\mathbf{W}) = \sum_{\mathbf{x}} x_i x_j \, P(\mathbf{x} \mid \mathbf{W}) = \langle x_i x_j \rangle_{P(\mathbf{x} \mid \mathbf{W})}$$

For us, this is a very familiar step in the MEP scheme of things,

$$\frac{\partial \ln Z}{\partial \lambda_j} = \langle F_j \rangle$$

MacKay then turns his attention to the first term. For that first term, if we recall that (see Exercises 42.9.1 and 42.9.2),

$$\frac{1}{2}\left[\mathbf{x}^T \mathbf{W} \mathbf{x}\right]$$

ends up as $\sum_{i,j} w_{ij} x_i x_j$, then,

$$\frac{\partial}{\partial w_{ij}} 1/2 \left[\mathbf{x}^{(n)T} \mathbf{W} \mathbf{x}^{(n)}\right] = x_i^{(n)} x_j^{(n)}$$

Substituting these findings into the log likelihood, MacKay derives,

$$\begin{aligned}
\frac{\partial \ln\left[\prod_{n=1}^{N} P(x^{(n)} \mid \mathbf{W})\right]}{\partial w_{ij}} &= \sum_{n=1}^{N} \left[x_i^{(n)} x_j^{(n)} - \langle x_i x_j \rangle_{P(\mathbf{x} \mid \mathbf{W})}\right] \\
&= \sum_{n=1}^{N} x_i^{(n)} x_j^{(n)} - \sum_{n=1}^{N} \langle x_i x_j \rangle_{P(\mathbf{x} \mid \mathbf{W})} \\
&= N \left[\frac{1}{N} \sum_{n=1}^{N} x_i^{(n)} x_j^{(n)} - \langle x_i x_j \rangle_{P(\mathbf{x} \mid \mathbf{W})}\right] \\
&= N \left[\langle x_i x_j \rangle_{\text{Data}} - \langle x_i x_j \rangle_{P(\mathbf{x} \mid \mathbf{W})}\right]
\end{aligned}$$

Rather than solving for the maximum likelihood by setting these equations to 0, MacKay stops here. The interpretation for Boltzmann machine learning is that the difference in the two terms inside the brackets is a *gradient*. The weights w_{ij} are adjusted by a numerical procedure that descends along this gradient.

The way we prefer to complete these steps is by the standard approach arriving at the maximum likelihood equations. Thus, we have a system of m equations,

$$\frac{\partial \ln[P(\mathcal{D}|\mathcal{M}_k)]}{\partial \lambda_j} = 0$$

$$\frac{\partial}{\partial \lambda_j} N \left(\sum_{j=1}^{m} \lambda_j \overline{F}_j - \ln Z \right) = 0$$

$$N(\overline{F}_j - \langle F_j \rangle) = 0$$

$$\langle F_j \rangle_{\text{ML}} = \overline{F}_j \qquad \text{for } j = 1 \text{ to } m$$

Despite the notation, the two derivations have arrived at the same result after these two identifications are made,

$$\langle x_i\, x_j \rangle_{\text{Data}} \equiv \overline{F}_j$$

$$\langle x_i\, x_j \rangle_{P(\mathbf{x}|\mathbf{W})} \equiv \langle F_j \rangle_{\text{ML}}$$

The model behind point p provides the information from the constraint function averages $\langle F_j \rangle_{\text{ML}}$. Do not forget, as was just demonstrated, that these constraint function averages must match the sample averages from the data. This model can definitely be called a complicated model because it includes the information from all 31 possible $\langle F_j \rangle$. Point $q^\star$ matches $\langle F_6 \rangle_{\text{ML}}$ through $\langle F_{15} \rangle_{\text{ML}}$. It is the Boltzmann machine that is "closest," in the sense of the distance measures defined in Information Geometry, to point p.

Implementing a learning procedure for a Boltzmann machine depends upon an elaborate procedure involving weight readjustments through the gradient and Monte Carlo techniques like simulated annealing. The annealing schedule is defined by introducing a temperature parameter T; thus, the one model parameter from the canonical distribution of statistical mechanics makes a reappearance.

But none of this kind of learning is required, at least not for the straightforward inferencing scenarios discussed in this Chapter. The optimal weights, that is, the correct Lagrange multipliers, are found immediately by the MEP algorithm for both the complicated model for point p and the Boltzmann machine behind point $q^\star$. There are none of the attendant difficulties wrapped up in Boltzmann machine learning, or perhaps more accurately, the numerical procedures involving the gradient are hidden within *Mathematica*'s black box function **NMinimize[ ]**.

42.9 Solved Exercises for Chapter Forty Two

Exercise 42.9.1: Discuss the energy function $E(x)$ for the Boltzmann machine as it is developed through the standard explanation for vector and matrix multiplications.

Solution to Exercise 42.9.1

We are given that the energy function for a Boltzmann machine is the product of a transposed vector $\mathbf{x^T}$ times a matrix $\mathbf{W}$ times the vector $\mathbf{x}$,

$$-E(\mathbf{x}) = 1/2\,[\,\mathbf{x^T W x}\,]$$

To enforce conformability in the requisite multiplications, the vector $\mathbf{x}$ must be a column vector. So, in our numerical example, $\mathbf{x^T}$ is a (1×5) row vector, $(x_1, x_2, x_3, x_4, x_5)$.

The matrix $\mathbf{W}$ of connection strengths is then a (5×5) matrix,

$$\mathbf{W} = \begin{pmatrix} w_{11} & w_{12} & w_{13} & w_{14} & w_{15} \\ w_{21} & w_{22} & w_{23} & w_{24} & w_{25} \\ w_{31} & w_{32} & w_{33} & w_{34} & w_{35} \\ w_{41} & w_{42} & w_{43} & w_{44} & w_{45} \\ w_{51} & w_{52} & w_{53} & w_{54} & w_{55} \end{pmatrix}$$

Multiplying the (5×5) matrix $\mathbf{W}$ by the (1×5) row vector $\mathbf{x^T}$ is a conformable operation resulting in a (1×5) row vector $\mathbf{x^T W}$.

Multiplying the first column of $\mathbf{W}$ by the first (and only) row of $\mathbf{x^T}$ yields the first element in the vector $\mathbf{x^T W}$,

$$w_{11}x_1 + w_{21}x_2 + w_{31}x_3 + w_{41}x_4 + w_{51}x_5$$

Following the same rule, multiplying the fifth column of $\mathbf{W}$ by the first row of $\mathbf{x^T}$ by yields the fifth element in the vector $\mathbf{x^T W}$,

$$w_{15}x_1 + w_{25}x_2 + w_{35}x_3 + w_{45}x_4 + w_{55}x_5$$

For the second multiplication, we have the (1×5) row vector $\mathbf{x^T W}$ multiplying the (5×1) column vector $\mathbf{x}$ which we know by now is going to result in a (1×1) scalar $\mathbf{x^T W x}$.

This sum is going to consist of 25 terms which look like,

$$w_{11}x_1^2 + w_{12}x_1x_2 + w_{21}x_2x_1 + \cdots + w_{45}x_4x_5 + w_{54}x_5x_4 + w_{55}x_5^2$$

However, because the w_{ii} elements in the matrix $\mathbf{W}$ will be set to 0, and matrix $\mathbf{W}$ is symmetric, the sum will be reduced to just ten terms when we compute $1/2\,\mathbf{x^T W x}$. For example, the first three terms above get reduced to,

$$w_{11}x_1^2 + w_{12}x_1x_2 + w_{21}x_2x_1 = 1/2\,(0 + w_{12}x_1x_2 + w_{21}x_2x_1) = w_{12}x_1x_2$$

The ten terms in the scalar $1/2\,[\,\mathbf{x}^T\mathbf{W}\mathbf{x}\,]$ are then,

$$w_{12}x_1x_2 + w_{13}x_1x_3 + w_{14}x_1x_4 + w_{15}x_1x_5 + w_{23}x_2x_3+$$

$$w_{24}x_2x_4 + w_{25}x_2x_5 + w_{34}x_3x_4 + w_{35}x_3x_5 + w_{45}x_4x_5$$

shown as the energy function, Equation (42.5), in section 42.4.

Exercise 42.9.2: Use *Mathematica* to construct an example of the results of the first exercise.

Solution to Exercise 42.9.2

First, construct the vector $\mathbf{x}^T$ as `vectorX = Array[x, 5]`. This produces the list `{x[1], x[2], x[3], x[4], x[5]}`. Notice that we don't have to be distracted by the distinction between row and column vectors.

Second, construct a matrix $\mathbf{W}$ containing actual numbers as opposed to the symbolic entries w_{ij}. *Mathematica* provides an easy way to do this with a command under the **Insert** Menu called **Table/Matrix** that allows the user to specify the number of rows and columns for a desired matrix.

Then, just fill in all 25 entries for the template of $\mathbf{W}$. Remember that the matrix will be symmetric and that the five diagonal entries will be 0. Here is the $\mathbf{W}$ matrix assigned as `matrixW` that I constructed to illustrate the general result found in the first exercise.

$$\mathtt{matrixW} = \begin{pmatrix} 0 & 1 & 2 & 3 & 4 \\ 1 & 0 & 5 & 6 & 7 \\ 2 & 5 & 0 & 8 & 9 \\ 3 & 6 & 8 & 0 & 10 \\ 4 & 7 & 9 & 10 & 0 \end{pmatrix}$$

Everything is now set up for the vector–matrix $\mathbf{x}^T\mathbf{W}$ and vector–vector $\mathbf{x}^T\mathbf{W}\mathbf{x}$ multiplications through the `Dot[ ]` command. The first multiplication is,

`Dot[vectorX, matrixW]`

and both together are,

`Dot[Dot[vectorX, matrixW], vectorX]`

To get the result in the form we want, a couple of more embellishments,

`1/2 Dot[Dot[vectorX, matrixW], vectorX] // Expand`

produces,

$$x_1x_2 + 2x_1x_3 + 3x_1x_4 + 4x_1x_5 + 5x_2x_3+$$

$$6x_2x_4 + 7x_2x_5 + 8x_3x_4 + 9x_3x_5 + 10x_4x_5$$

where $w_{12} = 1$, $w_{13} = 2$, $\cdots$, $w_{45} = 10$. This result conforms to the symbolic formula of the first exercise.

Are these applications of **Dot[ ]** to perform the vector–matrix multiplications equivalent to performing the actual low–level sums and multiplications? Let's ask *Mathematica* for the answer. Evaluating,

Equal[Expand[Dot[Dot[vectorX, matrixW], vectorX]],
        Sum[w[j, k] x[j] x[k], {k, 1, 5}, {j, 1, 5}]]

returns **True**.

Exercise 42.9.3: What kind of model produced the results as presented in section 42.5?

Solution to Exercise 42.9.3

Let's begin with a relatively simple model that, like the Boltzmann machine, will incorporate information from only double interactions. Suppose that the model discussed in section 42.5 inserted information only about two double interactions, *AB* and *AC*. Thus, there are correlations between intelligence and beer preference, as well as intelligence and hand preference in this model. The exact form of this information was specified as,

$$\langle F_6 \rangle = 0.50$$

$$\langle F_7 \rangle = 0.50$$

Since under the fair model, or point u, the same numerical assignment to each joint statement is $1/n = 1/32 = 0.03125$, the constraint function averages $\langle F_6 \rangle$ and $\langle F_7 \rangle$ would both be equal to $8/32 = 0.25$ under this model. Inserting the above information changes the assignments from $1/32$. In fact, the numerator and the partition function in the denominator for Q_{11} in the Bayes's Theorem computation discussed in that section work out to the values,

$$Q_{11} = \frac{\exp[\sum_{j=6}^{15} \lambda_j F_j(X = x_{11})]}{\sum_{i=1}^{32} \exp[\sum_{j=6}^{15} \lambda_j F_j(X = x_i)]} = \frac{5.2711}{123.651} = 0.04263$$

Decomposing the numerator in Q_{11} even further,

$$\exp\left[\sum_{j=6}^{15} \lambda_j F_j(X = x_{11})\right]$$

we find that,
$$F_j(X = x_{11}) = 1 \text{ for } j = 7, 9, \text{ and } 14$$
and,
$$F_j(X = x_{11}) = 0 \text{ for } j = 6, 8, 10, 11, 12, 13, \text{ and } 15$$

The Lagrange multipliers found by the MEP algorithm for this model must have λ_1 through λ_5 equal to 0, and λ_{16} through λ_{31} also equal to 0. There are potentially ten non–zero Lagrange multipliers λ_6 through λ_{15}. But, as we just discussed, only λ_7, λ_9, and λ_{14} will be multiplied by 1, and all of the remaining λ_j will be multiplied by 0.

$$\lambda_7 = +2.86889$$

$$\lambda_9 = -1.73182$$

$$\lambda_{14} = +0.52516$$

Thus, the numerator for Q_{11} is,

$$\exp[\,2.86889 + (-1.73182) + 0.52516\,] = 5.2711$$

Exercise 42.9.4: Examine some extreme models that serve as anchor points for any inference.

Solution to Exercise 42.9.4

What then are the implications for the fair model? The information under the fair model for the five marginal probabilities representing the main effects is that all five constraint function averages $\langle F_1 \rangle$ through $\langle F_5 \rangle$ equal 1/2.

The ten constraint function averages $\langle F_6 \rangle$ through $\langle F_{15} \rangle$ representing the marginal probabilities for the double interactions are all equal to 1/4. The ten constraint function averages $\langle F_{16} \rangle$ through $\langle F_{25} \rangle$ representing the marginal probabilities for the triple interactions are all equal to 1/8. The five constraint function averages $\langle F_{26} \rangle$ through $\langle F_{30} \rangle$ representing the marginal probabilities for the quadruple interactions are all equal to 1/16. The final constraint function average $\langle F_{31} \rangle$ represents the quintuple interaction, which, in this case, is not a marginal probability at all. It is simply Q_1, which is equal to 1/32. The dual parameters for this model, the Lagrange multipliers λ_1 through λ_{31}, are all equal to 0.

The upshot of this model is that *no active information* is inserted into the probability distribution. This model, point u, possesses the maximum amount of *missing information* of any probability distribution. It is not THE maximum entropy distribution, it is ONE maximum entropy distribution. Other maximum entropy distributions, like q^* for example, will get as close as they possibly can to u while still retaining all of their active information.

Point u will not be a Boltzmann machine because it cannot get as close to point p as point q^* can. Point u has no information about any double interactions, that is, all the weights in matrix $\mathbf{W}$ are equal to 0.

The joint probability distribution u will lead to inferences that the probability for any intelligence level and any beer preferences of any kangaroo are all the same and equal to 1/4. It doesn't make any difference where the kangaroo resides, its fur color, or hand preference; the probability of intelligence and beer preference remains unmoved at 1/4.

For example, Bayes's Theorem tells us that the probability that a kangaroo is above average intelligence and prefers Foster's given that it resides in Melbourne, has beige fur color, and is right handed, is,

$$P(A, B \mid C, D, E, \mathcal{M}_k) = \frac{P(A, B, C, D, E \mid \mathcal{M}_k)}{P(C, D, E \mid \mathcal{M}_k)}$$

$$= \frac{Q_1}{Q_1 + Q_2 + Q_3 + Q_4}$$

$$= \frac{1/32}{1/32 + 1/32 + 1/32 + 1/32}$$

$$= 1/4$$

Bayes's Theorem also tells us that the probability that a kangaroo is below average intelligence and prefers Corona given that it resides in Sydney, has sandy fur color, and is left handed, is the same,

$$P(\overline{A}, \overline{B} \mid \overline{C}, \overline{D}, \overline{E}, \mathcal{M}_k) = \frac{P(\overline{A}, \overline{B}, \overline{C}, \overline{D}, \overline{E} \mid \mathcal{M}_k)}{P(\overline{C}, \overline{D}, \overline{E} \mid \mathcal{M}_k)}$$

$$= \frac{Q_{32}}{Q_{29} + Q_{30} + Q_{31} + Q_{32}}$$

$$= \frac{1/32}{1/32 + 1/32 + 1/32 + 1/32}$$

$$= 1/4$$

What about an extreme model at the other end of the spectrum? What are the implications of models that permit some deductions to be made?

The following model might be considered somewhat bizarre. All of its main effects, all of its double interactions, all of its triple interactions, and all of its quadruple interactions are exactly the same as for the fair model. However, it does differ in one important respect.

The quintuple interaction is not $\langle F_{31} \rangle = 1/32$ as the fair model dictates, but, instead, the information in this bizarre model says that $\langle F_{31} \rangle = 0$. The difference is really not all that dramatic as the quintuple interaction under the fair model is only 0.03125 when compared to 0 under this "complicated" model. So one thing we know at the outset is that $Q_1 = 0$ under this model.

Hand the issue of the numerical assignments under this model over to the MEP algorithm. It returns with a very simple and symmetrical assignment, but one that is quite an interesting assignment nonetheless, as revealed in the joint probability table of Figure 42.4 below.

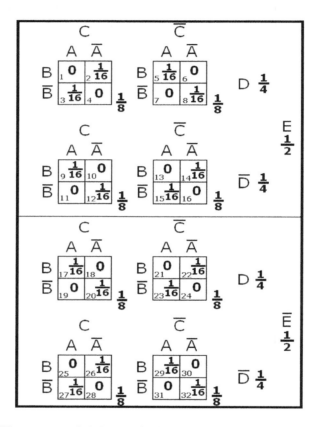

Figure 42.4: *The joint probability table for five traits of kangaroos as conditioned on the truth of a model that gives rise to strong deductions. This is a model behind a point p in the dually flat Riemannian manifold $\mathcal{S}^{32}$.*

You may verify at your leisure that all of the constraint function averages are satisfied. As an example, pick some quadruple interaction out of the blue. Say it is $\langle F_{28} \rangle$ which must equal $1/16$ according to what was just said about the model; that is, the fact that it matches all of the quadruple interactions from the fair model.

A quadruple interaction is the sum over two joint probabilities, and here this would be,

$$\langle F_{28}\rangle \equiv P(ABDE) = P(ABCDE) + P(AB\overline{C}DE) = Q_1 + Q_5 = 0 + 1/16 = 1/16$$

Select some triple interaction to examine. Suppose $\langle F_{23}\rangle$ is selected. This must equal 1/8 in the above model because that is the information in the fair model for any triple interaction. A triple interaction is the sum over four joint probabilities, and here this would be,

$$\langle F_{23}\rangle \equiv P(BCE)$$

$$P(BCE) = P(ABCDE) + P(\overline{A}BCDE) + P(ABC\overline{D}E) + P(\overline{A}BC\overline{D}E)$$

$$= Q_1 + Q_2 + Q_9 + Q_{10}$$

$$= 0 + 1/16 + 1/16 + 0$$

$$= 1/8$$

With 16 0s scattered throughout the joint probability table, it might be expected that a number of deductions could be reached. For example, what can be said about right handed, sandy fur colored kangaroos from Sydney? It is impossible for such a kangaroo to be below average in intelligence and prefer Foster's.

$$P(\overline{A}, B \mid C, D, \overline{E}, \mathcal{M}_k) = \frac{P(\overline{A}, B, C, D, \overline{E} \mid \mathcal{M}_k)}{P(C, D, \overline{E} \mid \mathcal{M}_k)}$$

$$= \frac{Q_{18}}{Q_{17} + Q_{18} + Q_{19} + Q_{20}}$$

$$= \frac{0}{1/16 + 0 + 0 + 1/16}$$

It is also impossible for a kangaroo with these three traits to be above average in intelligence and prefer Corona. This leaves only two possibilities.

There is a probability of 1/2 that such a kangaroo will be above average in intelligence and prefer Foster's, split evenly with a probability of 1/2 that such a kangaroo will be below average in intelligence and prefer Corona. We can quickly check that all bases are covered since all four possibilities for right handed, sandy fur colored kangaroos from Sydney add up to 1.

Exercise 42.9.5: What is the maximum value and minimum value of the information for the quintuple interaction for the type of models just discussed in the previous exercise?

Solution to Exercise 42.9.5

The minimum value of $\langle F_{31} \rangle = 0$ has been examined above. Since the information is the specification of a probability, and here it is just Q_1, we know that it must be bounded from below at 0 and from above at 1.

The maximum value of $\langle F_{31} \rangle = \frac{1}{16} = 0.06125$. It can not be any larger than this because any higher value would violate the feasible values for any quadruple interaction.

Information concerning quadruple interactions is a sum over two joint probabilities. For example, the information about the $BCDE$ quadruple interaction is $\langle F_{30} \rangle = \frac{1}{16}$ since it must match the quadruple interaction for the fair model.

$$P(BCDE \mid \mathcal{M}_k) = P(ABCDE \mid \mathcal{M}_k) + P(\overline{A}BCDE \mid \mathcal{M}_k)$$

$$= Q_1 + Q_2$$

$$= \frac{1}{16}$$

If $P(ABCDE \mid \mathcal{M}_k)$ were to be set any value higher than 0.06125, then obviously $P(BCDE \mid \mathcal{M}_k)$ could not satisfy the information requirement for these type of models. If the information about the quintuple interaction is set at its maximum value of $Q_1 = 1/16$, then $Q_2 = 0$.

Exercise 42.9.6: What kind of numerical assignment and deductions are made when the information about the quintuple interaction takes on its maximum value?

Solution to Exercise 42.9.6

When the information about the quintuple interaction takes on its maximum value of $\langle F_{31} \rangle = 0.06125$, Q_1 is assigned the value $\frac{1}{16}$, while Q_2 is assigned the value 0. For the situation where the information about the quintuple interaction took on its minimum value of $\langle F_{31} \rangle = 0$, Q_1 was assigned the value 0, while Q_2 was assigned the value $\frac{1}{16}$. The assignments of 0 and $\frac{1}{16}$ are interchanged throughout the joint probability table under the minimum and maximum value of the information for the quintuple interaction. Therefore, the pattern of conditional probabilities for intelligence and beer preference undergo the same symmetrical switch at the maximum value from what they were at the minimum value.

SOLVED EXERCISES FOR CHAPTER 42

Exercise 42.9.7: Conduct a "back of the envelope" rough analysis to get some feel for the formula for the ratio of the posterior probabilities of any two models.

Solution to Exercise 42.9.7

Let's start off by avoiding the relative weight for the maximum likelihood model be some enormous number as in section 42.7. Rather, we would prefer that the ratio of posterior probabilities have its maximum at a value of 1. Then, all other models other than the maximum likelihood model would have a relative weight less than 1.

If we want the ratio of the posterior probabilities have its maximum at 1, then the formula in Equation (42.17) gets rearranged to look like this,

$$\frac{P(\mathcal{M}_k|\mathcal{D})}{P(\mathcal{M}_{\text{ML}}|\mathcal{D})} = \exp\left\{N \times \left[\sum_{j=1}^{m}(\lambda_j^k - \lambda_j^{\text{ML}})\overline{F}_j(X=x_i) + \ln\left(\frac{Z_{\text{ML}}}{Z_k}\right)\right]\right\}$$

Replace model $\mathcal{M}_k$ in the numerator with model $\mathcal{M}_{\text{ML}}$. Now, of course, the right hand side must equal 1.

$$\frac{P(\mathcal{M}_{\text{ML}}|\mathcal{D})}{P(\mathcal{M}_{\text{ML}}|\mathcal{D})} = \exp\left\{N \times \left[\sum_{j=1}^{m}(\lambda_j^{\text{ML}} - \lambda_j^{\text{ML}})\overline{F}_j(X=x_i) + \ln\left(\frac{Z_{\text{ML}}}{Z_{\text{ML}}}\right)\right]\right\}$$

$$= \exp\left\{N \times \left[\sum_{j=1}^{m}[0 \times \overline{F}_j(X=x_i)] + \ln(1)\right]\right\}$$

$$= \exp\left\{N \times 0\right\}$$

$$= 1$$

This is the weight for the prediction from $\mathcal{M}_{\text{ML}}$ and is the maximum value for any weight in the overall averaging.

Any alternative model, like the Boltzmann machine model $\mathcal{M}_{\text{BM}}$, appearing in the numerator, if very, very close to $\mathcal{M}_{\text{ML}}$, must produce a ratio slightly less than 1. The argument to the exponential function must then be a small negative number, say,

$$N \times \left[\sum_{j=1}^{m}(\lambda_j^{\text{BM}} - \lambda_j^{\text{ML}})\overline{F}_j + \ln\left(\frac{Z_{\text{ML}}}{Z_{\text{BM}}}\right)\right] = -0.001$$

$$\exp\{-0.001\} = 0.999$$

The difference in the Lagrange multipliers between the two models $(\lambda_j^{\text{BM}} - \lambda_j^{\text{ML}})$ may be positive or negative, so the ratio of the two partition functions will have to be such that the entire term in the square brackets $[\cdots]$ does become negative.

Whatever negative number is in the square brackets $[\cdots]$ gets multiplied by N. If this number were once again -0.001, then $\mathcal{M}_{\text{BM}}$ has less weight relative to $\mathcal{M}_{\text{ML}}$ as N becomes large. If $N = 1$, then the ratio is 0.999, whereas if $N = 1000$, the ratio drops dramatically to $\exp\{-1\} = 0.3679$. This is as it should be. As the sample size becomes larger and larger, alternative models becomes far less competitive with the maximum likelihood model.

Exercise 42.9.8: Indicate how combinatorial explosion sets in with a vengeance to increase the size of the state space on both complicated models and their simplifications as Boltzmann machines.

Solution to Exercise 42.9.8

This Chapter's example looked at a Boltzmann machine with $N = 5$ nodes, or five units. The dimension of the state space consisting of these five units, each of which could take on only binary values, was $n = 2^5 = 32$.

The maximum number of thirty one constraint functions, $m = n - 1 = 31$, was broken down into five main effects, ten double interactions, ten triple interactions, five quadruple interactions, and one quintuple interaction. A complicated model like point p utilizing all m constraint functions was projected down to a simpler model of a Boltzmann machine at $q^\star$. The model behind $q^\star$ had only $m = 10$ constraint functions, and consequently only ten weights w_{ij}.

Increasing the number of nodes in the Boltzmann machine by just three from five to eight increases the dimension of the state space from 32 to $n = 2^8 = 256$. For the complicated model p utilizing all $m = n - 1 = 255$ constraint functions, there would be 70 quadruple interactions alone.

Use the combinatorial decomposition discussed before to systematically break down the number of interactions of each type,

$$2^N = 2^8 = \sum_{j=1}^{u} W_j(N) \rightarrow \sum_{j=1}^{9} W_j(8) = 256$$

The upper index u for the above summation is,

$$u = \frac{(N + n^\star - 1)!}{N!\,(n^\star - 1)!} = \frac{(8 + 2 - 1)!}{8!\,1!} = 9$$

where $n^\star = 2$ counts the number of states for any binary node. An alternative form is,

$$2^8 = \sum_{j=0}^{8} \binom{8}{j} = \binom{8}{0} + \binom{8}{1} + \cdots + \binom{8}{8} = 256$$

SOLVED EXERCISES FOR CHAPTER 42

Table 42.1 below shows the breakdown into all of the various interaction types. The final column shows how many of the lowest level joint probability terms, the Q_i, are included in the specified interaction.

Table 42.1: *Evidence of an incipient combinatorial explosion for a full decomposition of a complicated model consisting of just eight nodes.*

j	$\binom{8}{j}$	Interaction Type	Example	# Q_i involved
0	1	Zero order		
1	8	Main	$P(C)$	128
2	28	Double	$P(DG)$	64
3	56	Triple	$P(BEF)$	32
4	70	Quadruple	$P(ACFH)$	16
5	56	Quintuple	$P(BCEFG)$	8
6	28	Sixth order	$P(ABCDFH)$	4
7	8	Seventh order	$P(ABDEFGH)$	2
8	1	Eighth order	$P(ABCDEFGH)$	1
Sum	256			

The combinatorial explosion is especially onerous when trying to calculate the partition function of any complicated model Z_p.

$$Z_p = \sum_{i=1}^{256} \exp\left[\sum_{j=1}^{255} \lambda_j F_j(X = x_i)\right]$$

Even when the constraint functions map to 0 and 1, a rough sketch of the partition function is going to look something like,

$$Z_p = \overbrace{\exp[\lambda_1 + \cdots + \cdots] + \exp[\underbrace{\cdots + \lambda_j + \cdots}_{255 \text{ terms}}] + \cdots + \exp[0]}^{256 \text{ terms}}$$

The combinatorial explosion reflected in the increasingly difficult to compute partition function Z_p is the origin of Markov Chain Monte Carlo (MCMC) efforts to compute ratio of integrals within statistical mechanics which have the generic appearance,

$$\mathcal{I} = \frac{\int F(\theta) \exp\{-\frac{E(\theta)}{kT}\} d\theta}{\int \exp\{-\frac{E(\theta)}{kT}\} d\theta}$$

where the complicated partition function appears as the integral in the denominator.

Exercise 42.9.9: Identify the information in the double interaction BD under the model for the engineered system that is designed to correctly categorize kangaroos?

Solution to Exercise 42.9.9

The BD double interaction stands for any association between beer preference and fur color. Place all ten double interactions in the following order as shown below in Table 42.2. The index j in the first column refers to the specific constraint function for the double interaction $F_j(X = x_i)$. The final column shows MacKay's corresponding $x_i x_j$ terms in his energy function for the Boltzmann machine. Review the first exercise to refresh your memory.

Table 42.2: *All ten double interactions for a Boltzmann machine with 5 nodes.*

j	Interaction	Description	$x_i x_j$
6	AB	Intelligence x Beer Preference	$x_1 x_2$
7	AC	Intelligence x Hand Preference	$x_1 x_3$
8	AD	Intelligence x Fur Color	$x_1 x_4$
9	AE	Intelligence x Residence	$x_1 x_5$
10	BC	Beer Preference x Hand Preference	$x_2 x_3$
11	BD	Beer Preference x Fur Color	$x_2 x_4$
12	BE	Beer Preference x Residence	$x_2 x_5$
13	CD	Hand Preference x Fur Color	$x_3 x_4$
14	CE	Hand Preference x Residence	$x_3 x_5$
15	DE	Fur Color x Residence	$x_4 x_5$

The constraint function for the double interaction BD has the generic notation $F_{11}(X = x_i)$. It represents the mapping from all 32 joint statements to a vector of 1s and 0s. By definition, the marginal probability $P(BD)$ is the sum over the following eight joint probabilities,

$$P(BD) = P(ABCDE) + P(\overline{A}BCDE) + P(AB\overline{C}DE) + P(\overline{A}B\overline{C}DE) +$$

$$P(ABCD\overline{E}) + P(\overline{A}BCD\overline{E}) + P(AB\overline{C}D\overline{E}) + P(\overline{A}B\overline{C}D\overline{E})$$

Locate the eight matching Q_i in the joint probability table,

$$P(BD) = Q_1 + Q_2 + Q_5 + Q_6 + Q_{17} + Q_{18} + Q_{21} + Q_{22}$$

Since,
$$\langle F_{11}\rangle = \sum_{i=1}^{32} F_{11}(X = x_i)\, Q_i$$
we see that,
$$F_{11}(x_1) = F_{11}(x_2) = F_{11}(x_5) = F_{11}(x_6) = F_{11}(x_{17}) = F_{11}(x_{18}) = F_{11}(x_{21}) = F_{11}(x_{22}) = 1$$

with all the remaining $F_{11}(X = x_i) = 0$. The information inserted by the model for the BD interaction in the designed system is then,
$$\langle F_{11}\rangle = Q_1 + Q_2 + Q_5 + Q_6 + Q_{17} + Q_{18} + Q_{21} + Q_{22}$$
$$= 1/8 + 0 + 0 + 1/8 + 0 + 0 + 0 + 0$$
$$= 1/4$$

Exercise 42.9.10: Provide a more detailed description of some designed inferencing system.

Solution to Exercise 42.9.10

It is relevant to note that the designed system is based on the *single* model $\mathcal{M}_{\text{DS}}$. The designed system might have been inspired by an already existing understanding of the relationships among the variables, by a logical analysis accompanied by the appropriate logic expressions, or, even more abstractly, by Boolean functions as in some sort of cellular automata.

It is quite distinct then from an unsupervised system based on a point p, a maximum likelihood model $\mathcal{M}_{\text{ML}}$ derived from the sample averages of the data. There are no immediate data behind $\mathcal{M}_{\text{DS}}$, just as there is no immediate theory behind a model based solely on the empirical data.

As we said earlier, a judicious placement of the numerical assignments $1/8$ and 0 into all 32 cells of the joint probability table will allow this designed system to effect the correct "pattern completion" for any of the eight possible "clamped units." In fact, if these eight assignments are symmetrically placed into the joint probability table,
$$Q_1 = Q_6 = Q_{11} = Q_{16} = Q_{20} = Q_{23} = Q_{26} = Q_{29} = 1/8$$
with the remaining twenty four assignments all 0, then it is easy to see under what conditions of the "clamped variables" the designed inferencing system will output its correct categorizations of the intelligence and beer preference of any kangaroo with probabilities of 1 or 0.

For example, right handed sandy colored kangaroos from Melbourne and left handed beige colored kangaroos from Sydney are certain to be smart Foster's drinkers. It is impossible for a kangaroo to possess any other intelligence and beer preference trait given these known characteristics under this model. Both,

$$P(AB \,|\, CDE, \mathcal{M}_k) = \frac{Q_1}{Q_1 + Q_2 + Q_3 + Q_4} = 1$$

and,

$$P(AB \,|\, \overline{C}\,\overline{D}\,E, \mathcal{M}_k) = \frac{Q_{29}}{Q_{29} + Q_{30} + Q_{31} + Q_{32}} = 1$$

On the other hand, all right handed sandy colored kangaroos from Sydney and left handed beige colored kangaroos from Melbourne are certain to be dumb Corona drinkers. It is simply impossible for a kangaroo to be either a smart Foster's drinker, a dumb Foster's drinker, or a smart Corona drinker given these known traits under this model for the designed system. Both,

$$P(\overline{A}\,\overline{B} \,|\, CD\overline{E}, \mathcal{M}_k) = \frac{Q_{20}}{Q_{17} + Q_{18} + Q_{19} + Q_{20}} = 1$$

and,

$$P(\overline{A}\,\overline{B} \,|\, \overline{C}\,DE, \mathcal{M}_k) = \frac{Q_{16}}{Q_{13} + Q_{14} + Q_{15} + Q_{16}} = 1$$

while,

$$P(A\overline{B} \,|\, CD\overline{E}, \mathcal{M}_k) = \frac{Q_{19}}{Q_{17} + Q_{18} + Q_{19} + Q_{20}} = 0$$

But for this designed system, all of the information in the main effects and double interactions is the same as for the fair model. For example, we examined the information contained in the double interaction BD in the previous exercise. Since $\langle F_{11} \rangle = 1/4$, the information was the same as in the fair model. It turns out that the information in all ten double interactions is also equal to $1/4$.

The significant information in the designed inferencing system is contained in the two triple interactions where $\langle F_{20} \rangle$ and $\langle F_{24} \rangle$ both equal $1/4$ instead of $1/8$, in all five quadruple interactions where $\langle F_{26} \rangle$ through $\langle F_{30} \rangle$ are equal to $1/8$ instead of $1/16$, and in the one quintuple interaction $\langle F_{31} \rangle$ equal to $1/8$ instead of $1/32$.

Table 42.3 at the top of the next page takes a more detailed look at the five quadruple interactions of the complicated model behind the designed system, in other words, point p in the Riemannian manifold.

From the viewpoint of Information Geometry, we have another case where the Boltzmann machine $q^\star$ cannot serve as any reasonable approximation to the general inferencing system p because its projection down into the sub-manifold $\boldsymbol{E}(q)$, that is, the sub-manifold which matches all of the double interactions of the designed system, is no better than the fair model.

Table 42.3: *A more detailed look at the five quadruple interactions of a complicated model behind some designed inferencing system.*

Number	Information	Interaction	Cells	Probability
1	$\langle F_{26} \rangle$	$P(ABCD)$	$Q_1 + Q_{25}$	$1/8 + 0$
2	$\langle F_{27} \rangle$	$P(ABCE)$	$Q_1 + Q_{13}$	$1/8 + 0$
3	$\langle F_{28} \rangle$	$P(ABDE)$	$Q_1 + Q_5$	$1/8 + 0$
4	$\langle F_{29} \rangle$	$P(ACDE)$	$Q_1 + Q_3$	$1/8 + 0$
5	$\langle F_{30} \rangle$	$P(BCDE)$	$Q_1 + Q_2$	$1/8 + 0$

Exercise 42.9.11: Summarize all of the categorizations, or, if you prefer, the pattern completions, made by the designed system.

Solution to Exercise 42.9.11

Table 42.4 below lists all eight possible combinations of the input variables of hand preference, fur color, and place of residence. The output variables of intelligence and beer preference are shown wherever the conditional probability is equal to 1.

Table 42.4: *A summary of the categorizations, or pattern completions, made by a designed system for the intelligence and beer preference traits of kangaroos for all eight possible combinations of the input variables.*

Number	Probability	Output	Input
1	$P(AB \mid CDE) = 1$	Smart Foster's	Right Sandy Melbourne
2	$P(\overline{A}\,\overline{B} \mid CD\overline{E}) = 1$	Dumb Corona	Right Sandy Sydney
3	$P(A\overline{B} \mid C\overline{D}E) = 1$	Smart Corona	Right Beige Melbourne
4	$P(\overline{A}B \mid C\overline{D}\,\overline{E}) = 1$	Dumb Foster's	Right Beige Sydney
5	$P(\overline{A}B \mid \overline{C}DE) = 1$	Dumb Foster's	Left Sandy Melbourne
6	$P(A\overline{B} \mid \overline{C}D\overline{E}) = 1$	Smart Corona	Left Sandy Sydney
7	$P(\overline{A}\,\overline{B} \mid \overline{C}\,\overline{D}E) = 1$	Dumb Corona	Left Beige Melbourne
8	$P(AB \mid \overline{C}\,\overline{D}\,\overline{E}) = 1$	Smart Foster's	Left Beige Sydney

Exercise 42.9.12: What conclusions does the general inferencing scheme reach about any left handed beige colored kangaroo from Sydney?

Solution to Exercise 42.9.12

Let's compare these two probability expressions, the first conditioned on the model for the designed system, $P(A, B \,|\, \overline{C}, \overline{D}, \overline{E}, \mathcal{M}_{\mathrm{DS}})$, and the second conditioned on the data, $P(A_{N+1}, B_{N+1} \,|\, \overline{C}_{N+1}, \overline{D}_{N+1}, \overline{E}_{N+1}, \mathcal{D})$.

The first probability obviously uses the prediction from one model only, while the second probability is an average over all possible models. There is no subscript necessary in the first expression, while the $N+1$ subscripts in the second expression indicate that this is a probability for any $N+1^{st}$ kangaroo with N kangaroos comprising the data base.

The general inferencing scheme utilizes the formula in Equation (42.16) for any joint statement about the next kangaroo,

$$P(M_1 = 0, \cdots, M_i = 1, \cdots, M_n = 0 \,|\, \mathcal{D}) = \frac{N_i + 1}{N + n}$$

We will want to examine the prediction for all four possibilities of intelligence and beer drinking behavior given that the next kangaroo is definitely a left handed beige fur colored kangaroo from Sydney. Thus, the future frequency counts of $M_{29} = 1$ through $M_{32} = 1$ will be of interest. For example, checking the contingency table in Figure 42.3, we see that $N_{29} = 20$.

Thus, the probability that the next kangaroo is a smart, left handed, beige fur colored Foster's drinker from Sydney is,

$$P(M_1 = 0, \cdots, M_{29} = 1, \cdots, M_{32} = 0 \,|\, \mathcal{D}) = \frac{21}{640 + 32}$$

Relying on Bayes's Theorem, the probability that the next kangaroo is a smart Foster's drinking kangaroo when conditioned on the known facts of its other traits,

$$P(A_{641}, B_{641} \,|\, \overline{C}_{641}, \overline{D}_{641}, \overline{E}_{641}, \mathcal{D}) = \frac{21}{21 + 20 + 21 + 23}$$

The other three possibilities are also very close to $1/4$ as well.

Despite using the generality of probabilistic inferencing, the IP can not reach any definitive conclusions about how to categorize the kangaroos with these particular traits. It is not as successful as when it was dealing with the right handed sandy fur colored kangaroos from Sydney in section 42.6.3. But it does remain as the best inference that can be made in this case.

This case is an example of an unsupervised system, that is, a non–designed system based on the empirical data. However, a designed system is able to make deductions about the correct pattern completion for the kangaroos. As we saw in the last exercise,

$$P(AB\,|\,\overline{C}\,\overline{D}\,\overline{E}, \mathcal{M}_{\text{DS}}) = 1$$

$$P(A\overline{B}\,|\,\overline{C}\,\overline{D}\,\overline{E}, \mathcal{M}_{\text{DS}}) = 0$$

$$P(\overline{A}B\,|\,\overline{C}\,\overline{D}\,\overline{E}, \mathcal{M}_{\text{DS}}) = 0$$

$$P(\overline{A}\,\overline{B}\,|\,\overline{C}\,\overline{D}\,\overline{E}, \mathcal{M}_{\text{DS}}) = 0$$

Exercise 42.9.13: How would *Mathematica* ensure that the complement of double interactions is definitely the only the information being used by the Boltzmann machine?

Solution to Exercise 42.9.13

In the code for the MEP algorithm, make use of the allowed syntactical variation for constraining the parameters as documented in,

NMinimize[{*f, constraints*}**,** {*parameters*}**]**

```
solution = NMinimize[{Log[zsymbolic] - Dot[lsymbolic, cfa],
           Join[Table[λ_i == 0, {i, 1, 5 }],
                Table[λ_i == 0, {i, 16, 31 }]]}, lsymbolic]
```

All the coordinates, except for the ten model parameters λ_6 through λ_{15}, are constrained to be equal to 0 in the solution for the Boltzmann machine.

Chapter 43

Hidden Units for Boltzmann Machines

43.1 Introduction

Let's return to the theme of some contingent history associated with artificial neural networks. A by–product of such an accidental history is a certain idiosyncratic way of thinking about problems, together with an associated terminology and notation. I am going to couch the discussion in this Chapter by employing the Boltzmann machine's own unique *Gestalt* as it formed outside of the separate and distinct contingent history that gave rise to general inferencing.

Thus, this Chapter will tend to frame inferential explanations in terms of some neural network architecture, consisting of input units, hidden units, and output units. The input units and output units are labeled as visible units because they correspond to known observables in the real world, while the hidden units are latent variables created for the sole purpose of revealing some hidden pattern or structure not readily perceivable in the input units. The hidden units and output units are activated through weights connecting the units.

The input units are activated when environmental variables are clamped to the input units, which then in turn feed into activation of hidden units, and finally into the activation of output units. These output units are supposed to represent any target variables. This propagation is an asynchronous dynamic affair with each activation a momentary fleeting thing awaiting the next round of new activations. All of the weights connecting two units undergo a continuous change that depends on a gradient.

The main question here is the following: Will the introduction of hidden units help the Boltzmann machine overcome the impediment of relying on models with information solely from double interactions?

The most significant insight from the scientists who first explored Boltzmann machines was the nearly instantaneous realization that an architecture with weights connecting visible units could never hope to emulate the performance capabilities of complicated models available within general inferencing. A simple demonstration of the failure mode of Boltzmann machines was highlighted in the previous Chapter.

The weakness of the Boltzmann machine was revealed when a complicated model possessed the same information on all ten double interactions as the fair model, but nonetheless had important and relevant information on higher order interactions that enabled it to complete patterns successfully. The Boltzmann machine, by its very construction, was necessarily restricted to just the information in the double interactions and could only mimic the performance of the fair model. The ability of any fair model to perform pattern completion is of course non–existent.

Early on, though, a brilliant idea occurred to those seeking a way around this fundamental impediment. Perhaps the architecture of the basic Boltzmann machine could be augmented by introducing "hidden units" in addition to the already present input and output units. Perhaps the failure mode of Boltzmann machine could be circumvented by conceptualizing these hidden units as "feature detectors." The suggestion was that such hidden units, analogous to feature detectors in the visual part of the human brain, could distill out relevant implicit features inherent in any environmental variables clamped to the input units.

Quite naturally, it was taken for granted that the designer of such an inferential system didn't know much about these feature detectors, otherwise they would have been built in right from the start as visible units. Thus, the notion of "machine learning" became prevalent since the Boltzmann machine would have to "learn" what weights should connect the input units to the hidden units, as well as the weights from the hidden units to the output units.

In contrast, for a general inferential system designed to perform deductions, the appropriate weights under some model can always be found immediately. The MEP algorithm has no problem producing them after it becomes clear where the 0s should be placed within the joint probability table. There is no sense of the system forced to undergo a lengthy "learning process" to determine these weights, or model parameters.

It was recognized full well that a Boltzmann machine, even when augmented with hidden units, would never match the overpowering majesty of any ultimate complicated model. But the tradeoff in the number of parameters in the simpler model of the Boltzmann machine versus the full complicated model was deemed worthwhile.

For a numerical example, we are going to examine another simple Boltzmann machine, this time with three input units, three hidden units, and two output units. As we saw in Exercise 42.9.8, just by increasing the number of units from $N = 5$ to $N = 8$, we increased the dimension of the state space dramatically from $n = 32$ to $n = 256$.

Perhaps even more disconcerting is the fact that the complicated models we would like to construct in order to implement good inferences might involve 70 quadruple interactions, 56 quintuple interactions, and so on. This daunting prospect is enough incentive to consider what alternatives might exist.

A sketch of the linkages between the eight nodes in this particular Boltzmann machine is shown below as Figure 43.1. This network architecture indicates that the three input units are labeled as nodes C, D, and E. The three hidden units are labeled as nodes F, G, and H. The two output units are nodes A and B.

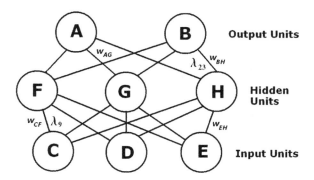

Figure 43.1: *A restricted Boltzmann machine which will try to classify kangaroos through the assistance of hidden units acting as feature detectors.*

A simplification is reflected in the fact that there are only nine connections between the input units and the hidden units, and a further six connections between the hidden units and the output units. These connections are labeled as weights with subscripts indicating a particular double interaction such as w_{CF}. With the voluntary restriction imposed by these limited connections, this kind of network architecture implements what is called a *restricted Boltzmann machine*.

Once again, this is easiest to understand by looking at it from the perspective of general inferencing. There are a total of $\binom{8}{2} = 28$ possible double interactions starting with AB, AC, and extending all the way through to GH. But in the restricted Boltzmann machine, the model will use the information from only 15 of the 28 total possible double interactions. Thus, the nine double interactions CF, CG, ..., EH connecting the input units to the hidden units will be used, together with the six double interactions AF, AG, ..., BH connecting the hidden units to the output units will also be used.

Possible double interactions that remain available from general inferencing, such as DE connecting two input units, or BD directly connecting an input and output unit, will not be employed. And, of course, any possible higher order interaction available to general inferencing such as $BCDGH$ can not be used by the restricted Boltzmann machine.

After this description, it should be clear that the overall inferencing objective remains the same. The IP would like to assess its degree of belief in the truth of any of the four joint statements about a kangaroo's intelligence and beer preference when conditioned on knowledge of its hand preference, fur color, and place of residence.

The only new element introduced into the plot is the appearance of three "feature detectors." It is hoped that these feature detectors can spot patterns in the predictor variables that enable the Boltzmann machine to do a better job at the correct classification of the kangaroos. We might not expect the Boltzmann machine to match the performance of a full blown general inferencing scheme. However, the trade–off in favor of the Boltzmann machine is the substantial reduction in the number of model parameters from $m = 31$ to $m = 15$.

43.2 Various Inferencing Scenarios

Let's establish an inferential scenario with a state space of dimension of $n = 256$. We envision $N = 8$ units that lead to this larger state space of dimension $n = 2^N = 256$. Remember that we are now using N to indicate the number of nodes in the network, and not to indicate the number of data points. The generality inherent in such a set up permits one to think about whichever kind of scenario happens to be most aesthetically pleasing. Here is a representative sampling of various popular scenarios.

Scenario #1 Arguably, at the most basic and abstract level, one might consider eight statements A, B, C, D, E, F, G, H. The only resolution for each of these statements is TRUE or FALSE.

Scenario #2 Or, we might choose to think in terms of observables $A, B, \cdots, H$ like coin flips having only binary measurement outcomes like HEADS or TAILS.
$$(A = a_1, B = b_2, \cdots, H = h_2) \equiv (A, \overline{B}, \cdots, \overline{H})$$

Scenario #3 Or, we might instead employ the language where eight binary nodes labeled as A through H can be in either an OFF state or an ON state. The ON state has a value of 1, and the OFF state has a value of 0.

Scenario #4 Or, getting closer to an actual physical situation, as eight atoms labeled as A through H in a lattice arrangement that could individually be in a SPIN UP state or a SPIN DOWN state. The SPIN UP takes on a value of $+1$, while the SPIN DOWN state takes on a value of -1.

Scenario #5 Or, in a thermodynamic setting or statistical mechanics setting where the nodes are eight Fermi oscillators. There is the inherent notion of physical activity taking place over time leading to a state of equilibrium controlled by a temperature parameter.

Scenario #6 Or, as used here to illustrate general inferencing, eight variables labeled as A through H to reflect a system designed to complete patterns for kangaroos. The partial pattern consists of three input variables C, D, and E standing for hand preference, fur color, and place of residence. The full pattern reaches completion when the correct categorization into the A and B variables of intelligence and beer preference can be made on the basis of the incomplete pattern represented by the above three variables. The three hidden variables of F, G, and H, by definition, cannot be given readily understandable labels like the other variables because we don't know what they are. All that can be said is that they pick out certain relevant features from the input variables that permit good pattern completions.

43.3 The Formal Probabilistic Start

To perform the pattern completion, we will need to compute the joint probability $P(A, B \,|\, C, D, E, \mathcal{M}_k)$ for the two output units A and B when the three input units C, D, and E are clamped to their given values. But in addition to these five visible units, we have three hidden variables F, G, and H that must be included in the probabilistic representation.

The hidden units F, G, and H are unknown, so the **Sum Rule** is used for the joint probability that would appear in the numerator of Bayes's Theorem,

$$P(A, B, C, D, E \,|\, \mathcal{M}_k) = \sum_{F,G,H} P(A, B, C, D, E, F, G, H \,|\, \mathcal{M}_k) \qquad (43.1)$$

Since F, G, and H can only assume two values each, the sum on the right hand side would consist of eight Q_i probabilities.

If we are looking at the consequences from some model $\mathcal{M}_k$ designed to classify kangaroos according to some criterion, then Bayes's Theorem becomes,

$$P(A, B \,|\, C, D, E, \mathcal{M}_k) = \frac{P(A, B, C, D, E \,|\, \mathcal{M}_k)}{P(C, D, E \,|\, \mathcal{M}_k)} \qquad (43.2)$$

$$= \frac{\sum_{F,G,H} P(A, B, C, D, E, F, G, H \,|\, \mathcal{M}_k)}{\sum_{A,B,F,G,H} P(A, B, C, D, E, F, G, H \,|\, \mathcal{M}_k)} \qquad (43.3)$$

where the sum in the numerator, as already mentioned, is over eight Q_i assignments, while the sum in the denominator is over thirty two Q_i assignments.

Thus, we will have to take care, for any particular inference, that the summations are being performed over the correct probabilities of joint statements as indicated in the above expressions.

43.4 How to Avoid Probability

But this rough overview of a restricted Boltzmann machine from a fundamental probabilistic grounding is not generally the description one finds in the literature. Instead, Boltzmann machines are viewed as a specific case of a type of artificial neural network called a *multilayer perceptron*. The primary emphasis is not on drawing out the consequences of the network architecture from probability theory, but rather as a somewhat different venture that sees the network as a means for approximating arbitrary functions.

Rather than trying to figure out how to compute the probability for some joint statement when conditioned on the truth of some number of predictor statements as in $P(A, B \,|\, C, D, E, \mathcal{M}_k)$, the goal is shifted to computing two *functions* $f_1(x)$ and $f_2(x)$ as output units where the arguments to the functions x are no longer statements, or mappings from statements, but simply "inputs" to the network. Likewise, any conditioning statements, or any statements to be summed over, are revamped conceptually into input units, or hidden units.

The specific role of Lagrange multipliers as model parameters is morphed into sets of weights connecting input units to hidden units, and hidden units to output units. Moreover, additional model parameters appear as "bias" terms at the hidden and output units.

It is very easy to lose track of what all of these "units" and "weights" were supposed to be mimicking. By not straying conceptually from within the confines of probability theory, it always remains clear that restricted Boltzmann machines never once deviate from very simple models involving the information from some subset of all the available double interactions. The weights are the associated Lagrange multipliers for whatever double interactions do enter into the models.

Other elements not part of the formal probabilistic approach also make their *deus ex machina* appearances. All k functions $f_k(x)$ are computed only after having inputs pass through some kind of *activation function* at the hidden layer. It is announced that the activation function should be something like the hyperbolic tangent function $\tanh(x)$ that converts large negative numbers to -1, large positive numbers to $+1$, and 0 to 0.

The physicist I. I. Rabi is purported to have querulously complained *"Who ordered that?"* when a particle called the muon was first broached as a necessary addition to the atomic zoo. We might utter a similar complaint when concepts like the $f_k(x)$ output functions, $\tanh(x)$ activation functions, and bias parameters make a sudden and mysterious appearance within what was thought to be a purely probabilistic enterprise.

There is a price to pay for this policy of excluding probability in the initial formulation of the inferential scenario. The output functions $f_k(x)$ now have to be somehow brought back into a probability expression. For example, in an *ad hoc* gesture, the $f_k(x)$ might appear as the means of independent Gaussian distributions.

Of course, it goes without saying that one never leaves the realm of probability in our preferred approach. So we spare ourselves the trouble of trying to justify this kind of *ad hoc* effort of trying to paste back functions into probability theory.

And, once one has finally gone to all the trouble of computing these $f_k(x)$, the job is still not finished. The $f_k(x)$ will be compared to some desired output of the network, and, presumably, there is at this beginning stage some discrepancy between the $f_k(x)$ just computed based on the some initial selection of arbitrary weights and biases, and what one would prefer the network to have computed. And so some long complicated routine must be discovered for adjusting all of the model parameters that extends over very many trials to reduce some error function, the measure of the discrepancy between the $f_k(x)$ on any given trial and what is ultimately desired from the network.

Operating from within the strictly formal probability approach, the answer from any model is immediately available. The prediction for some joint statement on intelligence and beer preference when given a kangaroo's status on hand preference, fur color, and place of residence, and moreover relying upon three feature detectors, is contained in $P(A, B \mid C, D, E, \mathcal{M}_k)$. This probability does not have be found by repeatedly changing weights and biases through some sort of "gradient descent" procedure over very many trials. We only have to use the MEP algorithm *once* to find the necessary Lagrange multipliers.

If only my advocacy of a strict adherence to the confines of a probabilistic framework did not lead to insurmountable computational problems. Alas, it is not to be!

The *ad hoc* approach, an approach I have maligned by levying several criticisms against it, has one distinct advantage over my preferred approach that relies strictly on formal probability. It may be that this approach provides the only pragmatic computational solution for very large networks that have to deal with problems of practical importance. We will see that even for so small a network comprised of eight nodes as discussed in this Chapter, forming the necessary constraint matrices and computing the partition function is subject to an incipient combinatorial explosion.

43.5 How Not to Avoid Probability

Despite the looming shadow of the inevitable combinatorial explosion, we can still rely on the MEP algorithm to find numerical assignments to all of the 256 joint probabilities required by the restricted Boltzmann machine. We manage to avoid computing the assignments from really complicated models because of the restricted Boltzmann machine's dependence on a subset of all possible double interactions.

For our example here, the models need include only the information from 15 of the possible 28 double interactions. A really complicated model from a network of eight nodes might include information from all eight main effects, some or all of the 28 double interactions, some or all of the 56 triple interactions, some or all of the 70 quadruple interactions, and so on. A daunting prospect indeed! But the competition is really between the restricted Boltzmann machine with three hidden units and complicated models from the five node network of the last Chapter.

Nonetheless, the state space cannot be reduced and is of dimension $n = 256$. Each of the 15 constraint functions $F_j(X = x_i)$ implementing the required double interactions will still have to assign values of 0 and 1 across all 256 joint statements $(X = x_i)$. Thus, the constraint matrix will consist of 15 lists with each of these lists containing 256 elements. Each of these 15 lists will have to reflect the mapping for a double interaction.

More explicitly, the first list will contain the mapping for the double interaction of CF representing the connection from the first input unit to the first hidden unit. The second list will contain the mapping for the double interaction of CG representing the connection from the first input unit to the second hidden unit. And so on, until the fifteenth list contains the mapping for the double interaction of BH that represents the connection from the last hidden unit to the second output unit.

The Lagrange multipliers for any of these models implementing the restricted Boltzmann machine will be labeled as λ_9 through λ_{23}. The constraint functions and constraint function averages will likewise be labeled as $F_9(X = x_i)$ through $F_{23}(X = x_i)$ and $\langle F_9 \rangle$ through $\langle F_{23} \rangle$. We would like to retain this indexing pattern to remind ourselves that there are many more potential terms in the MEP formula.

The numerators for the Q_i might be as simple as something like $\exp[\lambda_9]$ for the numerator of Q_{220}, or as complicated as,

$$\exp[\lambda_9 + \lambda_{10} + \lambda_{11} + \lambda_{12} + \lambda_{13} + \lambda_{14} + \lambda_{15} + \lambda_{16} + \lambda_{17} + \lambda_{18} + \lambda_{19} + \lambda_{20} + \lambda_{21} + \lambda_{22} + \lambda_{23}]$$

for the numerator of Q_1. Actually, it is easier to figure out that this last numerator must be the numerator for Q_1 since every single one of the fifteen lists must have a 1 as their first element.

The partition function Z_{q^*}, even for the relatively simple models of the restricted Boltzmann machine q^*, must however look quite complicated. By definition, Z_{q^*} consists of all possible 256 terms in the numerator of the Q_i, two of which were just shown above,

$$Z_{q^*} = \sum_{i=1}^{256} \exp\left[\sum_{j=9}^{23} \lambda_j^{q^*} F_j(X = x_i)\right]$$

Imagine the computational nightmare of finding the minimum value of the Legendre transformation in order to find the appropriate values for the model parameters in some complicated model p if the number of model parameters were $m = 255$.

$$H_{\max}(p) = \min_{\lambda_j^p} \left\{ \ln Z_p - \sum_{j=1}^{255} \lambda_j^p \langle F_j \rangle_p \right\}$$

MORE ON BOLTZMANN MACHINES

It's pretty obvious that, to make any further progress, we are going to have to lean rather heavily on *Mathematica's* computational prowess. Some of these details will be explored in Exercises 43.9.6, 43.9.7, 43.9.8, and 43.9.12.

There is an added benefit when the informational mind set is not abandoned. We can gauge how much active information the simpler models for the restricted Boltzmann machine are inserting into probability distributions, and, perhaps more importantly, how much residual information remains untapped.

For example, the maximum amount of missing information can be assessed for our current inferential scenario as 5.54518. Various restricted Boltzmann machines may reduce this missing information to around 5. This demonstrates that there is a large amount of potential information that these Boltzmann machines are not accessing because of the need to insert information only from a limited number of double interactions.

43.6 The Boltzmann Machine from the Pure Probability Approach

Let's take a look at an example of a restricted Boltzmann machine working on the kangaroo scenario. This presentation will concentrate exclusively on the purely probabilistic description as I outlined this philosophy in the last section. This restricted Boltzmann machine with three hidden variables will hopefully be an improvement over the woeful performance of the Boltzmann machine of the last Chapter which was bereft of any feature detectors.

Nevertheless, this Boltzmann machine under current consideration still remains a comparatively simple model with only 15 model parameters. These parameters incorporate the information from 15 double interactions. The Boltzmann machine can not take advantage of any of the potentially important information in more complicated models.

There does exist an advantageous trade-off for an IP which decides to approach the problem from the very same probabilistic perspective as employed for all of our other inferential scenarios. There is no need for a backpropagation scheme to adjust weights. There is no need to set up a huge number of Markov Chain Monte Carlo simulations. There is no need for all of the complications ensuing by treating the problem as, say, a multilayer perceptron problem, whose solution strays quite noticeably away from the strict confines of probability theory.

In contrast, the IP constructs a state space appropriate to the problem at hand, and then relies on the formal manipulation rules from probability theory, Bayes's Theorem being the most prominent of these. Thus, if the Boltzmann machine has a design specification to classify kangaroos as to their intelligence and beer drinking preferences given knowledge of known predictor variables like hand preference, fur color, and place of residence, and with the hopeful assistance of some feature detectors, formal probability expressions such as,

$$P(A, B \mid C, D, E, \mathcal{M}_k) = \frac{\sum_{F,G,H}^{8} P(A, B, C, D, E, F, G, H \mid \mathcal{M}_k)}{\sum_{A,B,F,G,H}^{32} P(A, B, C, D, E, F, G, H \mid \mathcal{M}_k)}$$

immediately present themselves as the desired solution under what I have labeled as the purely probabilistic approach.

The joint probability can be calculated only by conditioning on the truth of some model. This model $\mathcal{M}_k$ inserts information into the probability distribution over the joint statements. The MEP algorithm can now be relied upon to find numerical assignments to the probability for all of the joint statements in consonance with the information provided by this model.

Begin by examining the output from a small *Mathematica* program as shown in Figure 43.2 below. First, examine the information entropy of this restricted Boltzmann machine, considered as the point $q^\star$, an **m**–projection of some more complicated model point p that lives in the full Riemannian manifold.

| Entropy = 4.98515 | Z = 770.917 |

All 15 Lagrange multipliers. Constraint function averages in final row.

λ_9	λ_{10}	λ_{11}	λ_{12}	λ_{13}	λ_{14}	λ_{15}	λ_{16}	λ_{17}	λ_{18}	λ_{19}	λ_{20}	λ_{21}	λ_{22}	λ_{23}
-0.226595	-0.226595	-0.226595	-0.226595	-0.226595	-0.226595	0.444961	0.444961	0.444961	0.444961	0.444961	0.444961	0.444961	0.444961	0.444961
0.25	0.25	0.25	0.25	0.25	0.25	0.5	0.5	0.5	0.5	0.5	0.5	0.5	0.5	0.5

Probability for all four combinations of intelligence and beer preference conditioned on a sandy fur colored, right handed kangaroo from Melbourne.

P(A, B)	P(Not A, B)	P(A, Not B)	P(Not A, Not B)
Smart Foster's	Dumb Foster's	Smart Corona	Dumb Corona
0.492903	0.204649	0.204649	0.0977991

Figure 43.2: *The* Mathematica *output for a restricted Boltzmann machine.*

This point $q^\star$ must have inserted some active information into its probability distribution because the maximum possible information entropy of,

$$H_{\max}(u) = \ln n = \ln 256 = 5.54518$$

has, in fact, been reduced down to the maximum attainable information entropy of the distribution $q^\star$, which is $H_{\max}(q^\star) = 4.98515$. This result gives some indication of the amount of residual information from other models that is being ignored by the restricted Boltzmann machine.

Even though the partition function $Z_{q^\star}$ of the Boltzmann machine consists of 256 exponential terms,

$$Z_{q^\star} = \sum_{i=1}^{256} \exp\left[\sum_{j=9}^{23} \lambda_j^{q^\star} F_j(X = x_i)\right]$$

Mathematica has no problem calculating this sum as 770.917.

We next remark on the immediate availability of all 15 Lagrange multipliers from the MEP algorithm. As already mentioned as an advantage inherent to the strictly probabilistic approach, there was no need to rely upon a backpropagation algorithm, or extensive Monte Carlo simulations. These are the weights, the model parameters, enforcing the 15 double interaction constraint functions. In other words, this model is doing the same thing as the multilayer perceptron by setting up connections between input units and hidden units, and connections from hidden units to output units.

The most critical feature, in the sense of making the MEP algorithm work, is the specification of 15 constraint function averages. These averages are presented below the values for the Lagrange multipliers. These are the dual coordinates η_i that work in tandem with the θ^i coordinates if we want to indulge in the jargon of Information Geometry.

This is where the serious debate about the engineering design specifications of the Boltzmann machine would take place. I have made this Boltzmann machine especially transparent in specifying the 15 marginal probabilities evident in Figure 43.2 for all of the double interactions, like $P(CF)$, $P(CG)$, and so on, up through $P(BH)$.

On the other hand, if the Boltzmann machine were to be looked at from the perspective as an unsupervised system that tries to capture relevant associations in empirical data, these constraint function averages would represent sample averages based on all of the data. These sample averages could then represent the maximum likelihood model for some complicated model behind point p. Therefore, they would incorporate more information that is available from the selected double interactions in our restricted Boltzmann machine.

But the readily apparent stumbling block in a Boltzmann machine relying upon empirical data is that frequency counts are not available for the hidden units. That's why they are hidden. That is also why the various backpropagation schemes were devised in the first place. Some way was needed to adjust weights coming into and exiting from hidden units when no actual data could exist specifically for these units.

And finally the end goal of all of this effort, the actual probabilities for all four intelligence beer preference combinations are displayed. The highest degree of belief places a sandy colored right handed kangaroo from Melbourne into the category of an intelligent Foster's drinker. This restricted Boltzmann machine has overcome the limitation of not being able to categorize the kangaroos because probabilities for all the categories were the same.

The alternative I have discussed here was to try to design a formal probabilistic system that ensures that the hidden units recognize or detect patterns in the input units. This could be accomplished through logic expressions that try to replicate the feature detection capabilities inherent at the input and hidden layers levels.

43.7 A Critical Assessment

I have tried to maintain the sense of a consistent probabilistic framework in my approach to Boltzmann machines. And by that I mean consistency in comparison with the primitive, but general, inferencing scenarios that have been presented in my discussions so far.

However, the strains of enforcing this consistency are beginning to show cracks in the edifice. The most serious sign of a strain centers on the computational problem evident in the purely probabilistic approach as the dimension of the state space increases. For example, my joint probabilities are computed for a limited number of discrete statements, while really ambitious ANNs attempt to deal with hundreds, if not thousands, of units in potentially several layers of hidden units. These ANNs have the goal of computing continuous functions $f_k(x)$, not a few discrete probabilities.

Even with my preferred approach, there is no way of avoiding the incipient combinatorial explosion. The computational strategies in the two approaches are markedly different because of this problem. I persist in finding individual numerical assignments Q_i despite the size of the ever increasing state space. The number of model parameters for complicated models is growing rapidly; the partition function becomes ever more unwieldy. But, as a trade-off, I do find everything in one fell swoop through the MEP algorithm.

The ordinary non-probabilistic methods for Boltzmann machines, on the other hand, must undergo lengthy and complicated weight adjustment procedures. All sorts of sophisticated Markov Chain Monte Carlo techniques have to be researched and programmed. And, in the final analysis, no probability distribution exists over the underlying variables, and therefore there can be no truly fundamental rationale for any results from a probability perspective.

As alluded to in the last section, as an alternative to unsupervised systems relying solely on empirical data to tease out unknown relationships, a system might be designed using logic expressions to mimic feature detection. Otherwise, hidden units would have to be abandoned all together. This would take us back to visible units only, and investigation of complicated models with higher order interactions.

Nonetheless, I would not want to abandon the purely probabilistic analysis just yet. One reason is that I believe there is some merit in exploring logic inspired models for large dimensional state spaces. It seems reasonable that this tactic would be complementary to some empirical data based approach. The advantage here is that developing this one "good" model could circumvent the computational difficulties in the MEP algorithm.

Furthermore, many of the assignments would be 0 under a logic based model, while the remaining assignments could be found immediately. Thus, we circumvent the necessity for the computations inherent in the MEP algorithm. I provide some hints of this direction in Exercises 43.9.15 through 43.9.18.

43.8 Connections to the Literature

The seminal article that introduced the world to the idea of the Boltzmann machine was written by Hinton and Sejnowski [34, Chapter 7] in the mid–1980s. It is a clear, lucid, and still very readable introduction to what was then an exciting new paradigm for human information processing.

> The basic idea is simple: The weights on the connections between processing units encode knowledge about how things normally fit together in some domain and the initial states or external inputs to a subset of the units encode some fragments of a structure within the domain. These fragments constitute a problem: What is the whole structure from which they probably came? The network computes a "good solution" to the problem by repeatedly updating the states of the units that represent other possible parts of the structure until the network eventually settles into a stable state of activity that represents the solution.

From this description, one can easily discern that they envisioned such a network to be very much of a dynamic entity. The Boltzmann machine would be constantly updating unclamped nodes through some prescribed activation functions until the entire network system finally settled into something closely akin to a physical system that has found its thermodynamic equilibrium. The analogy to the energy of a real physical system was some sort of "constraint satisfaction function" or perhaps "goodness-of-fit function" that measured how well the Boltzmann machine could complete any number of "fragmented structures."

Since I have chosen to force this whole enterprise into the strict framework of a probabilistic inference, I have managed to avoid the basic requirement of a dynamic updating process that relaxes into an equilibrium over many simulations. The "whole structure of the constituent fragments" is then interpreted as simply a detailed description of a state space consisting of n joint statements. The weights, which were thought of as network connections as well as model parameters, now become the MEP's Lagrange multipliers, also model parameters. Completing the structure from the provided fragments now becomes the straightforward probability computation of a conditional probability like $P(A, B \mid C, D, E, \mathcal{M}_k)$.

They clearly recognized that Boltzmann machines would not work for problems where the significant structure was beyond the ken of the pairwise communication between units. This led to a network architecture that demanded the inclusion of "hidden units" that might circumvent this limitation. The strictly probabilistic approach would not have thought to do such a thing, but rather would have begun to investigate more and more complex models that used the information from triple interactions, quadruple interactions, and so forth.

> One way of using a parallel network is to treat it as a pattern completion device. A subset of the units are "clamped" into their on or off states and the weights in the network then complete the pattern by determining the states

of the remaining units. There are strong limitations on the sets of binary vectors that can be learned if the network has one unit for each component of the vector. These limits can be transcended by using extra units whose states do not correspond to components in the vectors to be learned. The weights of connections to these extra units can be used to represent complex interactions that cannot be expressed as pairwise correlations between the components of the vectors. We call these extra units *hidden units* ... and we call the units that are used to specify the patterns to be learned the *visible units*. ... The hidden units are where the network can build its own internal representations.

MacKay [29, pg. 524], in his treatment of Boltzmann machines and already mentioned as a useful reference in the last Chapter, also points out that simple Boltzmann machines need a fix in order to capture higher order structure.

> ... The real world, however, often has higher order correlations that must be included if our description of it is to be effective. Often the second order correlations in themselves carry little or no useful information. ...
> So, how can we develop such models? One idea might be to create models that directly capture higher order correlations, such as:
>
> $$P'(\mathbf{x} \mid \mathbf{W}, \mathbf{V} \ldots) = \frac{1}{Z'} \exp\left(\frac{1}{2} \sum_{ij} w_{ij} x_i x_j + \frac{1}{6} \sum_{ij} v_{ijk} x_i x_j x_k + \ldots \right)$$
>
> Such *higher order Boltzmann machines* are equally easy to simulate using stochastic updates, and the learning rule for the higher order parameters v_{ijk} is equivalent to the learning rule for w_{ij}.

Of course, MacKay's suggestion is exactly the one that I prefer to follow, but without the needless complication of stochastic updating and learning rules. His equation is simply a reworking of how the MEP incorporates information from higher order interactions. Models like these, together with even more complicated models, lie behind all the points p that live in the full Riemannian manifold before they get projected down into the sub–spaces of the restricted Boltzmann machines.

Despite the initial feeling of gloom brought about by Hinton and Sejnowski's introduction of these hidden units, any subsequent distraction to the formal analysis by probability proves unfounded. These "feature detectors" are added as part of the state space and, as a consequence, the number of joint statements to be considered is increased. By invoking the exceedingly simple **Sum Rule** axiom as one example of the formal manipulation rules of probability theory, the "pattern completion" for "fragments" is amended, as it was done in this Chapter's example, to,

$$P(A, B \mid C, D, E, \mathcal{M}_k) = \frac{\sum_{F,G,H}^{8} P(A, B, C, D, E, F, G, H \mid \mathcal{M}_k)}{\sum_{A,B,F,G,H}^{32} P(A, B, C, D, E, F, G, H \mid \mathcal{M}_k)}$$

I would like to make one more critical observation at this juncture. I think that probability theory's ability to generalize logic is often overlooked. I have spent a lot of time in my previous writings trying to demonstrate that probability theory handles things like Boolean functions and Classical Logic. Our brief foray into the intricacies of Boltzmann machines is, at least for me, another confirmation of the tendency to ignore the rewards to be gained from a strict adherence to the probability framework.

By drawing on the lessons from the examples that looked at logic functions as inspirations for probabilistic models, we should be willing to devote more effort into designing or engineering Boltzmann machines based on a single model. That is precisely what I did in this Chapter. The designer of the Boltzmann machine should examine more closely the ramifications of specifying feasible values for constraint functions, and the information contained in marginal probabilities like $P(CF \mid \mathcal{M}_k)$ connecting the input units and the hidden units, and marginal probabilities like $P(BH \mid \mathcal{M}_k)$ between hidden units and output units.

Neal [31, pp. 10–11] provides a wonderful introduction to *multilayer perceptrons*. I lean on his example for several of this Chapter's exercises. But, as any reader of my books might come to expect by now, the choices taken by individuals to go down one branch in the road versus another leads to surprising conceptual anomalies. One would naturally suppose, as I did for the longest time, that people possessed of overall sympathetic philosophies, would also tend to approach, and then analyze problems from a similarly sympathetic perspective.

But to cite just one curious example of where this all goes wrong, Neal chooses to think about artifical neural networks more as an artifice to construct arbitrary functions. Inferencing and probability seem to play no role at this level. Then, out of the blue, prior probabilities are attached to these functions rather than, say, prior probabilities over model space.

On the other hand, I choose, belaboring once again my main concern, to think about these networks as just another example of an inferential scenario. I end up using the formal manipulation rules of probability theory and the assignment of numerical values from the MEP algorithm in an entirely different way than Neal, even though we both started from complete agreement about every single Bayesian equation for probabilities. There is not one equation that Neal writes down that I would disagree with.

As just one example, Neal [31, pp. 5–6] presents an expression for the predictive distribution that is *almost* the same as mine.

> To predict the value of an unknown quantity, $x^{(n+1)}$, a Bayesian integrates the predictions of the model with respect to the posterior distribution of the parameters, giving
>
> $$P(x^{(n+1)} \mid x^{(1)}, \ldots, x^{(n)}) = \int P(x^{(n+1)} \mid \theta) \, P(\theta \mid x^{(1)}, \ldots, x^{(n)}) \, d\theta$$
>
> This *predictive distribution* for $x^{(n+1)}$ given $x^{(1)}, \ldots, x^{(n)}$ is the complete Bayesian inference regarding $x^{(n+1)}$ ...
>
> The Bayesian procedure avoids jumping to conclusions by considering not just the value of θ that explains the data best, but also other values of θ that explain the data reasonably well, and hence also contribute to the [above] integral ...

Instead, I write the predictive distribution as,

$$P(X_{N+1} = x_i \mid X_1, X_2, \ldots, X_N) = \sum_{k=1}^{\mathcal{M}} P(X_{N+1} = x_i \mid \mathcal{M}_k) \, P(\mathcal{M}_k \mid X_1, X_2, \ldots, X_N)$$

in order to emphasize that it is the statement that the model $\mathcal{M}_k$ asserts about the numerical assignments to all of the statements in the state space, and not the model parameters, which should be included in the probability expressions. As I keep trying to point out, there is no uncertainty about the parameters in a model, there is only uncertainty about the truth of the model's assignments.

Finally, I cannot resist the temptation to engage, as I have done in the past with Jaynes, MacKay, and Amari, in an almost word–by–word deconstruction of Neal's introductory remarks about probability. I seem to have a psychological fixation on this theme of trying to get myself inside the brains of people whom I respect for the quality of their arguments.

Some of the people relevant to the information processing concerns of my books who fall into this category are men like Frank Tipler, Stephen Wolfram, Douglas Hofstadter, Ed Jaynes, Sir Harold Jeffreys, Sun-ichi Amari, David J. C. MacKay, and Radford Neal.

It never ceases to amaze me that, despite my initial warm fuzzy feelings that I can make my life so much easier by following in their footsteps, after a few short steps, a fork in the road appears. They go down one path, but I cannot follow them there. Here is but one minor example where Neal goes off in one direction, and we must part ways [31, pp. 3–4].

> The statistical methodology of *Bayesian learning* is distinguished by its use of probability to express all forms of uncertainty. Learning and other forms of inference can then be performed by what are in theory simple applications of the rules of probability. The results of Bayesian learning are expressed in terms of a probability distribution over all unknown quantities. In general, these probabilities can be interpreted only as expressions of our degree of belief in various possibilities.

As I mentioned, everything starts out swimmingly. There is only a slight discord or quibble when the words *uncertainty* and *unknown quantities* are introduced. Does he mean the epistemological uncertainty that exists inside the consciousness of the IP, or some sort of ontological uncertainty? Mmmm ...

To me, language like, "a probability distribution over all unknown quantities" has a bit of a jarring ring to it. The joint statements in a state space are not unknown quantities. They must be, in fact, definite known assertions which can only be TRUE or FALSE. The statements themselves are in no way uncertain. There is no uncertainty surrounding the statement, "The coin will land with HEADS up." It is only in the mind of the IP, where probability exists as a degree of belief in the truth of any joint statement, where any uncertainty arises.

MORE ON BOLTZMANN MACHINES

However, Neal proceeds to indulge in some sloppy language reflecting an opinion that a probability is indeed a property of a physical object. I previously raised some strong objections to placing this kind of ontological cast to probabilities at the very beginning of Volume II [8, pg. 4].

> To illustrate the difference between Bayesian and frequentist learning, consider tossing a coin of unknown properties. There is an irreducible uncertainty regarding the outcome of each toss, which can be expressed by saying that *the coin has a certain probability of landing heads rather than tails* [my emphasis]. Since the properties of the coin are uncertain, however, we do not know what this probability of heads is (it might be one–half). A Bayesian will express this uncertainty using a *probability distribution over possible values for the unknown probability*[my emphasis] of the coin landing heads, and will update this distribution using the rules of probability theory as the outcome of each toss becomes known.

A coin is a definite physical object possessing however many certain and known physical properties we may care to measure. It is manufactured from chemical elements, say, Zinc and Copper, in certain ratios with a diameter of 20 mm and a thickness of 2 mm and a weight of 3 grams. It will be tossed at a room temperature of 27° C with a particular momentum, and so on, and so on. There are no unknown properties of a coin unless you want to go all the way down to the Heisenberg Uncertainty Principle, but I don't think that is what Neal meant.

Far more egregious is the highlighted phrase, "the coin has a certain probability of landing heads rather than tails." Coins, or for that matter, any physical objects, do not possess a substance called a probability. The coin is what ever it happens to be ontologically, while any notion of a probability must retreat to the consciousness of the IP who is contemplating the degree of belief it would like to repose in the truth of the statement, "The coin will land with HEADS up."

You don't need to be a "Bayesian" to assign probabilities; anybody is allowed to do it. Once again, we see some sloppy language in "a probability distribution over possible values for the unknown probability ..." Certainly, this must be a forbidden concept, emanating from the very axiomatic foundations of probability theory, which disallows language like "a probability of a probability." There is no concept in probability theory for constructing a probability distribution over *probabilities* of HEADS or TAILS.

We discussed this error once before. It conceivably might have been confusing to some when I introduced my standard probability expression for the assignment of a numerical value to a joint statement conditioned on the information resident within some model as,

$$Q_i \equiv P(X = x_i \mid \mathcal{M}_k)$$

Someone might object that this expression has no real content because what is really on the right hand side of the conditioned upon symbol (the solidus) is just $\{q_1, q_2, \cdots, q_n\}$. If this were true, then my expression would be subject to the very same criticism of the disallowed operation of wrapping one probability around another probability.

Recall that $\mathcal{M}_k$ must be a *statement*. A probability can be only be wrapped around statements. The statement $\mathcal{M}_k$ is something like, "I am model $\mathcal{M}_k$, and I am asserting that $\{q_1, q_2, \cdots, q_n\}$ is the correct numerical assignment to all n joint statements in the state space."

Now, the IP has every right to possess a degree of belief as to the truth of this statement. The IP has every right to believe that another assignment under the information resident in another model might, in fact, be the correct numerical assignment to all n joint statements in the state space. The probability attached to any of these statements, $P(\mathcal{M}_k)$, captures the IP's freedom in this regard.

That is why the expression $P(\mathcal{M}_k)$, the prior probability for models, is absolutely mandatory for the development of the Bayesian approach. We wouldn't be able to write down examples of the **Product Rule** for the joint occurrence of data and models, $P(\mathcal{D}, \mathcal{M}_k) = P(\mathcal{D} \mid \mathcal{M}_k) P(\mathcal{M}_k)$, if this were not the case.

Even Jaynes, of all people, once fell into this trap of constructing expressions involving probabilities of probabilities [16]. He first wrote out this expression for a binomial probability as,

> In the theory of Bernoulli trials, we calculate the probability that we shall obtain r successes in n trials as
>
> $$p(r \mid n) = \binom{n}{r} p^r (1-p)^{n-r}$$

You can not have the same symbol p doing double duty; in one sense on the left hand side for the probability of a statement about the number of successes, and then, in a second sense on the right hand side, for a particular numerical assignment. This error is fully disambiguated in my notation when I write,

$$P(N_1, N_2 \mid \mathcal{M}_k) = \frac{N!}{N_1! \, N_2!} Q_1^{N_1} Q_2^{N_2}$$

Just in case you thought it might have been a momentary slip of the pen, later in that same article, (and even in his book [22, pg. 164]) he writes down an even worse expression and explanation, where again a probability labeled as θ appears within a probability expression $p(\cdots \mid \cdots, \theta)$.

> As a third and less trivial example, where intuition did not anticipate the result, consider Bernoulli trials with an unknown probability of success. *Here the probability of success is itself the parameter θ to be estimated.* [Emphasis is mine.] Given θ, the probability that we shall observe r successes in n trials is
>
> $$p(r \mid n, \theta) = \binom{n}{r} \theta^r (1-\theta)^{n-r}$$

θ is both the parameter in a model, and the assigned probability? This would be like me writing an expression like,

$$P(N_1, N_2 \mid \lambda) = \frac{N!}{N_1! \, N_2!} \lambda^{N_1} (1-\lambda)^{N_2}$$

which is complete nonsense.

Here is Neal doing the same thing where a probability appears inside another probability expression [31, pp. 3–4],

> Consider a series of quantities, $x^{(1)}, x^{(2)}, \ldots$, generated by some process in which each $x^{(i)}$ is independently subject to random variation. We can define a *probabilistic model* for this random process, in which a set of unknown *model parameters*, θ, determine the probability distributions of the $x^{(i)}$. Such probabilities, or probability densities, will be written in the form $P(x^{(i)} \,|\, \theta)$. In the coin tossing example, the $x^{(i)}$ are the results of the tosses (heads or tails), *and θ is the unknown probability of the coin landing heads;* [The emphasis here is mine; it does not appear in the original] we then have $P(x^{(i)} \,|\, \theta) = [\theta$ if $x^{(i)} =$ heads; $1 - \theta$ if $x^{(i)} =$ tails]. [all other emphases in the original.]

The language is doubly objectionable because of the intimation that the data are generated by some probabilistic process. There is also the universal obligatory nod to "random variation," whatever mysterious thing that might be, impacting the data. Any probabilistic model should be set up for the benefit of the epistemological needs of the IP to express its uncertainty about the truth of the joint statements making up the data, not as some kind of alternative to an ontological physics model of the process.

The model parameters are never unknown. Taking the MEP as an example, it is quite the opposite. They happen to be completely defined by the mathematical procedure of undetermined Lagrange multipliers that finds the maximum of an objective function subject to side constraints. How can you have any uncertainty about a well defined mathematical procedure? Would you be using it if you did? The resulting model parameters λ_j must be known before any further progress can be made in assigning probabilities!

Although Jaynes never seemed to make a clear and repeated emphasis on this issue, I have found a few instances buried here and there [23, pg. 136] that leave no doubt as to his agreement that the concept of a probability distribution over model parameters is based on a false understanding of what the Lagrange multiplier is all about. I believe that after digesting these passages over an embarrassingly protracted period of time, it also dawned on me that any language in the literature that talked about prior or posterior probability distributions over parameters was total nonsense.

> In the maximum entropy problem, the quantity β had no previous existence; it is a Lagrange multiplier that is created only in the process of entropy maximization. But it appears only for mathematical convenience; the problem could be solved also by direct algebraic reduction without ever introducing it. β is not "estimated", but *defined,* by the PME formalism. That it is defined exactly and not approximately, far from being cause for complaint, merely indicates that our maximization problem was well posed. There would be cause to complain were it otherwise.
>
> It does not make sense, therefore, to speak of having prior knowledge of β, much less of honestly representing that knowledge. **A Lagrange multiplier does not have a probability distribution**; it is no different, in principle,

from a normalization constant that also appears in a probability distribution. That too is not estimated, but defined; and indeed, to infinite accuracy. ... [quotation marks and emphasis in the original, but I have made the critical statement **bold** which did not appear in the original]

Once one realizes that the model parameters are always known, then it becomes clear that the notion of a probability distribution for the model parameters is quite unnecessary! Any subsequent formal probability manipulation procedures that use anything like "$P(\theta)$ as a prior probability for the model parameters" is conceptually in error.

One must instead write out something like $P(\mathcal{M}_k)$ as the prior probability for the model space before any further derivation using the manipulation rules can take place. An IP *is allowed* to have a degree of belief about the truth of the dogmatic assertion made by any particular model that its assignments are absolutely correct because of the information it has inserted.

I realize that it becomes tiresome for the reader to continue on in this ultra-critical fashion. So I will stop here. But it is important, from time to time, to closely examine and think about the language that is used in our explanations.

Furthermore, in an apology to Neal, let me quote him on a very cogent issue where he is absolutely spot on. Very few people seem to recognize the distinction that exists between data and information [31, pg. 9].

> From a Bayesian perspective, adjusting the complexity of a model based on the amount of training data makes no sense. A Bayesian defines a model, selects a prior, collects data, computes the posterior, and then makes predictions. There is no provision in the Bayesian framework for changing the model or the prior depending on how much data was collected. If the model and prior are correct for a thousand observations, they are correct for ten observations as well ...
>
> For problems where we do not expect a simple solution, the proper Bayesian approach is therefore to use a model of a suitable type that is as complex as we can afford computationally, regardless of the size of the training set. ...

Neal is telling us that we may safely ignore the common warning to avoid the naive error of trying to "estimate more parameters in your model than the number of available data points." Such advice is conceptual nonsense of the highest order, but seriously promulgated to the unwary and naive.

I myself for the longest time believed in it as sage advice from the statistical *cognoscenti*. Now after having deprogrammed myself, I see that a model may have a million parameters with no data whatsoever, and inferencing proceeds unobstructed.

Even so august a personage as Sir Harold Jeffreys put himself on record as saying [24, pg. 245],

> We saw in Chapter I that if there are n observational data we cannot hope to determine more than n parameters.

And thus I repeat:

The data have nothing, nothing whatsoever, to do with the goal of establishing the information inserted into a probability distribution. The model parameters have that job.

A complicated model may have up to $n-1$ parameters if the state space is of dimension n with, again, all of this having nothing to do with any data. There may be no data, one data point, $n-1$ data points, or a million data points, and this has nothing to do with the model parameters.

There can be no notion of "overfitting" a model because of the number of data points. Whether there are no data, one data point, $n-1$ data points, or a million data points, these data can not change the information resident in that model; all these data can do is reorient the relative status in the degree of belief in the truth of all the models. Even after such a reorientation, the information in each model (and thus the parameters for each model) remains unaffected.

It is important to keep foundational concepts sacrosanct.

As a final reference, I recommend a very good overview of neural networks as looked at by a physicist interested in *complex systems* [37].

43.9 Solved Exercises for Chapter Forty Three

Exercise 43.9.1: What is the maximum amount of missing information for this Chapter's restricted Boltzmann machine?

Solution to Exercise 43.9.1

Since the dimension of the state space for the restricted Boltzmann machine is $n = 2^N = 2^8 = 256$, the quantitative measure of the missing information is the information entropy of the fair model where each $u_i = 1/256$,

$$\text{Maximum amount of missing information} \rightarrow -\sum_{i=1}^{256} u_i \ln u_i = \ln 256 = 5.54518$$

Every model other than the fair model must, by necessity, reduce this amount of missing information by the active information it inserts. Thus, all of these models will possess an information entropy less than 5.54518. The more active information a model inserts, the more the missing information is reduced. The quantitative measure of the amount of missing information reaches its absolute minimum value of $-\sum_{i=1}^{n} p_i \ln p_i = 0$ when, under the information from another model, one of the $p_i = 1$, and all the rest of the p_i equal 0.

The restricted Boltzmann machine must lie in the sub-manifold $E_k(q)$ where all of the coordinates λ_1 through λ_8, and all of the coordinates λ_{24} through λ_{255} are equal to 0. There are many, many points q in this sub-manifold, but only one q^* that is the **m**-projection of a point p residing in the full dimensional Riemannian manifold.

Point p is called the maximum likelihood model if the Boltzmann machine is unsupervised, that is, if we are letting the empirical data guide the inferences. If the Boltzmann machine is supervised, point p is the designed or engineered system with the correct classifications built into the system.

In either case, the missing information is reduced by the information resident in model q^* from the absolute maximum of 5.54518 to the maximum information entropy calculated by the Legendre transformation,

$$H_{\max}(q^*) = \min_{\lambda_j^{q^*}} \left\{ \ln Z_{q^*} - \sum_{j=9}^{23} \lambda_j^{q^*} \langle F_j \rangle_p \right\}$$

The fifteen constraint function averages $\langle F_j \rangle_p$ in the above expression are the averages computed from the probability distribution p. These averages are marginal probabilities like $P(CH)$ which capture the information in the double interaction between, say, an input unit C and a hidden unit H.

Exercise 43.9.2: Give an example of how a restricted Boltzmann machine is described as a multilayer perceptron.

Solution to Exercise 43.9.2

I shall illustrate the conventional description of a multilayer perceptron as explained by Neal in his book [31, pp. 10–11]. The point I wish to make here is that in this kind of conventional description, the probabilistic concepts which I have tried to place front and center in my presentation tend to be played down or obscured in some way or another. These alternative descriptions lean rather more heavily on the rationale of artificial neural networks as a novel way of approximating arbitrary functions, and from this way of approaching things reliance upon any probability framework is diminished.

To begin, the network architecture of the multilayer perceptron is the same as the restricted Boltzmann machine with the first layer the set of input units, the second layer, the set of hidden units, and the third layer, the set of output units. The notation stresses the idea of functions being computed with no appearance of probability expressions. Thus, the hidden layer computes j functions $h_j(x)$ as,

$$h_j(x) = \tanh\left(a_j + \sum_i u_{ij} x_i\right)$$

The hyperbolic tangent function serves as a so–called *activation function*. The x_i are the arguments to the functions and are the values clamped at the input units.

Notice that these x_i at the input units are not described as some subset of the statements making up the state space as I emphasize. We will talk more about the probabilistic origin of the hyperbolic tangent function in the next two Chapters. The u_{ij} are the weights connecting the input units to the hidden units, Once again, there is no mention of these weights being the same as the Lagrange multipliers for double interactions.

The next stage takes these $h_j(x)$ that have just been computed at the hidden layer level and passes them up to output layer level to form k functions $f_k(x)$. This time there is no activation function, but there is another set of weights v_{jk} modifying the $h_j(x)$.

$$f_k(x) = b_k + \sum_j v_{jk} h_j(x)$$

The multilayer perceptron has produced these $f_k(x)$, but since probability was excluded at the outset, it must perforce be nowhere evident at this stage of the solution. It has to be tacked on at the end in an *ad hoc* manner. Neal gives an example of the probability for a number of targets y_k conditioned on x,

$$p(y \,|\, x) = \prod_k \frac{1}{\sqrt{2\pi}\sigma_k} \exp\left(-(f_k(x) - y_k)^2 / 2\sigma_k^2\right)$$

where we see the functions $f_k(x)$ produced by the multilayer perceptron as the model parameters, the means μ_k, of Gaussian distributions. Now, to my way of thinking, this development is as *ad hoc* as it comes.

Compare this description of the multilayer perceptron with the description of the restricted Boltzmann machine from the purely probabilistic approach. The probability concepts are built in right from the beginning. In the end, we derive probabilities for statements at the output layer conditioned on knowing the statements at the input layer as in,

$$P(A, B \,|\, C, D, E, \mathcal{M}_k)$$

There are no mysteries as to the origin of things like activation functions, weights between units, and probabilities for targets.

Exercise 43.9.3: Fill in the details for the specific example that Neal provides for a multilayer perceptron.

Solution to Exercise 43.9.3

My Figure 43.3 below is taken directly from Neal's Figure 1.1 with a few minor changes [31, pg. 11]. It can be observed that there are $k = 1, 2$ output layer units, $j = 1, \ldots, 5$ hidden layer units, and $i = 1, \ldots, 4$ input layer units. Only some of the weights connecting the units are shown to unclutter the diagram.

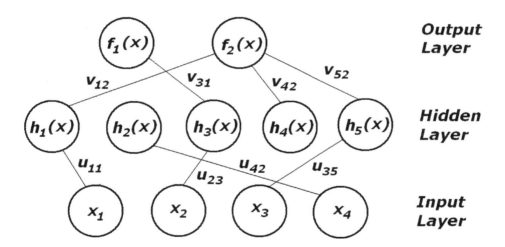

Figure 43.3: *Sketch of a multilayer perceptron consisting of an input layer, a hidden layer, and an output layer. Only a few of the weights are shown in order to keep the diagram uncluttered.*

SOLVED EXERCISES FOR CHAPTER 43

Let's start at the output layer level and work our way down. Compute the function $f_2(x)$ as our example.

$$f_2(x) = b_2 + \sum_{j=1}^{5} v_{j2}\, h_j(x)$$

$$= b_2 + v_{12}\, h_1(x) + v_{22}\, h_2(x) + v_{32}\, \overbrace{h_3(x)}^{\text{select}} + v_{42}\, h_4(x) + v_{52}\, h_5(x)$$

Now, select one of these hidden layer functions $h_j(x)$ in the above expression to see how it is expanded,

$$h_3(x) = \tanh\left(a_3 + \sum_{i=1}^{4} u_{i3}\, x_i\right)$$

$$= \tanh\left(a_3 + u_{13}\, x_1 + u_{23}\, x_2 + u_{33}\, x_3 + u_{43}\, x_4\right)$$

Substitute this back into the expression for $f_2(x)$ with the understanding that the other $h_j(x)$ would be similarly expanded,

$$f_2(x) = b_2 + v_{12}\, \overbrace{h_1(x)}^{\text{expand}} + v_{22}\, \overbrace{h_2(x)}^{\text{expand}} + v_{32}\, \overbrace{\tanh(a_3 + u_{13}x_1 + u_{23}x_2 + u_{33}x_3 + u_{43}x_4)}^{\text{already expanded}} +$$

$$v_{42}\, \overbrace{h_4(x)}^{\text{expand}} + v_{52}\, \overbrace{h_5(x)}^{\text{expand}}$$

Compare this generic description of a multilayer perceptron to our restricted Boltzmann machine. Matching things up, the network architecture indicates that the three input units are statements C, D, and E. These are the predictor, or explanatory variables of hand preference, fur color, and place of residence. The three hidden units are the feature detector statements F, G, and H. We can call these variables anything convenient. The two output units are whatever statements A and B stand for that need to be correctly categorized. Here, we would like a kangaroo's intelligence and beer preference to be correctly predicted.

Exercise 43.9.4: What is the numerical value of the third term when computing $f_2(x)$?

Solution to Exercise 43.9.4

This is just a primitive calculation to see what is happening at the nuts and bolts level. Let the four input units be clamped to the value of,

```
inputs = {-1, 0, 1, 2}
```

Let the four weights connecting all four input units to hidden unit 3 be,

```
iweights = {.1, .2, .3, .4}
```

Let the bias of the third hidden unit be $a_3 = -0.25$, `bias = -.25`. Let the weight connecting the third hidden unit to $f_2(x)$ be $v_{32} = 0.5$. Then,

```
.5 Tanh[bias + Dot[iweights, inputs]] // N
```

will evaluate to 0.317574. But it would be wrong to look at values like this and assume that some sort of probability has been computed.

Exercise 43.9.5: Building on the findings of the previous exercise, use *Mathematica* to compute both $f_1(x)$ and $f_2(x)$.

Solution to Exercise 43.9.5

First, set up a list of weights `uweights` from all four input units to all five hidden units. This selection of weights is arbitrary.

```
uweights = {{.2, .1, .25, .05}, {.05, .15, .2, .3 },
            {.1, .2, .3, .4}, {.1,.15,-.3,-.4}, {.1,.15,-.2,-.1}}
```

Next, set up five arbitrary bias parameters for the hidden layer,

```
biash = {.2, .3, -.25, .1, .3}
```

Put this through the activation function as the output at the hidden layer,

```
Tanh[biash + Dot[uweights, inputs]]
```

which returns with a list of five values as a result of this computation. Now, set up the list of weights from all five hidden units to both output units as,

```
vweights = {{.05,.06,.07,.08,.09}, {.09,.08,.07,.06,.05}}
```

and, as before, we need two bias parameters at the output layer,

```
biasf = {.01, .02}
```

We are now ready for the final evaluation returning the values at the output layer,

```
biasf + Dot[vweights, Tanh[biash + Dot[uweights, inputs]]]
```

as,

$$f_1(x) = 0.03638$$

$$f_2(x) = 0.09938$$

Once again, these are values of the functions $f_k(x)$ where neither the functions or their arguments involve any kind of probabilities.

SOLVED EXERCISES FOR CHAPTER 43

Exercise 43.9.6: Write some *Mathematica* code to begin construction of the constraint matrix needed for the purely probabilistic analysis of the restricted Boltzmann machine.

Solution to Exercise 43.9.6

The constraint matrix will be represented in *Mathematica* as a list of 15 lists. The basic structure looks like this,

$$\text{cm} = \{\{list~1~CF\}, \{list~2~CG\}, \ldots, \{list~15~BH\}\}$$

Each one of these fifteen lists implements one of the fifteen double interactions used in the restricted Boltzmann machine. This information ranges from the CF double interaction of the first input unit to the first hidden unit through the BH double interaction of the last hidden unit to the last output unit. Each list will itself consist of 256 elements, all either 0 or 1. These are the mappings $F_j(X = x_i)$ from the joint statements to numbers on the real line.

The lists for the eight main effects A through H will be built first because any of the required double interactions can be constructed from them by multiplication. The easy beginning **Clear[a, b, c, d, e, f, g, h]** clears all of the variables, and then creates each 256 element list for the main effects from the user created function **makeconstraintfunction[ ]**,

```
h = makeconstraintfunction[1]
g = makeconstraintfunction[2]
           ...
a = makeconstraintfunction[8]
```

The job of this function is to discern the pattern of 0s and 1s dictated by the pattern of the variables as they are laid out in the joint probability table.

```
makeconstraintfunction[j_] :=
            Flatten[Table[Join[Table[1, {2^(8 - j)}],
                Table[0, {2^(8 - j)}]], {2^(j - 1)}]]
```

For an example, consider the construction of the list,

```
g = makeconstraintfunction[2]
```

Given the hierarchical pattern of the joint probability table where statement A is at the lowest level and statement H is at the highest level, statement G will consist of the pattern of 64 1s, 64 0s, 64 1s, and 64 0s. **Table[1, {2^6}]** will create the first 64 1s, and then **Table[0, {2^6}]** will create the next 64 0s. These will be merged into a list of 128 elements through **Join[ ]**.

This task will be accomplished through **Table[..., {2^1}]** two times which manufactures a list of lists to complete the 256 elements. The **Flatten[ ]** command gets rid of the excess list structure and turns the expression into just one list with the required pattern of 64 1s, 64 0s, 64 1s, and 64 0s,

$$\{\underbrace{1,1,\ldots}_{64\ 1s},\underbrace{0,0,\ldots}_{64\ 0s},\underbrace{1,1,\ldots}_{64\ 1s},\underbrace{0,0,\ldots}_{64\ 0s}\}$$

Now that the main effects lists have been constructed, it is easy to arrive at our end goal of creating all fifteen lists of the double interactions.

```
cf = {Times[c, f]}
cg = {Times[c, g]}
    ...
bh = {Times[b, h]}
```

Finally, all 15 lists are merged and the constraint matrix is created with,

cm =
Join[cf, cg, ch, df, dg, dh, ef, eg, eh, af, ag, ah, bf, bg, bh]

The **Times[ ]** outputs above were wrapped in a list in order that the structure for **cm** from the **Join[ ]** was directly in the form of a matrix, that is, the required list of 15 lists.

Exercise 43.9.7: Verify that nothing seems to be seriously amiss with these manipulations described in the previous exercise.

Solution to Exercise 43.9.7

We might perform some of the necessary vector–matrix multiplications to see if anything has gone wrong in the developments of the last exercise. Start out with the symbolic expressions for the fifteen Lagrange multipliers,

lsymbolic = Map[Subscript[λ, #]&, Range[9, 23]]

which creates the list $\{\lambda_9, \lambda_{10}, \cdots, \lambda_{23}\}$. Does the vector–matrix multiplication involving the matrix **cm** work out to the correct length?

Length[Dot[lsymbolic, cm]]

returns the correct length of 256. Since this much seems OK, go ahead and form the symbolic expression for the partition function,

zsymbolic = Total[Exp[Dot[lsymbolic, cm]]]

Check to see whether the Legendre transformation is working properly by returning the information entropy and the actual numerical values of the Lagrange multipliers from some feasible, but otherwise arbitrarily selected model. The selected model parameters are contained in the list for all fifteen constraint function averages **cfa**.

Evaluating the Legendre transformation with,

NMinimize[Log[zsymbolic] - Dot[lsymbolic, cfa], lsymbolic]

returns with the expected nested list structure. The first element is the maximized information entropy for the restricted Boltzmann machine q^* as dictated by the above selection of model parameters in **cfa**, followed by a list of the numerical values for λ_9 through λ_{23}. These values for the Lagrange multipliers will, of course, be used to compute the numerical assignments to all 256 Q_i.

Exercise 43.9.8: How can you select the appropriate cells in the joint probability table for the summation over variables required by Bayes's Theorem?

Solution to Exercise 43.9.8

The example of Bayes's Theorem used in the probabilistic approach for the restricted Boltzmann machine, as illustrated in section 43.6, was to find the probability of all four combinations of intelligence and beer preference when conditioned on the known status of a kangaroo's hand preference, fur color, and place of residence. Normally, the IP would compute Bayes's Theorem in the form,

$$P(A, B \mid C, D, E \mid \mathcal{M}_k) = \frac{P(A, B, C, D, E \mid \mathcal{M}_k)}{P(C, D, E \mid \mathcal{M}_k)}$$

for this inferential scenario.

But the restricted Boltzmann machine includes three hidden variables F, G, and H to act as feature detectors in addition to the already mentioned five visible variables of A, B, C, D, and E. Bayes's Theorem must be amended to include the summation over these hidden variables.

$$P(A, B \mid C, D, E, \mathcal{M}_k) = \frac{\sum_{F,G,H}^{8} P(A, B, C, D, E, F, G, H \mid \mathcal{M}_k)}{\sum_{A,B,F,G,H}^{32} P(A, B, C, D, E, F, G, H \mid \mathcal{M}_k)}$$

For the numerator in Bayes's Theorem, we must pick out eight appropriate Q_i, and for the denominator 32 appropriate Q_i from the 256 Q_i available to us from the full joint probability table. Let's focus on the first of the four probabilities the restricted Boltzmann machine will compute at the output units: the probability of an intelligent Foster's drinking kangaroo, (A, B), given that the input variables are clamped to a right handed sandy fur colored kangaroo from Melbourne, (C, D, E).

Once again, it is a matter of trying to find some discernible pattern to the cell numbers in the joint probability table. When we look at the joint statement $ABCDEFGH$, $ABCDE$ will be fixed, and FGH will vary over eight possibilities. Starting with H and working back to F, all eight possibilities and their cell locations within the joint probability table are shown in Table 43.1 below.

Table 43.1: *All eight joint statements involved in the numerator of Bayes's Theorem used to calculate $P(A, B \mid C, D, E, \mathcal{M}_k)$.*

Number	Joint Statement	Cell Location	Joint Statement	Reordered
1	$ABCDEFGH$	1	$ABCDEFGH$	1
2	$ABCDEFG\overline{H}$	129	$ABCDE\overline{F}GH$	33
3	$ABCDEF\overline{G}H$	65	$ABCDEF\overline{G}H$	65
4	$ABCDEF\overline{G}\,\overline{H}$	193	$ABCDE\overline{F}\,\overline{G}H$	97
5	$ABCDE\overline{F}GH$	33	$ABCDEFG\overline{H}$	129
6	$ABCDE\overline{F}G\overline{H}$	161	$ABCDE\overline{F}G\overline{H}$	161
7	$ABCDE\overline{F}\,\overline{G}H$	97	$ABCDEF\overline{G}\,\overline{H}$	193
8	$ABCDE\overline{F}\,\overline{G}\,\overline{H}$	225	$ABCDE\overline{F}\,\overline{G}\,\overline{H}$	225

Notice that if we rearrange these joint statements in increasing numerical order, a pattern is easily found. They simply start out at 1 and increase by 32 until the final cell location at 225 is reached. Thus, it becomes pretty straightforward to create a *Mathematica* function,

```
bt[j_] := Total[Table[qi[[j + i]], {i, 0, 224 + j, 32}]]
```

to sum over the eight appropriate Q_i for the numerator.

Then **bt[1]** is used for (A, B), **bt[2]** for $(A, \overline{B})$, **bt[3]** for $(\overline{A}, B)$, and **bt[4]** for $(\overline{A}, \overline{B})$. The denominator is also easy because the 32 Q_i are found as,

```
bt[1] + bt[2] + bt[3] + bt[4]
```

and Bayes's Theorem is evaluated by the function,

```
bayes[k_] := bt[k] / (bt[1] + bt[2] + bt[3] + bt[4])
```

for all four possibilities through **Table[bayes[i], {i, 1, 4}]**.

For example, when the output in Figure 43.2 was being discussed, we observed that this restricted Boltzmann machine gave the highest probability at nearly 50% that a right handed, sandy fur colored kangaroo from Melbourne would be classified as an intelligent Foster's beer drinker.

SOLVED EXERCISES FOR CHAPTER 43

Exercise 43.9.9: Does the formal probability expression for a pattern completion in the Boltzmann machine conditioned on data provide a reasonable answer even when there are no data?

Solution to Exercise 43.9.9

Yes. It provides more than a reasonable answer. It provides the answer that one would want and expect.

This Chapter's focus on the probabilistic analysis of the Boltzmann machine ended up with the probability expression for a pattern completion conditioned on one model only. It was implicit that this one model implemented a system designed to complete patterns according to some predefined criterion.

All right handed, sandy colored kangaroos from Melbourne ought to be classified as smart Foster's drinkers. Intelligent kangaroos who prefer to drink Foster's beer completed the full pattern from the partial fragment provided through their hand preference, fur color, and place of residence.

There was no mention of any data in this discussion because no data were required. I chose to characterize our current effort to produce one model as more akin to what was done before in trying to find one model that would implement some logic function, or perhaps some elementary cellular automata.

We have seen previously that working through the formal rules when data *are* present results in this probability expression for the very next joint observation,

$$P(A_{N+1}, B_{N+1}, C_{N+1}, D_{N+1}, E_{N+1}, F_{N+1}, G_{N+1}, H_{N+1} \mid \mathcal{D}) = \frac{N_i + 1}{N + n}$$

Since we have no data, all $N_i = 0$ and $N = 0$, leading to the probability of the first joint observation of,

$$P(A_1, B_1, C_1, D_1, E_1, F_1, G_1, H_1 \mid \mathcal{D}) = \frac{1}{256}$$

By definition, we cannot observe the hidden variables F, G, and H, so we must sum over them to find the numerator in Bayes's theorem,

$$P(A, B, C, D, E \mid \mathcal{D}) = \sum_{F,G,H}^{8} P(A_1, B_1, C_1, D_1, E_1, F_1, G_1, H_1 \mid \mathcal{D}) = \frac{8}{256}$$

In like manner, to find the denominator in Bayes's theorem, we must sum over A and B as well,

$$P(C, D, E \mid \mathcal{D}) = \sum_{A,B,F,G,H}^{32} P(A_1, B_1, C_1, D_1, E_1, F_1, G_1, H_1 \mid \mathcal{D}) = \frac{32}{256}$$

We arrive at the entirely reasonable answer of a probability equal to 1/4 for the pattern completion for each of the four possible intelligence and beer preference traits when no data are available,

$$P(A, B \mid C, D, E, \mathcal{D}) = \frac{P(A, B, C, D, E \mid \mathcal{D})}{P(C, D, E \mid \mathcal{D})} = \frac{1}{4}$$

Exercise 43.9.10: What if the IP could collect frequency counts for each cell of the contingency table?

Solution to Exercise 43.9.10

In contrast to the last exercise, the IP would now have values for all N_i based on the observed data. Some of these N_i might still be 0, though. I must apologize here for making my notation do double duty. I have used N in two different and definitely confusing ways.

Earlier, N indicated the number of nodes in a Boltzmann machine. For example, here in this Chapter, $N = 8$ consisting of three input units, three hidden units, and two output units.

Now, I am reverting back to using N as I have done in the past. It refers to the data $\mathcal{D}$, the total number of observations of joint statements from the defined state space. The N_i refer to the actual frequency counts in each of the $2^N = 2^8 = 256$ cells of the contingency table where the conflicting notation is illustrated on purpose, and with $\sum_{i=1}^{256} N_i = N$.

Suppose that $N = 10,000$, and we want to calculate $P(A, B \,|\, C, D, E, \mathcal{D})$. There are eight cells in the contingency table that are relevant for the numerator. These are the frequency counts, (refer back to Table 43.1 in Exercise 43.9.8 for the appropriate cell locations),

$$N_1 = 183, N_{33} = 76, N_{65} = 76, N_{97} = 31, N_{129} = 76, N_{161} = 31, N_{193} = 31, N_{225} = 13$$

Thus, we have for the numerator of Bayes's Theorem, and the first term in the denominator,

$$\sum^{8}(N_i + 1) = 517 + 8 = 525$$

In similar fashion, for the second term in the denominator,

$$N_2 = 48, N_{34} = 31, N_{66} = 31, N_{98} = 20, N_{130} = 31, N_{162} = 20, N_{194} = 20, N_{226} = 13$$

and summing,

$$\sum^{8}(N_i + 1) = 214 + 8 = 222$$

SOLVED EXERCISES FOR CHAPTER 43

When all is said and done,

$$P(A, B \mid C, D, E, \mathcal{D}) = \frac{525}{525 + 222 + 222 + 112} = \frac{525}{1081}$$

$$P(\overline{A}, B \mid C, D, E, \mathcal{D}) = \frac{222}{525 + 222 + 222 + 112} = \frac{222}{1081}$$

$$P(A, \overline{B} \mid C, D, E, \mathcal{D}) = \frac{222}{525 + 222 + 222 + 112} = \frac{222}{1081}$$

$$P(\overline{A}, \overline{B} \mid C, D, E, \mathcal{D}) = \frac{112}{525 + 222 + 222 + 112} = \frac{112}{1081}$$

The probability for an intelligent Foster's drinking kangaroo has the highest probability, at close to 50%, for completing the full pattern when given the fragment of a right handed, sandy fur colored kangaroo residing in Melbourne, and based as well on this vast amount of unsupervised empirical data.

In fact, the above example frequency counts were generated by multiplying the Q_i from the designed system by 10,000 and rounding. For example, under the information in the model for the designed Boltzmann machine as discussed in this Chapter, the assigned numerical value to the joint probability was,

$$P(A, B, C, D, E, \overline{F}, \overline{G}, \overline{H} \mid \mathcal{M}_k) \equiv Q_{225} = 0.00130$$

The most critical feature in the probability expression $P(A, B \mid C, D, E, \mathcal{D})$ is that, in its derivation, an average was taken over all conceivable models. Therefore, the above probabilities are slightly lower or slightly higher than those calculated for the same conditions under the single model. However, when N is very large, this difference is not very pronounced.

Exercise 43.9.11: How do you complete the pattern for a kangaroo's correct intelligence and beer preference traits given the partial fragment of knowing that the kangaroo is left handed, has sandy fur color, and resides in Melbourne?

Solution to Exercise 43.9.11

This is the same as Exercise 43.9.8, but with a different set of values clamped to the input units of the restricted Boltzmann machine. Now the IP must compute the probabilities for the four joint statements (1) (A, B), (2) $(\overline{A}, B)$, (3) $(A, \overline{B})$, and (4) $(\overline{A}, \overline{B})$ conditioned on $\overline{C}$, D, and E. The joint statement with the highest conditional probability will complete the pattern.

For example, the computation of the third possibility through Bayes's Theorem becomes,

$$P(A, \overline{B} \mid \overline{C}, D, E, \mathcal{M}_k) = \frac{P(A, \overline{B}, \overline{C}, D, E \mid \mathcal{M}_k)}{P(\overline{C}, D, E \mid \mathcal{M}_k)}$$

Notice that this is the same model for the Boltzmann machine as before, so the 256 Q_i assignments don't have to be recalculated. What does change is that assignments from different cell locations will be involved in Bayes's Theorem.

Just as before, the numerator and denominator in Bayes's Theorem will involve sums over the appropriate eight and thirty two cell locations in the joint probability table,

$$P(A, \overline{B} \,|\, \overline{C}, D, E, \mathcal{M}_k) = \frac{\sum_{F,G,H}^{8} P(A, \overline{B}, \overline{C}, D, E, F, G, H \,|\, \mathcal{M}_k)}{\sum_{A,B,F,G,H}^{32} P(A, \overline{B}, \overline{C}, D, E, F, G, H \,|\, \mathcal{M}_k)}$$

The numerical assignment for $P(A\overline{B}\,\overline{C}DEFGH \,|\, \mathcal{M}_k)$ conditioned on the truth of the current model is located in cell 7 of the joint probability as Q_7. Its value is $Q_7 = 0.00949$. The remaining seven probabilities for the numerator are found successively 32 cell locations away each time as $Q_{39}, Q_{71}, \ldots, Q_{231}$.

Q_{39} is the assignment to $P(A\overline{B}\,\overline{C}DEFGH \,|\, \mathcal{M}_k)$, and so on, following the same pattern as laid down in Table 43.1. These will be summed as the summation over the three hidden variables F, G, and H. The denominator in Bayes's Theorem demands a sum over 32 Q_i assignments. These are the eight assignments just accounted for in $A\overline{B}$, plus the eight assignments each for the summation over the remaining three conditions AB, $\overline{A}B$, and $\overline{A}\,\overline{B}$.

As far as the actual values are concerned,

$$P(A\overline{B}\,\overline{C}DE \,|\, \mathcal{M}_k) = Q_7 + Q_{39} + Q_{71} + \cdots + Q_{231} = 0.03301$$

calculated with **bt[7]** where,

bt[j_] := Total[Table[qi[[j + i]], {i, 0, 224 + j, 32}]]

Continuing on, calculate the other three terms in the denominator of Bayes's theorem with **bt[5]**, **bt[6]** and **bt[8]**. Summarizing these calculations, we have,

$$P(AB\overline{C}DE \,|\, \mathcal{M}_k) = Q_5 + Q_{37} + Q_{69} + \cdots + Q_{229} = 0.08485$$

$$P(\overline{A}B\overline{C}DE \,|\, \mathcal{M}_k) = Q_6 + Q_{38} + Q_{70} + \cdots + Q_{230} = 0.03301$$

$$P(A\overline{B}\,\overline{C}DE \,|\, \mathcal{M}_k) = Q_7 + Q_{39} + Q_{71} + \cdots + Q_{231} = 0.03301$$

$$P(\overline{A}\,\overline{B}\,\overline{C}DE \,|\, \mathcal{M}_k) = Q_8 + Q_{40} + Q_{72} + \cdots + Q_{232} = 0.01466$$

We have everything in place for the calculation of Bayes's Theorem to determine the probability of each of the four intelligence beer preference combinations given the new clamped inputs, but with the same information as the previous model. The first combination is,

$$P(AB \,|\, \overline{C}DE, \mathcal{M}_k) = \frac{0.08485}{0.08485 + 0.03301 + 0.03301 + 0.01466} = 0.5126$$

SOLVED EXERCISES FOR CHAPTER 43

Summarizing,

$$P(AB\,|\,\overline{C}DE, \mathcal{M}_k) = \frac{0.08485}{0.08485 + 0.03301 + 0.03301 + 0.01466} = 0.5126$$

$$P(\overline{A}B\,|\,\overline{C}DE, \mathcal{M}_k) = \frac{0.03301}{0.08485 + 0.03301 + 0.03301 + 0.01466} = 0.1994$$

$$P(A\overline{B}\,|\,\overline{C}DE, \mathcal{M}_k) = \frac{0.03301}{0.08485 + 0.03301 + 0.03301 + 0.01466} = 0.1994$$

$$P(\overline{A}\,\overline{B}\,|\,\overline{C}DE, \mathcal{M}_k) = \frac{0.01466}{0.08485 + 0.03301 + 0.03301 + 0.01466} = 0.0886$$

The conditional probability for the (A, B) condition,

$$P(AB\,|\,\overline{C}DE, \mathcal{M}_k) = 0.5126$$

is the largest probability of the four. Apparently it does not make any difference whether a kangaroo is right handed or left handed; as long as they have sandy fur color and hail from Melbourne, they will be classified by this restricted Boltzmann machine as smart Foster's drinkers.

Exercise 43.9.12: Identify all of the numerical assignments that go to make up any constraint function average.

Solution to Exercise 43.9.12

What we want are those Q_i that form the average $\langle F_j \rangle$. By definition,

$$\langle F_j \rangle = \sum_{i=1}^{256} F_j(X = x_i)\, Q_i$$

But the only allowable information for a restricted Boltzmann machine is some subset of all 28 possible double interactions. For our situation here, these are the 15 CF through BH double interactions. The index j will run from $j = 9$ through $j = 23$ and where, in addition, as is clearly indicated, the index i must run from $i = 1$ through $i = 256$.

Take as an example the CF double interaction. This is represented as a list of 256 1s and 0s within the constraint matrix **cm**. It was the first double interaction constructed from the C and F main effects.

This constraint function is thus $F_9(X = x_i)$. Use the code below to pick out the 64 elements in this list where $F_9(X = x_i) = 1$.

```
Clear[x]; x = {}; For[i = 1, i < 257, i++,
         If[Flatten[cf][[i]] == 1, x = Append[x, i]]]; x
```

The list,

$$\{1, 2, 3, 4, 9, 10, 11, 12, \cdots, 217, 218, 219, 220\}$$

is returned identifying the 64 numerical assignments $Q_1, Q_2, \cdots, Q_{220}$ as the average $\langle F_9 \rangle$. This sum of Q_i is the marginal probability $P(CF)$.

As another example, what are the Q_i that constitute the marginal probability $P(BH)$, the information affecting the third hidden unit and the second output unit? The BH double interaction was the last list inserted into the constraint matrix **cm** through **bh = {Times[b, h]}**. Substitute **bh** into the above code replacing **cf** where it returns the list,

$$\{1, 2, 5, 6, 9, 10, 13, 14, \cdots, 121, 122, 125, 126\}$$

These 64 Q_i form $P(BH)$ through,

$$\langle F_{23} \rangle = \sum_{i=1}^{256} F_{23}(X = x_i) \, Q_i$$

$$= Q_1 + Q_2 + Q_5 + Q_6 + \cdots + Q_{121} + Q_{122} + Q_{125} + Q_{126}$$

Exercise 43.9.13: Show all the details of the MEP calculation for the numerical assignment of the probability for any arbitrarily selected joint statement handled by the restricted Boltzmann machine.

Solution to Exercise 43.9.13

Arbitrarily select the joint statement in the 79th cell of the joint probability table. This is the joint statement $A\overline{B}\,\overline{C}\,\overline{D}EF\overline{G}H$. The assigned probability under the given model to the joint statement, "This is a smart left handed beige fur colored kangaroo from Melbourne who drinks Corona with the three feature detectors being ON, OFF, and ON." is then,

$$Q_{79} \equiv P(A\overline{B}\,\overline{C}\,\overline{D}EF\overline{G}H \mid \mathcal{M}_k)$$

The MEP algorithm calculates the numerical value as,

$$Q_{79} = \frac{\exp\left[\sum_{j=9}^{23} \lambda_j F_j(X = x_{79})\right]}{\sum_{i=1}^{256} \exp\left[\sum_{j=9}^{23} \lambda_j F_j(X = x_i)\right]}$$

The partition function was calculated to have the value,

$$Z_{q^*} = \sum_{i=1}^{256} \exp\left[\sum_{j=9}^{23} \lambda_j F_j(X = x_i)\right] = 770.917$$

as shown in the *Mathematica* output of Figure 43.2.

The numerator for Q_{79} will have some number of the $\lambda_j^{q^*}$ as the arguments to the exponential function. We could go to the 79th element in the constraint matrix and "read down" all 15 $F_j(X = x_{79})$ 0s and 1s to see which specific $\lambda_j^{q^*}$ make up the arguments.

We will make *Mathematica* take care of this tedious chore for all the Q_i,

`Column[Table[{i, Part[Dot[lsymbolic, cm], i]}, {i, 1, 256}]]`

Inspecting this output, we see that the numerator for Q_{79}, and eventually Q_{79} itself, is,

$$\exp\left[\sum_{j=9}^{23} \lambda_j F_j(X = x_{79})\right] = \exp\left[\lambda_{15} + \lambda_{17} + \lambda_{18} + \lambda_{20}\right]$$

$$= \exp\left[0.444961 + 0.444961 + 0.444961 + 0.444961\right]$$

$$= 5.9289$$

$$Q_{79} = \frac{5.9289}{770.917}$$

$$= 0.00769$$

Exercise 43.9.14: Create a "dendrogram" to orient yourself as to the locations of the cells in the joint probability table.

Solution to Exercise 43.9.14

The sketch of the dendrogram appears below in Figure 43.4.

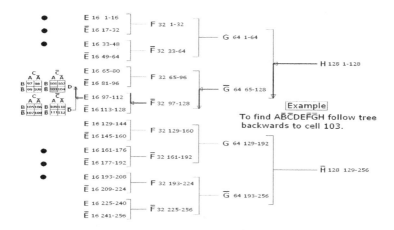

Figure 43.4: *Locating a cell in the joint probability table with a "dendrogram."*

Exercise 43.9.15: Before beginning a discussion of logic functions serving as models for Boltzmann machines, clear up some infrastructure details.

Solution to Exercise 43.9.15

We begin with a *Mathematica* built–in function called **Tuples[** *arg1, arg2* **]**. The first argument to **Tuples[ ]** will be the list **{True, False}**, and the second argument will be an integer $2, 3, 4, \ldots$

The reason why we need to understand the output from **Tuples[ ]** is because it underlies how Boolean functions create the particular *order* of the orthogonal basis functions for, say, Boole's Expansion Theorem.

Let's first examine some output from **Tuples[** *arg1, arg2* **]**. To understand the ordering for two variables, **Tuples[{True, False}, 2]** returns with,

{{True, True}, {True, False}, {False, True}, {False, False}}

This ordering matches up with the orthogonal logic basis functions of AB, $A\overline{B}$, $\overline{A}B$, and $\overline{A}\,\overline{B}$. And thus it becomes important to pay attention to the details of how the cells of the joint probability table were constructed. I have usually constructed it such that AB is cell 1, $A\overline{B}$ is cell 3, $\overline{A}B$ is cell 2, and $\overline{A}\,\overline{B}$ is cell 4.

When **BooleanTable[BooleanConvert[Implies[A,B]]]** returns the list **{True, False, True, True}**, the result from **Tuples[ ]** will tell us that the functional assignment **True** goes with AB, the functional assignment **False** goes with $A\overline{B}$, the functional assignment **True** goes with $\overline{A}B$, and the functional assignment **True** goes with $\overline{A}\,\overline{B}$. And, of course, this is the correct functional assignment for the implication logic function $f_{13}(A, B)$. The IP would place a $P(A\overline{B} \mid \mathcal{M}_k) = 0$ in cell 3 of the joint probability table in order to have probability theory generalize logic when the implication logic function serves as a model.

With this as background, it is easy to see how things generalize. Move up to **Tuples[{True, False}, 3]** which returns with,

**{{True, True, True}, {True, True, False}, {True, False, True},
{True, False, False}, {False, True, True}, {False, True, False},
{False, False, True}, {False, False, False}}**

Match up this pattern with the orthogonal logic basis functions,

(1) ABC, (2) $AB\overline{C}$, (3) $A\overline{B}C$, (4) $A\overline{B}\,\overline{C}$,
(5) $\overline{A}BC$, (6) $\overline{A}B\overline{C}$, (7) $\overline{A}\,\overline{B}C$, (8) $\overline{A}\,\overline{B}\,\overline{C}$

Suppose we choose our canonical inference of $(A \to B) \wedge (B \to C)$ as the model. Then,

BooleanTable[BooleanConvert[And[Implies[A,B], Implies[B,C]]]]

returns the list **{True, False, False, False, True, False, True, True}**.

Here is where we have to be careful in matching up with the cells of the joint probability table however it may have been constructed. Figure 43.5 shows a joint probability table for the three logic variables A, B, and C constructed such that the first four cells index C, and the last four cells index $\overline{C}$. Note that this layout of the joint probability table is different than that shown in Figure 37.1.

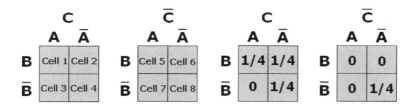

Figure 43.5: *A joint probability table for three variables laid out with the C variable at the highest level.*

Matching things up, we see that the four **False** go to cells 5, 3, 7, and 6, while the four **True** go to cells 1, 2, 4, and 8. One numerical assignment under this logic model would place 0s in cells 3, 5, 6, and 7, and 1/4 in cells 1, 2, 4, and 8.

$$AB\overline{C} \to \text{Cell 5}$$

$$A\overline{B}C \to \text{Cell 3}$$

$$A\overline{B}\,\overline{C} \to \text{Cell 7}$$

$$\overline{A}B\overline{C} \to \text{Cell 6}$$

Before we depart this example, we should double-check to see whether Bayes's Theorem, as a generalization of logic, makes sense. It seems that it does.

$$P(C\,|\,A,\mathcal{M}_k) = \frac{P(AC)}{P(A)} = \frac{1/4 + 0}{1/4 + 0 + 0 + 0} = 1$$

$$P(A\,|\,C,\mathcal{M}_k) = \frac{P(AC)}{P(C)} = \frac{1/4 + 0}{1/4 + 1/4 + 0 + 1/4} = 1/3$$

$$P(\overline{A}\,|\,C,\mathcal{M}_k) = \frac{P(\overline{A}C)}{P(C)} = \frac{1/4 + 1/4}{1/4 + 1/4 + 0 + 1/4} = 2/3$$

$$P(A\,|\,\overline{B},\overline{C},\mathcal{M}_k) = \frac{P(A\overline{B}\,\overline{C})}{P(\overline{B}\,\overline{C})} = \frac{0}{0 + 1/4} = 0$$

Exercise 43.9.16: Apply the lessons learned in the previous exercise to a simple Boltzmann machine with two hidden units.

Solution to Exercise 43.9.16

We are still deep into the notion of logic functions that serve as inspiration for models of Boltzmann machines. Suppose, for our current example, we have four logic variables A, B, C, and D. A is the one output unit, B is the one input unit, while C and D are two hidden units. The IP possesses somewhat vague criteria about the role of the hidden units that any machine learning should somehow implement. These are then the three logical constraints: (1) $A \to B$, (2) $C \to \overline{B}$, and (3) $D \to B$.

Since we are using probability theory to generalize logic, we don't reason in the direction dictated by the above logic expressions. Doing so would dictate that knowledge about A should lead to a *deduction* about B. Our reasoning is always in the other direction, the inferential direction, where B is the input and A is the desired output. This is even more so in the case where we want to reason about a hidden unit like D from a visible unit like B, even though the implication goes from D to B.

For an initial glance at where we are heading, peruse Table 43.2 below where variable D is at the highest level in the joint probability table.

Table 43.2: *Summarizing the placement of the numerical assignments in the joint probability table based on a model that implements certain logical constraints.*

Order	Tuples []	Joint Statement	Cell	Assignment	Probability
1	$TTTT$	$ABCD$	1	F	0
2	$TTTF$	$ABC\overline{D}$	9	F	0
3	$TTFT$	$AB\overline{C}D$	5	T	1/6
4	$TTFF$	$AB\overline{C}\,\overline{D}$	13	T	1/6
5	$TFTT$	$A\overline{B}CD$	3	F	0
6	$TFTF$	$A\overline{B}C\overline{D}$	11	F	0
7	$TFFT$	$A\overline{B}\,\overline{C}D$	7	F	0
8	$TFFF$	$A\overline{B}\,\overline{C}\,\overline{D}$	15	F	0
9	$FTTT$	$\overline{A}BCD$	2	F	0
10	$FTTF$	$\overline{A}BC\overline{D}$	10	F	0
11	$FTFT$	$\overline{A}B\overline{C}D$	6	T	1/6
12	$FTFF$	$\overline{A}B\overline{C}\,\overline{D}$	14	T	1/6
13	$FFTT$	$\overline{A}\,\overline{B}CD$	4	F	0
14	$FFTF$	$\overline{A}\,\overline{B}C\overline{D}$	12	T	1/6
15	$FFFT$	$\overline{A}\,\overline{B}\,\overline{C}D$	8	F	0
16	$FFFF$	$\overline{A}\,\overline{B}\,\overline{C}\,\overline{D}$	16	T	1/6

Make *Mathematica* tell us the logic function of our four units that implements these logic constraints.

```
BooleanTable[BooleanConvert[And[Implies[A, B],
                Implies[C, Not[B]], Implies[D, B]]]]
```

which returns the list of sixteen `True` and `False` we will need in order to fill in the 16 cells of the joint probability table.

But before we can place these assignments into the correct cells of the joint probability table, we have to run a `Tuples[{True, False}, 4]` to return the list which shows us the order of the orthogonal logic functions. Then we can match this up with whatever manner we might have constructed the table. With this table as a guide, we can fill in the numerical assignments for all 16 cell locations of the joint probability table. See Figure 43.6 below.

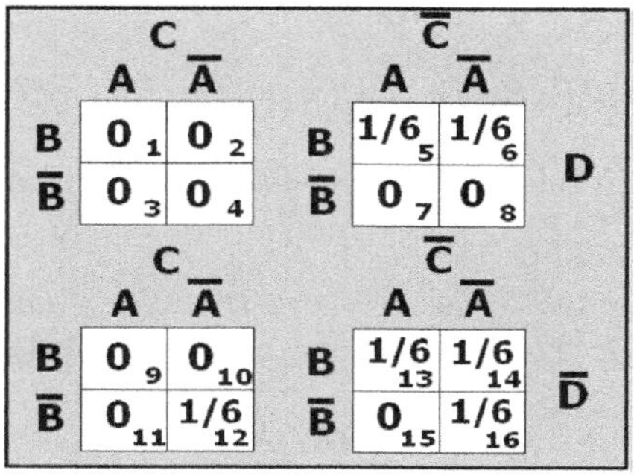

Figure 43.6: *The joint probability table for all four variables of a Boltzmann machine following some logical constraints.*

It is always an interesting exercise to see if our understanding of the implication functions jibes with the placement of the 0s in the table. For example, if A is TRUE, then B can not be FALSE. Check cells 3, 7, 11, and 15 to verify that any joint statement involving $A\overline{B}$ has an assignment of 0. If C is TRUE, B can not be TRUE. Check cells 1, 2, 9, and 10 to verify that any joint statement involving CB has an assignment of 0. If D is TRUE, then B can not be FALSE. Sure enough, checking cells 3, 4, 7, and 8 verifies that any joint statement involving $D\overline{B}$ has an assignment of 0.

There is another way to check that we have not made any errors. This method involves the use of **De Morgan's axiom** introduced in Volume I [7, Chapter 5]. Besides the full DNF expansion of the implication logic function as,

$$A \to B \equiv AB \vee \overline{A}B \vee \overline{A}\,\overline{B}$$

there is the shorter version expressed as,

$$A \to B \equiv \overline{A} \vee B$$

By **De Morgan's axiom**, we have the following logical equivalency,

$$\overline{A} \vee B = T \equiv A \wedge \overline{B} = F$$

Thus, our joint probability table must confirm that,

$$P(A\overline{B}) = P(A\overline{B}CD) + P(A\overline{B}\,\overline{C}D) + P(A\overline{B}C\overline{D}) + P(A\overline{B}\,\overline{C}\,\overline{D}) = 0$$

These probabilities are located in cells 3, 7, 11, and 15, and they are all assigned a value of 0.

We can verify the second implication $C \rightarrow \overline{B}$ in exactly the same way. By **De Morgan's axiom**, we have the following logical equivalency,

$$\overline{C} \vee \overline{B} = T \equiv C \wedge B = F$$

Thus, our joint probability table must confirm that,

$$P(BC) = P(ABCD) + P(\overline{A}BCD) + P(ABC\overline{D}) + P(\overline{A}BC\overline{D}) = 0$$

These probabilities are located in cells 1, 2, 9, and 10, and they also are all assigned a value of 0.

Now it is time to address the inferences that can be made about output unit A from knowledge of input unit B. To start, relying as always on Bayes's Theorem, we simply have,

$$P(A \mid B, \mathcal{M}_k) = \frac{P(AB \mid \mathcal{M}_k)}{P(B \mid \mathcal{M}_k)}$$

From the purely probabilistic perspective, we must sum over the hidden variables C and D.

$$P(A \mid B, \mathcal{M}_k) = \frac{\sum_{CD}^{4} P(ABCD \mid \mathcal{M}_k)}{\sum_{ACD}^{8} P(ABCD \mid \mathcal{M}_k)}$$

The four cells contributing to the numerator $P(AB \mid \mathcal{M}_k)$ are then,

$$P(ABCD) \rightarrow \text{Cell 1}$$

$$P(AB\overline{C}D) \rightarrow \text{Cell 5}$$

$$P(ABC\overline{D}) \rightarrow \text{Cell 9}$$

$$P(AB\overline{C}\,\overline{D}) \rightarrow \text{Cell 13}$$

The eight cells contributing to the denominator $P(B \mid \mathcal{M}_k)$ are found by summing over A in addition to C and D. These are the four cells above and cells 2, 6, 10, and 14. It is even easier to see from the joint probability table that $P(B \mid \mathcal{M}_k)$ consists of the cells 1, 2, 5, 6, 9, 10, 13, and 14.

It becomes a simple matter of substituting the numerical assignments from the model dictated by the logical constraints of the Boltzmann machine to find the probabilistic answer for the output of unit A given that input unit B was clamped to a value of TRUE.

$$P(A \,|\, B, \mathcal{M}_k) = \frac{0 + 1/6 + 0 + 1/6}{(0 + 0 + 1/6 + 1/6) + (0 + 0 + 1/6 + 1/6)} = 1/2$$

The axioms of probability theory already tell us that $P(\overline{A}\,|\,B, \mathcal{M}_k)$ must also be equal to 1/2. But let's check anyway. By Bayes's Theorem, we have,

$$P(\overline{A}\,|\,B, \mathcal{M}_k) = \frac{P(\overline{A}B\,|\,\mathcal{M}_k)}{P(B\,|\,\mathcal{M}_k)}$$

The denominator will be the same as just calculated, while the numerator consists of cells 2, 6, 10, and 14. It is clear then that the correct value of 1/2 is returned,

$$P(\overline{A}\,|\,B, \mathcal{M}_k) = \frac{0 + 1/6 + 0 + 1/6}{(0 + 0 + 1/6 + 1/6) + (0 + 0 + 1/6 + 1/6)} = 1/2$$

What happens if the input unit is clamped to a value of FALSE?

$$P(A\,|\,\overline{B}, \mathcal{M}_k) = \frac{P(A\overline{B}\,|\,\mathcal{M}_k)}{P(\overline{B}\,|\,\mathcal{M}_k)}$$

The numerator in Bayes's Theorem then consists of the four cells 3, 7, 11, and 15. But we have determined that these four cells must sum to 0. Remember that the implication function $A \rightarrow B$ demands this,

$$A \rightarrow B \equiv P(A\overline{B}) = 0$$

Thus,

$$P(A\,|\,\overline{B}, \mathcal{M}_k) = 0 \text{ and } P(\overline{A}\,|\,\overline{B}, \mathcal{M}_k) = 1$$

The output A is certain to be OFF if the input B is OFF. The IP is ambivalent about the output if the input is ON since

$$P(A\,|\,B, \mathcal{M}_k) = P(\overline{A}\,|\,B, \mathcal{M}_k) = 1/2$$

Exercise 43.9.17: The model behind this Boltzmann machine is a Boolean function. Which one is it?

Solution to Exercise 43.9.17

There are 2^{2^k} possible Boolean functions for k variables. We have already examined the $2^{2^2} = 16$ logic functions, and the $2^{2^3} = 256$ Boolean functions behind the elementary cellular automata. There are $2^{2^4} = 65,536$ possible Boolean functions for our Boltzmann machine, one of which we have explored in the previous exercises.

We can find out that our Boolean function is the $12,341^{st}$ function of four variables out of the possible 65,536 functions with the *Mathematica* expression,

```
FromDigits[Boole[BooleanTable[BooleanConvert[
   And[Implies[A, B],Implies[C, Not[B]], Implies[D, B]]]]], 2]
```

Boole[ ] converts the list of **True** and **False** to a list of corresponding 1s and 0s, which then allows **FromDigits**[*list, base*] to treat this list of 0s and 1s as a number in base 2.

So now construct this $12,341^{st}$ function with,

```
bf = BooleanFunction[12341, 4]
```

and then test for a tautology with,

```
TautologyQ[Equivalent[bf[A, B, C, D],
    And[Implies[A, B], Implies[C, Not[B]], Implies[D, B]]]]
```

This evaluates to **True** confirming that our Boltzmann machine is, in fact, the $12,341^{st}$ Boolean function of four variables. A further confirmation exists in how *Mathematica* returns the DNF for **bf[A, B, C, D]**,

```
BooleanConvert[bf[A, B, C, D]] // FullForm
```

as,

```
Or[And[Not[A], Not[B], Not[D]], And[B, Not[C]]]
```

Distributing the probability operator over the **Or** results in,

$$P(\overline{A}\,\overline{B}\,\overline{D} \vee B\overline{C}) = P(\overline{A}\,\overline{B}\,\overline{D}) + P(B\overline{C}) - P(\overline{A}\,\overline{B}\,\overline{D}B\overline{C})$$

$$= P(\overline{A}\,\overline{B}\,\overline{D}) + P(B\overline{C}) - 0$$

$$P(\overline{A}\,\overline{B}\,\overline{D}) = P(\overline{A}\,\overline{B}C\overline{D}) + P(\overline{A}\,\overline{B}\,\overline{C}\,\overline{D})$$

$$P(B\overline{C}) = P(AB\overline{C}D) + P(\overline{A}B\overline{C}D) + P(AB\overline{C}\,\overline{D}) + P(\overline{A}B\overline{C}\,\overline{D})$$

Refer back to Table 43.2 to verify that these six joint statements are the very same ones that the DNF for the logic expression,

```
And[Implies[A, B], Implies[C, Not[B]], Implies[D, B]]
```

assigned a **True**. The probabilities assigned to these six joint statements must add up to 1.

SOLVED EXERCISES FOR CHAPTER 43   427

Exercise 43.9.18: How does the actual Boltzmann machine compare to the purely probabilistic approach presented in the last few exercises?

Solution to Exercise 43.9.18

The most complicated model would possess 15 constraint functions $F_j(X = x_i)$ consisting of the four main effects, A, B, C, and D, the six double interactions, AB, AC, AD, BC, BD, and BC, the four triple interactions, ABC, ABD, ACD, and BCD, and the one quadruple interaction, $ABCD$.

The actual restricted Boltzmann machine must be a simpler model than this, constructed as it is from just the two double interactions between the hidden units and the output units, AC and AD, and the two double interactions from the input units to the hidden units, BC and BD. We will also allow the Boltzmann machine to use the four marginal probabilities, $P(A)$ through $P(D)$, as "biases." Thus, there are a total of eight model parameters in the Boltzmann machine as compared to the maximum allowable 15 model parameters for the most complicated models p.

The Boltzmann machine $q^\star$ is one of the less complicated models because it is an **m**–projection of the more complicated model in p down into a sub–manifold. For our current inferential scenario, the model behind p is the model dictated by the set of logic constraints. The **m**–projection of p down to the sub–manifold with only eight coordinates forces $q^\star$ to match p on those corresponding η_i.

From the joint probability table as shown in Figure 43.6, we can easily form the required eight marginal probabilities for $q^\star$. Notice that the ordering of the constraint function averages as given above means that $\eta_5 \equiv \langle F_5 \rangle \equiv P(AB)$ is skipped in the sequence.

$$\eta_1 = P(A) = 1/3$$

$$\eta_2 = P(B) = 2/3$$

$$\eta_3 = P(C) = 1/6$$

$$\eta_4 = P(D) = 1/3$$

$$\eta_6 = P(AC) = 0$$

$$\eta_7 = P(AD) = 1/6$$

$$\eta_8 = P(BC) = 0$$

$$\eta_9 = P(BD) = 1/3$$

When these model parameters are used in the MEP algorithm, the numerical assignments to all 16 cells of the joint probability table are found as shown in Figure 43.7 below. The first 12 assignments exactly match those from the model implementing the logic constraints. However, in the final four cells of cells 13, 14, 15, and 16, the assignments are slightly different. The complicated model p is utilizing more information than is available to q^*.

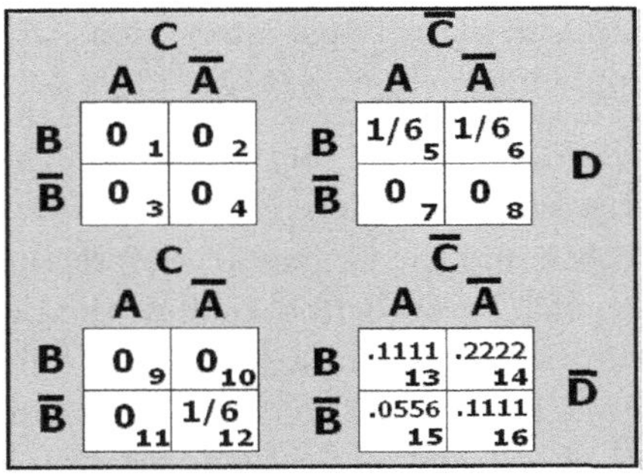

Figure 43.7: *The joint probability table for all four units of a simplified Boltzmann machine as the **m**-projection from a more complicated model.*

We have already investigated the performance of the logic inspired model p. It gave us the probabilities for the output unit A to be ON or OFF given the ON or OFF status of the input unit B,

$$P(A \mid B, \mathcal{M}_p) = 1/2$$

$$P(\overline{A} \mid \overline{B}, \mathcal{M}_p) = 1$$

The restricted Boltzmann machine q^* does respond in a comparable manner, although its performance is somewhat degraded as might be expected from a more limited model with less information.

$$P(A \mid B, \mathcal{M}_{\text{BM}}) = \frac{5}{12}$$

$$P(\overline{A} \mid \overline{B}, \mathcal{M}_{\text{BM}}) = \frac{5}{6}$$

The Boltzmann machine has a lesser degree of belief that A will be ON or that A will be OFF when compared to the more informative model.

More detail shows the cell locations for both the numerator and the denominator in Bayes's Theorem,

$$P(\overline{A} \mid \overline{B}, \mathcal{M}_{\text{BM}}) = \frac{4 + 8 + 12 + 16}{3 + 4 + 7 + 8 + 11 + 12 + 15 + 16}$$

$$= \frac{0 + 0 + 1/6 + 1/9}{0 + 0 + 0 + 0 + 0 + 1/6 + 1/18 + 1/9}$$

$$= \frac{5}{6}$$

Chapter 44

Canonical Inferences

44.1 Introduction

In the upcoming Chapters, I am going to introduce two topics, *variational Bayes* and *mean field theory*, which, like Boltzmann machines, have been treated as rather exotic concepts in the literature. I intend to shed some light on these issues by leveraging some of our new found knowledge about IG. But I would also like to adhere as closely as possible to the strict confines of the probabilistic framework as constructed over the first two Volumes.

I have discovered to my dismay that as more complicated inferential scenarios are broached in the literature, any previously touted virtues of the simplicity of the probability approach are cast adrift at the first signs of trouble. The prospect of immersing oneself into these fascinating new details is very tempting indeed. Indulging ourselves in this manner has the unfortunate tendency for the big picture, as they say, to get lost in the shuffle.

This Chapter is my attempt at an antidote to such overindulgence. So, for me, the big picture will take the form of what I am going to label as *canonical inferences*. The effort undertaken here is just a continuation of the theme developed in the first two Volumes concerning the amazing analytical formulas that can be derived when averaging over model space.

Furthermore, this Chapter continues the discussion on Boltzmann machines as begun in the last two Chapters. These so-called canonical inferences are no more, and no less, than the "pattern completions" that the Boltzmann machines were attempting to implement.

"Clamping" a partial pattern onto the Boltzmann machine hopefully resulted in a correct classification of some target, that is, a successful pattern completion. A successful pattern completion might be achieved through supervised or unsupervised learning. Our canonical inferencing scheme tends to rely more on empirical data, and thus falls more into the unsupervised learning category of Boltzmann machines.

Some of the characteristic features of canonical inferences that I would like to emphasize are the following: In the upcoming, but familiar, probability expressions, we will observe a reliance on empirical data, no hidden or latent variables, lack of appearance of model statements, no MEP calculations with attendant difficulties computing the partition function for complicated models, and, finally, freedom to include or exclude at will any of the variables in the joint statements.

Canonical inferences are based on the posterior predictive distribution which, in turn, is based on the adoption of a uniform prior probability over models. An ancillary goal of this effort at describing canonical inferences, but nonetheless a very important one, is to introduce some of Harold Jeffreys's objections to adopting a uniform prior probability over model space. These objections ultimately led to the Bayesian cottage industry of manufacturing and explaining the dubious necessity for a *Jeffreys's prior*.

44.2 Symbolic Expressions

We begin with the primary symbolic probability expressions used throughout the Chapter. First up is the probability of any number of generic variables, labeled as A, B, C, ... when conditioned on knowing some number of additional generic variables labeled as ... X, Y, Z, as well as the data, $\mathcal{D}$. For more clarity, we call the A, B, C, ... variables, the "target" variables. The ... X, Y, Z variables are called the "predictor" variables. Always remember, though, that these "variables" are in reality *statements*.

A subscript $N+1$ is attached to all of the variables to clearly indicate that this is a prediction about the *next* joint occurrence of the target variables A_{N+1}, B_{N+1}, C_{N+1}, ... when given the predictor variables ... X_{N+1}, Y_{N+1}, Z_{N+1}. It also implies that the data $\mathcal{D}$ have consisted of the previous N joint occurrences of both target and predictor variables. Thus, the emphasis is on the number of *future* frequency counts of joint occurrences. When we talk about the *very next* occurrence, we specify this with the notation of $M = 1$.

44.2.1 Primary expression

After these preliminaries, it is the special conditional probability, the so–called posterior predictive probability, that we are interested in,

$$P(A_{N+1}, B_{N+1}, C_{N+1}, \ldots \mid \ldots, X_{N+1}, Y_{N+1}, Z_{N+1}, \mathcal{D})$$

Notice particularly that the model statement $\mathcal{M}_k$ does NOT appear anywhere in the above conditional probability expression. This notation should really be expanded to show that each statement, measurement, or observation, must result in something that has already been defined as part of the state space. Thus, a symbolic joint statement should show an actual detailed observation at trial t as something like,

$$(A_t = a_i, B_t = b_j, C_t = c_k, \ldots, Z_t = z_l)$$

CANONICAL INFERENCES

44.2.2 Bayes's Theorem

Next up is the solution to this primary probability expression as provided by Bayes's Theorem. Only those who place a continued reliance on this theorem should be called "Bayesians."

$$P(A_{N+1}, B_{N+1}, C_{N+1}, \ldots \mid \ldots, X_{N+1}, Y_{N+1}, Z_{N+1}, \mathcal{D}) =$$

$$\frac{P(A_{N+1}, B_{N+1}, C_{N+1}, \ldots, X_{N+1}, Y_{N+1}, Z_{N+1} \mid \mathcal{D})}{\sum_{n_A, n_B, n_C, \ldots} P(A_{N+1}, B_{N+1}, C_{N+1}, \ldots, X_{N+1}, Y_{N+1}, Z_{N+1} \mid \mathcal{D})} \quad (44.1)$$

44.2.3 The amazingly simple answer

From the detailed derivation of the probability for the next joint occurrence when conditioned on all of the past data, we have Laplace's *Rule of Succession*,

$$P(A_{N+1}, B_{N+1}, C_{N+1}, \ldots, X_{N+1}, Y_{N+1}, Z_{N+1} \mid \mathcal{D}) = \frac{N_i + 1}{N + n} \quad (44.2)$$

where N_i is the observed frequency count in the ith cell of the contingency table indexing the joint statement $(A, B, C, \ldots, X, Y, Z)$, and $\sum_{i=1}^{n} N_i = N$. As usual, n is the dimension of the state space and here, $n = n_A \times n_B \times \cdots \times n_Z$.

The solution via Bayes's Theorem then works out to,

$$P(A_{N+1}, B_{N+1}, C_{N+1}, \ldots \mid \ldots, X_{N+1}, Y_{N+1}, Z_{N+1}, \mathcal{D}) =$$

$$\frac{\frac{N_i+1}{N+n}}{\frac{N_i+1}{N+n} + \frac{N_j+1}{N+n} + \frac{N_k+1}{N+n} + \cdots +} =$$

$$\underbrace{\frac{N_i + 1}{(N_i + 1) + (N_j + 1) + (N_k + 1) + \cdots}}_{n_A \times n_B \times n_C \times \cdots \text{ terms}} \quad (44.3)$$

44.2.4 Summing over variables

As we observed in the last Chapter on the restricted Boltzmann machine, Bayes's Theorem undergoes a slight modification due to the **Sum Rule** when some given variables are not explicitly included as part of the posterior predictive equation even though they *are* part of the defined state space. In the same spirt of a generic labeling, let F, G, and H stand for these variables that do not appear as either target variables or predictor variables even though they help define the n joint statements in the state space.

Through the **Sum Rule** then, a joint probability over the target and predictor variables becomes,

$$P(ABCXYZ) = \sum_{n_F, n_G, n_H} P(ABCFGHXYZ)$$

Bayes's Theorem in Equation (44.1) is then modified to look like,

$$P(A_{N+1}, B_{N+1}, C_{N+1}, \ldots \mid \ldots, X_{N+1}, Y_{N+1}, Z_{N+1}, \mathcal{D}) =$$

$$\frac{\sum_{n_F, n_G, n_H, \ldots} P(A_{N+1}, B_{N+1}, \ldots, F_{N+1}, G_{N+1}, H_{N+1}, \ldots, Y_{N+1}, Z_{N+1} \mid \mathcal{D})}{\sum_{n_A, n_B, n_C, n_F, n_G, n_H, \ldots} P(A_{N+1}, B_{N+1}, \ldots, F_{N+1}, G_{N+1}, H_{N+1}, \ldots, Y_{N+1}, Z_{N+1} \mid \mathcal{D})}$$

(44.4)

44.2.5 Primitive confirmations

Let's start by applying the formula in Equation (44.4) to the very first cases one would encounter in order to make sure that the known results in these specific cases can be confirmed.

About the most primitive case one can imagine would be the two–dimensional state space with no data, no predictor variables, and no hidden variables. Thus, we have that $n = n_A = 2$, $N = 0$, and $M = 1$. There is one target variable A, but no predictor variables X, Y, Z, or hidden variables F, G, H. The formula yields as the probability at the very first trial, $t = 1$, for the second kind of possible measurement for statement A, $(A_1 = a_2)$,

$$P(A_1 = a_2 \mid \mathcal{D}) = \frac{P(A_1 = a_2 \mid \mathcal{D})}{P(A_1 = a_1 \mid \mathcal{D}) + P(A_1 = a_2 \mid \mathcal{D})}$$

Substitute Laplace's result, the *Rule of Succession*, for each term appearing in Bayes's Theorem,

$$P(A_1 = a_1 \mid \mathcal{D}) = \frac{N_1 + 1}{N + n} = \frac{0 + 1}{0 + 2} = \frac{1}{2}$$

and

$$P(A_1 = a_2 \mid \mathcal{D}) = \frac{N_2 + 1}{N + n} = \frac{0 + 1}{0 + 2} = \frac{1}{2}$$

Finally,

$$P(A_1 = a_2 \mid \mathcal{D}) = \frac{N_2 + 1}{\underbrace{(N_1 + 1) + (N_2 + 1)}_{n_A = 2 \text{ terms}}}$$

$$= \frac{1}{1 + 1}$$

$$= 1/2$$

CANONICAL INFERENCES

Augment this most primitive condition ever so slightly by adding one binary predictor variable X to the existing target variable A. The constraint of having observed no previous data still applies. Now, we have $n = n_A \times n_X = 4$, $N = 0$, and $M = 1$.

The formula again yields the probability for the very first trial, say, this time for the first kind of possible observation ($A_1 = a_1$) when conditioned on knowledge that the predictor variable was in its second state $X_1 = x_2$, as,

$$P(A_1 = a_1 \mid X_1 = x_2, \mathcal{D}) = \frac{P(A_1 = a_1, X_1 = x_2 \mid \mathcal{D})}{P(X_1 = x_2 \mid \mathcal{D})}$$

$$= \frac{P(A_1 = a_1, X_1 = x_2 \mid \mathcal{D})}{P(A_1 = a_1, X_1 = x_2 \mid \mathcal{D}) + P(A_1 = a_2, X_1 = x_2 \mid \mathcal{D})}$$

Assume that the 2×2 contingency table has been laid out such that the columns are labeled by the target variable and the rows by the predictor variable. Each term in Bayes's Theorem above is a joint conditional probability, so substitute Laplace's result,

$$\frac{N_i + 1}{N + n}$$

The posterior predictive probability becomes,

$$P(A_1 = a_1 \mid X_1 = x_2, \mathcal{D}) = \frac{1/4}{1/4 + 1/4}$$

$$= \frac{N_3 + 1}{\underbrace{(N_3 + 1) + (N_4 + 1)}_{n_A = 2 \text{ terms}}}$$

$$= \frac{1}{1 + 1}$$

$$= 1/2$$

It's pretty obvious that the probability of a binary target variable at the first trial remains at $1/2$ no matter how many predictor variables are added. There are still only two joint statements, $A \cdots XYZ$ and $\overline{A} \cdots XYZ$, involved in the denominator of Bayes's Theorem,

$$P(A_1 \mid \cdots X_1, Y_1, Z_1, \mathcal{D}) = \frac{N_i + 1}{\underbrace{(N_i + 1) + (N_j + 1)}_{n_A = 2 \text{ terms}}}$$

$$= \frac{0 + 1}{(0 + 1) + (0 + 1)}$$

$$= 1/2$$

We want to avoid an unnecessarily hasty rush to judgment in our conjectures. It is always advisable to take baby steps before taking the plunge. Up the ante with two binary target variables, A and B, but retain the one binary predictor variable X. Also, let the minimum amount of data enter the fray by supposing that a frequency count of 1 has been recorded in the contingency table for some joint occurrence.

Now, we have $n = n_A \times n_B \times n_X = 8$, $N = 1$, $N_3 = 1$, and $M = 1$. The formula yields the joint probability for the second observation that the target variables are $(A_2 = a_1)$ and $(B_2 = b_2)$ when conditioned on knowledge that the predictor variable is observed to be $(X = x_2)$. The one previous data point recorded in cell 3 of the eight cell contingency table is $\mathcal{D} = (A_1 = a_1, B_1 = b_2, X_1 = x_1)$.

Bayes's Theorem now reads,

$$P(A_2 = a_1, B_2 = b_2 \mid X_2 = x_2, \mathcal{D}) = \frac{P(A_2 = a_1, B_2 = b_2, X_2 = x_2 \mid \mathcal{D})}{P(X_2 = x_2 \mid \mathcal{D})}$$

where the four terms arising from the explicit expansion by the **Sum Rule** in the denominator are,

$$P(X_2 = x_2 \mid \mathcal{D}) = P(A_2 = a_1, B_2 = b_1, X_2 = x_2 \mid \mathcal{D}) +$$

$$P(A_2 = a_1, B_2 = b_2, X_2 = x_2 \mid \mathcal{D}) +$$

$$P(A_2 = a_2, B_2 = b_1, X_2 = x_2 \mid \mathcal{D}) +$$

$$P(A_2 = a_2, B_2 = b_2, X_2 = x_2 \mid \mathcal{D})$$

Suppose the lay out of the eight cell contingency table has the target variables at the lowest levels and the predictor variable at the highest level. Then, Bayes's Theorem translates into,

$$P(A_2 = a_1, B_2 = b_2 \mid X_2 = x_2, \mathcal{D}) = \frac{N_7 + 1}{\underbrace{(N_5 + 1) + (N_7 + 1) + (N_6 + 1) + (N_8 + 1)}_{n_A \times n_B = 4 \text{ terms}}}$$

$$= \frac{0 + 1}{(0 + 1) + (0 + 1) + (0 + 1) + (0 + 1)}$$

$$= 1/4$$

CANONICAL INFERENCES

However, the minimal data of a single frequency count $N_3 = 1$ for the joint occurrence of $(A_1 = a_1, B_1 = b_2, X_1 = x_1) \equiv \mathcal{D}$ does makes a difference if the predictor variable were to be observed as $(X = x_1)$ at the next trial,

$$P(A_2 = a_1, B_2 = b_2 \mid X_2 = x_1, \mathcal{D}) = \frac{N_3 + 1}{\underbrace{(N_1 + 1) + (N_3 + 1) + (N_2 + 1) + (N_4 + 1)}_{n_A \times n_B = 4 \text{ terms}}}$$

$$= \frac{1 + 1}{(0 + 1) + (1 + 1) + (0 + 1) + (0 + 1)}$$

$$= 2/5$$

The probabilities for the other three possibilities of the target variables then become equal to $1/5$ when $(X_2 = x_1)$ is known at trial 2.

44.3 Logistic Regression

An example of logistic regression was treated in some depth in Chapter Twenty Three of Volume II [8]. Here is an overview of that previous example from the perspective of canonical inferencing.

The target variable A was the binary variable of Cardiac Heart Disease (CHD) with the two possible observations being $(A = a_1) \equiv$ CHD Present and $(A = a_2) \equiv$ CHD Absent. There were two predictor variables X and Y standing for Age and Lab Test with three possible observations for each predictor variable. Thus, the dimension of the state space was $n = n_A \times n_X \times n_Y = 2 \times 3 \times 3 = 18$.

Data were collected on $N = 1000$ patients with all eighteen frequency counts, the N_i, allocated to the appropriate cell of the contingency table. If the IP wanted to categorize any new patient, not already in the data base, by the probability for CHD when the patients's age and results on the test are known, then M is taken equal to 1.

For the specific predictors of a patient's age between 40 and 60, and a high score on the lab test, the posterior predictive probability for cardiac heart disease is calculated as,

$$P(A_{1001} = a_1 \mid X_{1001} = x_2, Y_{1001} = y_3, \mathcal{D}) = \frac{N_8 + 1}{\underbrace{(N_8 + 1) + (N_{17} + 1)}_{n_A = 2 \text{ terms}}}$$

$$= \frac{41 + 1}{(41 + 1) + (125 + 1)}$$

$$= 1/4$$

For the Boltzmann machines, the decision was made to place the target variable into the category with the highest probability. For example, a right handed kangaroo with beige fur color living in Melbourne was placed into the target category of being a Foster's drinker of above average intelligence because this posterior predictive probability was larger than any of the other three categories.

In other situations, like the current medical scenario, other costs and risks come into play. Perhaps a threshold probability of 0.20 has been established in the medical community such that a probability for CHD higher than this demands some further attention. Thus, this new patient would be subject to closer scrutiny because of his posterior predictive probability for CHD.

I think it must be emphasized here that this posterior predictive probability for the target variables does not arise from the information resident in any one single model. It is, in fact, an average over the posterior probability of all models.

As we have already intimated, and shall mention again, the rapidly increasing dimension of the state space as inferential problems become more realistic is a serious computational issue at the level of a single model. Basically, since the partition function consists of a sum of n exponential functions, as new variables are added, combinatorial explosion sets in.

Amazingly though, as was evident in these introductory examples, any such computational difficulties that might exist at the level of the single individual model disappear completely. Averaging over all models in canonical inferencing alleviates the combinatorial explosion existing at the individual model level.

44.4 Implications for More Realistic Scenarios

Conduct a rough "back of the envelope" look at some of the interesting implications of canonical inferencing when applied to more realistic scientific scenarios. We are indeed curious about the implications due to an increasing number of variables, both target variables and predictor variables. In other words, what happens when dealing with a much larger state space, and thus a much larger n. Supposing that data are expensive to collect, N is restricted to be of moderate size. Finally, the number of categories for the target variables is also becoming quite a bit larger.

The kangaroos have begun to wonder whether correct categorizations can be made about their less obvious behavioral traits like beer preference, hand preference, and intelligence, when based on more obvious physical facts like fur color, place of residence, height, and weight.

There are a total of seven variables, divided between the three target variables, A, B, and C, and four predictor variables W, X, Y, and Z. There are two possible observations for beer and hand preference, and seven possible measurements of intelligence. There are two possible observations for fur color and place of residence, and seven possible measurements for both height and weight.

CANONICAL INFERENCES 437

The dimension of the state space is therefore,

$$n = n_A \times \times n_B \times n_C \times n_W \times n_X \times n_Y \times n_Z$$

$$= 2 \times 2 \times 7 \times 2 \times 2 \times 7 \times 7$$

$$= 5,488$$

The denominator in Bayes's Theorem will be a sum over,

$$n_A \times n_B \times n_C = 2 \times 2 \times 7 = 28 \text{ terms}$$

So perhaps the rejoicing over the disappearance of n at the end of the last section was a little premature. There are still 28 target categories to be kept track of.

Since it was claimed that gathering data was a bit expensive, suppose that a limited sample size of $N = 200$ kangaroos was queried and tested on all seven variables. For example, the 127th kangaroo tested is part of the data base $\mathcal{D}$,

$$(A_{127} = a_1, B_{127} = b_3, C_{127} = c_6, W_{127} = w_2, X_{127} = x_1, Y_{127} = y_4, Z_{127} = z_7)$$

Any canonical inferencing prediction can be made for any kangaroo not in this data base. This is the *next* kangaroo with $M = 1$; so it will be labeled with the subscript $N + M = N + 1 = 201$.

The probability that will be the basis for any decision is calculated according to the canonical expression,

$$P(A_{201}, B_{201}, C_{201}, | W_{201}, X_{201}, Y_{201}, Z_{201}, \mathcal{D}) =$$

$$\frac{P(A_{201}, B_{201}, C_{201}, W_{201}, X_{201}, Y_{201}, Z_{201} \mid \mathcal{D})}{\sum_{n_A \times n_B \times n_C = 28} P(A_{201}, B_{201}, C_{201}, W_{201}, X_{201}, Y_{201}, Z_{201} \mid \mathcal{D})}$$

$$= \frac{N_i + 1}{\underbrace{(N_i + 1) + (N_j + 1) + \cdots}_{n_A \times n_B \times n_C = 28 \text{ terms}}}$$

But with 5,488 cells in the contingency table and only $N = 200$ frequency counts, it is likely there will be many 0s, something less than 200 1s, and a very few 2s, 3s, and so forth as frequency counts. The most likely probabilities for specific targets then are probabilities like $1/28, 2/29, 3/30, \ldots$

For example, focus on a beige kangaroo of average height and weight living in Sydney, that is, with predictor variables $(W = w_1, X = x_2, Y = y_4, Z = z_4)$. Why might you guess that a kangaroo who is a left handed, Foster's drinking genius, that is, with target variables $(A = a_1, B = b_2, C = c_7)$ has a probability of $1/10$? If the frequency count in that one particular cell of the contingency table, the ith cell, indexing the joint occurrence of,

left handed beige fur colored Foster's drinking geniuses of average height and weight living in Sydney

is $N_i = 2$ from the overall total of $N = 200$ frequency counts scattered over the 5,488 cells of the contingency table, and there are 0s in all the other 27 cells indexing beige kangaroos of average height and weight living in Sydney, then the probability is equal to 3/30.

44.5 Relevance for Jeffreys's Prior?

I will be discussing the topic of Jeffreys's prior in more detail later on in Chapter Forty Seven. But the current topic of canonical inferencing offers an intriguing backdrop for discussions of this kind. Bayesians have a special, parochial interest in debating Jeffreys's argument in favor of abandoning the uniform prior over model space. For example, the derivation of the posterior predictive formula depended crucially on ignoring Jeffreys on this point, by making explicit the assumption of a uniform prior over models.

Jeffreys began to lay down the general motivation for his displeasure with the uniform prior by treating the issue of sampling from a finite population. Jeffreys seemed to settle on unnecessarily obfuscating notation to develop his arguments, and, not surprisingly, his discussion of sampling from a finite population also suffers from this flaw.

Despite this notational handicap, and what you might initially think would be the introduction of an alternative to the uniform prior, he does, in fact, end up essentially rederiving Laplace's result encapsulated in the *Rule of Succession*. *A fortiori*, these formulas as derived by Jeffreys are also the same as the posterior predictive formulas when they are compared to the particular way in which I have developed them.

He then presents a summary of the application of the *Rule of Succession* to the finite sampling without replacement scenario. This summary is an accurate portrayal of the consequences of adopting the uniform prior over model space. I quote Jeffreys below on this matter in the next section.

Nevertheless, as the central tenet of the debate, Jeffreys subsequently refused to accept the uniform distribution for the prior probability of models $P(\mathcal{M}_k)$. To refresh our memories, what technically is that prior probability over model space? It is a probability density function, namely the Dirichlet probability density function, defined by,

$$\text{pdf}(\mathcal{M}_k) = C_D \, q_1^{\alpha_1 - 1} \, q_2^{\alpha_2 - 1} \cdots q_n^{\alpha_n - 1}$$

This "conjugate" form of the Dirichlet pdf, that is, conjugate to the probability expression for the data under a model, is absolutely vital to simplifying the ensuing analytical expression that ultimately results in the posterior predictive formula.

CANONICAL INFERENCES

As already mentioned *ad nauseam*, the canonical inferencing formula crucially depends on adopting the uniform prior form of pdf $(\mathcal{M}_k)$, in other words, in setting all the α_i parameters in the Dirichlet distribution equal to 1. Jeffreys opted instead to set all the α_i parameters in the Dirichlet distribution equal to $1/2$. He did this in order to avoid the uniform probability, and to place a "lump of probability" at both extremes of 0 and 1 for any q_i.

I find it vitally important to take notice of the fact that the above Dirichlet probability distribution is over the assigned q_i, and **NOT** a probability distribution over **THE PARAMETERS IN THE MODEL**. If this is not recognized, it becomes the source of almost all of the confusion that surrounds Jeffreys's prior.

The implication for IG is that the function of the metric tensor g, so intimately involved in the definition of Jeffreys's prior, *must be found in the manifold holding the Dirichlet distributions*. If, instead, the metric tensor is defined on the space of the $P(X = x_i \,|\, \mathcal{M}_k)$, in other words, on the structure (S, g) we have been examining so far, mass confusion will ensue.

Consider how we would formulate Jeffreys's "animals with feathers" problem from the standpoint of a canonical inference. We have one binary target variable A and one binary predictor variable X. The target variable A is a statement about whether an animal has a beak with,

$$(A = a_1) \equiv \text{"The animal has a beak."}$$

$$(A = a_2) \equiv \text{"The animal does not have a beak."}$$

and the predictor variable X is a statement about whether an animal has feathers,

$$(X = x_1) \equiv \text{"The animal has feathers."}$$

$$(X = x_2) \equiv \text{"The animal does not have feathers."}$$

Suppose that we have made some extensive observations on animals with feathers and beaks so that the probabilistic answer is seen to address an empirical question. See Exercise 44.7.10 for the solution to the question: What is the probability that the *next* animal the IP sees with feathers will have a beak?

The general answer from canonical inferencing will have the form,

$$P(A_{N+1} = a_1 \,|\, X_{N+1} = x_1, \mathcal{D}) = \frac{N_1 + 1}{N_1 + N_3 + 2}$$

where N_1 is the actual frequency count in the data base for feathered animals with beaks, and N_3 is the actual frequency count for feathered animals without beaks. So you are not surprised when canonical inferencing arrives at the simple answer that the probability that the *next* animal with feathers has a beak is,

$$P(\text{has a beak} \,|\, \text{has feathers}, \mathcal{D}) = \frac{\text{Number with beaks and feathers} + 1}{\text{Number with beaks and feathers} + 2}$$

44.6 Connections to the Literature

Let me set the scene for Jeffreys's displeasure with both Bayes and Laplace and their decision to use a uniform, or "flat," prior distribution over the models. After making this critical commentary, Jeffreys illustrated what he was getting at with his "animals with feathers" example. [24, pg. 128]

> It follows that with the uniform distribution of the prior probability (1) a large homogeneous sample will establish a high probability that the next member will be of the same type, and a moderate probability that a further sample comparable in size with the first sample will be of the type, (2) sampling will never give a high probability that the whole population is homogeneous unless the sample constitutes a large fraction of the whole population.

This is all quite correct as I endeavored to show in my amplification of Jeffreys's example in Volume I [7, pp. 337–338]. In my numerical example, if Jeffreys had sampled, say, 5000 birds and they all had beaks, the probability for the next bird to have a beak was 0.9998, the next 5000 to have a beak was around $1/2$, and the whole population of birds to have beaks, a small probability comparable to the ratio of the number in the sample to the whole population.

I construct some similar numerical examples in the Exercises to this Chapter. See Exercises 44.7.8 through 44.7.14.

This follow on quote from Jeffreys [24, pg. 128] sums up in a nutshell what was bothering him about the uniform prior probability. It is the source for constructing all those infamous "animals with feathers" examples.

> The last result was given by Broad [...] and was the first clear recognition, I think, of the need to modify the uniform assessment if it was to correspond to actual processes of induction. ... The rule of succession had been generally appealed to as a justification of induction; what Broad showed was that it was no justification whatsoever for attaching even a moderate probability to a general rule if the possible instances of the rule are many times more numerous than those already investigated. If we are ever to attach a high probability to a general rule, on any practicable amount of evidence, it is necessary that it must have a moderate probability to start with. Thus, I may have seen 1 in 1,000 of the 'animals with feathers' in England; on Laplace's theory the probability of the proposition 'all animals with feathers have beaks', would be about 1/1000. This does not correspond to my state of belief or anybody else's.

I would counter with this remark: You can not establish with 100% certainty a *logical deduction* based on the inferencing methods of probability theory that also embraces empirical data unless you start out by believing that logical deduction to the exclusion of anything else. Exercises 44.7.12 and 44.7.13 explore in more detail what I mean by this.

44.7 Solved Exercises for Chapter Forty Four

Exercise 44.7.1: Review the posterior predictive formula assuming the uniform distribution for the prior probability over model space.

Solution to Exercise 44.7.1

Because this formula will be in constant use for the upcoming numerical examples, we will repeat it here at the outset. Notice that the expression involves 1) n, the dimension of the state space, 2) the *future* frequency counts, that is, the M_i and $M = \sum_{i=1}^{n} M_i$, and 3) the past data $\mathcal{D} \equiv N_i$ and $N = \sum_{i=1}^{n} N_i$. Any model $\mathcal{M}_k$ does not explicitly appear in the formula because the critically important averaging over $P(\mathcal{M}_k \mid \mathcal{D})$ has already taken place.

$P(M_1, M_2, \cdots, M_n \mid N_1, N_2, \cdots, N_n) =$

$$\frac{M!\,(N+n-1)!}{\prod_{i=1}^{n} N_i!\,(M+N+n-1)!} \times \frac{\prod_{i=1}^{n}(M_i+N_i)!}{\prod_{i=1}^{n} M_i!}$$

Exercise 44.7.2: How drastically does the probability of TAILS change after adopting an initial state of complete ignorance about the coin, and then observing one HEADS?

Solution to Exercise 44.7.2

In section 44.2.4, we found the correct answer that, under a state of complete ignorance about the causal nature of some unknown coin and its manner of being flipped, the probability of TAILS is equal to 1/2. But flipping the coin once and observing a HEADS will change this probability to 1/3.

There is only one binary target variable A, a statement about what the coin will show after it has been tossed,

$$(A = a_1) \equiv \text{``The coin will show HEADS.''}$$

$$(A = a_2) \equiv \text{``The coin will show TAILS.''}$$

There are no predictor variables. There is one piece of data that consists of having observed the coin land HEADS on the first trial. The canonical formula tells us that the probability for TAILS on the second toss is,

$$P(A_2 = a_2 \mid \mathcal{D} = \{A_1 = a_1\}) = \frac{N_2 + 1}{N + n} = \frac{0+1}{1+2} = 1/3$$

We have the settings of $n = n_A = 2$, $N = 1$, $N_1 = 1$, $N_2 = 0$, and $M = 1$, $M_1 = 0$, and $M_2 = 1$.

Now, this result might clash with your intuition. I suspect you might judge dropping the probability for TAILS to 1/3 after just one trial a bit harsh.

Exercise 44.7.3: How drastically does the probability of TAILS change after adopting an initial state of complete ignorance about the coin, and then observing two HEADS?

Solution to Exercise 44.7.3

In the last exercise, we found that flipping the coin once and observing a HEADS changed the probability of TAILS on the next trial to 1/3.

The problem has not changed with regard to any predictor variables. However, there are now two pieces of data consisting of having observed the coin land HEADS on the first two trials. The posterior predictive formula tells us that,

$$P(A_3 = a_2 \,|\, \mathcal{D} = \{A_1 = a_1, A_2 = a_1\}) = \frac{N_2 + 1}{N + 2} = \frac{0 + 1}{2 + 2} = 1/4$$

We have the settings of $n = n_A = 2$, $N = 2$, $N_1 = 2$, $N_2 = 0$, and $M = 1$, $M_1 = 0$, and $M_2 = 1$.

Now, I persist in claiming that this result might also clash with your intuition. Just as in the previous case, I suspect you might judge dropping the probability for TAILS to 1/4 after just two trials a bit extreme.

Exercise 44.7.4: How drastically does the probability of TAILS change after adopting an initial state of complete ignorance about the coin, and then observing N HEADS?

Solution to Exercise 44.7.4

In the previous exercise, we found that flipping the coin twice and observing two HEADS changed the probability of TAILS on the next trial to 1/4.

There are now N pieces of data consisting of having observed the coin land HEADS on all N trials. The posterior predictive formula now tells us that,

$$P(A_{N+1} = a_2 \,|\, \mathcal{D} = \{A_1 = a_1, A_2 = a_1, \cdots A_N = a_1\}) = \frac{N_2 + 1}{N + n} = \frac{1}{N + 2}$$

We have the settings of $n = n_A = 2$, N, $N_1 = N$, $N_2 = 0$, and $M = 1$, $M_1 = 0$, and $M_2 = 1$.

Suppose that $N = 100$. Now, this result might also clash with your intuition, *but in exactly the opposite direction* as in the previous two situations! I suspect you might judge dropping the probability for TAILS to *only* 1/102 after all 100 trials have revealed all HEADS is a bit lenient. It was this kind of quantitative answer that prompted Harold Jeffreys to have misgivings about the uniform prior.

SOLVED EXERCISES FOR CHAPTER 44

Exercise 44.7.5: What is the probability for two HEADS and one TAILS on the first three coin flips?

Solution to Exercise 44.7.5

What does the canonical expression say for the next M future occurrences? So far, we have concentrated exclusively on the *next* occurrence where $M = 1$, and not where M could equal any number of future occurrences. We are not stymied, however, because we can refer back to the original predictive formula for future occurrences. This formula explicitly mentions a general M so we can use it here. Since the dimension of the state space still equals 2, the formula simplifies to,

$$P(M_1, M_2 \mid N_1, N_2) = \frac{M!\,(N+1)!}{N_1!\,N_2!\,(M+N+1)!} \times \frac{(M_1+N_1)!\,(M_2+N_2)!}{M_1!\,M_2!}$$

Since we are asking about the *first* three coin flips, there are no data. Thus, $N = 0$, $N_1 = 0$ and $N_2 = 0$. We still have just the one target variable A and no predictor variables, so $n = n_A = 2$. The new interesting change is that $M = 3$, $M_1 = 2$ and $M_2 = 1$.

Plugging these values into the above formula,

$$P(M_1 = 2, M_2 = 1 \mid N_1 = 0, N_2 = 0) = \frac{3!\,1!}{0!\,0!\,4!} \times 1$$

$$= \frac{3!}{4!}$$

$$= 1/4$$

There are only four possibilities for the frequency counts in the first three trials.

Possibility 1. All three HEADS.

Possibility 2. Two HEADS, one TAILS.

Possibility 3. Two TAILS, one HEADS.

Possibility 4. All three TAILS.

Each of these four possibilities has exactly the same probability of $1/4$ to occur during the first three trials. Any multiplicity factor has been nullified. This is the undeniable consequence of adopting a uniform prior over model space.

The probability for Possibilities 2 and 3 would be three times that of Possibilities 1 and 4 if the one model assigning $Q_1 = Q_2 = 1/2$ had been in place because a multiplicity factor of three would have been in effect for the three ways that two HEADS and one TAILS, or two TAILS and one HEADS, could have occurred.

Putting it, I think, even more forcefully, is the fact that these probabilities are the undeniable consequence of frankly admitting that the IP does not know anything concerning the physical causality of the coin. Not knowing anything, it is conceivable that the coin just *might* be a coin constructed with two TAILS. Not knowing anything, it is conceivable that the coin just *might* be a coin constructed with two HEADS. Not knowing anything, it is conceivable that the coin just *might* be a fair coin. Not knowing anything, it is conceivable that the coin *might* have a bias for HEADS, and so on. Allowances have to be made in the model space for *every conceivable peculiarity of the coin!*

I find it a fascinating outcome that the formal rules of probability manipulation, when handed this conundrum, decide to split up the probability evenly over all of the possibilities for the future frequency counts. It reaches this conclusion in a very efficient way, and, moreover, in a way that I don't think I would have ever stumbled upon however hard and long I thought about the problem.

Nonetheless, this simple and satisfying resolution continues to provoke serious head scratching even today. There seems to be an enduring conflict between our ingrained intuition, and the above consequences ensuing from the formal rules of probability manipulation. Ground zero centers on the adoption of the uniform prior probability for models to represent a state of complete ignorance on the part of the IP. Sir Harold Jeffreys is one person in particular who found this outcome discomfiting. And he is by no means alone in those wanting to find some rationale to replace Laplace's argument.

Exercise 44.7.6: What is the probability for any possibility on the first M coin flips?

Solution to Exercise 44.7.6

It is easy to generalize the last exercise to cover the case of any number of future frequency counts. Remember that there are no previous data, so not only does $N = 0$, all of the N_i must also equal 0.

$$P(M_1, M_2 \mid N_1, N_2) = \frac{M! \, (N+1)!}{N_1! \, N_2! \, (M+N+1)!} \times \frac{\prod_{i=1}^{2}(M_i + N_i)!}{\prod_{i=1}^{2} M_i!}$$

$$= \frac{1}{M+1}$$

During the first $M = 99$ coin flips, every possibility from all HEADS, to all TAILS, to 50 HEADS and 49 TAILS, and everything in between, has the same probability of 1/100. This result, undeniably correct by applying the formal manipulation rules of probability, can serve as a serious conversation starter with those who have never thought beyond the adoption of a single model.

SOLVED EXERCISES FOR CHAPTER 44   445

Exercise 44.7.7: What happens when the target variable is not binary?

Solution to Exercise 44.7.7

The above result is generalized to,

$P(M_1, M_2, \cdots, M_n \mid N_1, N_2, \cdots, N_n) =$

$$\frac{M!\,(N+n-1)!}{N_1!\cdots N_n!\,(M+N+n-1)!} \times \frac{\prod_{i=1}^{n}(M_i+N_i)!}{\prod_{i=1}^{n} M_i!}$$

$$= \frac{M!\,(n-1)!}{(M+n-1)!} \times \frac{\prod_{i=1}^{n} M_i!}{\prod_{i=1}^{n} M_i!}$$

$$= \frac{M!\,(n-1)!}{(M+n-1)!}$$

Consider the case of rolling a die. Here, the target variable A can be in one six possible states, $n = n_A = 6$. What is the probability of any outcome in the first six rolls of the die?

For example, the probability of seeing three ONEs, no TWOs, FOURs, or FIVEs, one THREE and two SIXes in the first six rolls of the die is,

$$P(M_1 = 3, M_2 = 0, M_3 = 1, M_4 = 0, M_5 = 0, M_6 = 2 \mid \mathcal{D}) = \frac{M!\,(n-1)!}{(M+n-1)!}$$

$$= \frac{6!\,5!}{11!}$$

$$= \frac{1}{462}$$

This is the same probability of seeing every face once, of seeing all SIXes, of seeing all THREEs, and anything in between in the first six rolls of the die. There must be a total, then, of,

$$\frac{(M+n-1)!}{M!\,(n-1)!} = \frac{11!}{6!\,5!} = 462$$

possible frequency counts in the first six rolls of the die, all of which have the same probability of $1/462$.

We can't help but notice that the formula for the posterior predictive probability depends solely on the fixed values of n and M in this case. Thus, no matter what pattern of conceivable future frequency counts M_i is desired, its probability will be the same as any other set of M_i.

Once again, these equal probabilities are the consequences when one adopts the uniform prior probability over model space. We follow the advice of Laplace who said that when one cannot distinguish among the possible *causes*, then all possible future events are equiprobable.

Exercise 44.7.8: Jeffreys began his discussion of finite sampling with a variation on the coin flip scenario. By restricting the problem to "animals with feathers," he addressed the question of the probability for such animals to have beaks. Approach this same problem from within our framework of canonical inferencing.

Solution to Exercise 44.7.8

What is the probability that the rest of some defined finite population is of one type after a sample of size N is taken, and it turns out that the entire sample belongs to that type. Get comfortable with the calculations by solving a problem involving a small population of feathered animals. Suppose that the entire population consists of 100 animals with feathers and these are birds, turkeys, chickens, ducks, geese, swans, and so on.

We have taken a sample of $N = 50$ of these feathered animals, and duly recorded whether they had beaks or not. To no one's surprise, the data consist of $N_1 = 50$ and $N_2 = 0$ where $(A = a_1) \equiv$ "It had a beak." and $(A = a_2) \equiv$ "It didn't have a beak.". The number in the entire population is $N + M = 100$, so the rest of the population consists of the $M = 50$ unsampled feathered animals. The sought for probability is then,

$$P(M_1 = 50, M_2 = 0 \,|\, N_1 = 50, N_2 = 0)$$

The posterior predictive probability formula tells us that,

$$P(M_1 = 50, M_2 = 0 \,|\, \mathcal{D}) = \frac{M!\,(N+n-1)!}{N_1!\,N_2!\,(M+N+n-1)!} \times \frac{(M_1+N_1)!\,(M_2+N_2)!}{M_1!\,M_2!}$$

After plugging in the numbers for the stated conditions for this problem,

$$P(M_1 = 50, M_2 = 0 \,|\, \mathcal{D}) = \frac{50!\,51!}{50!\,0!\,(50+50+1)!} \times \frac{100!\,0!}{50!\,0!}$$

$$= \frac{51!}{101!} \times \frac{100!}{50!}$$

$$= \frac{51}{101}$$

$$\approx 1/2$$

After sampling half the entire population, and discovering that all of the animals with feathers had beaks, the degree of belief that the remainder of the unsampled population also had beaks is only about $1/2$. According to Jeffreys, this was clearly a failure on the part of probability theory to adequately reach a sensible induction. Surgery had to be performed on the offending uniform prior probability for models.

SOLVED EXERCISES FOR CHAPTER 44

Exercise 44.7.9: Continue the previous exercise by considering a larger population.

Solution to Exercise 44.7.9

Let's enlarge the size of the population to $M+N = 10,000$, but retain the condition that half the population was sampled. After plugging in the numbers for the stated conditions for this problem,

$$P(M_1 = 5000, M_2 = 0 \mid \mathcal{D}) = \frac{5000!\,5001!}{5000!\,0!\,(5000+5000+1)!} \times \frac{10000!\,0!}{5000!\,0!}$$

$$= \frac{5001!}{10001!} \times \frac{10000!}{5000!}$$

$$= \frac{5001}{10001}$$

$$\approx 1/2$$

The same outcome holds as in the previous exercise even when we consider a much larger population. The lesson is clear. No matter the size of the population, if we sample about half the population, the probability that the remaining unsampled portion is all of the same type hovers around 1/2.

Exercise 44.7.10: What about a large population, but the IP only wants to know the probability that the *next* animal has a beak?

Solution to Exercise 44.7.10

The size of the population is still 10,000, and half the population have been sampled, all with beaks. Now, though, since we are asking only about the next animal in feathers, M, M_1 and M_2 reduce back down to the very small numbers, $M = 1$, $M_1 = 1$ and $M_2 = 0$.

$$P(M_1 = 1, M_2 = 0 \mid \mathcal{D}) = \frac{1!\,5001!}{5000!\,0!\,(1+5000+2-1)!} \times \frac{5001!\,0!}{1!\,0!}$$

$$= \frac{5001!}{5000!\,5002!} \times 5001!$$

$$= \frac{5001}{5002}$$

$$\approx 0.9998$$

If all the IP desires is the probability that the next feathered animal has a beak, and I maintain that this is overwhelmingly the inferential question that is most often asked, then a large sample will provide a very high probability for that event. Despite the argued for shortcomings of the uniform prior probability over model space, it has returned a very reasonable answer.

Exercise 44.7.11: What happens to this large probability for the very next animal as we ask for the probability that the next two, the next three, and so on, up to the next 5,000 animals all have beaks?

Solution to Exercise 44.7.11

It's straightforward to figure out the simple formula for determining the probability for the next M animals in this situation.

$$P(M_1 = M, M_2 = 0 \mid N_1 = N, N_2 = 0) =$$

$$\frac{M!\,(N+n-1)!}{N_1!\,N_2!\,(M+N+n-1)!} \times \frac{(M_1+N_1)!\,(M_2+N_2)!}{M_1!\,M_2!}$$

$$= \frac{M_1!\,(N_1+1)!}{N_1!\,(M_1+N_1+1)!} \times \frac{(M_1+N_1)!}{M_1!}$$

$$= \frac{(N_1+1)!}{N_1!\,(M_1+N_1+1)!} \times (M_1+N_1)!$$

$$= \frac{N_1+1}{M_1+N_1+1}$$

At first, nothing too serious seems to be happening for the next few unsampled animals.

$$P(\text{Next two} \mid \mathcal{D}) = \frac{5001}{5003} = 0.9996$$

$$P(\text{Next three} \mid \mathcal{D}) = \frac{5001}{5004} = 0.9994$$

$$P(\text{Next ten} \mid \mathcal{D}) = \frac{5001}{5011} = 0.9980$$

$$P(\text{Next hundred} \mid \mathcal{D}) = \frac{5001}{5101} = 0.9804$$

The probability, even out to as many as the next 100 animals, remains very high that all will have beaks. The probability, however, that we will observe more and more animals in feathers with beaks is slowly, but steadily, decreasing,

$$P(\text{Next thousand} \,|\, \mathcal{D}) = \frac{5001}{6001} = 0.8334$$

$$P(\text{Next three thousand five hundred} \,|\, \mathcal{D}) = \frac{5001}{8501} = 0.5883$$

$$P(\text{Remaining five thousand} \,|\, \mathcal{D}) = \frac{5001}{10001} \approx 1/2$$

We can put our finger on the source of the trouble by using the binomial probability as an approximation.

$$P(\text{Next } M) \approx W(M)\, Q_1^{M_1}\, Q_2^{M_2}$$

The binomial is an approximation to the posterior predictive distribution because the Q_i are the same at every trial. The posterior predictive is dealing with the case where we have sampling without replacement, while the binomial assumes sampling with replacement.

Nevertheless, the approximation remains pretty good even out to the next 1,000 unsampled animals,

$$P(\text{Next two}) \approx \binom{2}{0}(0.9998)^2 (0.0002)^0 = 0.9996$$

$$P(\text{Next three}) \approx \binom{3}{0}(0.9998)^3 (0.0002)^0 = 0.9994$$

$$P(\text{Next ten}) \approx \binom{10}{0}(0.9998)^{10} (0.0002)^0 = 0.9980$$

$$P(\text{Next hundred}) \approx \binom{100}{0}(0.9998)^{100} (0.0002)^0 = 0.9802$$

$$P(\text{Next thousand}) \approx \binom{1000}{0}(0.9998)^{1000} (0.0002)^0 = 0.8188$$

But more important, by far, is the insight stemming from the appearance of the multiplicity factor $W(M)$ within the binomial distribution. There is only one way, $W(M) = 1$, that all M animals can appear as one type.

But as M increases, there are more and more ways that some mixture of M_1 feathered animals with beaks and M_2 feathered animals without beaks might just happen to appear, even though the probability for a feathered animal without a beak must be very low based on the sample data. The multiplicity factor becomes very large for large M, in other words, whenever a large number of unsampled animals still remains.

This is the rationale for why the probability must decrease to what seems like an unreasonably low value like $1/2$ when the number of unsampled animals is roughly the same size as in the sample. Using this rationale, we can even ferret out roughly where the components to the decreasing probability in going from, say, $M = 1$ to $M = 1000$, occur.

Even for a fantastically large probability like 0.9998 on each trial, the probability of seeing this type occur 1000 times in a row is 0.8188. The probability for seeing 999 of this type and 1 of its negation is 0.1637. The probability for seeing 998 of this type and 2 of its negation is 0.0164. The probability for seeing 997 of this type and 3 of its negation is 0.0011. The probability for seeing four or more of the other type becomes vanishingly small. However, it does becomes increasingly likely to see at least one occurrence of the other type as we ask about a larger and larger unsampled population.

For example, and keeping in mind that we are using the binomial probability merely as a rough approximation, we see that even with the aforementioned large probability of 0.9998 at each trial, the probability of observing *one* beakless feathered animal somewhere in 5000 future trials is a significantly large probability of around 0.37,

$$P(M_1 = 4999, M_2 = 1) \approx \frac{5000!}{4999!\,1!} \times (0.9998)^{4999}(0.0002)^1 = 0.3679$$

due to the multiplicity factor having a value of 5,000.

That one beakless feathered animal might happen to be found at the next data sample, the one after that, ..., or at the 5000th and last observation of the yet unsampled animals. Of course, there are far more ways,

$$\frac{5000!}{4998!\,2!} = 12,497,500$$

to find *two* beakless feathered animals in the next 5000 sampled, but the probability is not as high as for one simply because the $(0.9998)^{4998}(0.0002)^2$ terms more than compensate for the increase in the multiplicity factor.

It is my opinion that the most important lesson from these exercises is that the almost universal condemnation directed at Laplace's *Rule of Succession* for using a uniform prior probability is unfounded. Bayes, Laplace, Jaynes, and my treatment over these books, all followed the formal manipulation rules of probability. People want an *inductive* rule that allows them to make a *deductive* conclusion based on some small sample.

The rules of probability tell one exactly what degree of belief various statements are entitled to. The critic's ire should be leveled at the rules of probability theory, and not at Bayes or Laplace for adopting a uniform prior, if he doesn't like these consequences. The rules of probability theory tell us that even with a degree of belief as high as 0.9998 that the next animal with feathers will also have a beak, there is a rather large probability that at least one out of the next five thousand will *not* have a beak.

SOLVED EXERCISES FOR CHAPTER 44

Exercise 44.7.12: Discuss Jeffreys's "animals with feathers" problem in a desultory fashion from the perspective of canonical inferencing.

Solution to Exercise 44.7.12

As we have seen, Jeffreys proposed the very straightforward inferential scenario of finding the probability that an animal has a beak given that it has feathers. We have therefore one binary target variable A where A is a statement about whether an animal has a beak,

$$(A = a_1) \equiv \text{"The animal has a beak."}$$

$$(A = a_2) \equiv \text{"The animal does not have a beak."}$$

But the solution demands that we also consider the other statement in the problem that will serve as the predictor variable. There is just one of these predictor variables X, a statement about whether an animal has feathers,

$$(X = x_1) \equiv \text{"The animal has feathers."}$$

$$(X = x_2) \equiv \text{"The animal does not have feathers."}$$

The dimension of the state space is thus $n = n_A \times n_X = 4$.

The conditional probability that the next animal has a beak given that it does have feathers, and conditioned as well on some amount of previous data, is written out as usual by the expression $P(A_{N+1} = a_1 \mid X_{N+1} = x_1, \mathcal{D})$.

Start out from scratch by considering probabilities when no data are available. Thus, $N = 0$ and all $N_i = 0$. The above expression for the probability of the next animal reduces to the probability that the very first animal has a beak given that it has feathers. By Bayes's Theorem,

$$P(A_1 = a_1 \mid X_1 = x_1) = \frac{P(A_1 = a_1, X_1 = x_1)}{P(X_1 = x_1)}$$

$$= \frac{P(A_1 = a_1, X_1 = x_1)}{P(A_1 = a_1, X_1 = x_1) + P(A_1 = a_2, X_1 = x_1)}$$

The posterior predictive probability for each of the four joint statements is,

$$P(A_1 = a_i, X_1 = x_j \mid \mathcal{D}) = \frac{N_i + 1}{N + n} = \frac{0 + 1}{0 + 4} = \frac{1}{4}$$

and for the particular case of a feathered animal with a beak,

$$P(A_1 = a_1 \mid X_1 = x_1, \mathcal{D}) = \frac{N_1 + 1}{(N_1 + 1) + (N_3 + 1)} = \frac{1}{(0 + 1) + (0 + 1)} = 1/2$$

Move slowly away from this most primitive case by acknowledging that one piece of data exists. Suppose the animal observed was beakless and featherless. Now, $N = 1$, $N_1 = N_2 = N_3 = 0$ and $N_4 = 1$. The posterior predictive probability now becomes 1/5 for the first three joint statements and 2/5 for the last joint statement. The probability that the next animal has a beak given that it has feathers remains, however, at 1/2.

$$P(A_2 = a_1 \mid X_2 = x_1, \mathcal{D} = \{A_1 = a_2, X_1 = x_2\}) = \frac{N_1 + 1}{(N_1 + 1) + (N_3 + 1)}$$

$$= \frac{1}{(0 + 1) + (0 + 1)}$$

$$= 1/2$$

After having whetted our appetite with these primitive cases, analyze a more realistic situation where data has been gathered on $N = 7000$ animals. These data are laid out in the contingency table in Figure 44.1 below.

	X feathers	**X̄** no feathers	
A beak	N_1 500	N_2 2	502
Ā no beak	N_3 0	N_4 6498	6498
	500	6500	7000

Figure 44.1: *A fictitious contingency table to illustrate Jeffreys's "animals with feathers" problem.*

The N_i frequency counts are presented in each of the four cells with $N_1 = 500$, $N_2 = 2$, $N_3 = 0$, and $N_4 = 6,498$. The probability that the next animal has a beak given that it has feathers jumps up to near certainty with a value of 0.9982.

$$P(A_{7001} = a_1 \mid X_{7001} = x_1, \mathcal{D} = \{A_1, X_1, \cdots, A_{7000}, X_{7000}\}) = \frac{N_1 + 1}{(N_1 + 1) + (N_3 + 1)}$$

$$= \frac{500 + 1}{(500 + 1) + (0 + 1)}$$

$$= 501/502$$

$$= 0.9982$$

SOLVED EXERCISES FOR CHAPTER 44   453

Exercise 44.7.13: Suppose that some of these data from the previous exercise were collected in Australia.

Solution to Exercise 44.7.13

Look at the new contingency table in Figure 44.2 below. There are now eight cells in the contingency table to reflect the addition of a new binary predictor variable H. H is the statement that the observations about the animals were made in Australia. As evident from the labeling of this new predictor variable, we will eventually issue probabilities by marginalizing over H.

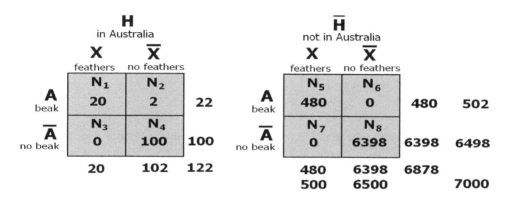

Figure 44.2: *Another fictitious contingency table to illustrate Jeffreys's "animals with feathers" problem. A new binary predictor variable H has been added.*

But first, these data will serve as another elementary illustration of a canonical inference. The symbolic conditional probability to be calculated is,

$$P(A_{N+1} = a_1 \mid X_{N+1} = x_2, H_{N+1} = h_1, \mathcal{D})$$

which for this problem is translated more specifically into,

The degree of belief that the following statement is true: "The next animal to be observed has a beak given that it has no feathers and it was observed in Australia. Also, all of the previous observations of animals whether in Australia or not are taken into account."

We rely on Bayes's Theorem to find the answer,

$$P(A_{N+1} = a_1 \mid X_{N+1} = x_2, H_{N+1} = h_1, \mathcal{D}) = \frac{P(A_{N+1} = a_1, X_{N+1} = x_2, H_{N+1} = h_1 \mid \mathcal{D})}{P(X_{N+1} = x_2, H_{N+1} = h_1 \mid \mathcal{D})}$$

Using the *Rule of Succession* for each of the terms in Bayes's Theorem,

$$P(A_{N+1} = a_1, X_{N+1} = x_2, H_{N+1} = h_1 \mid \mathcal{D}) = \frac{N_2 + 1}{N + n}$$

$$P(X_{N+1} = x_2, H_{N+1} = h_1 \mid \mathcal{D}) = \frac{(N_2 + 1) + (N_4 + 1)}{N + n}$$

$$P(A_{N+1} = a_1 \mid X_{N+1} = x_2, H_{N+1} = h_1, \mathcal{D}) = \frac{N_2 + 1}{(N_2 + 1) + (N_4 + 1)}$$

$$= \frac{2 + 1}{(2 + 1) + (100 + 1)}$$

$$= \frac{3}{104}$$

Compare this probability to the same situation with the difference that the animal was not observed in Australia. Then,

$$P(A_{N+1} = a_1 \mid X_{N+1} = x_2, H_{N+1} = h_2, \mathcal{D}) = \frac{1}{6400}$$

We are left with the completely sensible result of a lower probability that the next animal has a beak given that it has no feathers when it is not observed in Australia.

If the IP wants probabilities collapsed over the location where the animals were observed, then canonical inferencing uses the **Sum Rule** to re–express Bayes's Theorem as shown in the following condensed notation,

$$P(A_{N+1} \mid X_{N+1}, \mathcal{D}) = \frac{\sum_{n_H} P(AHX \mid \mathcal{D})}{\sum_{n_A, n_H} P(AHX \mid \mathcal{D})}$$

Thus, there will be two $(N_i + 1)$ terms in the numerator, and four $(N_i + 1)$ terms in the denominator.

Substituting the frequency counts from the contingency table in Figure 44.2, the probability that the next animal has a beak given that it has no feathers is,

$$P(A_{N+1} = a_1 \mid X_{N+1} = x_2, \mathcal{D}) = \frac{(N_2 + 1) + (N_6 + 1)}{(N_2 + 1) + (N_4 + 1) + (N_6 + 1) + (N_8 + 1)}$$

$$= \frac{(2 + 1) + (0 + 1)}{(2 + 1) + (100 + 1) + (0 + 1) + (6398 + 1)}$$

$$= \frac{4}{6504}$$

Once again, this is an eminently reasonable answer that has been returned to the IP when it relied strictly on the formal rules of probability. It is higher than the previous probability of $1/6400$ when it was given that the animal was not observed in Australia, but much lower than the previous probability of $3/104$ when it was given that the animal was observed in Australia.

It doesn't make any difference in the final probabilities if the statement H is placed to the left of the conditioned upon symbol along with the statement A instead of on the right along with $\overline{X}$. Thus, the condensed probability expression $P(AH \mid \overline{X}, \mathcal{D})$ would be calculated as usual by,

$$P(AH \mid \overline{X}, \mathcal{D}) = \frac{P(AH\overline{X} \mid \mathcal{D})}{\sum_{n_A, n_H} P(AH\overline{X} \mid \mathcal{D})}$$

Four probabilities will be calculated on the left hand side for every combination of A and H and the fixed value of $(X = x_2)$. There will be one joint probability in the numerator and four joint probabilities in the denominator. These four joint probabilities in the denominator of Bayes's Theorem are,

$$P(A_{N+1} = a_1, H_{N+1} = h_1, X_{N+1} = x_2 \mid \mathcal{D}) = 3/7008$$

$$P(A_{N+1} = a_2, H_{N+1} = h_1, X_{N+1} = x_2 \mid \mathcal{D}) = 101/7008$$

$$P(A_{N+1} = a_1, H_{N+1} = h_2, X_{N+1} = x_2 \mid \mathcal{D}) = 1/7008$$

$$P(A_{N+1} = a_2, H_{N+1} = h_2, X_{N+1} = x_2 \mid \mathcal{D}) = 6399/7008$$

So, for example,

$$P(A_{N+1} = a_2, H_{N+1} = h_1 \mid X_{N+1} = x_2, \mathcal{D}) = \frac{101}{3 + 101 + 1 + 6399} = \frac{101}{6504}$$

To replicate the results found in the previous exercise, sum over the two probabilities where H was explicitly mentioned,

$$P(A_{N+1} = a_1 \mid X_{N+1} = x_2, \mathcal{D}) = \frac{3}{6504} + \frac{1}{6504} = \frac{4}{6504}$$

$$P(A_{N+1} = a_2 \mid X_{N+1} = x_2, \mathcal{D}) = \frac{101}{6504} + \frac{6399}{6504} = \frac{6500}{6504}$$

Exercise 44.7.14: Revisit the Rule of Succession as used in the previous Exercise to find the probability that the very next animal to be observed would be a beakless non–feathered animal from Australia.

Solution to Exercise 44.7.14

Since the state space for this problem is dimension $n = 8$, the posterior predictive probability is,

$$P(M_1, M_2, \cdots, M_8 \mid N_1, N_2, \cdots, N_8) = \frac{M!(N+n-1)!}{\prod_{i=1}^{8} N_i!(M+N+n-1)!} \times \frac{\prod_{i=1}^{8}(M_i + N_i)!}{\prod_{i=1}^{8} M_i!}$$

Since the IP is concerned with the very next animal, the above formula immediately simplifies with $M = 1$ to,

$$P(M_1, M_2, \cdots, M_8 \mid N_1, N_2, \cdots, N_8) = \frac{(N+n-1)!}{\prod_{i=1}^{8} N_i! \, (N+n)!} \times \prod_{i=1}^{8} (M_i + N_i)!$$

$$= \frac{1}{\prod_{i=1}^{8} N_i! \, (N+n)} \times \prod_{i=1}^{8} (M_i + N_i)!$$

$$= \frac{1}{(N+n)} \times \frac{\prod_{i=1}^{8} (M_i + N_i)!}{\prod_{i=1}^{8} N_i!}$$

The IP wants to find the joint posterior probability for the next beakless non–feathered animal from Australia,

$$P(A_{N+1} = a_2, H_{N+1} = h_1, X_{N+1} = x_2 \mid \mathcal{D})$$

Consulting the contingency table, we see that this requires that $M_4 = 1$ and the remaining seven M_i to be equal to 0. This leads to,

$$P(M_1 = 0, \cdots, M_4 = 1, \cdots, M_8 = 0 \mid N_1, N_2, \cdots, N_8) = \frac{1}{N+n} \times (N_4 + 1)$$

$$P(A_{N+1} = a_2, H_{N+1} = h_1, X_{N+1} = x_2 \mid \mathcal{D}) = \frac{101}{7008}$$

The denominator value of 7008 is common to all of the terms in Bayes's Theorem, so it cancels out in the end, as illustrated in the previous exercise.

Exercise 44.7.15: Define a *Mathematica* function to calculate binomial distribution probabilities as discussed in Exercise 44.7.11.

Solution to Exercise 44.7.15

The very straightforward function,

binom[qi_List, Ni_List]:=
        Apply[Multinomial, Ni] Apply[Times, Power[qi, Ni]]

computes a probability of 0.183958 for observing two beakless feathered animals in the next 5000 observations with **binom[{0.9998, 0.0002}, {4998, 2}]**. The correct probability for this future observation using the general posterior predictive formula shown in Exercise 44.7.1 is,

$$P(M_1 = 4998, M_2 = 2 \mid N_1 = 5000, N_2 = 0) = 0.125$$

The interesting part of this function definition resides in the built–in function **Apply[*f, expr*]**. It replaces the *head* of the expression *expr* with *f*. Thus, it takes the list of the frequency counts and replaces the head **List** with the new head **Multinomial**. It does exactly the same thing with the list of the assigned numerical values of the probabilities raised to the power of the frequency counts and replaces the old head **List** with the new head **Times**.

Exercise 44.7.16: What was Jeffreys really looking for in the inferential problem of the animals with feathers?

Solution to Exercise 44.7.16

I introduced the idea of probability theory generalizing Classical Logic in Chapter Seven of Volume I [7, pg. 169]. We looked first at the classical case of *modus ponens*, that is, the implication $A \to B$, together with the assertion of the truth of the premise A leading to the deduction that B is true.

Probability theory can reproduce this certain deduction,

$$P(B \mid A, \{\mathcal{M}_k \equiv A \to B\}) = 1$$

through Bayes's Theorem. Not only that, but probability theory is able to generalize to the situation that logic deems invalid, that is, finding the degree of belief in the truth of the premise A when only the consequence B is known,

$$P(A \mid B, \{\mathcal{M}_k \equiv A \to B\}) = \frac{P(AB \mid \mathcal{M}_k)}{P(B \mid \mathcal{M}_k)}$$

Thus, in the end, we see that what Jeffreys really wanted was to invoke Classical Logic to achieve a confirmed deduction that if an animal had feathers, then it must also have a beak. The probabilistic version of this is the same as above with,

$$P(A \mid X, \{\mathcal{M}_k \equiv X \to A\}) = 1$$

It is interesting to calculate what the probabilistic generalization finds for the reverse situation of the probability that an animal has feathers where it is known that it has a beak,

$$P(X \mid A, \{\mathcal{M}_k \equiv X \to A\}) = \frac{P(AX \mid \mathcal{M}_k)}{P(AX \mid \mathcal{M}_k) + P(A\overline{X} \mid \mathcal{M}_k)} = \frac{Q_1}{Q_1 + Q_2}$$

Take for the numerical assignments of Q_1 and Q_2 under the single model to be the normed frequency counts shown of the contingency table of Figure 44.1,

$$Q_1 = \frac{N_1}{N} = \frac{500}{7000} \text{ and } Q_2 = \frac{N_2}{N} = \frac{2}{7000}$$

(it is conceptually wrong, of course, to think that probabilities are frequencies) to arrive at,

$$P(X \mid A, \{\mathcal{M}_k \equiv X \to A\}) = \frac{500}{500 + 2}$$

So the resolution to Jeffreys's discomfort and the motivation for introducing his "Jeffreys's prior" is, to my way of looking at it, very straightforward. Jeffreys wanted a *deduction* based on *one* model, that model, of course being the logical implication $X \to A$, "If an animal has feathers, then it must have a beak."

Probability theory then provides Jeffreys with his desired outcome. Every time he sees an animal in feathers, then with probability 1, it will also have a beak. No data are necessary for logical deductions, only the verification that the premise X is indeed correct.

If the IP already knows the one, single, exclusive model $\mathcal{M}_k$ that it wants to employ, then this is the way to proceed. However, I think everyone, even Jeffreys, would concede that, in making inferences as opposed to making deductions, the IP must take into account every conceivable model behind the numerical assignments to the statements in the state space. And, if the IP admits to the fact that it is very ill–informed about the causal nature of the phenomenon under investigation, it would be best to adopt that uniform prior probability over model space.

The only price you have to pay, is 1) empirical support for your prediction, and 2) probabilities like 0.99 instead of 1.00. The advantage is that all of the models are all still lurking in the background, no matter how low their probabilities might go. If something unforeseen crops up, the IP hasn't closed off all the avenues for escape as it has when it adopted that one model of logical implication.

Exercise 44.7.17: Revisit the details of the implicational model.

Solution to Exercise 44.7.17

It is fun to go back now and again to revisit the arguments tying together Classical Logic and probability theory. Every logic function can be decomposed according to its disjunctive normal form (DNF). Applying Boole's Expansion Theorem to the two variable implication logic function,

$$A \to B = f(T,T)\, AB \vee f(T,F)\, A\overline{B} \vee f(F,T)\, \overline{A}B \vee f(F,F)\, \overline{A}\,\overline{B}$$

we may observe that the coefficients of the expansion,

$$f(T,T) = T$$

$$f(T,F) = F$$

$$f(F,T) = T$$

$$f(F,F) = T$$

lead to the full DNF expression for the implication function as,

$$A \to B = AB \vee \overline{A}B \vee \overline{A}\,\overline{B}$$

The coefficients in the function expansion tell us the unambiguous meaning of any particular logic function. For the implication logic function, given that the coefficient of the basis function $A\overline{B}$ is $f(T,F) = F$, means that a probability of

0 must be assigned to that cell of the joint probability table indexing the joint statement A and not B. Therefore, the essence of implication is that it is impossible for A to be true and B false.

But anything else is OK. For example, since the coefficient of the basis function $\overline{A}B$ is $f(F,T) = T$, A can be false and B true without violating the implication. Any legitimate probability other than 0 can be assigned to the joint statement of A is false and B is true.

Make use of the generality inherent in Boolean Algebra to make the connection with probabilities even closer. Suppose that the carrier set for the Boolean Algebra is enlarged from the bare bones carrier set,

$$\mathbf{B} = \{T, F\}$$

to something like,

$$\mathbf{B} = \{a, a', b, b', c, c', T, F\}$$

Then, as shown in [7, pg. 98], the four cells of the joint probability table can be assigned the values $f(T,T) = a$, $f(F,T) = c'$, $f(T,F) = F$, and $f(F,F) = b'$ as the Boolean Algebra analogs to actual probability assignments.

Jeffreys's animals with feathers scenario demands that we be a bit careful in the pattern matching between the statements in the generic example just presented and how the statements were defined in Jeffreys's problem. Referring back to the contingency table in Exercise 44.7.12, the label X matches up with the old A, and the label A matches with the old B. So the expansion becomes,

$$X \to A = f(T,T)\, XA \vee f(T,F)\, X\overline{A} \vee f(F,T)\, \overline{X}A \vee f(F,F)\, \overline{X}\,\overline{A}$$

The coefficient $f(T,F) = F$ is attached to the joint statement (the joint statement now impossible by implication),

$$X\overline{A} \equiv \text{"An animal has feathers, but it has no beak."}$$

The 0 gets placed into cell 3 of the joint probability table, that is, $Q_3 = 0$. The conditional probability that any animal showing feathers must have a beak now becomes a certainty,

$$P(A = a_1 \,|\, X = x_1, \{\mathcal{M}_k \equiv X \to A\}) = \frac{Q_1}{Q_1 + Q_3} = 1$$

The clearly obvious differences with this solution compared to the one suggested by Laplace is: 1) there is only one single model that is accepted as universally true to the exclusion of all other models, and 2) no empirical data are required.

The IP is, contrary to Laplace's criterion of total ignorance on the part of the IP, very well–informed indeed when it adopts the implication logic function as the basis for an inference. It also enjoys the luxury of eschewing data based inferencing, because logically the mere truth of the premise is enough by itself to guarantee the truth of the consequence.

Exercise 44.7.18: Construct a table of probabilities as calculated by the posterior predictive formula for observing six or fewer beakless feathered animals in, say, the remaining 5000 feathered animals not yet sampled.

Solution to Exercise 44.7.18

The probability is very high, at over 99%, that there are six or fewer beakless feathered animals in the 5,000 remaining feathered animals in the population. However, the contentious result is the probability of close to 50% for at least one beakless feathered animal. As in the previous exercises, we have made the assumption that the total population consists of 10,000 feathered animals with 5,000 of these already sampled and all found to possess beaks.

The probabilities are shown below in Table 44.1.

Table 44.1: *The posterior predictive probabilities for seeing future frequency counts of feathered animals with and without beaks in the remaining 5,000 that were left unsampled in the population.*

M_1	M_2	M	$P(M_1, M_2 \mid N_1 = 5000, N_2 = 0)$
5000	0	5000	0.5000500
4999	1	5000	0.2500250
4998	2	5000	0.1250000
4997	3	5000	0.0624875
4996	4	5000	0.0312344
4995	5	5000	0.0156109
4994	6	5000	0.0078016
	Sum		0.9922094

This result, of course, runs counter to the intuitive inductive response that we, and most especially, Harold Jeffreys would expect after observing one half of the entire population with every single one of the feathered animals possessing a beak. If you can't see the penny drop after being hit between the eyes with such force, then how are you ever going to make any successful induction?

But, just to indulge me, mull this "paradox" over in your mind for a while. It may eventually dawn on you that an *absolutely complete* description of how to conduct the sampling procedure was never specified. It may very well be, and this is exactly what Laplace's criterion of a uniform prior probability over *causes* allows for, that the sampling procedure employed to date for that first 5000 samples is not really, in some sense, a sampling procedure with universal coverage. Perhaps that first sample, although large in number, was pragmatically limited to a definite geographic region over a prescribed time period in a particular order, as well as potentially many other constraining circumstances to the sampling procedure.

SOLVED EXERCISES FOR CHAPTER 44

Harold Jeffreys's experience with feathered animals took place in one country, England, during one time period, the 1930s. It was further limited by his contacts with other like–minded people who observed feathered animals as well as the books that he read about the biology of feathered animals, and so forth.

Even though we currently believe that there could be no feathered animals without beaks here on the plant Earth at the present time, that doesn't rule out that earlier in the evolutionary history of feathered animals there might have been some without beaks. Or, perhaps in some remote place on Earth where sampling has not yet taken place, or more likely on another planet with feathered animal life forms, there might be some without beaks.

Unlike Jeffreys's implicit sampling assumptions restricted to, let's say, the planet Earth at a specific geographical location and specific time and no prescription as to exactly how the sampling was to proceed, Laplace did not, in his uniform prior over models, place any such restrictions on how a population might be sampled.

The essential point is simply that any sampling procedure is never specified in enough detail to exhaustively rule out an exception. And that is what the posterior predictive probability formula is trying to tell us when it uses the uniform prior probability over model space.

In addition, and to correct the almost universally accepted error about Laplace, he most certainly realized that "total ignorance on the part of the IP for the causes of events" for which the uniform prior *was* appropriate, did not apply to physically understood situations like chemistry, or the sun's appearance each morning. These kind of situations were the ones that first aroused Jeffreys's ire. Although perhaps Laplace might have chosen a better example in this latter case in order to avoid the undeserved ridicule to which he has since been subjected.

Laplace's result, encapsulated in his *Rule of Succession*, and modernized in the posterior predictive probability formula, makes allowances for every bizarre contingency that has not been specified in the sampling procedure. So perhaps the outcome from his formula such as about a 50% probability for seeing at least one beakless feathered animal when half have been sampled is not as objectionable as Jeffreys would have liked us to believe.

But these remarks have to be tempered if we are considering the case of a *fixed population*. Why would the probability for at least one beakless feathered animal continue to drop as the sample size becomes larger and larger given my argument? It should remain appreciable no matter how large the initial sample size if we are to make allowances for the supposed bizarre contingencies left unspecified in the sampling procedure.

The resolution is easy enough. For the fixed population size of supposedly, $N + M = 10,000$, the number used for the purposes of this exercise, as the sample size N becomes larger and larger, the number of future observations M must become smaller and smaller, until at the end, the probability for a future observation of a beakless feathered animal must approach 0.

In essence, the problem of the ambiguity over what was the exact sampling procedure, and did it in fact exhibit universal coverage, becomes a moot point. When $N + M$ is the fixed population size, then, by definition, there can be no further mystery surrounding any bizarre circumstances that the sampling procedure has neglected because as N becomes larger and larger, and M becomes smaller and smaller, the total population, however it may have been defined, is getting closer and closer to being fully accounted for. The argument for the potentially possible beakless feathered animal due to a myopic sampling procedure withers away as N approaches $N + M$.

Chapter 45

Variational Bayes

45.1 Introduction

The topic of *variational free energy minimization*, together with its application to inferencing called *variational Bayes*, is broached in this Chapter. This concept is usually introduced as a specialized, perhaps even as an advanced, technique within statistical mechanics. Further elaboration of the core ideas behind variational free energy minimization typically leads one on to another advanced topic in statistical mechanics traveling under the rubric of *mean field theory*.

One focus in this Chapter is to explore whether, in fact, variational free energy minimization can be treated coherently from the perspective of general inferencing. Topics like this crop up all the time in other disciplines, the motivation stemming from the desire for some internal consistency with the path the contingent history of the subject has taken. Given their acknowledged success in these other areas, they are often transported over to inferencing problems as if no further thought or analysis were demanded.

And so, this notion of *variational Bayes* has been applied in an attempt at finding some simpler approximation to the posterior probability distribution for models. But is this even necessary? I have shown that a perfectly acceptable analytical solution exists that takes into account *the probability for every single conceivable model*. Why then should we ever want to invest any effort in techniques that purportedly short–circuit this perfectly fine Bayesian solution?

We have achieved some familiarity with how Information Geometry approaches the general question of approximating a complicated probability distribution by a less complicated model. With this as background knowledge, variational free energy minimization becomes somewhat less daunting. Information Geometry does provide a nice rationale for this esoteric, rather specialized technique, illustrating how it arises quite naturally even when we purposefully limit ourselves to inferences made within the confines of our standard probabilistic framework.

We have been examining the notion of α–projecting some point p living in a dually flat Riemannian manifold down to another point $q^\star$ in a sub–manifold. And, relying upon geometric intuition, we found these projections as minimum distances.

It turns out that the **e**–projection of point p, and not the **m**–projection we have recently been concentrating on, **is exactly the same as variational free energy minimization**. As usual, a point p is taken to represent an impossibly complicated model with a computationally intractable partition function, while point $q^\star$ is some sort of approximation to p that *is* computable.

I am forced to take on a somewhat negative tone for this Chapter. All of the effort put into understanding this technique of variational free energy minimization turns out, in my opinion, to be wasted effort. It is all very confusing and, in the final analysis, one is saddled with a result that is not particularly satisfying.

But there is always something to be gained even if one has gone down the wrong path and has to eventually backtrack. It forces us to closely examine the differences in these two distance measures from Information Geometry, the **m**–projection and the **e**–projection, and why one of them is valuable and the other not so much.

As I mentioned earlier, a rationale for some computational technique is often developed in an environment far removed from any grounding in basic probability theory. Then, because of some assumed success, they are touted as beneficial for inferencing because of their pedigree from this other highly respected discipline.

This negative exercise of trying to explain something that, in the end, I feel must be rejected is personally relevant to me. Why? Because it serves as a wonderful example of where the mathematics is all very correct, but the point of it all is completely misdirected. It is the classic red herring that some feel compelled to foist off on us on a regular basis.

It is why I included the following commentary in my *Apologia* to Volume I.

> Because this is such a contentious issue, let me expand a little in my own defense. The mistakes I object to are rarely computational or mathematical errors. These people are too smart to commit such blunders. That would have been beaten out of them in graduate school.
>
> No, more often than not, the mistakes are of the "conceptual" variety. They bungle some conceptual notion at the outset, and then proceed to build a house of (mathematical) straw, oblivious to the fact that everything they have done starting at the point of their initial confusion is a red herring.

I think that these misguided attempts to find simpler approximations to the posterior probability for models should be bounced off the lessons learned in the previous Chapter. There, simple Bayesian formulas for making the kind of inferences ultimately desired were derived from averaging over *all* models. The question of whether these models represented complicated or uncomplicated models made no difference in the resulting inference.

45.2 What Becomes Computationally Intractable?

Consider a larger state space than what we have typically dealt with in the past. Suppose that such a state space has a dimension of $n = 512$. Why this particular size of the state space will become clear in a moment. All I want to highlight now is that the partition function must be a sum over 512 terms. Moreover, each one of these 512 terms involves an exponential that itself might have a potentially large number of arguments.

Nonetheless, it is a fact that even simple models must have the same number of terms in their normalizing factor as complicated models. State spaces as "large" as $n = 512$ are laughingly minuscule by the standards of statistical mechanics. Inevitably, within statistical mechanics, a grudging realization dawned that some kind of resolution must be found for the computationally intractable nature of the partition function.

As the *entrée* to some basic conceptual notions, consider an assembly of $N = 9$ binary systems. Each system can be in only two states, which, for notational simplicity, are labeled as $+$ or $-$. These are conventionally called *spin systems*. Thus, the entire assembly can be in any one of $n = 2^N = 2^9 = 512$ states.

Write out a probability for any arbitrarily chosen state from the available 512 states in the following standard fashion,

$$P(X = x_i \mid \mathcal{M}_k) = P(-, +, +, -, +, -, +, +, + \mid \mathcal{M}_k) \tag{45.1}$$

Suppose that this state is, say, state number 297.

Each state of the assembly, as in the above state ($X = x_{297}$), has an energy function $E(X = x_{297})$ attached to it, and an average energy is found by averaging over all of the states.

$$\langle E \rangle = \sum_{i=1}^{512} E(X = x_i) \, P(X = x_i \mid \mathcal{M}_k) \tag{45.2}$$

A simplistic energy function $E(\mathbf{x})$ might be something like a sum over the assembly of the nine systems where a $-$ condition for any system contributes $S = -1$ and a $+$ condition for any system contributes a $S = +1$. Then state 297 of the assembly has an energy of $E(X = x_{297}) = \sum_{t=1}^{9} S_t = +3$ attached to it.

Supposing furthermore that this is a type of problem that should be dealt with by statistical mechanics, the probability assigned to any state is dictated by the canonical probability distribution with $\beta = -\frac{1}{k_B T}$,

$$P(X = x_{297} \mid \mathcal{M}_k) = \frac{e^{\beta E(X=x_{297})}}{\sum_{i=1}^{512} e^{\beta E(X=x_i)}} \tag{45.3}$$

$$= \frac{e^{3\beta}}{\underbrace{e^{9\beta} + e^{7\beta} + \cdots + e^{3\beta} + \cdots + e^{-9\beta}}_{512 \text{ terms}}} \tag{45.4}$$

The partition function $Z(\beta)$ is explicitly shown as a sum over 512 terms. If the energy function is more complicated than the illustrative easy one just chosen, then the computation of $Z(\beta)$ obviously becomes that much harder to compute.

Now, let's switch the notation from that traditionally employed by statistical mechanics over to the MEP notation we prefer to use for general inferencing.

$$P(X = x_i \mid \mathcal{M}_k) = \frac{\exp\left[\sum_{j=1}^{m} \lambda_j F_j(X = x_i)\right]}{\sum_{i=1}^{n} \exp\left[\sum_{j=1}^{m} \lambda_j F_j(X = x_i)\right]} \tag{45.5}$$

We will eventually want to compare formulas derived within the context of statistical mechanics with those derived from general inferencing. However, switching notation doesn't offer any respite from our troubles with the partition function. There is still that sum over n exponential terms with each argument of the exponential containing m terms.

For example, in so simple an inferential problem with $n = 8$, say, the kangaroo scenario that included the three variables of beer preference, hand preference, and fur color, the partition function Z_p for a complicated model looked like,

$$Z_p = \underbrace{e^{\lambda_1 + \lambda_2 + \cdots + \lambda_7} + e^{\lambda_1 + \lambda_3 + \lambda_5} + \cdots + e^{\lambda_2} + 1}_{8 \text{ terms}} \tag{45.6}$$

when the constraint functions $F(X = x_i)$ took on values of 0 and 1.

Moving up to $n = 512$ in the general MEP notation, the partition function will follow a similar pattern,

$$Z_p = \underbrace{e^{\lambda_1 + \lambda_2 + \cdots + \lambda_{511}} + e^{\lambda_1 + \lambda_3 + \lambda_5 + \cdots} + \cdots + 1}_{512 \text{ terms}} \tag{45.7}$$

with many more λ_j available. The more complicated the model, the more λ_j will be required in order to include information from higher order interactions. For example, from the 512 terms there could be a total of,

$$\binom{9}{4} = \frac{9!}{5!\,4!} = 126 \tag{45.8}$$

possible quadruple interactions to be accounted for. This potential flexibility to include information about higher order interactions leads to many more λ_j than would otherwise be contemplated in simpler models.

It has been a common practice in statistical mechanics to rely on simplified models that invoke some sort of additivity of energy functions at the level of sub–systems. For example, Schrödinger [35, pg. 18] tells us that,

> First a simple, but useful, remark. We have stated that [the partition function] is 'multiplicative', [...] and all other thermodynamical functions are additive, if the system in question is made up of two or more systems in loose–energy contact, so that its *levels* ε_l are the sums of any level of the first system (α_k), and any level of the second system (β_m) ...

Copying this idea from statistical mechanics represents one obvious way out of this conundrum. The optimistic hope is that invoking independence at the level of the individual sub–systems might work out for the larger system constituted from those sub–systems.

Supposing now that the individual sub–systems are coins, then by adopting the independence assumption, an assembly of $N = 9$ coins (now thought of as just one larger linked system) would be said to have a partition function,

$$Z_q = N \ln\left(1 + e^\lambda\right) \tag{45.9}$$

when the mapping of the states of each binary system goes to 0 and 1. If an alternative mapping were to be adopted of -1 and $+1$ to maintain the analogy with a spin system, then,

$$Z_q = N \ln\left(e^\lambda + e^{-\lambda}\right) \tag{45.10}$$

I no longer use the pedantic qualification in the wording, "an assembly of N binary systems." The former "assembly" is now just a (larger) system composed of smaller sub–systems. The assembly of N coins can itself be called a system composed of a subsystem of individual coins.

If these many sub–systems can be thought of as independent of one another (the "loose–energy contact"), the larger system has a partition function,

$$Z_{\text{system}} = Z_{\text{coin 1}} \times Z_{\text{coin 2}} \times \cdots \times Z_{\text{coin N}} \tag{45.11}$$

This is the reason why what should be a complicated partition function for a physical system at some fundamental level is instead always broken down into a multiplication of partition functions for the simpler component sub–systems. The best known example of this is in physical chemistry where, as Schrödinger states [35, pg. 18]

> ... if the system is one gas molecule whose energy is the sum of its translational, rotational, and vibrational energy, all the thermodynamic functions are made up additively of a translational, rotational, and a vibrational contribution—the mathematical situation being the same as if these three types of energy belonged to three independent systems in juxtaposition.

45.3 Review of α–projections

45.3.1 m–projections

The $\alpha = -1$ projection has been examined in previous Chapters. The **m**–projection of a complicated model p down to the sub–manifold of a less complicated model q is the minimization of the Kullback–Leibler relative entropy,

$$KL(p,q) \equiv D^{(-1)}(p \parallel q) = \sum_{i=1}^{n} p_i \ln\left(\frac{p_i}{q_i}\right) \qquad (45.12)$$

Manipulate this expression into a form resembling the Legendre transformation used in the MEP algorithm,

$$D^{(-1)}(p \parallel q) = \sum_{i=1}^{n} p_i \ln\left(\frac{p_i}{q_i}\right)$$

$$= \sum_{i=1}^{n} p_i \ln p_i - \sum_{i=1}^{n} p_i \ln q_i$$

$$= -H(p) - \sum_{i=1}^{n} p_i \left[\sum_{j=1}^{m} \lambda_j^q F_j(X = x_i) - \ln Z_q \right]$$

$$= -H(p) - \left[\sum_{j=1}^{m} \lambda_j^q \sum_{i=1}^{n} p_i F_j(X = x_i) - \sum_{i=1}^{n} p_i \ln Z_q \right]$$

$$= \ln Z_q - H(p) - \sum_{j=1}^{m} \lambda_j^q \langle F_j \rangle_p \qquad (45.13)$$

The relative entropy becomes zero just for that case where $p \equiv q$. Then, the last expression above becomes,

$$H_{\max}(p) = \ln Z_p - \sum_{j=1}^{m} \lambda_j^p \langle F_j \rangle_p \qquad (45.14)$$

the source of the MEP algorithm for finding the numerical assignments to p.

Since $-H(p)$ will be some negative value, minimizing $D^{(-1)}(p \parallel q)$ requires the minimization of $\ln Z_q - \sum_{j=1}^{m} \lambda_j^q \langle F_j \rangle_p$. By utilizing the information from some number m of constraint function averages from p, $\langle F_j \rangle_p$, the MEP algorithm will find the numerical assignment for some point q through,

$$\text{Maximum entropy of some } q_i = \min_{\lambda_j^q} \left[\ln Z_q - \sum_{j=1}^{m} \lambda_j^q \langle F_j \rangle_p \right] \qquad (45.15)$$

45.3.2 e–projections

We haven't talked much about the $\alpha = 1$ projection. But it is the basis for the variational free energy minimization procedure. The **e**–projection of a complicated model p down to a less complicated model q is the dual of the Kullback–Leibler relative entropy,

$$KL(q,p) \equiv D^{(1)}(p \parallel q) = \sum_{i=1}^{n} q_i \ln\left(\frac{q_i}{p_i}\right) \qquad (45.16)$$

The difference compared to the **m**–projection is in,

$$D^{(1)}(p \parallel q) = \sum_{i=1}^{n} q_i \ln\left(\frac{q_i}{p_i}\right)$$

$$= \sum_{i=1}^{n} q_i \ln q_i - \sum_{i=1}^{n} q_i \ln p_i$$

$$= -H(q) - \sum_{i=1}^{n} q_i \left[\sum_{j=1}^{m} \lambda_j^p F_j(X = x_i) - \ln Z_p\right]$$

$$= \ln Z_p - H(q) - \sum_{j=1}^{m} \lambda_j^p \langle F_j \rangle_q$$

$$D^{(1)}(p \parallel q) - \ln Z_p = -\sum_{j=1}^{m} \lambda_j^p \langle F_j \rangle_q - H(q) \qquad (45.17)$$

Just as above, $-H(q)$ will be some negative value. Minimizing $D^{(1)}(p \parallel q)$ requires the minimization of $\ln Z_p - \sum_{j=1}^{m} \lambda_j^p \langle F_j \rangle_q$. By utilizing the information from some number m of constraint function averages from q, $\langle F_j \rangle_q$, the MEP algorithm will find the numerical assignment for some point p through,

$$\text{Maximum entropy of some } p_i = \min_{\lambda_j^p} \left[\ln Z_p - \sum_{j=1}^{m} \lambda_j^p \langle F_j \rangle_q\right]$$

But the *raison d'être* for entering into this enterprise in the first place was the supposed intractable nature of Z_p, namely, the partition function of the complicated distribution. Thus, we can not find the e–projection with this expression if we can not compute Z_p.

45.4 Defining Free Energy

Within thermodynamics, there is a concept called the *Helmholtz free energy*. This concept was introduced, without naming it as such, when I got around to examining Schrödinger's thermodynamic equations in Chapter Twenty Seven of Volume II.

The attempt was made there to couch expressions that Schrödinger had written in the traditional language of statistical thermodynamics back into the kind of MEP notation we had been developing.

So, for example, Schrödinger wrote down his Equation (2.20) to eventually define the free energy as,

$$k \log Z \equiv \Psi = S - \frac{U}{T}$$

where the concepts of free energy, the partition function, entropy, average energy, and temperature all make an appearance. It's easy enough to see that this may manipulated into,

$$-kT \log Z = U - S$$

where the Helmholtz free energy appears on the left hand side of the equation. So, in words, apparently we are looking to minimize an average energy minus the entropy.

The MEP version of the free energy equation comes directly from the Legendre transformation for point q,

$$H(q) = \ln Z_q - \sum_{j=1}^{m} \lambda_j^q \langle F_j \rangle_q \qquad (45.18)$$

$$-\ln Z_q = -\sum_{j=1}^{m} \lambda_j^q \langle F_j \rangle_q - H(q) \qquad (45.19)$$

$$-kT \log Z = U - S \qquad (45.20)$$

This is the MEP version of a "free energy" defined as a constraint function average minus the information entropy. One must always pay attention to whether a "+" or "−" sign is required in front of the term $\sum_{j=1}^{m} \lambda_j^q \langle F_j \rangle_q$. Given the way we have defined the parameters in the MEP notation, we always have a "−" sign. Also, we have labeled this distribution as a point q.

But we pose the question: In variational Bayes, is it not the expression arrived at in Equation (45.17)

$$D^{(1)}(p \parallel q) - \ln Z_p = -\sum_{j=1}^{m} \lambda_j^p \langle F_j \rangle_q - H(q)$$

that is the true free energy we seek to minimize? With this labeling, we seem to be focusing on the "approximation" to the distribution of point p. However, due to the ubiquitous difficulty in computing anything associated with the complicated distribution p, we shall have to define an approximation to the "true free energy" $-\ln Z_p$ by the free energy of the approximating distribution $-\ln Z_q$.

Begin to expand this initial expression with an eye towards turning it into something that might look like a relative entropy. Since the negative of the information entropy is for point q,

$$-H(q_i) = \sum_{i=1}^{n} q_i \ln q_i \tag{45.21}$$

substitute this into the Legendre transformation for q,

$$-\ln Z_q = \sum_{i=1}^{n} q_i \ln q_i - \sum_{j=1}^{m} \lambda_j^q \langle F_j \rangle_q \tag{45.22}$$

$$= \sum_{i=1}^{n} q_i \ln q_i - \sum_{i=1}^{n} q_i \left\{ \ln \exp\left[\sum_{j=1}^{m} \lambda_j^q F_j(X = x_i) \right] \right\} \tag{45.23}$$

$$= \sum_{i=1}^{n} q_i \ln \left\{ \frac{q_i}{\exp\left[\sum_{j=1}^{m} \lambda_j^q F_j(X = x_i) \right]} \right\} \tag{45.24}$$

Now, if we are willing to consider an argument that the denominator could be turned into p with these manipulations,

$$-\ln Z_q = \sum_{i=1}^{n} q_i \ln \left\{ \frac{q_i}{\exp\left[\sum_{j=1}^{m} \lambda_j^q F_j(X = x_i) \right]} \right\} \tag{45.25}$$

$$= \sum_{i=1}^{n} q_i \ln \left\{ \frac{q_i}{\exp\left[\sum_{j=1}^{m} \lambda_j^p F_j(X = x_i) - \ln Z_p + \ln Z_p \right]} \right\} \tag{45.26}$$

$$= \sum_{i=1}^{n} q_i \ln \left(\frac{q_i}{p_i} \right) - \ln Z_p \tag{45.27}$$

$$-\ln Z_q = KL(q, p) - \ln Z_p \tag{45.28}$$

$$KL(q, p) = \ln Z_p - \ln Z_q \tag{45.29}$$

we see the appearance of the divergence function that figured prominently in the e–projection,

$$D^{(1)}(p \parallel q) \equiv KL(q, p)$$

Once again, given the obvious geometric notion that if q and p are the same point, then $KL(q, p) = 0$. Also, $\ln Z_p = \ln Z_q$. Otherwise, the dual of the relative entropy will be some positive number, and the free energy of the approximating distribution q will be less than the "true free energy" of the intractable distribution p.

45.5 Connections to the Literature

My development of variational methods as portrayed in this Chapter was inspired by the treatment rendered by MacKay in his book [29, Chapter 33]. At the outset, let me say that both of our presentations spring from the same general probabilistic motivation. For example, reliance on the relative entropy as a quantitative measure of the similarity of two probability distributions is critical. However, where we differ is my tendency to emphasize an overall consistency with the MEP notation and concepts.

MacKay labels his approximating distribution as $Q(\mathbf{x}; \boldsymbol{\theta})$ where $\boldsymbol{\theta}$ stands for all of the adjustable parameters. I would write my equivalent approximating distribution as the point $q^\star$ in a Riemannian manifold,

$$q^\star \equiv P(X = x_i \mid \mathcal{M}_k)$$

where, of course, any of the model parameters λ_j or $\langle F_j \rangle$ are adjustable under the information in different models. Because $q^\star$ is supposed to be a much simpler approximation to p, the number of model parameters λ_j for $q^\star$ will be far fewer than for p.

The complicated distribution that $Q(\mathbf{x}; \boldsymbol{\theta})$ is trying to approximate is a generic canonical distribution from statistical mechanics written as,

$$P(\mathbf{x} \mid \beta, \mathbf{J}) = \frac{1}{Z(\beta, \mathbf{J})} \exp\left[-\beta E(\mathbf{x}; \mathbf{J})\right]$$

My equivalent complicated distribution is p.

However, I choose not to place any restrictions on p such that it must follow some energy function as defined by statistical mechanics. These energy functions can usually be interpreted as some number of double interactions and main effects, but my complicated model p may be very much more complicated because it is allowed to include all possible types of higher order interactions.

$$p \equiv P(X = x_i \mid \mathcal{M}_k) = \frac{\exp\left[\sum_{j=1}^{m} \lambda_j F_j(X = x_i)\right]}{\sum_{i=1}^{n} \exp\left[\sum_{j=1}^{m} \lambda_j F_j(X = x_i)\right]}$$

where both n and m are large numbers.

MacKay commences with a definition for a *variational free energy* $\beta \tilde{F}(\boldsymbol{\theta})$,

> The objective function chosen to measure the quality of the approximation is the *variational free energy*
>
> $$\beta \tilde{F}(\boldsymbol{\theta}) = \sum_{\mathbf{x}} Q(\mathbf{x}; \boldsymbol{\theta}) \ln \frac{Q(\mathbf{x}; \boldsymbol{\theta})}{\exp\left[-\beta E(\mathbf{x}; \mathbf{J})\right]} \qquad (33.5)$$

We recognize here in MacKay's Equation (33.5) something that is not quite the relative entropy, but pretty close. All that is missing is the partition function that should accompany $\exp[-\beta E(\mathbf{x};\mathbf{J})]$. He rectifies this a short time later with this explanation,

Second, we can use the definition of $P(\mathbf{x}\,|\,\beta,\mathbf{J})$ to write:

$$\beta\tilde{F}(\boldsymbol{\theta}) = \sum_\mathbf{x} Q(\mathbf{x};\boldsymbol{\theta})\ln\frac{Q(\mathbf{x};\boldsymbol{\theta})}{P(\mathbf{x}\,|\,\beta,\mathbf{J})} - \ln Z(\beta,\mathbf{J}) \qquad (33.8)$$

$$= D_{KL}(Q \parallel P) + \beta F \qquad (33.9)$$

where F is the true free energy defined by

$$\beta F \equiv -\ln Z(\beta,\mathbf{J}) \qquad (33.10)$$

and $D_{KL}(Q \parallel P)$ is the relative entropy between the approximating distribution $Q(\mathbf{x};\boldsymbol{\theta})$ and the true distribution $P(\mathbf{x}\,|\,\beta,\mathbf{J})$. Thus by Gibbs' inequality, the variational free energy $\tilde{F}(\boldsymbol{\theta})$ is bounded below by F and only attains this value for $Q(\mathbf{x};\boldsymbol{\theta}) = P(\mathbf{x}\,|\,\beta,\mathbf{J})$.

Our strategy is thus to vary $\boldsymbol{\theta}$ in such a way that $\beta\tilde{F}(\boldsymbol{\theta})$ is minimized. The approximating distribution then gives a simplified approximation to the true distribution that may be useful, and the value of $\beta\tilde{F}(\boldsymbol{\theta})$ will be an upper bound for βF. Equivalently, $\tilde{Z} \equiv e^{-\beta\tilde{F}(\boldsymbol{\theta})}$ is a lower bound for Z.

MacKay's Equation (33.8) is the same as my,

$$-\ln Z_q = KL(q,p) - \ln Z_p$$

as derived in the last section. I altered the flow of MacKay's presentation because of the desire to immediately identify where the relative entropy occurred in his expressions. After presenting his Equation (33.5), MacKay manipulates it into the following two forms, both of which are quite familiar to us from our exposure to the MEP.

$$\beta\tilde{F}(\boldsymbol{\theta}) = \sum_\mathbf{x} Q(\mathbf{x};\boldsymbol{\theta})E(\mathbf{x};\mathbf{J}) - \sum_\mathbf{x} Q(\mathbf{x};\boldsymbol{\theta})\ln\frac{1}{Q(\mathbf{x};\boldsymbol{\theta})} \qquad (33.6)$$

$$\equiv \beta\langle E(\mathbf{x};\mathbf{J})\rangle_Q - S_Q, \qquad (33.7)$$

where $\langle E(\mathbf{x};\mathbf{J})\rangle_Q$ is the average of the energy function under the distribution $Q(\mathbf{x};\boldsymbol{\theta})$, and S_Q is the entropy of the distribution $Q(\mathbf{x};\boldsymbol{\theta})$ ...

Adhering strictly to the notational formalism as it has evolved in our study of the MEP, I developed the following equivalent expression in section 45.3.2.

$$D^{(1)}(p \parallel q) - \ln Z_p = -\sum_{j=1}^m \lambda_j^p \langle F_j \rangle_q - H(q)$$

What sub-manifold $\boldsymbol{E}_k$ does this q inhabit? For really simplified approximations to complicated p, q lives in the space of independent distributions.

45.6 Solved Exercises for Chapter Forty Five

Exercise 45.6.1: Demonstrate the equivalency with the way that MacKay expresses the information entropy and my way.

Solution to Exercise 45.6.1

MacKay writes,

$$S_Q \equiv \sum_{\mathbf{x}} Q(\mathbf{x}) \ln \frac{1}{Q(\mathbf{x})}$$

for the information entropy, while I always express it in this form,

$$H(Q_i) \equiv -\sum_{i=1}^{n} Q_i \ln Q_i$$

They are equivalent expressions because of the easy series of manipulations,

$$\sum_{\mathbf{x}} Q(\mathbf{x}) \ln \frac{1}{Q(\mathbf{x})} = \sum_{\mathbf{x}} Q(\mathbf{x})[\ln 1 - \ln Q(\mathbf{x})]$$

$$= \sum_{\mathbf{x}} Q(\mathbf{x})[0 - \ln Q(\mathbf{x})]$$

$$= \sum_{\mathbf{x}} Q(\mathbf{x})[-\ln Q(\mathbf{x})]$$

$$= -\sum_{\mathbf{x}} Q(\mathbf{x}) \ln Q(\mathbf{x})$$

$$= -\sum_{i=1}^{n} Q_i \ln Q_i$$

In the final line, where MacKay simply uses $\mathbf{x}$ for any probability in the discrete state space, I prefer to explicitly subscript the probability with the index i where i runs from 1 to n, the dimension of the state space.

But far more important than this equivalency in notation, is the identification of exactly which state space is involved in the computation of the missing information. This is, in fact, one of the central issues in this Chapter. Is the IP trying to maximize the missing information at the level of the individual "system," or at the level of the larger "assembly" of many systems? Is the dimension of the state space n defined at the lowest level of the system, or at the higher level of the assembly of N systems? The larger state spaces considered here have a dimension of, say, $n = 2^N$, when they are thought of as constructed from lower level binary systems.

SOLVED EXERCISES FOR CHAPTER 45

Exercise 45.6.2: Reconstruct in a very general way how Schrödinger solved the problem of computing the partition function for extremely large state spaces.

Solution to Exercise 45.6.2

The following exercise is a sketch of how Schrödinger solved the "computationally intractable" nature of the partition function in statistical mechanics [35, pg. 19]. There are actually two places in this statistical mechanics scenario of the "ideal monatomic gas" where large numbers have to be finessed.

The first opportunity to face large numbers crops up in finding the partition function for just *one* atom in the gas. In the classical, pre–quantum era, state spaces for thermodynamics were constructed based on the concept of a "phase space." The definition began with the x, y, z coordinates of the atom located within the ordinary three dimensional space we all inhabit. The phase space was completed with the addition of three momentum variables, p_x, p_y, and p_z available to the atom at each space coordinate.

The canonical Boltzmann distribution, expressed in our standard MEP notation is the very familiar looking,

$$P(X = x_i \mid \mathcal{M}_k) = \frac{e^{\lambda F(X=x_i)}}{Z(\lambda)}$$

$$= \frac{e^{\lambda F(X=x_i)}}{\sum_{i=1}^{n} e^{\lambda F(X=x_i)}}$$

where n is an enormously large number since the atom can be anywhere in space and have any momentum.

Imagining our single atom inside a box, one instantiation of the phase space might be the statement, say, $(X = x_{293,478,064,...})$, that, "The atom is located 10 cm from the left edge along the x–axis, 15 cm from the front edge along the y–axis, and 5 cm towards the top edge along the z–axis. Additionally, the velocity of the atom at this particular location inside the box along each of the three dimensions is such and such ..."

The symbolic expression $P(X = x_i \mid \mathcal{M}_k)$ then reflects the IP's degree of belief in this statement when given some temperature and the acceptance of the physical ontological rationale behind an ideal monatomic gas, in other words, the information behind the model $\mathcal{M}_k$.

Translating the above generic MEP notation into a form more appropriate for statistical mechanics,

$$P(\text{state of atom} \mid \mathcal{M}_k) = \frac{\exp\left[-\frac{1}{k_B T} E(\boldsymbol{x})\right]}{Z(T)}$$

where the Lagrange multiplier λ has been replaced with Boltzmann's constant and the temperature T, and the constraint function $F(X = x_i)$ with the defined energy function $E(x)$. Classically, the energy function was defined as,

$$E(x) = \frac{p_x^2 + p_y^2 + p_z^2}{2m}$$

where m is the mass of the atom.

$Z(T)$ would initially be thought of as a sum over a discrete partitioning of the phase space, the *cells* of the phase space. But the analytical tactic is chosen to convert the sum into an integral in order to calculate the denominator of the canonical distribution.

$$Z(T) = \int_x \cdots \int_{p_z} \exp\left[-\frac{1}{k_B T} \frac{p_x^2 + p_y^2 + p_z^2}{2m}\right] dx\, dy\, dz\, dp_x\, dp_y\, dp_z$$

It is at this juncture, that although conceptually things have not gotten out of hand, many of us, myself included, would be left scratching their heads as to what to do next. Herr Doktor Professor Schrödinger's Nobel laureate facility with math comes to our rescue.

He recommends, first of all, to simplify from an integration over six dimensions to one over three dimensions by considering the triple integration over the x, y, and z coordinates to be the volume V. We are left with a partition function looking like,

$$Z(T) = V \int_{p_x} \int_{p_y} \int_{p_z} \exp\left[-\frac{1}{k_B T} \frac{p_x^2 + p_y^2 + p_z^2}{2m}\right] dp_x\, dp_y\, dp_z$$

Next, he tells us that there is "an obvious transformation of variables" for the integrand that takes the partition function to,

$$Z(T) = V \left(\frac{2m}{1/k_B T}\right)^{3/2} \int_{-\infty}^{+\infty} \int_{-\infty}^{+\infty} \int_{-\infty}^{+\infty} \exp\left[\xi^2 + \eta^2 + \zeta^2\right] d\xi\, d\eta\, d\zeta$$

We must recall that earlier Schrödinger has instructed us that in statistical mechanics everything of importance boils down to calculating his Ψ function where $\Psi = k_B \log Z$. We are quite happy to discover that this makes the next manipulation easy,

$$\Psi = k_B \ln V + \frac{3k}{2} \ln T + \text{constant}$$

because the integral has been judged to be a constant.

All of this effort has resulted in finding the partition function that determines the probability for the state of just *one* atom constituting the ideal monatomic gas. We now face a second imbroglio involving extremely large numbers. But here is where Schrödinger's rationale for invoking independence, or additivity, becomes critical.

SOLVED EXERCISES FOR CHAPTER 45

The atom is one system. Schrödinger is concerned with an assembly of N such systems in a statistical mechanics treatment of an ideal gas. If the assembly is in "a loose energy contact" with all of its constituent systems, then Schrödinger says that we can invoke independence with the happy consequence for calculating the partition function of the overall assembly that,

$$Z_1 \times Z_2 \times \cdots \times Z_N \to \sum_{t=1}^{N} \ln Z_t = N \ln Z$$

The Ψ function for the assembly (the entire volume of gas) becomes,

$$\Psi = N k_B \ln V + \frac{3}{2} N k_B \ln T + N \times \text{constant}$$

As discussed in Volume II [8, Chapter 27], the important MEP feature involving the average constraint function, the partition function, and the model parameter,

$$\langle F \rangle = \frac{\partial \ln Z}{\partial \lambda}$$

becomes for Schrödinger the average energy,

$$U = T^2 \frac{\partial \Psi}{\partial T}$$

Given the above development for Ψ relating to an assembly in loose energy contact, together with the assumption of independence, we arrive at Schrödinger's final result for the average energy of the ideal monatomic gas,

$$U = \frac{3}{2} N k_B T$$

As a conceptual check, Boltzmann's constant k_B has T in its denominator, and ergs (or joules) in its numerator, so that $k_B T$ cancels out to an energy as U must be.

This exercise was an interesting overview of how Schrödinger in particular, and statistical mechanics in general, tries to finesse the computationally intractable nature of the combinatorial explosion when n becomes very large. Carefully note that we started out with an n, though itself very large in the beginning for the single system, that had to grow enormously larger for the assembly of N such systems.

Exercise 45.6.3: Spell out in detail the consequences of independence when considering simplified models for an assembly composed of $N = 9$ coins.

Solution to Exercise 45.6.3

We start out with a binary system of $n = 2$. Then we graduate to an assembly of $N = 9$ such binary systems so that the final state space has dimension $n = 512$. A spin system example was presented in section 45.2, and the abstract probability for one possibility from the total of 512 states available to the assembly was,

$$P(X = x_i \mid \mathcal{M}_k) = P(-, +, +, -, +, -, +, +, + \mid \mathcal{M}_k)$$

A completely analogous assembly would be an assembly of nine coins where the IP would be interested in the probability of the disposition of the coins as,

$$P(X = x_i \mid \mathcal{M}_k) = P(\text{T, H, H, T, H, T, H, H, H} \mid \mathcal{M}_k)$$

Note that this statement is NOT about the same coin being sequentially tossed nine times in a row. Rather, it describes a layout, so to speak, of nine possibly different coins that may have what is face up determined by the faces showing on the other coins in the layout.

The whole thrust of the variational free energy minimization scheme is that since it is an intolerable burden to calculate Z_p for any complicated model $\mathcal{M}_k$, some simpler approximating model will have to be found so that a Z_q can be computed.

For example, the most complicated model behind point p would have a partition function Z_p with the following pattern,

$$Z_p = \underbrace{e^{\lambda_1 + \lambda_2 + \cdots + \lambda_{511}} + e^{\lambda_2 + \lambda_3 + \lambda_4 + \cdots} + \cdots + 1}_{512 \text{ terms}}$$

where 511 model parameters λ_1 through λ_{511} would have to be found.

Clearly, having to cope with an alternative log partition function $\ln Z_q$,

$$\ln Z_q = N \ln\left(e^{\lambda} + e^{2\lambda}\right)$$

with only one model parameter λ would be an astounding simplification. The constraint functions $F(X = x_1) = 1$ and $F(X = x_2) = 2$ are used here.

From the fundamental principles of probability theory, the probability for some assembly when the results of a trial are independent of all previous trials, or, more accurately, when the probability for the face of each coin in the layout is independent of the disposition of every other coin, is given by the familiar series of manipulations as,

$$P(X = x_i \mid \mathcal{M}_k) = P(Y_9 = \text{H} \mid Y_8, Y_7, \cdots, \mathcal{M}_k) \times P(Y_8 = \text{H} \mid Y_7, Y_6, \cdots, \mathcal{M}_k) \times \cdots$$

$$= P(Y_9 = \text{H} \mid \mathcal{M}_k) \times P(Y_8 = \text{H} \mid \mathcal{M}_k) \times \cdots \times P(Y_1 = \text{T} \mid \mathcal{M}_k)$$

$$= \underbrace{Q_1 \times Q_1 \times \cdots \times Q_2}_{9 \text{ terms}}$$

$$= \underbrace{\frac{e^\lambda}{e^\lambda + e^{2\lambda}} \times \frac{e^\lambda}{e^\lambda + e^{2\lambda}} \times \cdots \times \frac{e^{2\lambda}}{e^\lambda + e^{2\lambda}}}_{9 \text{ terms}}$$

$$Z_q = (e^\lambda + e^{2\lambda}) \times (e^\lambda + e^{2\lambda}) \times \cdots \times (e^\lambda + e^{2\lambda})$$

$$= (e^\lambda + e^{2\lambda})^9$$

$$\ln Z_q = 9 \ln(e^\lambda + e^{2\lambda})$$

where $Y_t = y_l$ stands for an outcome at the level of an individual coin (the system) at either the tth trial or t^{th} position in the layout.

The energy $E(\mathbf{x})$ of any state in the large system (the assembly) must be in the range from 9 to 18 if the energy had been defined as $E(\mathbf{x}) = \sum_{t=1}^{N} F(X = x_i)$. Suppose that the information consists of an average energy $U = 15$ for the assembly of nine coins, or $\langle F \rangle = 1\frac{2}{3}$ at the system level of the individual coin. This is where we can now apply Schrödinger's lesson about additivity from the previous exercise.

Since the constraint function average, or average energy U, for the nine member system of coins under independence (or the "loose energy contact" where the face of one coin in one position doesn't impact the face of another coin in another position) is,

$$\langle F \rangle = \frac{\partial N \ln Z_q}{\partial \lambda}$$

$$= \frac{\partial \left[9 \ln(e^\lambda + e^{2\lambda}) \right]}{\partial \lambda}$$

$$= 9 \left(\frac{e^\lambda + 2e^{2\lambda}}{e^\lambda + e^{2\lambda}} \right)$$

$$= 18 - \frac{9}{1 + e^\lambda}$$

The interesting last line comes from *Mathematica* evaluating,

```
D[9 Log[Exp[λ] + Exp[2 λ]], λ] // Apart
```

The *Mathematica* built-in function `Apart[ ]` splits the original fraction arising from the partial differentiation in the next to last line into two terms. Sometimes this provides some additional insight as it does in this case. In this format for the average energy, it is clear that when the model parameter $\lambda \to \infty$, the average energy approaches 18, and when $\lambda \to -\infty$, the average energy approaches 9.

The value of the model parameter at the level of the individual system, namely at the level of one coin, is $\lambda = 0.693147$. This value is found with,

`Solve[D[9 Log[Exp[`λ`] + Exp[2 `λ`]], `λ`] == 15, `λ`, Reals] // N`

This is easy enough to check because the expression below returns 15,

`D[9 Log[Exp[`λ`] + Exp[2 `λ`]], `λ`] /. `$\lambda \to$` 0.693147`

Now let's find that the partition function at the level of the individual system, Z_1 through Z_9, is equal to 6.

`Exp[`λ`] + Exp[2 `λ`] /. `$\lambda \to$` 0.693147`

This makes the probability for HEADS equal to 1/3 and probability for TAILS equal to 2/3 at the level of the binary system.

$$P(\text{HEADS} \mid \mathcal{M}_k) = \frac{e^\lambda}{e^\lambda + e^{2\lambda}}$$

$$= \frac{e^{0.693147}}{e^{0.693147} + e^{2(0.693147)}}$$

$$= \frac{2}{2+4}$$

$$= \frac{1}{3}$$

To find the probability for the particular assembly that started off this exercise becomes an easy calculation under independence,

$$P(X = x_i \mid \mathcal{M}_k) = 1/3 \times 1/3 \times \cdots \times 2/3$$

$$= \frac{1 \times 1 \times \cdots \times 2}{3^9}$$

$$= \frac{8}{19683}$$

SOLVED EXERCISES FOR CHAPTER 45

From this we see that Z_q, the partition function under the approximation of a simpler model q, has the value of,

$$Z_q = Z_1 \times Z_2 \times \cdots \times Z_9 = 3^9 = 19683$$

Since we carried through with the division by $e^\lambda = 2$ everywhere, $\ln Z_q$ changes from,

$$\ln Z_q = N \ln(e^\lambda + e^{2\lambda}) = 9 \ln 6 \to Z_q = 6^9$$

into,

$$\ln Z_q = N \ln \left(\frac{e^\lambda + e^{2\lambda}}{e^\lambda} \right) = N \ln (1 + e^\lambda) = 9 \ln 3 \to Z_q = 3^9$$

With these results in hand, the probability for any selected coin assembly can be readily calculated.

$$P(\text{All HEADS} \,|\, \mathcal{M}_k) = \frac{9!}{9!\,0!}(1/3)^9 \,(2/3)^0 = \frac{1}{19683}$$

$$P(\text{All TAILS} \,|\, \mathcal{M}_k) = \frac{9!}{0!\,9!}(1/3)^0 \,(2/3)^9 = \frac{512}{19683}$$

$$P(\text{6 HEADS, 3 TAILS} \,|\, \mathcal{M}_k) = \frac{9!}{6!\,3!}(1/3)^6 \,(2/3)^3 = \frac{672}{19683}$$

To check that there are 512 possibilities for the assembly of nine coins, and that the denominator in the probabilities for these possibilities is equal to 19683, evaluate,

```
Total[Table[Multinomial[9 - j, j] Power[2, j],{j, 0, 9}]]
```

To check that the average energy over all of the above probabilities does equal 15, evaluate (notice the need for parentheses),

```
Total[Table[ (Multinomial[9 - j, j])
        (Power[2, j]) (j + 9), {j, 0, 9}]] / 19683
```

Exercise 45.6.4: After digesting the results from the last exercise, what would be a plausible direction in which to generalize?

Solution to Exercise 45.6.4

We might retain the notion of independence at the level of the individual coin system, but allow each coin to have its own distinct probability for HEADS or TAILS. Thus, each coin has its own private, and possibly different, λ_j.

For example, Coin 1 might have a probability for HEADS=0.25 and TAILS=0.75 with $\lambda_1 = 1.098612$, but Coin 2 is allowed to set $\lambda_2 = 0$ with a probability for HEADS=0.50 and TAILS=0.50, and so on, for the remaining seven coins. We no longer have quite as simple a model as before where only one λ had to be found, but still one where computing the overall probability for the assembly is doable.

The probability for the example assembly under this newer approximating model would look like,

$$P(X = x_i \,|\, \mathcal{M}_k) = \underbrace{\frac{e^{\lambda_1}}{e^{\lambda_1} + e^{2\lambda_1}} \times \frac{e^{\lambda_2}}{e^{\lambda_2} + e^{2\lambda_2}} \times \cdots \times \frac{e^{2\lambda_9}}{e^{\lambda_9} + e^{2\lambda_9}}}_{9 \text{ terms}}$$

$$= \underbrace{\frac{1}{1 + e^{\lambda_1}} \times \frac{1}{1 + e^{\lambda_2}} \times \cdots \times \frac{e^{\lambda_9}}{1 + e^{\lambda_9}}}_{9 \text{ terms}}$$

For the sake of the numerical example, let's first just arbitrarily assign possibly different probabilities for HEADS to each of the nine coins as shown in the list below.

$$P(\text{HEADS Coin 9}) = 1/3$$

$$P(\text{HEADS Coin 8}) = 1/2$$

$$P(\text{HEADS Coin 7}) = 3/4$$

$$P(\text{HEADS Coin 6}) = 1/10$$

$$P(\text{HEADS Coin 5}) = 1/2$$

$$P(\text{HEADS Coin 4}) = 2/3$$

$$P(\text{HEADS Coin 3}) = 4/5$$

$$P(\text{HEADS Coin 2}) = 1/6$$

$$P(\text{HEADS Coin 1}) = 1/2$$

As was to be expected, the probability for the particular assembly of six HEADS and three TAILS of the current example must change when compared to the previous approximation.

$$P(X = x_i \,|\, \mathcal{M}_k) = P(\text{HEADS Coin 9}) \times P(\text{HEADS Coin 8}) \times \cdots \times P(\text{TAILS Coin 1})$$

$$= 1/3 \times 1/2 \times 3/4 \times 9/10 \times 1/2 \times 1/3 \times 4/5 \times 1/6 \times 1/2$$

$$= 0.00125$$

Let's see what this looks like in the MEP notation for the probabilities,

$$P(X = x_i \mid \mathcal{M}_k) = \frac{1}{1+e^{\lambda_1}} \times \frac{1}{1+e^{\lambda_2}} \times \cdots \times \frac{e^{\lambda_9}}{1+e^{\lambda_9}}$$

$$= \frac{1}{1+e^{0.6931472}} \times \frac{1}{1+e^0} \times \cdots \times \frac{e^0}{1+e^0}$$

$$= 1/3 \times 1/2 \times 3/4 \times 9/10 \times 1/2 \times 1/3 \times 4/5 \times 1/6 \times 1/2$$

$$= 0.00125$$

Notice that we are still taking advantage of the independence assumption that, for example, allows the IP to assign the probability of HEADS to Coin 9 independent of whether HEADS or TAILS are showing on any of the other eight coins. But we did have to abandon the multiplicity factor since the coins might have different probabilities depending on their position in the layout.

Exercise 45.6.5: What conceptual niceties have we glibly passed over?

Solution to Exercise 45.6.5

We could continue to vary the nine λ_j parameters to our heart's content, just as variational free energy minimization instructs us to do, and calculate different probabilities for all 512 possible assemblies. But what is the criterion for judging one of these approximations as superior to any other? Apparently, from MacKay's expression,

$$\beta \tilde{F}(\boldsymbol{\theta}) = \beta \langle E(\mathbf{x}; \mathbf{J}) \rangle_Q - S_Q$$

we are to calculate some average energy function with respect to our approximating model. However, that energy function is still a pre-defined $E(\mathbf{x}; \mathbf{J})$ appropriate to some complicated model p, and thus does not change when simpler approximating models are being considered.

We take home the same general lesson by reverting to the variational free energy minimization analog as it appears in our preferred inferential notation,

$$D^{(1)}(p \parallel q) - \ln Z_p = -\sum_{j=1}^{m} \lambda_j^p \langle F_j \rangle_q - H(q)$$

The sum on the right hand side involves some number of m parameters λ_j^p together with m constraint functions $F_j(X = x_i)$ from the complicated model p. It is only specified that the approximating distribution q is to be used to form the average of all of these constraint functions.

Thus, the defined energy function $E(\mathbf{x}; \mathbf{J})$ from statistical mechanics, as well as some of the $F_j(X = x_i)$ in the general inferencing scheme, are more complicated than the example presented in the previous exercise. The m model parameters λ_j^p come from the unknown complicated model, and m must also refer to a larger number of constraint functions.

We see that, although we are indeed taking an average with respect to a simpler approximating distribution $Q(\mathbf{x}; \boldsymbol{\theta})$, the energy function does not become simpler, and might involve interactions between neighboring spins. Alternatively, although we are taking an average with respect to a simpler approximating distribution q, the constraint functions do not become simpler, and might involve higher order interactions.

There is also another unresolved mystery here. As far as statistical mechanics is concerned, the Boltzmann canonical probability distribution only ever has the one model parameter β and the one energy function $E(\mathbf{x}; \mathbf{J})$. In general inferencing, we have m model parameters λ_j^p and m constraint functions $F_j(X = x_i)$.

In the numerator of the MEP formula, the argument to the exponential function is the sum $\sum_{j=1}^m \lambda_j^p F_j(X = x_i)$ which seems to mimic any complicated energy function. If that is the case, then what is the analogous role of β in general inferencing? An argument can be made that the temperature in statistical mechanics plays the role of an additional and independent annealing parameter in general inferencing.

Exercise 45.6.6: Compare the m–projection and the e–projection in the simple kangaroo scenario.

Solution to Exercise 45.6.6

For this simple inferential scenario, $n = 4$ and the complicated models p have three Lagrange multipliers because they include information about one double interaction. Our numerical examples have taken a particular p to be the maximum likelihood model with,

$$p_i = (0.70, 0.05, 0.05, 0.20)$$

When this point p is **m**–projected to that specific sub–manifold containing only independent distributions q, the shortest distance is from p to a $q^\star$. This $q^\star$ has the same information, in the form of constraint function averages, as the first two averages $\langle F_1 \rangle_p = 3/4$ and $\langle F_2 \rangle_p = 3/4$, from p. Thus,

$$q_i^\star = (0.5625, 0.1825, 0.1825, 0.0625)$$

For the e–projection, and thus also for **variational Bayes**, we have to once again find a minimum distance, but now that squared distance is defined as the dual to the Kullback–Leibler relative entropy, namely $D^{(1)}(p \parallel q)$,

$$D^{(1)}(p \parallel q) = \sum_{i=1}^{4} q_i \ln\left(\frac{q_i}{p_i}\right)$$

$$= \sum_{i=1}^{4} q_i \ln q_i - \sum_{i=1}^{4} q_i \ln p_i$$

$$= -H(q) - \sum_{i=1}^{4} q_i \left[\sum_{j=1}^{3} \lambda_j^p F_j(X = x_i) - \ln Z_p\right]$$

$$D^{(1)}(p \parallel q) - \ln Z_p = -H(q) - \sum_{j=1}^{3} \lambda_j^p \langle F_j \rangle_q$$

On the right hand side of this last step, the three λ_j^p and the three $F_j(X = x_i)$ must already be known. This is analogous to knowing $\mathbf{E}(\mathbf{x}; \mathbf{J})$. The q_i are unknown at this juncture, but we do know that only two λ_j^q can be varied so that ultimately an $H(q)$ and all three $\langle F_j \rangle_q$ can be computed.

Exercise 45.6.7: Set up three points in a triangular relationship to further examine the e–projection.

Solution to Exercise 45.6.7

Continue with the usual labeling where point p is some complicated distribution living in a dually flat Riemannian manifold, point q is some simpler distribution living in a sub–manifold, and point u is the fair model distribution also living in the same sub–manifold as point q. The special point q^* refers to that q that is closest to p when squared distance is measured by the divergence function $D^{(1)}(p \parallel q)$.

From Chapter Forty, we have Amari's formula, Equation (40.7), that allows us to look at the triangular relationship among these three points p, q, and u,

$$D^{(1)}(p \parallel q) + D^{(1)}(q \parallel u) - D^{(1)}(p \parallel u) = \{\theta^i(p) - \theta^i(q)\}\{\eta_i(u) - \eta_i(q)\}$$

When the right hand side is equal to 0, the triangle is a right triangle, and we have found that point q^* closest to p.

Exercise 45.6.8: Begin the computational exploration of this e–projection.

Solution to Exercise 45.6.8

Perform the exploration within the easy computational realm of the initial kangaroo scenario. Set up a quantitative measure of the departure from orthogonality of a triangle consisting of three points p, q, and u,

$$\text{orthogonality measure} = D^{(1)}(p \parallel q) + D^{(1)}(q \parallel u) - D^{(1)}(p \parallel u)$$

Compute each of the three divergence functions on the right hand side. Begin with the longest distance between the complicated model behind point p and the fair model behind point u,

$$D^{(1)}(p \parallel u) = \sum_{i=1}^{4} u_i \ln\left(\frac{u_i}{p_i}\right)$$

$$= 0.25 \ln\left(\frac{0.25}{0.70}\right) + 0.25 \ln\left(\frac{0.25}{0.05}\right) + 0.25 \ln\left(\frac{0.25}{0.05}\right) + 0.25 \ln\left(\frac{0.25}{0.20}\right)$$

$$= 0.6031$$

The goal of the variational free energy minimization procedure is to find some suitable q that is the minimum **e**–projection from p. Therefore, we don't know what q^* might be at this juncture. We will vary the q_i to achieve some sense of where q^* is located. What we are looking for is that q^* that makes the orthogonality measure equal to 0.

Let's arbitrarily start with the q that was the **m**–projection. Then we have,

$$D^{(1)}(p \parallel q) = \sum_{i=1}^{4} q_i \ln\left(\frac{q_i}{p_i}\right)$$

$$= 0.5625 \ln\left(\frac{0.5625}{0.70}\right) + 0.1875 \ln\left(\frac{0.1875}{0.05}\right)$$

$$+ 0.1875 \ln\left(\frac{0.1875}{0.05}\right) + 0.0625 \ln\left(\frac{0.0625}{0.20}\right)$$

$$= 0.2999$$

For the final term in the orthogonality measure,

$$D^{(1)}(q \parallel u) = \sum_{i=1}^{4} q_i \ln\left(\frac{q_i}{u_i}\right)$$

$$= 0.5625 \ln\left(\frac{0.5625}{0.25}\right) + 0.1875 \ln\left(\frac{0.1875}{0.25}\right)$$

$$+ 0.1875 \ln\left(\frac{0.1875}{0.25}\right) + 0.0625 \ln\left(\frac{0.0625}{0.25}\right)$$

$$= 0.2877$$

It is easy to ascertain that this q that arose as the **m**–projection of point p does not participate in forming a right triangle by the dual relative entropy criterion since the orthogonality measure is not 0.

$$0.2999 + 0.2877 - 0.6031 = -0.0155$$

Thus, there must be a different q that is closer to p by this alternative distance measure, and also further from u. If a new trial q were to have the assignment of,

$$Q_i = (0.576283, 0.182850, 0.182850, 0.058017)$$

then the orthogonality measure does becomes 0. Now, the Pythagorean theorem *is* satisfied because,

$$0.2903 + 0.3128 = 0.6031$$

Notice that this new q, and in fact it is q^*, is closer to p with $D^{(1)}(p \parallel q^*) = 0.2903$, and farther from u than the q we originally started from since $D^{(1)}(q^* \parallel u) = 0.3128$.

Exercise 45.6.9: Does Equation (40.7) as shown in Exercise 45.6.7 provide the same answer?

Solution to Exercise 45.6.9

Equation (40.7) gives us the orthogonality measure in terms of the dual coordinates for the three points,

$$D^{(1)}(p \parallel q^*) + D^{(1)}(q^* \parallel u) - D^{(1)}(p \parallel u) = \{\theta^i(p) - \theta^i(q^*)\} \{\eta_i(u) - \eta_i(q^*)\}$$

This expression should also compute to a value of 0 when the q^* proposed above is inserted.

The $\theta^i(p)$, the three Lagrange multipliers from the maximum likelihood model containing the information from the one double interaction, and the $\eta_i(u)$, the three constraint function averages from the fair model, are well–known to us. The $\theta^i(q^*)$, the two Lagrange multipliers from the **e**–projection of p to the sub–manifold with no double interaction must be found. The third Lagrange multiplier for q^* must be 0 by definition. Also, all three constraint function averages for $\eta_i(q^*)$ can be calculated once q^* has been found.

Orthogonality measure $= \sum_{i=1}^{3} [\,(\theta^i(p) - \theta^i(q^\star))\,] \times [\,(\eta_i(u) - \eta_i(q^\star))\,]$

$= [\,(-1.3862944 - 1.1479323) \times (0.50 - 0.759133)\,]$

$+ [\,(-1.3862944 - 1.1479323) \times (0.50 - 0.759133)\,]$

$+ [\,(4.0253517 - 0) \times (0.25 - 0.576283)\,]$

≈ 0

Exercise 45.6.10: Perform a final numerical check by calculating the squared distance function developed in section 45.3.2.

Solution to Exercise 45.6.10

The minimum squared distance from p to $q^\star$ as the **e**–projection has been computed as $\min[KL(q,p)] = 0.2903$. The same value should be calculated from the function derived as,

$$D^{(1)}(p \parallel q) = \ln Z_p - H(q) - \sum_{j=1}^{3} \lambda_j^p \langle F_j \rangle_q$$

The partition function for p is $Z_p = 5$. The three Lagrange multipliers λ_j^p are known,

$$\lambda_1^p = -1.3862944$$

$$\lambda_2^p = -1.3862944$$

$$\lambda_3^p = +4.0253517$$

The assigned distribution for the $q^\star$ that is the **e**–projection was determined as,

$$q_1^\star = 0.576283$$

$$q_2^\star = 0.182850$$

$$q_3^\star = 0.182850$$

$$q_4^\star = 0.058017$$

With this much determined, the three constraint function averages, $\langle F_j \rangle_{q_*}$,

$$\langle F_1 \rangle_{q_*} = \sum_{i=1}^{4} F_1(X = x_i)\, q_i^* = 0.576283 + 0.182850 + 0 + 0 = 0.759133$$

$$\langle F_2 \rangle_{q_*} = \sum_{i=1}^{4} F_2(X = x_i)\, q_i^* = 0.576283 + 0 + 0.182850 + 0 = 0.759133$$

$$\langle F_3 \rangle_{q_*} = \sum_{i=1}^{4} F_3(X = x_i)\, q_i^* = 0.576283 + 0 + 0 + 0 = 0.576283$$

and the information entropy,

$$H(q_i^*) = -\sum_{i=1}^{4} q_i^* \ln q_i^* = 1.10415$$

can be computed.

Reweave this detail in order to compute the minimum **e**–projection distance,

$$D^{(1)}(p \parallel q^*) = \ln Z_p - H(q^*) - \sum_{j=1}^{3} \lambda_j^p \langle F_j \rangle_{q^*}$$

$$= \ln 5 - 1.10415 - [\,(-1.3862944 \times 0.759133) +$$

$$(-1.3862944 \times 0.759133) + (+4.0253517 \times 0.576283)\,]$$

$$= 0.2903$$

Exercise 45.6.11: Show Amari's canonical divergence function and its MEP analog.

Solution to Exercise 45.6.11

Amari defines [2, pg. 61, Eq. (3.44)] what he calls a canonical divergence function for two points p and q in a dually flat Riemannian manifold as,

$$D(p \parallel q) \stackrel{\text{def}}{=} \psi(p) + \varphi(q) - \theta^i(p)\, \eta_i(q)$$

The left hand side $D(p \parallel q)$, the canonical divergence function, is more specifically $KL(q,p)$, the dual of the Kullback–Leibler relative entropy. When this squared distance between p and q is minimized, we then have the **e**–projection of p to q.

When we make the translation from Amari's IG notation back to our MEP notation through the following equivalencies,

$$\psi(p) \equiv \ln Z_p$$

$$\varphi(q) \equiv -H(q)$$

$$\theta^i(p)\,\eta_i(q) \equiv \sum_{j=1}^{m} \lambda_j^p \,\langle F_j \rangle_q \qquad \text{(Einstein summation convention)}$$

we have the analog of the canonical divergence function in the MEP format,

$$D^{(1)}(p \parallel q) \equiv KL(q,p) = \sum_{i=1}^{n} q_i \ln\left(\frac{q_i}{p_i}\right) = \ln Z_p - H(q) - \sum_{j=1}^{m} \lambda_j^p \,\langle F_j \rangle_q$$

And this final expression can be matched up quite easily with Amari's equivalent canonical divergence expression with which we started this exercise,

$$D(p \parallel q) = \psi(p) + \varphi(q) - \theta^i(p)\,\eta_i(q)$$

Exercise 45.6.12: But how was the point $q^\star$ actually found?

Solution to Exercise 45.6.12

An exploratory *Mathematica* program was written that would allow the Lagrange multipliers, that is, the model parameters for the candidate point $q^\star$, to be varied. As the model parameters were varied, a function, described below, assessing the departure from orthogonality was monitored. As the model parameters converged onto an orthogonality measure of 0, the resulting numerical assignment could be computed from these optimal parameters.

In this manner, it was discovered that the model parameters with values of,

$$\lambda_1^{q^\star} = 1.14793229294$$

$$\lambda_2^{q^\star} = 1.14793229294$$

$$\lambda_3^{q^\star} = 0$$

led to an MEP assignment of,

$$q_1^\star = 0.576283$$

$$q_2^\star = 0.182850$$

$$q_3^\star = 0.182850$$

$$q_4^\star = 0.058017$$

The program, of course, fixed the third model parameter at $\lambda_3^{q^\star} = 0$ because any candidate $q^\star$, just like any other q in the sub–manifold, had to ignore the double interaction. This goes back to how simple and complicated models were defined.

A subtlety in the above solution that might have been easily overlooked, but is nonetheless vital to understanding the variational free energy minimization, is the role of m in the summation $\sum_{j=1}^{m} \lambda_j^p \langle F_j \rangle_q$. The value of m must be set at that value appropriate to the complicated model p that is being approximated by q. Even though averages $\langle F_j \rangle_{q^\star}$ are being found with respect to the simpler model $q^\star$, there are still $m = 3$ constraint functions from the complicated model p that must be accounted for.

For example, in the solution to Exercise 45.6.10, the third constraint function, $F_3(X = x_i) = (1, 0, 0, 0)$, from the complicated model p had to be taken into account when,

$$\langle F_3 \rangle_{q^\star} = 0.576283 + 0 + 0 + 0 = 0.576283$$

was calculated.

Exercise 45.6.13: How was the orthogonality measure mentioned above implemented in *Mathematica*?

Solution to Exercise 45.6.13

In Exercise 40.7.6, the dual version of Amari's formula for detecting how close three points were to forming a right triangle was derived as,

$$KL(q,p) + KL(r,q) - KL(r,p) = \sum_{i=1}^{n}(q_i - r_i) \ln\left(\frac{q_i}{p_i}\right)$$

Point u, the fair model, now takes the place of point r.

Then, it is relatively straightforward to implement this in *Mathematica* with a function **dualVersion[** *arg1, arg2, arg3* **]** of three arguments. Each argument is a list containing the assigned probabilities p_i, q_i, and u_i.

```
dualVersion[p1_List, p2_List, p3_List] :=
                        Total[ (p2 - p3) Log[p2 / p1]]
```

dualVersion[pi, qi, ui] eventually reached a value of $7.03 \times 10^{-13} \approx 0$ with **qi** at the previously stated values for $q_i^\star$.

Exercise 45.6.14: Find the point q^* that is the e–projection of point p from the $n=8$ kangaroo scenario.

Solution to Exercise 45.6.14

Use the same exploratory program described in Exercises 45.6.12 and 45.6.13 to vary the first three Lagrange multipliers until the orthogonality measure decreases to 0. The **e**–projection of p is to a sub–manifold where independent distributions reside. Therefore, the coordinates λ_4 through λ_7 will be set to 0. It turned out that setting the first three coordinates to the same value,

$$\lambda_1^{q^*} = 0.779020285$$

$$\lambda_2^{q^*} = 0.779020285$$

$$\lambda_3^{q^*} = 0.779020285$$

led to an MEP assignment for q^* of,

$$q_1^* = 0.322080$$

$$q_2^* = 0.147788$$

$$q_3^* = 0.147788$$

$$q_4^* = 0.067813$$

$$q_5^* = 0.147788$$

$$q_6^* = 0.067813$$

$$q_7^* = 0.067813$$

$$q_8^* = 0.031117$$

This assignment to q^* resulted in the shortest squared distance from point p in the full manifold to any point q in the sub–manifold holding all of the independent distributions. The actual value was $D(p \parallel q^*) = 0.099916$.

Exercise 45.6.15: Perform a quick validity check of the above value of the divergence function using the MEP version.

Solution to Exercise 45.6.15

Since,

$$D(p \parallel q^*) = \ln Z_p - H(q^*) - \sum_{j=1}^{7} \lambda_j^p \langle F_j \rangle_{q^*}$$

plugging in the values for the partition function for p, the information entropy for $q^\star$, the Lagrange multipliers for p, and the constraint function averages with respect to $q^\star$ should reproduce the value of 0.099916. The *Mathematica* expression,

`Log[33.3333] + Total[qi Log[qi]] - Dot[lambdap, Dot[cm, qi]]`

does indeed return the same value as $D(p \parallel q^\star)$.

The individual divergence functions for the points p, $q^\star$, and u also satisfy the requirement to form a right triangle since,

$$D(p \parallel u) = D(p \parallel q^\star) + D(q^\star \parallel u)$$

$$0.321961 = 0.099916 + 0.222045$$

Exercise 45.6.16: Sketch out the approach for finding the point $q^\star$ that is the e–projection of a point p in a scenario involving three coins.

Solution to Exercise 45.6.16

Building on the findings of the last exercise, consider an inferential scenario involving three coins flipped simultaneously. Let the point p represent a rather unlikely situation of some sort of linkage involving the three coins. The state space would consist of $2^N = 2^3 = 8$ statements indicating the disposition of each of the three coins.

Under the information from the most complicated models, p would have seven model parameters reflecting the marginal probabilities for the three coins, the three double interactions between coins 1 and 2, coins 2 and 3, and coins 1 and 3, and finally the one triple interaction among all three coins. To mimic "nearest neighbor" interactions in statistical mechanics, the complicated model behind p considered here will only include the double interaction between coins 1 and 2, and the double interaction between coins 2 and 3 in addition to a coin's individual propensity for HEADS or TAILS.

A distribution p will be set up with five model parameters in order to investigate the inner workings of its **e**–projection to a point $q^\star$ in a sub–manifold E consisting of independent distributions and only three model parameters.

Exercise 45.6.17: Return to the example of the "assembly" of nine coins. Discuss possible energy functions more in tune with statistical mechanics. What kind of conceptual as well as practical computational issues ensue?

Solution to Exercise 45.6.17

In the previous exercises about the assembly of nine coins, the energy function was simply the sum of the individual "spins" of each coin, $E(\mathbf{x}) = \sum_l^9 S_l$, where S_l

could equal 1 or 2. The energy of any assembly then had to be in the range of 9, where all nine coins would show HEADS face up, through 18, where all nine coins would show TAILS face up.

Exercise 45.6.4 examined the case where an average energy $\langle E(\mathbf{x}) \rangle_q$ was defined with respect to some approximating distribution q in the sub–manifold consisting of all independent distributions. An independent distribution clearly means that the probability for HEADS or TAILS of any coin does not depend on what face is showing on any other coin.

An alternative energy function oriented more towards how statistical mechanics conceptualizes things is one that involves interactions between neighboring spins. This is, in turn, an attempt by physics and chemistry to simplify the computational effort in explicating what might be "really" happening at the "system" level of atoms and molecules when "assembled" to form a large scale object. MacKay gives a good example of typical energy functions in his discussion of variational Bayes for "spin systems." These energy functions assume the general appearance of MacKay's energy function definition for spin systems,

$$E(\mathbf{x}; \mathbf{J}, \mathbf{h}) = -\frac{1}{2} \sum_m \sum_n J_{mn} x_m x_n - \sum_n h_n x_n$$

Imagine the nine coins laid out in a 3×3 lattice arrangement where Coin 1 is at the upper left corner and Coin 9 is at the lower right corner. Following MacKay, construct an energy function $E(\mathbf{x})$ that not only includes the S_l of an individual coin, but also some selected double interactions between two nearest neighbors. Thus, the face showing on Coin 1 might be influenced by the faces appearing on Coins 2 and 4, these being defined as the "nearest neighbors" in the lattice.

Let's not deviate too much in this discussion from our standard MEP notation. A "spin" S_l is a specific instance of a constraint function $F_j(X = x_i)$ taking on the values of 1 or 2 at the level of the individual system. We first looked at the easiest independent case where there was just one model parameter λ providing the same numerical assignment of 1/3 to HEADS and 2/3 to TAILS to all nine coins.

Then, this starting situation was generalized to the more interesting case where there were nine model parameters λ_1 through λ_9. This permitted each coin to have its own private probability assignment for HEADS. For example, Coin 1 was assigned a probability of 1/2, Coin 2 was assigned a probability of 1/6, and so on.

With the introduction of interactions between neighboring coins, we are adding more constraint functions $F_j(X = x_i)$ in order to implement $S_k \times S_l$. The only possible values for these interaction terms contributing to $E(\mathbf{x})$ are a 1 where both coins show HEADS, a 2 where one of the two coins shows HEADS and the other shows TAILS, or a 4 where both coins show TAILS. There would be a total of 12 such interaction terms as in 1) $S_1 \times S_2$, 2) $S_1 \times S_4$, $\cdots$, 12) $S_8 \times S_9$.

If our revised energy function includes these 12 new interaction terms between coins in addition to the 9 old constraint functions, we have 21 constraint functions $F_j(X = x_i)$ and 21 model parameters λ_1 through λ_{21}. Although quite expanded from our original considerations, this still represents a significant distillation from, say, a really complicated model consisting of 511 parameters. The new energy function now can range from a minimum energy value of $E_{\min}(\mathbf{x}) = 21$ when all nine coins show HEADS to a maximum energy value of $E_{\max}(\mathbf{x}) = 66$ when all nine coins show TAILS.

To obtain a listing of all 512 statements in the state space $\mathbf{x}$, evaluate,

```
Grid[Table[{j, Tuples[{H, T}, 9][[j]]}, {j, 1, 512}]]
```

The layout where Coins 1 and 2 show TAILS and the remaining seven coins all show HEADS occurs at position 385 in the listing. The energy of this state is calculated as 11 due to $\sum_{l=1}^{9} S_l$ of the individual spins plus 18 due to $\sum_{12} S_k \times S_l$ of the twelve spin interactions for a total energy $E(\mathbf{x}) \equiv E(X = x_{385}) = 29$. We don't know the probability for this particular state ($X = x_{385}$), or for that matter, the probabilities for any of the 512 states, so no average energy can be computed.

The variational Bayes technique demands knowledge of the energy function for the complicated model. Constructing the matrix of constraint functions, even for models with just the information from these few interaction terms, is not a trivial task. This matrix for a model p would consist of 21 rows and 512 columns.

The first constraint function $F_1(X = x_i)$ would have a 1 wherever HEADS appears for Coin 1, and 2 otherwise. Thus, the first 256 columns would contain a 1, and the remaining 256 columns a 2. $F_2(X = x_i)$ would have a 1 wherever HEADS appears for Coin 2, and a 2 otherwise. Thus, $F_1(X = x_{385}) = 2$ and $F_2(X = x_{385}) = 2$, while $F_3(X = x_{385})$ through $F_9(X = x_{385})$ would equal 1.

The constraint functions for the double interactions would take account of any $S_l \times S_k$. Thus, $F_{10}(X = x_{385}) = 4$, $F_{11}(X = x_{385}) = 2$, $F_{12}(X = x_{385}) = 2$, $F_{13}(X = x_{385}) = 2$, and $F_{15}(X = x_{385}) = 2$. The remaining seven constraint functions through $F_{21}(X = x_{385})$ all equal 1. The MEP numerator for Q_{385} would look like,

$$Q_{385} \text{ numerator} = \exp\left[2\lambda_1 + 2\lambda_2 + \cdots + \lambda_9 + 4\lambda_{10} + 2\lambda_{11} + \cdots + \lambda_{21}\right]$$

with the partition function Z_p consisting of a sum of 512 similar looking terms. This approach is quickly getting out of hand.

However, revisiting MacKay's definition of variational minimization,

$$\beta \tilde{F}(\boldsymbol{\theta}) = \beta \langle E(\mathbf{x}; \mathbf{J}) \rangle_Q - S_Q$$

we see that we can begin to get purchase on this issue by picking any old distribution q from the sub–manifold of independent distributions. The new energy function

$E(\mathbf{x}; \mathbf{J})$ taking nearest neighbor interactions into account can be readily evaluated for any assembly. For example, the energy for the 385th assembly is calculated to be $E(X = x_{385}) = 26$.

Pick the distribution q discussed earlier in Exercise 45.6.4 manufactured from nine probabilities at the system level. One term in the average energy calculation then would be,

$$\langle E(\mathbf{x}; \mathbf{J}) \rangle_Q = \sum_{i=1}^{512} E(X = x_i)\, q_i$$

$$= \cdots + (1/3 \times 1/2 \times 3/4 \times 1/10 \times 1/2 \times 2/3 \times 4/5 \times 5/6 \times 1/2) \times 26 + \cdots$$

In this calculation of $\sum_{i=1}^{512} E(X = x_i)\, q_i$, any J_{mn} and h_n values were ignored. Thus, the analog to MacKay's,

$$E(\mathbf{x}; \mathbf{J}, \mathbf{h}) = \sum_{mn} J_{mn} x_m\, x_n + \sum_n h_n x_n$$

would be the MEP summation over the Lagrange multipliers λ_j and the constraint functions $F_j(X = x_i)$,

$$E(\mathbf{x}; \mathbf{J}, \mathbf{h}) \equiv \sum_{j=1}^{21} \lambda_j\, F_j(X = x_i)$$

In order to calculate the information entropy S_Q, the other 511 q_i would have to be found in a similar manner from the nine independent probabilities. The amount of missing information could then be computed,

$$S_Q \equiv H(q) = -\sum_{i=1}^{512} q_i \ln q_i$$

But there is a lingering mystery left unresolved so far. In statistical mechanics and the Boltzmann canonical distribution, there is only *one* model parameter, here with the notation of β. In statistical mechanics, the energy function always assumes the burden of any inherent physical complexity. The temperature remains as the only "parameter" that forces the probability for physical states to change.

Inevitably, we are led to a violation of our fundamental framework arising from this necessity of parsing out two levels of parameters, the λ_j and β. One resolution just discussed is that the λ_j become involved in the definition of a more complicated energy function, leaving β as the sole model parameter. Only if $\beta = 1$, do we return to our fundamental framework, but then one has to reconceptualize the issue of whether the λ_j are part of the definition of the constraint function, or are they model parameters?

SOLVED EXERCISES FOR CHAPTER 45

Under the canonical Boltzmann probability distribution of statistical mechanics, the probability of the state $\mathbf{x}$ is,

$$P(\mathbf{x}) = \frac{e^{-E(\mathbf{x})/k_B T}}{Z(T)}$$

The most obvious consequence is that as the temperature becomes very high, or, in other words, as β approaches 0, all of the states approach the same probability whatever their energy.

Conversely, as the temperature becomes very low, or as β approaches ∞, the ground state of all nine HEADS with the minimum energy becomes increasingly probable. The probability for all 512 assemblies is being determined by β, not by the λ_j which, in this view, are part of the complicated definition of the energy function,

$$E(\mathbf{x}; \mathbf{J}, \mathbf{h}) = \sum_{mn} J_{mn} x_m x_n + \sum_n h_n x_n$$

Now, to implement some general strategy for variational Bayes, β must already have been defined. Not only that, $E(\mathbf{x}; \mathbf{J}, \mathbf{h})$ must have already been defined. For inferencing, variational Bayes implies that *all* λ_j^p and *all* $F_j(X = x_i)$ must have been previously defined.

Then, the starting configuration for any arbitrary q_i could be altered slightly through a change in the 21 model parameters, a new average energy computed with respect to the newly altered q distribution, a new information entropy of the q distribution computed, and the approach to a minimum,

$$\beta \tilde{F}(\boldsymbol{\theta}) = \beta \langle E(\mathbf{x}; \mathbf{J}) \rangle_Q - S_Q$$

monitored through an iterative process.

I have tried to emulate in the above rambling argument some of the conceptual confusion that always seems to insinuate itself in these advanced applications of inferential problem solving. Here, of course, variational Bayes is an example of what I have in mind.

Chapter 46

Mean Field Theory

46.1 Introduction

We would like to see how long this framework for general inferencing can hold up when subjected to more and more difficult problems. So far, it seems to work exactly as we would want for coins, dice, and kangaroos.

Physics, though, might be expected to provide interesting challenges. In this Chapter, we examine approximations that have been developed over the years for hard problems in condensed matter physics. First, we will look at *Ising models*, and then *mean field theories*. The groundwork laid down in the previous Chapter on variational Bayes will come in handy as we delve into this area. The fact that Information Geometry has recently addressed these topics is additional motivation.

As you might have surmised, my objective is not to explain Physics. My concern is to remark on any parallels that might exist between the twin pillars of probability theory and Physics. There are some obvious similarities existing alongside some confusing conceptual mysteries when the formal manipulation rules and the MEP are juxtaposed next to the standard treatment physicists employ.

I am guilty of the sin of which Einstein accused many of his contemporaries. He mockingly accused them, in their choice of problems to work on, of drilling holes where the wood was thinnest. I will leave all of the hard problems to the Einsteins of the world, and do readily confess in my discussions of Ising models and mean field theory in trying to find the thinnest part of the board in which to drill.

The potentially hard problems have been avoided in order to select the easier versions because 1) they appear to be readily amenable to the tools at hand, and 2) because many other people have worked on them and have made the job a whole lot easier. However, I have found that there are interesting surprises even when one purposefully chooses the easier problem.

If the hard Physics problem is approached from the very outset as a general inferencing problem, then a lot of the obscure reasoning and the results generated from these suspect arguments become a whole lot clearer. I tried to give an inkling of this productive attitude in my discussion of Schrödinger's development of statistical thermodynamics in Chapter Twenty Seven of Volume II.

My overriding compulsion, whenever some resolution has finally been achieved for an inferencing scenario, is that there should be *no mysteries whatsoever* left as toxic residue in the mind of the IP. My contention is that our analysis of coin tossing, dice rolling, or correlations between traits of kangaroos leaves no pesky lingering doubts. We can honestly claim that we have grasped the essence of the solution to the problem. Every base should have been touched whenever the solution is rightly critiqued from the criteria laid down within the formal abstract framework.

For example, I hope you would agree with me that when the following question is posed: "What is the probability for HEADS on the first toss of a coin when nothing is known about the coin, or the manner in which it will be tossed?", the single, unreservedly correct resolution must be the answer $1/2$. There are no lingering unresolved issues with this answer. The IP is not constantly battling a lurking suspicion that something has been overlooked. The IP can proudly proclaim to the world: "There are no residual mysteries left for me in this answer."

But there always seems to be something to doubt when we depart from that pristine realm of general inferencing where we do understand everything about the answer. There most certainly are some reservations when we try to transfer that understanding over to an analogous answer in Physics. Are there unresolved mysteries and lingering doubts when we examine the Ising models and the mean field theory? There always are.

Fortunately, the solution to the abstract general inferencing problem can offer some guidance on judging solutions that *are* offered since they are not beset by any mysteries. The IP, in the end, is not really looking for a particular numerical answer, but always searching for some deeper insight into the problem.

Thus, it is important to examine, once again, the subtle interplay involved in the various concepts of how the state space is defined, the dimension of the state space, the probability of any possible state in the state space, the models, the parameters in the models, the data, and so on. This concern is reflected in the various atomic expressions appearing in our equations like n, N, 2^N, Z, Z^N, $P(\mathbf{x})$, $E(\mathbf{x})$, $W(N)$, $\mathcal{D}$, $\mathcal{M}_k$, and so on.

Finally, to end this **Introduction**, and to give a hint as to where we are heading, would you be at all perplexed when Physics comes up with a particular mathematical expression that happens to involve an hyperbolic trigonometric function? What is the origin for such an expression? Can we identify within a general inferencing problem where and why such an expression would be introduced? Need I add, in conclusion, that the IP would like the resolution to this little problem to leave no *mysteries* in its wake.

46.2 Binary Statements

Let's begin by reviewing elementary facts about statements where just two possible measurements are defined. Of course, this is the pedagogical coin flip scenario.

We introduced the MEP formula in Chapter Seventeen of Volume II with an assignment to the probabilities for HEADS and TAILS. As with every assignment made by the MEP, conditioning on the truth of the information under some specific model $\mathcal{M}_k$ must be indicated to the right of the conditioned upon symbol.

$$P(\text{HEADS} \,|\, \mathcal{M}_k) \equiv P(X = x_1 \,|\, \mathcal{M}_k) = \frac{e^\lambda}{e^\lambda + e^{2\lambda}} \tag{46.1}$$

$$P(\text{TAILS} \,|\, \mathcal{M}_k) \equiv P(X = x_2 \,|\, \mathcal{M}_k) = \frac{e^{2\lambda}}{e^\lambda + e^{2\lambda}} \tag{46.2}$$

In order for any information to be given an operational meaning under a model for the coin flip scenario, a constraint function, or a mapping from two statements to numbers, was set up as $F(X = x_1) = 1$ and $F(X = x_2) = 2$. The parameter for the model was the Lagrangian multiplier λ.

The coordinate systems for any point in the dually flat Riemannian manifold are represented by either the Lagrange multipliers $\lambda_j \equiv \theta^i$, or the constraint function averages $\langle F_j \rangle \equiv \eta_i$. Here, there is only one λ, and one $\langle F \rangle$.

A dually flat Riemannian manifold $(S, g, \nabla^{(e)}, \nabla^{(m)})$ must provide these kind of coordinate systems for a collection of points belonging to the manifold,

$$\{p, q, r, \cdots\} \in \mathcal{S}^n$$

Each point $p, q, r, \cdots$ is a separate probability distribution. As λ is varied between $-\infty$ and $+\infty$, or $\langle F \rangle$ is varied between 1 and 2, points associated with these varying coordinates are probability distributions under different models. The relationship between these two dual parameters when either is changed is reflected in the metric tensor, or its inverse.

Later on in Chapter Seventeen, the constraint function mappings were changed to $F(X = x_1) = 0$ and $F(X = x_2) = \epsilon$ to make an analogy with the Fermi oscillator. Under this alternative mapping, the MEP formula changed to,

$$P(\text{ground state} \,|\, \mathcal{M}_k) \equiv P(X = x_1 \,|\, \mathcal{M}_k) = \frac{1}{1 + e^{\lambda \epsilon}} \tag{46.3}$$

$$P(\text{excited state} \,|\, \mathcal{M}_k) \equiv P(X = x_2 \,|\, \mathcal{M}_k) = \frac{e^{\lambda \epsilon}}{1 + e^{\lambda \epsilon}} \tag{46.4}$$

Using Schrödinger's terminology of "assemblies" and "systems," we asserted that the probability for an assembly of N Fermi oscillator systems could be considered as analogous to N independent coin tosses. The denominator for the probability of N tosses became,

$$\prod_{t=1}^{N} Z_t = \sum_{t=1}^{N} \ln Z_t = N \ln Z = N \ln\left(1 + e^{\lambda \epsilon}\right) \qquad (46.5)$$

because the partition function Z_t at each trial was the same Z. The subscript t indicates the tth flip of the coin within the overall number of N flips.

When the scenario turns to physical situations like Ising models or spin glasses, the constraint functions may be expressed as SPIN UP, $F(X = x_1) = +1$, and SPIN DOWN, $F(X = x_2) = -1$. The MEP assignment changes in concert with this alternative mapping to,

$$P(\text{SPIN UP} \mid \mathcal{M}_k) \equiv P(X = x_1 \mid \mathcal{M}_k) = \frac{1}{1 + e^{-2\lambda}} \qquad (46.6)$$

$$P(\text{SPIN DOWN} \mid \mathcal{M}_k) \equiv P(X = x_2 \mid \mathcal{M}_k) = \frac{e^{-2\lambda}}{1 + e^{-2\lambda}} \qquad (46.7)$$

What happens to the dual coordinates, the constraint function averages, when this choice is made for Ising models? The universal constraint function average must remain at $\langle F_0 \rangle = 1$. But under this alternative mapping, $\langle F_1 \rangle \equiv \langle F \rangle = \tanh(\lambda)$. The above MEP formula and the derivation of the constraint function average is discussed in more detail in this Chapter's first two Exercises.

46.3 The Ising Model

Let's now focus on more of the details in this arbitrary, but perfectly acceptable, Ising model mapping and the resulting MEP assignment. It is important to do so because of the relevance to physical models that rely on statistical mechanics and Boltzmann's canonical distribution.

The Ising model is part of Physics. It attempts to describe magnetic properties of matter. Furthermore, any notion of probability makes an appearance solely because it comes along for the ride with statistical mechanics. It is definitely not something that had its origin in any kind of purely inferential setting.

Nonetheless, it is not too difficult to recognize that the specialized contingent history, as it has evolved through Physics, is just a special case of general inferencing. Of course, we must attribute this insight directly to Jaynes who said as much when he first laid out the MEP in 1957.

In laying out these details, I will periodically weave back and forth between our standard probabilistic interpretation, and the contingent history that Physics has provided for us in the form of the two dimensional Ising model. Placing these two interpretations side by side should prove interesting.

MEAN FIELD THEORY

Let's commence by once again revisiting that most primitive of all the general inferencing scenarios, the coin flip. The coin flip scenario had a state space of dimension $n = 2$ together with the arbitrary, but perfectly acceptable, mappings from the two statements to numbers of,

$$(X = x_1) \equiv \text{"HEADS"} \text{ with } F(X = x_1) = 1$$

and,

$$(X = x_2) \equiv \text{"TAILS"} \text{ with } F(X = x_2) = 2$$

The partition function for the coin flip scenario must then be expressed as,

$$Z(\lambda) = \exp[\lambda F(X = x_1)] + \exp[\lambda F(X = x_2)] = e^\lambda + e^{2\lambda}$$

where λ, the Lagrange multiplier, is the one parameter defining any model.

A subtle conceptual transition takes place if we equate N flips of one coin, or N coins flipped once, with an assembly of N binary systems. Here, you can see that we are reverting to Schrödinger's terminology so that the problem can be viewed in the light of statistical thermodynamics. The dimension of this new state space is $n = 2^N$. A reorientation is required that shifts the thinking about the *data* of N repeated tosses of a coin to a static assembly that is a joint statement existing in the newly enlarged $n = 2^N$ state space.

For example, the Ising model envisions N atomic sites arranged in a regular order on a lattice like a crystal. For the smallest slice of an Ising model, consider a 3×3 square lattice so that there are $N = 9$ atomic sites. If each of these N sites can be in either a SPIN UP or SPIN DOWN condition, then a typical state $\mathbf{x}$ in the $n = 2^N = 2^9 = 512$ dimensional state space might look like,

$$\mathbf{x} = \{\text{SPIN DOWN}, \text{SPIN UP}, \cdots, \text{SPIN DOWN}\} \equiv \overbrace{\{-1, +1, +1, \cdots, -1\}}^{N=9}$$

A corresponding typical state $\mathbf{x}$ for the assembly of $N = 9$ coins might look like,

$$\mathbf{x} = \{T, H, H, \cdots, T\} \equiv \overbrace{\{2, 1, 1, \cdots, 2\}}^{N=9}$$

Because Physics treats this as a problem in statistical mechanics, an energy $E(\mathbf{x})$ must be defined for every state $\mathbf{x}$. In a tremendous simplification of possibly very complex models, the energy might be defined as something like,

$$E(\mathbf{x}) = \sum J_{kl} S_k S_l - \sum h_l S_l$$

But even this turns out to be a little too hard to handle, so, for example, in the Ising model, this energy is redefined as something simpler,

$$E(\mathbf{x}) = J \sum S_k S_l - h \sum S_l$$

where S_l is the $+1$ or the -1 "spin" value assigned to the lth site.

Even physicists simplify the Ising model problem even further by setting $J = 0$. With this stipulation, the spins in the lattice arrangement are not coupled to any other spin. The energy function then reduces to,

$$E(\mathbf{x}) = h \sum_{l=1}^{N} S_l$$

Compare this to an analogous energy function for an assembly of N coins, say, where $N = 100$,

$$E(\mathbf{x}) = \sum_{t=1}^{100} \lambda F(X_t = x_i)$$

where $F(X_t = H) = 1$ and $F(X_t = T) = 2$. Suppose the assembly looks like this,

$$\mathbf{x} = \overbrace{\{T, H, H, \cdots, T\}}^{N=100}$$

then the total energy might be defined as,

$$E(\mathbf{x}) = \lambda \sum_{t=1}^{100} F(X_t = x_i) = \lambda(2 + 1 + \cdots + 2)$$

where λ was made equivalent to h and set to 1. If there are, say, 25 HEADS and 75 TAILS in the assembly of the 100 coins, then as Schrödinger would have it, the total "energy" is,

$$E(\mathbf{x}) = 175$$

The total energy for an assembly can range from a minimum of 100 to a maximum of 200. If the total energy over the assembly is divided by N, then this becomes the information for an individual coin, the information at the system level,

$$U = \frac{E}{N} = \langle F \rangle = 1.75$$

With the information in a model specified by the parameter $\langle F \rangle = 1.75$, and with a constraint function mapping of $F(\text{HEADS}) = 1$ and $F(\text{TAILS}) = 2$, then the numerical assignment to the probability for HEADS becomes $Q_1 = 0.25$ and the numerical assignment to the probability for TAILS becomes $Q_2 = 0.75$.

Substitute the MEP assignment for Q_1 and Q_2 in the overall probability for the state of the 100 coins,

$$P(\mathbf{x}) = \left(\frac{e^\lambda}{e^\lambda + e^{2\lambda}}\right)^{N_1} \left(\frac{e^{2\lambda}}{e^\lambda + e^{2\lambda}}\right)^{N_2}$$

The denominator involved in all of the N multiplications is the same Z value for Q_1 and Q_2, which is $\exp(\lambda) + \exp(2\lambda)$. Thus, we have the normalizing factor for $P(\mathbf{x})$ as,

$$Z^N = (e^\lambda + e^{2\lambda})^N$$

$$N \ln Z = N \ln(e^\lambda + e^{2\lambda})$$

MEAN FIELD THEORY

Because we have in our possession, compliments of Schrödinger, the following basic relationship between concepts in statistical mechanics of total energy E, N, and average energy U, and the average constraint function from general inferencing, the "average share of energy of one system,"

$$\frac{E}{N} = U = \langle F \rangle = \frac{\partial \ln Z}{\partial \lambda}$$

we can go ahead and find,

$$E = N \frac{\partial \ln Z}{\partial \lambda}$$

$$= \frac{\partial N \ln Z}{\partial \lambda}$$

$$= \frac{\partial N \ln (e^\lambda + e^{2\lambda})}{\partial \lambda}$$

$$= N \frac{e^\lambda + 2e^{2\lambda}}{e^\lambda + e^{2\lambda}}$$

$$N \frac{e^\lambda + 2e^{2\lambda}}{e^\lambda + e^{2\lambda}} = 175$$

With $N = 100$, the solution for the Lagrange multiplier is found to be $\lambda = 1.09861$ which, of course, must be the value of the model parameter for the assignment of $Q_1 = 0.25$ and $Q_2 = 0.75$ at the system level of each coin.

This is an **e**–projection of a potentially complicated model for the assembly down to the sub–manifold of independent distributions. It is also a "naive mean field approximation" to what in reality might be a very complicated relationship amongst all 100 coins. The mean field approximation is an tremendous simplification because each coin in the assembly of 100 coins has the same probability of 0.25 to show HEADS, and each coin has "a loose energy contact" with every other coin in the assembly.

Schrödinger has told us that the partition function Z_q for an assembly whose constituent systems are in a "loose energy contact" may be calculated by the convenient expediency of multiplying by N the simple partition functions existing at the system (each individual coin) level.

You might want to look at Exercise 46.7.3 at this juncture for further rumination about the conceptual differences between Schrödinger's approach to the **MOST PROBABLE DISTRIBUTION** and the tack that I take in trying to adhere strictly to a general inferencing scheme.

Proceed through precisely the same steps for the different mapping of the states of the Ising model, but with the same rationale of independent distributions at the system level,

$$P(\mathbf{x}) = \left(\frac{e^\lambda}{e^\lambda + e^{-\lambda}}\right)^{N_1} \left(\frac{e^{-\lambda}}{e^\lambda + e^{-\lambda}}\right)^{N_2}$$

$$Z^N = (e^\lambda + e^{-\lambda})^N$$

$$N \ln Z = N \ln(e^\lambda + e^{-\lambda})$$

$$\frac{e^x + e^{-x}}{2} = \cosh(x)$$

$$e^\lambda + e^{-\lambda} = 2\cosh(\lambda)$$

$$N \ln Z = N \ln[2\cosh(\lambda)]$$

Suppose that the magnetization M for an assembly of $N = 100$ atoms with the different constraint function reflecting a spin state for an atom is -50,

$$\langle M \rangle = \frac{\partial N \ln Z}{\partial \lambda}$$

$$= \frac{\partial N \ln[2\cosh(\lambda)]}{\partial \lambda}$$

$$= N \tanh(\lambda)$$

$$N \tanh(\lambda) = -50$$

$$\lambda = -0.549306$$

leading to the same assignment of $Q_1 = 0.25$ and $Q_2 = 0.75$ found for the coin assembly. Of course, $M = 25(+1) + 75(-1) = -50$.

46.4 Adding Complexity

This effort to describe Ising models, as surveyed in the last section, was about as stripped down a version as one could possibly wish for. But it *was* instructive to begin the comparison of a basic general inferencing scenario side by side with a Physics problem. And my overall message has been to harp on the lack of any outstanding mysteries in any aspect of the general inferencing scenario. There is nothing in the inferences made about coin tossing which lead to any head scratching over unresolved mysteries.

On the other hand, the assumptions underpinning the Ising model left a lot to be desired. There could be no association whatsoever between any of the N sites.

Furthermore, the probability at each of the N sites had to remain the same and unchanged. The essence of any model of a physical phenomenon like the magnetic properties of matter surely must involve both short, medium, and long range mutual interactions at the huge number of sub–atomic, atomic, and molecular sites that compose the substance in question.

So we begin ever so slowly to add complexity to the basic scenario as just discussed. In statistical mechanics, additional complexity is enforced through a more complicated energy function as in the familiar,

$$E(\mathbf{x}) = -1/2 \sum J_{kl} S_k S_l - \sum h_l S_l$$

For us though it is better to adhere to the general MEP formula where there will be a summation over some number of constraints taking into account the individual spins, augmented by a summation over some further number of constraints taking into account interactions between spin sites.

Thus, as the analog to the more complicated energy function $E(\mathbf{x})$ above, we would construct more constraint functions and Lagrange multipliers to implement individual spin sites and double interactions between "nearest neighbor" spin sites.

$$E(\mathbf{x}) \longrightarrow \sum_{j=1}^{9} \overbrace{\lambda_j F_j(X = x_i)}^{h_l S_l} + \sum_{j=10}^{21} \overbrace{\lambda_j F_j(X = x_i)}^{J_{kl} S_k S_l}$$

In this equation, we have reverted back to the $N = 9$ sites example for an easier understanding of the values presented for the upper indices in the two summations.

It's not too difficult to see how the calculation of the total energy is immediately impacted when we start to include spin interactions. Suppose that the assembly of $N = 100$ atomic spin sites actually consisted of $N_1 = 20$ SPIN UP sites and $N_2 = 80$ SPIN DOWN sites at the system level. With no spin interactions, the magnetization is calculated as -60 instead of -50.

However, if the model includes spin interactions at ten of the sites with these ten sites all SPIN UP, then the magnetization reverts back to -50. We would be wrong to select the "most probable distribution" as $N_1 = 25$ and $N_2 = 75$ as we did earlier based on the simpler energy model.

Nonetheless, the mean field approximation is still restricted to models that live in the sub–manifold of the encompassing full Riemannian flat manifold where all coordinates associated with any kind of higher order interaction must have a value of zero. This example of an α–projection in Information Geometry to implement a mean field approximation can now be viewed as a specific application of orthogonal decomposition and foliation as was broached in Chapter Forty One.

46.5 Comparing e–projection Notation

To facilitate the translation between Amari's argument for mean field approximation as an **e**–projection, and my MEP notation for the dual of the Kullback–Leibler relative entropy, I will repeat the following transformation sequence. Recall that, for me, model p is the complicated model and model q is the simpler approximation.

$$KL(q,p) \equiv D^{(1)}(p \parallel q)$$

$$= \sum_{i=1}^{n} q_i \ln\left(\frac{q_i}{p_i}\right)$$

$$= \sum_{i=1}^{n} q_i \left(\ln q_i - \ln p_i\right)$$

$$= \sum_{i=1}^{n} q_i \ln q_i - \sum_{i=1}^{n} q_i \ln p_i$$

$$= E_q\left[\ln q - \ln p\right]$$

$$= E_q\left[\sum_{j=1}^{m} \lambda_j^q F_j(X = x_i) - \ln Z_q\right] - E_q\left[\sum_{j=1}^{m} \lambda_j^p F_j(X = x_i) - \ln Z_p\right]$$

Take a breather at this step to take note that the index m for the summations refers to the total number of constraints as defined by the complicated model p. Pick up the transformation at the point where the expectation of the terms in brackets is taken with respect to probability distribution q,

$$D^{(1)}(p \parallel q) = \left[\sum_{j=1}^{m} \lambda_j^q \langle F_j\rangle_q - \ln Z_q\right] - \left[\sum_{j=1}^{m} \lambda_j^p \langle F_j\rangle_q - \ln Z_p\right]$$

$$KL(q,p) = \ln\left(\frac{Z_p}{Z_q}\right) + \sum_{j=1}^{m}(\lambda_j^q - \lambda_j^p)\langle F_j\rangle_q$$

Compare this expression in my MEP notation to Amari's equivalent expression (more detail in next section),

$$D_1\left[p : q\right] = E_q\left[\overline{\boldsymbol{h}} \cdot \boldsymbol{x} - \psi_q - (WX + \boldsymbol{h} \cdot \boldsymbol{x} - \psi_p)\right]$$

$$= \overline{\boldsymbol{h}} \cdot \boldsymbol{m} - \psi_q - W \cdot M + \boldsymbol{h} \cdot \boldsymbol{m} + \psi_p$$

46.6 Connections to the Literature

Amari clearly states that a mean field approximation is the same as an **e**–projection from some complicated model to a simpler model in a sub–manifold [5].

> If we use the e–projection of q to $\boldsymbol{E}$ instead of the m–projection, we have the naive mean field approximation ... To show this, we calculate the e–projection (1–projection) of q to $\boldsymbol{E}$. For $\alpha = 1$,
>
> $$D_1[q: p] = D_{-1}[p: q] = E_p[\overline{\boldsymbol{h}} \cdot \boldsymbol{x} - \psi_p - (WX + \boldsymbol{h} \cdot \boldsymbol{x} - \psi_q)]$$
>
> $$= \overline{\boldsymbol{h}} \cdot \boldsymbol{m} - \psi_p - W \cdot M - \boldsymbol{h} \cdot \boldsymbol{m} + \psi_q$$
>
> because of $M = E_p[X] = \boldsymbol{m}\boldsymbol{m}^T$.

Unfortunately for the notation, in this article Amari labels q as the complicated distribution and p as the simpler approximation, just the opposite of my standard labeling. Thus, Amari's divergence function $D_{-1}[p: q]$ becomes my dual relative entropy function $KL(q,p)$. The sub–manifold $\boldsymbol{E}$ is the sub–manifold containing all of the independent distributions. We found out in the previous Chapter on variational Bayes that approximations to models with complicated energy functions also came from the class of independent distributions.

It's a comfort to quickly ascertain that the right hand side of Amari's expression for the divergence function involved in the **e**–projection is exactly what we would expect despite the lack of standardization with the MEP notation. This beginning manipulation of the dual relative entropy is one we have carried out numerous times,

$$KL(q,p) = \sum_{i=1}^{n} q_i \ln\left(\frac{q_i}{p_i}\right)$$

$$= E_q[\ln q_i - \ln p_i]$$

remembering the reversal in roles where q is the simpler distribution approximating the complicated distribution p. Matching things up, but always keeping in mind the reversal between p and q,

$$E_q[\ln q_i] \equiv E_p[\overline{\boldsymbol{h}} \cdot \boldsymbol{x} - \psi_p]$$

$$\sum_{j=1}^{m} \lambda_j^q \langle F_j \rangle_q - \ln Z_q \equiv \overline{\boldsymbol{h}} \cdot \boldsymbol{m} - \psi_p$$

$$E_q[\ln p_i] \equiv E_p[WX + \boldsymbol{h} \cdot \boldsymbol{x} - \psi_q]$$

$$\sum_{j=1}^{m} \lambda_j^p \langle F_j \rangle_q - \ln Z_p \equiv W \cdot M - \boldsymbol{h} \cdot \boldsymbol{m} - \psi_q$$

Earlier in the article, Amari had discussed that the distributions in $\boldsymbol{E}$ were from the exponential family, which for us means that these distributions will also appear in the familiar MEP format.

Since $\boldsymbol{E}$ consists of all the independent distributions, it is easy to show the geometry of $\boldsymbol{E}$. Moreover, $\boldsymbol{E}$ itself is an exponential family.

$$p(\boldsymbol{x}, \overline{\boldsymbol{h}}) = \exp\{\overline{\boldsymbol{h}} \cdot \boldsymbol{x} - \psi_0(\overline{\boldsymbol{h}})\}$$

where

$$e^{\psi_0(\overline{\boldsymbol{h}})} = \prod_i (e^{\overline{h}_i} + e^{-\overline{h}_i})$$

or

$$\psi_0(\overline{\boldsymbol{h}}) = \sum_i \log(e^{\overline{h}_i} + e^{-\overline{h}_i})$$

This $\overline{\boldsymbol{h}} = (\overline{h}_1, \cdots, \overline{h}_n)$ is [sic] the e–affine coordinates of $\boldsymbol{E}$. Its m–affine coordinates are given by

$$m = E_p[\boldsymbol{x}] = \frac{\partial}{\partial \overline{\boldsymbol{h}}} \psi_0(\overline{\boldsymbol{h}})$$

which is easily calculated as

$$m_i = E_p[x_i] = \frac{\partial}{\partial \overline{h}_i} \psi_0(\overline{\boldsymbol{h}}) = \frac{e^{\overline{h}_i} - e^{-\overline{h}_i}}{e^{\overline{h}_i} + e^{-\overline{h}_i}} = \tanh h_i$$

This final result of the mean at each spin site under an independent distribution as shown by Amari will be examined in more detail in Exercise 46.7.4. It is clear that each site may have its own private mean value $\tanh h_i \equiv \tanh \lambda_j$.

To wrap up Amari's prescription on how to implement mean field theory as an e–projection, we finish the quote as begun above where he tells us that he will perform a partial differentiation of the divergence function with respect to all of the constraint function averages under the independent distribution and find a critical point by equating this result to 0.

Hence,

$$\frac{\partial D_1}{\partial \boldsymbol{m}} = \frac{\partial \overline{\boldsymbol{h}}}{\boldsymbol{m}} \frac{\partial (\overline{\boldsymbol{h}} \cdot \boldsymbol{m} - \psi_p)}{\overline{\boldsymbol{h}}} - W\boldsymbol{m} - \boldsymbol{h}$$

$$= \tanh^{-1} \boldsymbol{m} - W\boldsymbol{m} - \boldsymbol{h}$$

This gives

$$\boldsymbol{m}_1[q] = \tanh[W\boldsymbol{m}_1[q] + \boldsymbol{h}]$$

known as the "naive" mean field approximation. In the component form, this is

$$m_i = \tanh\left(\sum w_{ij} m_j + h_i\right)$$

This equation can have a number of solutions. It is necessary to check which solution minimizes $D_1[q:p]$. The minimum may be attained at the boundary of $m_i = \pm 1$.

MEAN FIELD THEORY

He finds, in his notation, that,

$$\tanh^{-1} \boldsymbol{m} - W\boldsymbol{m} - \boldsymbol{h} = 0$$

$$\tanh^{-1} \boldsymbol{m} = W\boldsymbol{m} - \boldsymbol{h}$$

$$\boldsymbol{m} = \tanh\left[W\boldsymbol{m} - \boldsymbol{h}\right]$$

MacKay [29, pg. 426] also has a good explanation of the Ising model from a fundamental probabilistic perspective. His example of mean field theory actually occurs as part of his Chapter on variational free energy minimization. I would just take the generalization one step further to point out that both variational Bayes and mean field theory can be addressed as specific applications of e–projection. These α–projections, as Amari has taken pains to emphasize, are not only relevant in some abstract mathematical sense to Differential Geometry, but can often help to elucidate our primary focus of concern on inferential questions from the standpoint of Information Geometry.

It is not too difficult to match up MacKay's terminology with our preferred MEP inspired approach. He begins with,

> In the simple Ising model ..., every coupling J_{mn} is equal to J if m and n are neighbors and zero otherwise. There is an applied field $h_n = h$ that is the same for all spins. A very simple approximating distribution is one with just a single variational parameter a, which defines a separable distribution
>
> $$Q(\mathbf{x}; a) = \frac{1}{Z_Q} \exp\left(\sum_n a x_n\right)$$
>
> in which all spins are independent and have the same probability
>
> $$q_n = \frac{1}{1 + \exp(-2a)}$$
>
> of being up. The mean magnetization is
>
> $$\bar{x} = \tanh(a)$$
>
> and the equation $[a_m = \beta\left(\sum_n J_{mn}\bar{x}_n + h_m\right)]$ which defines the minimum of the variational free energy becomes
>
> $$a = \beta(CJ\bar{x} + h)$$
>
> where C is the number of couplings that a spin is involved in, $C = 4$ in the case of a rectangular two–dimensional Ising model. We can solve [the last two equations above] for $\bar{x}$ numerically — in fact, it is easiest to vary $\bar{x}$ and solve for β — and obtain graphs of the free energy minima and maxima as a function of temperature ...

For us, an assigned probability at the binary system level (each atom), while adhering strictly to the MEP formula, is,

$$Q_i \equiv P(X = x_i \mid \mathcal{M}_k) = \frac{\exp\left[\lambda F(X = x_i)\right]}{\sum_{i=1}^{2} \exp\left[\lambda F(X = x_i)\right]}$$

For an atom with spin up, where $F(X = x_1) = +1$ and $F(X = x_2) = -1$, this becomes,

$$Q_1 = \frac{e^\lambda}{e^\lambda + e^{-\lambda}} = \frac{1}{1 + e^{-2\lambda}}$$

the same as MacKay's expression with his variational parameter a being our model parameter, the Lagrange multiplier λ.

Exercise 46.7.2 goes into all of the details of how the MEP also finds, like MacKay, that the mean magnetization at the system level must be,

$$\langle F \rangle = \tanh(\lambda)$$

Although it is certainly not immediately obvious, we also have a strong suspicion that MacKay's equations for his free energy minimization solution are the same as Amari's **e**–projection solution, as just quoted previously,

$$m_i = \tanh\left(\sum w_{ij} m_j + h_i\right)$$

Recall that Amari's w_{ij} are MacKay's J_{mn}, Amari's m_i are MacKay's $\bar{x}_n$, while for us, adhering strictly to the MEP formalism, the w_{ij}, the J_{mn}, and the h_i are all just Lagrange multipliers λ_j for either uncoupled spins, or for coupled spins defined within some neighborhood. Any averages like m_i or $\bar{x}_n$ for us must be any constraint function averages $\langle F_j \rangle$ in the definition of the complicated model p taken with respect to the simpler probability distribution of the independent model q.

46.7 Solved Exercises for Chapter Forty Six

Exercise 46.7.1: Provide more detail for the final form of Equations (46.6) and (46.7).

Solution to Exercise 46.7.1

We begin by establishing some elementary facts about exponential functions for the upcoming manipulations.

$$e^\lambda \times e^{-\lambda} = e^{(\lambda-\lambda)} = e^0 = 1$$

$$e^{-\lambda} \times (e^\lambda + e^{-\lambda}) = e^{(-\lambda+\lambda)} + e^{(-\lambda-\lambda)} = 1 + e^{-2\lambda}$$

The probability assigned to the SPIN UP and SPIN DOWN states conditioned on model $\mathcal{M}_k$ where $F(X = x_1) = +1$, and $F(X = x_2) = -1$ is given by the MEP formula,

$$P(\text{SPIN UP} \,|\, \mathcal{M}_k) = \frac{e^{\lambda(+1)}}{e^{\lambda(+1)} + e^{\lambda(-1)}}$$

$$= \frac{e^\lambda}{e^\lambda + e^{-\lambda}}$$

$$\frac{e^\lambda}{e^\lambda + e^{-\lambda}} \times \frac{e^{-\lambda}}{e^{-\lambda}} = \frac{1}{1 + e^{-2\lambda}}$$

$$P(\text{SPIN DOWN} \,|\, \mathcal{M}_k) = \frac{e^{\lambda(-1)}}{e^{\lambda(+1)} + e^{\lambda(-1)}}$$

$$= \frac{e^{-\lambda}}{e^\lambda + e^{-\lambda}}$$

$$\frac{e^{-\lambda}}{e^\lambda + e^{-\lambda}} \times \frac{e^{-\lambda}}{e^{-\lambda}} = \frac{e^{-2\lambda}}{1 + e^{-2\lambda}}$$

If you evaluate the expression **Exp[- λ] (Exp[λ] + Exp[- λ])** hoping to see the answer $1 + e^{-2\lambda}$ returned, you will be treated to an unexpected and quirky feature of *Mathematica*. Instead, the built–in function **Expand[**expr**]** must be wrapped around the above expression to force *Mathematica* to return the expected answer. Alternatively, one could ask *Mathematica* to simplify the expression with **Simplify[**expr**]**.

This is similar to the confusion experienced by the *Mathematica* novice (someone like me for example) over the seeming non–evaluation of elementary expressions like **Sqrt[Power[x, 2]]** or **Log[Power[a, b]]** when I expected to see the answers x or $b \ln a$.

In these cases, `PowerExpand[`*expr*`]` will return the expected answers. However, `Simplify[`*expr*`]` as an alternative will not work in this situation. Curiously, a *Mathematica* strength is the culprit here because it always treats the fully general case where complex functions might be involved.

Just as a somewhat crude heuristic, I find that I will often wrap `Simplify[`*expr*`]` or even `FullSimplify[`*expr*`]` as a matter of course around expressions involving `Exp[ ]` to see if a more acceptable answer emerges, and by more acceptable I mean one that I more easily recognize.

In addition, the `TrigToExp[`*expr*`]` and `ExpToTrig[`*expr*`]` functions are often helpful. For example, `ExpToTrig[Exp[x] + Exp[-x]]` returns `2 Cosh[x]`.

Exercise 46.7.2: Provide more detail about the information inserted by the implicit universal constraint function average and the explicit first constraint function average in the previous exercise.

Solution to Exercise 46.7.2

The information inserted into the probability distribution by the implicit universal constraint function average is simply $\langle F_0 \rangle = (1 \times Q_1) + (1 \times Q_2) = 1$,

$$\frac{1}{1+e^{-2\lambda}} + \frac{e^{-2\lambda}}{1+e^{-2\lambda}} = 1$$

The information inserted into the probability distribution by the specification of the only explicit constraint function $F(X = x_1) = +1$ and $F(X = x_2) = -1$ is that $\langle F \rangle = F(X = x_1) Q_1 + F(X = x_2) Q_2$.

$$\langle F \rangle = \left([+1] \times \frac{1}{1+e^{-2\lambda}}\right) + \left([-1] \times \frac{e^{-2\lambda}}{1+e^{-2\lambda}}\right)$$

$$= \frac{1}{1+e^{-2\lambda}} - \frac{e^{-2\lambda}}{1+e^{-2\lambda}}$$

$$= \tanh \lambda$$

Although in this case, it is better to keep the probabilities in their original MEP format so that the hyperbolic tangent function can be immediately recognized,

$$\langle F \rangle = \frac{e^{\lambda}}{e^{\lambda}+e^{-\lambda}} - \frac{e^{-\lambda}}{e^{\lambda}+e^{-\lambda}} = \tanh \lambda$$

Mathematica confirms this because an evaluation of the above expression wrapped within a `FullSimplify[`*expr*`]` returns `Tanh[λ]`.

SOLVED EXERCISES FOR CHAPTER 46

Exercise 46.7.3: Disambiguate the title of Schrödinger's Chapter II with the coin assembly example of section 46.3.

Solution to Exercise 46.7.3

The title of Chapter II in Schrödinger's *Statistical Thermodynamics* is,

THE METHOD OF THE MOST PROBABLE DISTRIBUTION

If you were to just casually assume that he (and Boltzmann) are talking about a *probability distribution*, you would be mistaken.

He is, in fact, referring to that one particular *distribution of frequency counts* that possesses the greatest probability. For example, the *distribution* of 25 HEADS and 75 TAILS has the maximum value of the probability,

$$P(N_1 = 25, N_2 = 75) = \frac{100!}{25!\,75!} (0.25)^{25} (0.75)^{75} = 0.0917997$$

when compared to the probability for any other possible frequency count in an assembly of $N = 100$ coins, given some average energy and assuming independence at the system level.

The *distribution of probabilities* is clearly a distinct concept because it refers to the probabilities over all possible frequency counts, where here this would consist of a distribution of 101 probabilities,

$$P(N_1 = 0, N_2 = 100) = \frac{100!}{0!\,100!} (0.25)^0 (0.75)^{100} = 3.2072 \times 10^{-13}$$

$$P(N_1 = 1, N_2 = 99) = \frac{100!}{1!\,99!} (0.25)^1 (0.75)^{99} = 1.0691 \times 10^{-11}$$

$$\ldots = \ldots$$

$$P(N_1 = 24, N_2 = 76) = \frac{100!}{24!\,76!} (0.25)^{24} (0.75)^{76} = 0.09059$$

$$P(N_1 = 25, N_2 = 75) = \frac{100!}{25!\,75!} (0.25)^{25} (0.75)^{75} = 0.09180$$

$$P(N_1 = 26, N_2 = 74) = \frac{100!}{26!\,74!} (0.25)^{26} (0.75)^{74} = 0.08827$$

$$\ldots = \ldots$$

$$P(N_1 = 50, N_2 = 50) = \frac{100!}{50!\,50!} (0.25)^{50} (0.75)^{50} = 4.5073 \times 10^{-8}$$

$$\ldots = \ldots$$

$$P(N_1 = 100, N_2 = 0) = \frac{100!}{100!\,0!} (0.25)^{100} (0.75)^0 = 6.2230 \times 10^{-61}$$

This entire distribution of probabilities is easily obtained from *Mathematica* by first creating the function,

prob[N1_, N2_] :=
    Multinomial[N1, N2] Power[.25, N1] Power[.75, N2]

followed by constructing a simple table of all 101 probabilities,

TableForm[Table[{j, 100 - j, prob[j, 100 - j]}, {j, 0, 100}]]

So you see that Schrödinger is very literally not trying to find a probability distribution, but rather finding the disposition of frequency counts with the highest probability. This would be, of course, the very same frequency counts as revealed by the data especially so as $N \to \infty$. It all seems so very circular!

Of course, in this binary case, either maximization of entropy or maximization of the multiplicity factor is superfluous because satisfaction of the specified information is enough by itself to assign Q_1 and Q_2, or find N_1 and N_2. The first piece of information is that $N = N_1 + N_2 = 100$ and the second piece of information is that $N_1 F(X = x_1) + N_2 F(X = x_2) = 175$. The only N_1 and N_2 that can satisfy this information is $N_1 = 25$ and $N_2 = 75$.

If there happened to be three or more statements in the state space as possible measurements of a coin flip, then we would have to find those frequency counts that satisfied the constraints exactly in the same way as above, but with the additional requirement that we would choose those frequency counts that could happen in the greatest number of ways.

For example, let's switch over to a system (an atom) which can be in one of *three* spin conditions, UP, DOWN, or NEUTRAL, $+1$, -1, or 0, instead of just the binary spin conditions. Since Schrödinger's derivation depends explicitly on letting $N \to \infty$, let $N = 900,000$, with, as usual, $N_1 + N_2 + N_3 = N$, and with the total energy of the entire assembly of 900,000 atoms at $E = 0$.

Now, the occupancy counts for the three spins of $N_1 = 0$, $N_2 = 900,000$, and $N_3 = 0$ with all atoms in the NEUTRAL spin state satisfies this information.

$$N_1 + N_2 + N_3 = 0 + 900,000 + 0 = N$$

$$N_1(+1) + N_2(0) + N_3(-1) = [0 \times (+1)] + [900,000 \times 0] + [0 \times (-1)] = 0$$

However, this disposition of frequency counts can happen in only *one* way!

Another set of occupancy counts which also satisfies the above information is,

$$N_1 + N_2 + N_3 = 300,000 + 300,000 + 300,000 = N$$

$$N_1(+1) + N_2(0) + N_3(-1) = [300,000 \times (+1)] + [300,000 \times 0] + [300,000 \times (-1)] = 0$$

but the multiplicity factor for these frequency counts is incomprehensibly larger at,

$$W(N) = \frac{900000!}{300000!\,300000!\,300000!}$$

Schrödinger's method of the "most probable distribution" of frequency counts says to pick the set of frequency counts that, while still able to satisfy the given information, can happen in the most number of ways. Therefore, from Schrödinger's Equation (2.6), since $N_i = a_l = 300,000$,

$$a_l = N \frac{e^{-\mu \epsilon_l}}{\sum e^{-\mu \epsilon_l}}$$

$$\frac{e^{-\mu \epsilon_l}}{\sum e^{-\mu \epsilon_l}} = 1/3$$

$$\mu = 0$$

Since $\mu \equiv 1/k_B T$, the temperature of the assembly must be very high in order for the atoms to be equally distributed to their three spin states.

Exercise 46.7.4: Discuss the MEP relationship between the constraint function average and the partition function as it applies to the Ising model.

Solution to Exercise 46.7.4

The familiar generic MEP relationship that shows the constraint function average to be the partial differentiation of the log of the partition function with respect to the Lagrange multiplier is,

$$\langle F \rangle = \frac{\partial \ln Z}{\partial \lambda}$$

Eventually, we are going to allow a possibly different λ_j at the level of each binary system. But for now, just go ahead and directly perform the partial differentiation at the system level,

`FullSimplify[D[Log[Exp[`λ`] + Exp[-` λ`]], `λ`]]`

in order to arrive at the expected answer of `Tanh[`λ`]`.

It is instructive to look back at Exercise 36.9.15. It was here that we asked *Mathematica* to evaluate the log of any function $f(x)$ through,

`D[Log[f[x]], x] // FullForm`

It provided us with a typically idiosyncratic *Mathematica* symbolic expression,

```
Times[Power[f[x], -1], Derivative[1][f][x]]
```

Such symbolic expressions do possess the virtue that they are subject to no false interpretation. Do you remember what I said in my *Apologia* to Volume I? Any question of interpretation of a symbolic expression must ultimately be resolved by subjecting the issue in question to a *Mathematica* evaluation.

Translating this into a more familiar notation, we have,

$$\frac{\partial \ln f(x)}{\partial x} = \frac{1}{f(x)} \times \frac{\partial f(x)}{\partial x}$$

After this bit of a digression, we can verify the above relationship through a slightly different series of manipulations beginning with,

$$\frac{\partial \ln Z}{\partial \lambda} = \frac{1}{Z} \times \frac{\partial Z}{\partial \lambda}$$

Recall that at the binary system level where $n = 2$, the partition function Z is,

$$Z = \sum_{i=1}^{2} e^{\lambda F(X=x_i)}$$

$$= e^{\lambda(+1)} + e^{\lambda(-1)}$$

$$= e^{\lambda} + e^{-\lambda}$$

$$= 2\cosh(\lambda)$$

With this much accomplished, we can proceed to,

$$\frac{\partial Z}{\partial \lambda} = \frac{\partial [2\cosh(\lambda)]}{\partial \lambda}$$

$$= 2\sinh(\lambda)$$

$$\frac{1}{Z} = \frac{1}{e^{\lambda} + e^{-\lambda}}$$

$$\frac{1}{Z} \times \frac{\partial Z}{\partial \lambda} = \frac{2\sinh(\lambda)}{2\cosh(\lambda)}$$

$$\langle F \rangle = \tanh(\lambda)$$

SOLVED EXERCISES FOR CHAPTER 46

As a quick verification,

$$\langle F \rangle = \sum_{i=1}^{n} F(X = x_i) \, Q_i$$

$$= F(X = x_1) \, Q_1 + F(X = x_2) \, Q_2$$

$$= \left([+1] \times \frac{e^\lambda}{e^\lambda + e^{-\lambda}} \right) + \left([-1] \times \frac{e^{-\lambda}}{e^\lambda + e^{-\lambda}} \right)$$

$$= \frac{e^\lambda}{e^\lambda + e^{-\lambda}} - \frac{e^{-\lambda}}{e^\lambda + e^{-\lambda}}$$

$$= \frac{e^\lambda - e^{-\lambda}}{e^\lambda + e^{-\lambda}}$$

$$= \frac{2\sinh(\lambda)}{2\cosh(\lambda)}$$

$$= \tanh(\lambda)$$

Exercise 46.7.5: What is the probability for an assembly of N atomic spin sites if independence no longer holds between sites?

Solution to Exercise 46.7.5

For the sake of a concrete numerical example, consider one of the 512 possible states for the $N = 9$ assembly of spins arranged in a 3×3 lattice,

$$\mathbf{x} = \begin{matrix} + & - & + \\ - & + & - \\ + & - & + \end{matrix}$$

Under independence, the probability for this state is the multiplication of the separate probabilities at each site Y_l,

$$P(\mathbf{x} \mid \mathcal{M}_k) = P(Y_9) \times P(Y_8) \times \cdots \times P(Y_1)$$

When all probabilities are the same at the binary system level,

$$P(Y_l = \text{SPIN UP} \mid \mathcal{M}_k) = \frac{1}{1 + e^{-2\lambda}}$$

$$P(Y_l = \text{SPIN DOWN} \mid \mathcal{M}_k) = \frac{e^{-2\lambda}}{1 + e^{-2\lambda}}$$

Thus,

$$P(\mathbf{x} \mid \mathcal{M}_k) = \frac{1}{1 + e^{-2\lambda}} \times \frac{e^{-2\lambda}}{1 + e^{-2\lambda}} \times \cdots \times \frac{1}{1 + e^{-2\lambda}}$$

In our ongoing numerical example where $\lambda = -0.549306$, then,
$$P(\mathbf{x}\,|\,\mathcal{M}_k) = (0.25)^5(0.75)^4 = 0.000309$$

If the probabilities at the binary system level are permitted to vary, then
$$P(\mathbf{x}\,|\,\mathcal{M}_k) = \frac{1}{1+e^{-2\lambda_9}} \times \frac{e^{-2\lambda_8}}{1+e^{-2\lambda_8}} \times \cdots \times \frac{1}{1+e^{-2\lambda_1}}$$

When we begin to allow correlations between sites, full independence no longer holds. Start off with a model where there is a correlation between sites 1 and 2. Then,
$$P(\mathbf{x}\,|\,\mathcal{M}_k) = P(Y_9) \times P(Y_8) \times \cdots \times P(Y_2\,|\,Y_1) \times P(Y_1)$$
$$= P(Y_9) \times P(Y_8) \times \cdots \times P(Y_2, Y_1)$$

Now, an MEP formula needs to be developed for a $n = 4$ joint probability table to assign probabilities to $P(Y_2, Y_1)$. Instead of two model parameters λ_1 and λ_2 for sites 1 and 2, the model must be augmented with another parameter because we require three model parameters to describe the correlation at these two sites.
$$P(\mathbf{x}\,|\,\mathcal{M}_k) = \frac{1}{1+e^{-2\lambda_{10}}} \times \frac{e^{-2\lambda_9}}{1+e^{-2\lambda_9}} \times \cdots \times \frac{e^{\sum^3 \lambda_j F_j(X=x_i)}}{\sum^4 e^{\sum^3 \lambda_j F_j(X=x_i)}}$$

Continuing on in the same vein, suppose there is a correlation among sites 1, 2, and 3. Then, we have,
$$P(\mathbf{x}\,|\,\mathcal{M}_k) = P(Y_9) \times P(Y_8) \times \cdots \times P(Y_3, Y_2, Y_1)$$

Instead of three model parameters for sites 1, 2, and 3, the model must be augmented with seven parameters in order to describe all possible interactions at these three sites.
$$P(\mathbf{x}\,|\,\mathcal{M}_k) = \frac{1}{1+e^{-2\lambda_{13}}} \times \frac{e^{-2\lambda_{12}}}{1+e^{-2\lambda_{12}}} \times \cdots \times \frac{e^{\sum^7 \lambda_j F_j(X=x_i)}}{\sum^8 e^{\sum^7 \lambda_j F_j(X=x_i)}}$$

In the end, we arrive at the necessity to assign probabilities to the full blown joint probability table where no independence exists anywhere.
$$P(\mathbf{x}\,|\,\mathcal{M}_k) = P(Y_9, Y_8, \cdots, Y_3, Y_2, Y_1)$$

Generalize to arbitrary N, while permitting all levels of possible interactions,
$$P(\mathbf{x}\,|\,\mathcal{M}_k) = P(Y_N, Y_{N-1}, \cdots, Y_3, Y_2, Y_1)$$

The assigned probabilities are still provided through the fully general MEP formula,
$$P(\mathbf{x}\,|\,\mathcal{M}_k) = \frac{e^{\sum_{j=1}^{(2^N-1)} \lambda_j F_j(X=x_i)}}{\sum_{i=1}^{2^N} e^{\sum_{j=1}^{(2^N-1)} \lambda_j F_j(X=x_i)}}$$

However, contemplate a very large number of atomic sites, say, $N = 10{,}000$. In a significant understatement, the IP is faced with an insurmountable computational issue due to the combinatorial explosion.

SOLVED EXERCISES FOR CHAPTER 46

Exercise 46.7.6: Use *Mathematica* to help find the value for the model parameter that satisfies magnetization as a total energy function for the $N = 100$ atomic spin sites example.

Solution to Exercise 46.7.6

$$\text{Total } E(\mathbf{x}) = N_1\, F(X = x_1) + N_2\, F(X = x_2)$$

$$= 25(+1) + 75(-1)$$

$$= -50$$

$$\frac{\partial N \ln Z}{\partial \lambda} = -50$$

$$100 \tanh(\lambda) = -50$$

Use the `Solve[ ]` function to find that,

$$\texttt{Solve[100 Tanh[}\lambda\texttt{]} == \texttt{-50,} \lambda\texttt{]}$$

returns the answer that $\lambda = -0.549306$.

Exercise 46.7.7: Plot the dual parameters in the Ising model with the constraint function average a function of the Lagrange multiplier.

Solution to Exercise 46.7.7

The Lagrange multiplier parameter λ can vary between $-\infty$ and $+\infty$, while the dual constraint function average parameter $\langle F \rangle$ can only vary between -1 and $+1$. This is exactly what the plot of $\tanh(\lambda)$ shows us in Figure 46.1 at the top of the next page. The value of λ doesn't have to go anywhere near these end points before Q_1 is very close to a numerical assignment of 1. For example, at $\lambda = 4$, $\tanh(4) = 0.9993293$.

$$Q_1 = \frac{e^\lambda}{e^\lambda + e^{-\lambda}} = \frac{e^4}{e^4 + e^{-4}} = 0.9996646$$

Thus,

$$\langle F \rangle = \tanh(4)$$

$$F(X = x_1)\, Q_1 + F(X = x_2)\, Q_2 = Q_1 - Q_2$$

$$\tanh(4) = 0.9996646 - 0.0003354$$

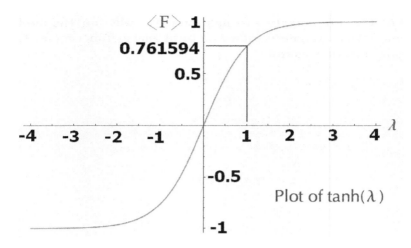

Figure 46.1: *The non–linear relationship between the dual parameters λ and $\langle F \rangle$ of the Ising model. A plot of* $\tanh(\lambda)$ *as a function of* λ. *At* $\lambda = 1$, $\tanh(1) = 0.761594$.

This plot of $\tanh(\lambda) \equiv \langle F \rangle$ vs. λ is simply another demonstration of the highly non–linear relationship between dual coordinates. A previous example discussed in Exercise 31.10.25 of Volume II illustrated another non–linear relationship that involved the dual coordinates of the Cauchy distribution. This is where Information Geometry is helpful because it provides us a means of assessing the quantitative changes in the dual coordinates through the metric tensor.

Exercise 46.7.8: Use *Mathematica* and the Legendre transformation to find the Lagrange multiplier and the information entropy in the previous exercise.

Solution to Exercise 46.7.8

The core of the MEP algorithm is the Legendre transformation, which for this easy scenario where $n = 2$ reduces to,

$$H_{\max}(Q_i) = \min_{\lambda} [\ln Z - \lambda \langle F \rangle]$$

The expression in brackets is to be minimized by varying λ and keeping $\langle F \rangle$ fixed at the information provided by the model $\mathcal{M}_k$. The information in this model is that $\langle F \rangle = \tanh(4)$. Let *Mathematica* evaluate the expression,

```
NMinimize[Log[Exp[λ] + Exp[-λ]] - 0.9993292997 λ, λ]
```

which returns {0.00301821, {λ → 4.}}. The information entropy is the first element in the returned list, thus,

$$H_{\max}(Q_i) = -(Q_1 \ln Q_1 + Q_2 \ln Q_2)$$

$$= -(0.999665 \ln 0.999665 + 0.000335 \ln 0.000335)$$

$$= 0.003018$$

The extended precision for tanh(4) was found by **N[Tanh[4],10]**. Since Q_1 is so very close to 1, the missing information reflected by the information entropy must be very close to 0.

Exercise 46.7.9: From the possible spin assemblies, pick out one and find its energy function. What is the probability for this particular chosen assembly?

Solution to Exercise 46.7.9

All of the points representing the probability distributions for the assembly of $N = 9$ spin sites live in a manifold $\mathcal{S}^{512}$. We label as points p those distributions based on the information from complicated models. We distinguish as points q those particular distributions living in a sub–manifold $\boldsymbol{E}$ based on information permitting only independent distributions.

Any point p reflecting a complicated model can be **e**–projected down into $\boldsymbol{E}$. A point q^* in $\boldsymbol{E}$ that has a minimum value for the dual relative entropy between the two distributions is the mean field solution for physical scenarios like the Ising model.

The points p in the Ising model are themselves relatively simple when judged by the potential complexity they might possess. They are characterized by coordinates in a vector for the individual spins and coordinates in a matrix for some selected set of double interactions between spin sites, that is, the set that has been defined as representing sites that are "nearest neighbors." In Amari's notation, these model parameters are in the vector $\boldsymbol{h}$ and matrix W. In the MEP notation, these model parameters are some relatively small number of Lagrange multipliers λ_j^p.

But the **e**–projection of a point p down to a q^* can, by definition, only be a model with information represented in a vector $\overline{\boldsymbol{h}}$, or, in a smaller number of $\lambda_j^{q^*}$.

Arbitrarily select the 171st assembly which is the same one with the alternating + and − spin sites as was shown in Exercise 46.7.5. The numbering of the assembly comes from how the **Tuples[ ]** function orders the assemblies. The probability of this particular assembly from the total of the 512 available assemblies in MEP notation is,

$$P(X = x_{171} | \mathcal{M}_k) = \frac{e^{\sum_{j=1}^{21} \lambda_j^p F_j(X=x_{171})}}{\sum_{i=1}^{512} e^{\sum_{j=1}^{21} \lambda_j^p F_j(X=x_{171})}}$$

This is the probability under the information in a point p where the "complicated" model looked at the spins at all nine sites together with twelve double interactions between nearest neighbor spin sites. For example, given the layout in the 3×3 lattice, site 1's nearest neighbors were site 2 and site 4, site 2's nearest neighbors were site 1, site 3, and site 5, but the interaction between site 1 and site 2 is only counted once. And so on. The MEP's λ_1^p through λ_9^p is Amari's $\boldsymbol{h}$, and the MEP's λ_{10}^p through λ_{21}^p is Amari's W.

However, the actual formula used to calculate this particular assembly must derive from a point q^* that is the **e**-projection of a point p, not from the above formula directly concerning a point p. So,

$$P(X = x_{171} \mid \mathcal{M}_k) = \frac{e^{\sum_{j=1}^{9} \lambda_j^{q^*} F_j(X = x_{171})}}{\sum_{i=1}^{512} e^{\sum_{j=1}^{9} \lambda_j^{q^*} F_j(X = x_{171})}}$$

The possibly different $\lambda_1^{q^*}$ through $\lambda_9^{q^*}$ are equivalent to Amari's $\overline{\boldsymbol{h}}$.

Nevertheless, to implement both the **e**-projection and mean field theory, the energy function $E(\mathbf{x})$ from the more complicated model behind p must be known and used even though its average is going to be taken with respect to the distribution q^*. Remember the term $\langle E(\mathbf{x}) \rangle_Q$ in MacKay's formulation?

In our numerical example, the energy of the 171st assembly is $E(\mathbf{x}) = -11$ reflecting a $+1$ contribution from the individual spins $\sum_{l=1}^{9} S_l = +1$ and a -12 contribution from the spin site interactions $\sum^{12} S_k S_l = -12$. But take care to notice that this means we must have taken an easier path by assuming that all 21 $\lambda_j^p = 1$ since $E(X = x_{171}) = \sum_{j=1}^{21} \lambda_j^p F_j(X = x_{171})$. In other words, the J_{mn} and h_n values in MacKay's version were all set to 1.

The most basic application of the above would be to find the probability of the assembly as the multiplication of the nine independent probabilities at each of the nine sites, and, moreover, let the probability at each site be the same. The same average constraint function at each site is,

$$\langle F \rangle = \tanh \lambda = \tanh \overline{h} = \tanh 1 = 0.761594$$

leading to a numerical assignment of a probability for a $+$ site as $Q_1 = 0.880797$ and a numerical assignment of a probability for a $-$ site as $Q_2 = 0.119203$. The probability for the assembly of five $+$ sites and four $-$ sites is then,

$$P(X = x_{171} \mid \mathcal{M}_k) = \underbrace{Q_1 \times Q_2 \times \cdots \times Q_1}_{9 \text{ terms}}$$

$$= Q_1^5 \times Q_2^4$$

$$= \left(\frac{e^1}{e^1 + e^{-1}}\right)^5 \times \left(\frac{e^{-1}}{e^1 + e^{-1}}\right)^4$$

$$= 1.07 \times 10^{-4}$$

This is indeed some point q in E, with $\lambda_1^q = \lambda_2^q = \cdots = \lambda_9^q = 1$, but we don't know whether it is the point $q^\star$ that is the minimum squared distance by the criterion of the dual relative entropy measure.

Exercise 46.7.10: Find another independent model that permits the nine λ_j^q to vary.

Solution to Exercise 46.7.10

Find another point q in the sub–manifold E where the nine coordinates λ_j^q are allowed to vary, but because this point q lives in E all of the remaining coordinates must have a value of 0. In order to compute the partition function Z_q, independence must still be assumed at the system level. Arbitrarily, assign the nine λ_j^q from 0.6 to 1.4 in increments of 0.1. Thus, Q_1 at site 1 is,

$$Q_1^{\text{Site 1}} = \frac{e^{0.6}}{e^{0.6} + e^{-0.6}} = 0.768525$$

Q_2 at site 2 is,

$$Q_2^{\text{Site 2}} = \frac{e^{-0.7}}{e^{0.7} + e^{-0.7}} = 0.197816$$

and Q_1 at site 9 is,

$$Q_1^{\text{Site 9}} = \frac{e^{1.4}}{e^{1.4} + e^{-1.4}} = 0.942676$$

Therefore, the different probability for the alternating SPIN UP and SPIN DOWN assembly under the information contained in this model is,

$$P(X = x_{171} \mid \mathcal{M}_k \equiv \text{ new } q) = 0.768525 \times 0.197816 \times \cdots \times 0.942676$$

Exercise 46.7.11: What fundamental conceptual issue has arisen from this discussion of e–projections when it has been applied to problems within statistical mechanics and a comparison with general inferencing?

Solution to Exercise 46.7.11

In statistical mechanics, the energy function $E(\mathbf{x})$ is assumed to be defined somehow from a prior understanding of the physics or chemistry principles underlying the problem. As we have seen in the Ising model for magnetization, an energy function is defined by something like,

$$E(\mathbf{x}) = \sum J_{kl}\, S_k\, S_l + \sum h_l\, S_l$$

The matrix J_{kl} and vector h_l are conceived as something that should be known from physical or chemical knowledge. The energy function bears "all the burden," so to speak, in setting up a complicated model.

Thus, from the perspective of statistical mechanics with the "slack" taken up by the magical energy function, all that remains in order to determine the probability of a state is the question of just one model parameter β in the canonical Boltzmann distribution,

$$P(\mathbf{x}) = \frac{e^{-\beta E(\mathbf{x})}}{\sum e^{-\beta E(\mathbf{x})}}$$

For general inferencing problems, the whole difficulty of complicated models is in specifying the information by constraint functions $F_j(X = x_i)$ together with their associated Lagrange multipliers λ_j. In the MEP formalism, the analog to J_{kl} and h_l are, say, the λ_j for $j = 1$ to 9, for the uncoupled spins, and λ_j for $j = 10$ to 21 for the coupled spins. This is indeed a verification of Jaynes's off the cuff remark that in setting up these $F_j(X = x_i)$ and their associated λ_j, an IP is inserting some physics or chemistry into a general inferencing problem in order to transform it into a problem for statistical mechanics.

But if the IP takes the same approach as in statistical mechanics, these λ_j^p for a complicated model must be known, because otherwise how could a minimum "squared distance" ever be computed between this p with its λ_j^p and some $q^\star$ with its different and smaller set of $\lambda_j^{q^\star}$?

Also, unless one takes $\beta = 1$, there is no place for another "meta–parameter" β within the MEP formalism. All of the model parameters for either the complicated model p or the simpler approximating model $q^\star$ have already been defined through their respective Lagrange multipliers.

Of course, the absolute necessity of finding some solution to the computational intractability of the partition function for the complicated model is still front and center. For general inferencing concerning arbitrarily complicated models this seems to remain a more difficult problem when the full complement of possible interactions is taken into account. Even the most complicated models examined for the Ising model of magnetization are relatively simple because even the most difficult problems look only at a limited number of double interactions.

Chapter 47

Jeffreys's Prior

47.1 Introduction

A confusing mess has grown up within the heart of what most people consider the Bayesian approach to inferencing. The general consensus, among Bayesians and non–Bayesians alike, is that there is some fundamental unresolved issue with "Bayesian priors."

This disconcerting state of affairs surrounding these so–called Bayesian priors does not center on any one well–defined topic, but insinuates itself in a rather vague manner over several overlapping issues. Among the more prominently mentioned "problematic issues" are *non–informative priors*, *entropic priors*, *improper priors*, and *Jeffreys's priors*, the latter, of course, named after the famous (at least to Bayesians) British geophysicist Sir Harold Jeffreys.

It seems to be a rather common lament among Bayesians that if this thorny issue could be resolved in some way, then even the most benighted would have the scales slip from their eyes. In consequence of this fortuitous event, people would live their lives happily ever after computing probabilities as confirmed Bayesians.

This situation has been an unpleasant interlude interrupting progress within the field. Strangely, however, it is not that uncommon within all scientific disciplines to experience lengthy fallow periods resulting from conceptual distinctions being ignored.

Physicists have a rather brutal way of characterizing some of the ill–conceived conceptual gyrations that arise during these confused times. One of the founders of quantum mechanics, the witty and acerbic Wolfgang Pauli, has been attributed with the famous put–down of conceptual groaners with the phrase: **"That's not even wrong!"**

I can think of no better summary of the state of affairs concerning this whole cottage industry of "problematic priors" than Pauli's scathing reaction. In this Chapter, we will return to some of the fundamental concepts in probability theory in an attempt to avoid the wrong turns that have been taken over the years.

The first question one might think to ask is: What prior probability are we talking about? Very few commentators will come right out and tell you at the outset because the problem might very well melt away before our very eyes. The opportunity to dazzle you has escaped. Let's try not to fall into the same trap.

We are talking about the *prior probability for models*. Letting the cat barely peek out of the bag, would you expect that an assignment to some joint statement based on a model already assumed true, that is, $P(X = x_i \,|\, \mathcal{M}_k)$, should ever be considered as some prior probability? Would you expect that conditioning on already observed data, that is, $P(\mathcal{M}_k \,|\, \mathcal{D})$, should ever be considered as some prior probability? Would you expect that an assignment to the next occurrence, that is, $P(X_{N+1} \,|\, X_1, X_2, \cdots X_N)$ should ever be considered as some prior probability?

I don't think you would advocate for any of these expressions as a definition for a *prior probability*. Then what is it? It is simply $P(\mathcal{M}_k)$, the probability assigned to all models *prior* to any data. And I repeat, because it is so very important, that $\mathcal{M}_k$ must be viewed as a **statement** to the effect: "The correct assignment of probabilities to all of the statements in the state space is $q_1, q_2, \cdots, q_n$."

I intend to make some modest progress in clearing up the mysteries surrounding "Bayesian priors" in this Chapter. As a very important starting example, consider Jeffreys's prior. Since the late 1940s when Jeffreys introduced the idea, an enormous number of man hours has gone into the creation of a huge extant literature on Jeffreys's prior alone.

To commence, we will pick up where we left off in Chapter Fifteen of Volume I by disambiguating where Jeffreys's prior fits into the grand scheme of things. The numerical context will be the kangaroo contingency table. I include several exercises at the end of the Chapter for a review of these notions.

It turns out, rather ironically given the contentious history between these two men, that Jeffreys's prior will eventually depend on the Fisher information matrix for its definition. We intend to take advantage of our initial forays into Information Geometry for help on this topic. But that discussion will have to wait until the next Chapter.

To give you an inkling of the conceptual confusion rampant in this area, you may recall that the Fisher information matrix is computed from the probability of statements *given the truth of some model*. You might have naively assumed, if you had just happened to be introduced to this issue of a prior probability, that you might instead have to wrestle with the meaning of a degree of belief for all models before you had observed any data, and not on the degree of belief in joint statements conditioned on the truth of some model.

47.2 Back to Basics

I am going to assume that everyone accepts the axiomatic nature of the **Sum Rule** and **Product Rule** as they apply to the formal manipulation of probabilities. Thus, in a generic notation, one *joint* probability is transformed into the multiplication of *two* probabilities, a *conditional* probability and a *marginal* probability, by the **Product Rule**.

$$P(A = a_i, B = b_j) = P(A = a_i \mid B = b_j) P(B = b_j) \qquad (47.1)$$

If the IP is interested in the probability of statement $(A = a_i)$, then an application of the **Sum Rule** results in,

$$P(A = a_i) = \sum_{j=1}^{n_B} P(A = a_i, B = b_j) \qquad (47.2)$$

$$= \sum_{j=1}^{n_B} P(A = a_i \mid B = b_j) P(B = b_j) \qquad (47.3)$$

where n_B represents the number of statements involving just B.

However, these expressions really only apply at the level of abstract probabilities where the practical concern of assigning legitimate numerical values to the abstract probabilities $P(A = a_i \mid B = b_j)$ and $P(B = b_j)$ never arises. If not quite an axiom, we take it as an imperative that a third type of statement, a statement about models, must also enter the picture. Volume II was devoted to showing how *information* resident in these models is one very good way to assign numerical values to abstract probabilities.

In this expanded picture that includes the statement about a model, while still relying upon the **Product Rule** and the **Sum Rule**, we now have,

$$P(A = a_i, B = b_j, \ldots, \mathcal{M}_k) = P(A = a_i, B = b_j, \ldots \mid \mathcal{M}_k) P(\mathcal{M}_k) \qquad (47.4)$$

$$P(A = a_i, B = b_j, \ldots) = \sum_{k=1}^{\mathcal{M}} P(A = a_i, B = b_j, \ldots \mid \mathcal{M}_k) P(\mathcal{M}_k) \qquad (47.5)$$

After translating any complicated joint statement $(A = a_i, B = b_j, \ldots)$ into the less cumbersome notation generally employed in my exposition, we have the explicit appearance of *one* marginal probability on the left hand side of Equation (47.6) decomposed into *two* separate probabilities on the right hand side.

$$P(X = x_i) = \sum_{k=1}^{\mathcal{M}} P(X = x_i \mid \mathcal{M}_k) P(\mathcal{M}_k) \qquad (47.6)$$

The first term on the right hand side of Equation (47.6), $P(X = x_i \mid \mathcal{M}_k)$, shows that the probability for some statement is conditioned upon a model, while the second term $P(\mathcal{M}_k)$ is the *prior probability for a model*. Thus, the derived *marginal*, not *prior*, probability $P(X = x_i)$ resulting from these formal manipulation rules happens to be constructed from *two* separate probabilities, a conditional probability $P(X = x_i \mid \mathcal{M}_k)$, and a prior probability $P(\mathcal{M}_k)$.

Now, $P(X = x_i \mid \mathcal{M}_k)$ is, in some sense, pre–determined because of the assumed truth of model $\mathcal{M}_k$. Remember that model $\mathcal{M}_k$ is a statement to the effect: **"The correct numerical values of probability to assign to the statements in the state space are $q_1, q_2, \cdots, q_n$."**

Not only is it permissible, but, in fact, mandatory to attach the probability operator around this statement $\mathcal{M}_k$, $P(\mathcal{M}_k)$, to express the IP's degree of belief that this above statement concerning the correct q_i is, in fact, TRUE.

When model $\mathcal{M}_k$ *is* assumed true, then $P(X = x_i \mid \mathcal{M}_k)$ becomes Q_i. If we desire to invoke the MEP rationale that information can be inserted into a probability distribution under the auspices of model $\mathcal{M}_k$, then the MEP formula provides us with the Q_i,

$$Q_i = \frac{\exp\left[\sum_{j=1}^{m} \lambda_j F_j(X = x_i)\right]}{Z(\lambda_j)} \tag{47.7}$$

where the λ_j are the *parameters* of the model appearing *only* in the above MEP formula for the conditional probability $P(X = x_i \mid \mathcal{M}_k)$. These parameters DO NOT appear in the prior probability $P(\mathcal{M}_k)$.

Now comes the critical conceptual distinction. The probability $P(\mathcal{M}_k)$ has no linkage to the probability $P(X = x_i \mid \mathcal{M}_k)$. It is a separate and distinct probability. It is very convenient for analytical purposes to adopt this conjugate probability density function for the models,

$$\text{pdf}(\mathcal{M}_k) = \int \cdots \int_{\sum q_i = 1} \frac{\Gamma(\sum_{i=1}^{n} \alpha_i)}{\prod_{i=1}^{n} \Gamma(\alpha_i)} q_1^{\alpha_1 - 1} q_2^{\alpha_2 - 1} \cdots q_n^{\alpha_n - 1} \, dq_i \tag{47.8}$$

These parameters labeled α_i appearing in the probability density function for $\mathcal{M}_k$ have nothing to do with the parameters λ_j appearing in the MEP formula. In order to further emphasize the distinction between the two probabilities, only the first term in Equation (47.6), $Q_i \equiv P(X = x_i \mid \mathcal{M}_k)$, has ever been mentioned in connection with the MEP formula.

Invoking Equation (47.8) for the prior probability of a model seems rather *ad hoc* because its form is very convenient when it becomes necessary to process some data. Nonetheless, it too, like any pdf, possesses an MEP rationale for its derivation. But I thought it just might add even more confusion to an already impossibly complicated situation to mention this fact earlier.

So to argue that $P(\mathcal{M}_k)$ should be derived from $P(X = x_i \mid \mathcal{M}_k)$ is fallacious conceptually right out of the starting gate. But amazingly that is, in fact, exactly what is done in the argument for Jeffreys's prior when it is defined as a function of Fisher's information matrix,

$$P(\mathcal{M}_k) \propto \mid g_{rc}(p) \mid^{1/2} \qquad (47.9)$$

We now know that the metric tensor $g_{rc}(p)$ is computed as,

$$g_{rc}(p) \equiv E_p \left[\frac{\partial \ln P(X = x_i \mid \mathcal{M}_k)}{\partial \lambda_r} \times \frac{\partial \ln P(X = x_i \mid \mathcal{M}_k)}{\partial \lambda_c} \right] \qquad (47.10)$$

$$= \sum_{i=1}^{n} \{ [F_r(X = x_i) - \langle F_r \rangle] \times [F_c(X = x_i) - \langle F_c \rangle] \} Q_i \qquad (47.11)$$

A serious conceptual error arises if you try to use Equation (47.11) to find $P(\mathcal{M}_k)$.

The Fisher information matrix is found by differentiating with respect to the model parameters λ_j in,

$$P(X = x_i \mid \mathcal{M}_k) \equiv Q_i = \frac{\exp\left[\sum_{j=1}^{m} \lambda_j F_j(X = x_i)\right]}{Z(\lambda_j)} \qquad (47.12)$$

The parameters α_i that appear in the probability density function for $P(\mathcal{M}_k)$ are completely different than the parameters λ_j appearing in the MEP formula.

So, my purpose here has been to cleanly separate out the conceptual distinction between $P(X = x_i \mid \mathcal{M}_k)$ and $P(\mathcal{M}_k)$. Repeating, attempting to form a prior probability for models on the basis of the λ_j is a misplaced enterprise, but that in fact is what is done.

Jeffreys, along with every one else, states flat out that the major issue here is the notion of a probability distribution over *parameters*. I quote him extensively about his thoughts on this matter in the upcoming **Connections to the Literature**. However, nowhere in my formal manipulation rules do we see any expressions like $P(\lambda_j)$ or $P(\alpha_i)$. What is the reason for this omission?

A model $\mathcal{M}_k$ and the completely separate concept of a *degree of belief* about model $\mathcal{M}_k$, captured as $P(\mathcal{M}_k)$, exist at different levels in a hierarchy. If you ask me how a model can provide numerical assignments of probability to statements, I will tell you that the parameters λ_j are inserting particular information under that model.

No uncertainty exists at this level of a purely mathematical problem; the MEP algorithm solves this purely mathematical problem through standard variational techniques. No probability is required. There is no degree of belief that the Lagrange multipliers λ_j are correct or not. They either solve the problem as a mathematics problem or they do not.

If the IP should move up one level in the hierarchy, attaching a probability to a model is no cause for alarm. The IP, as an epistemological creature, has every right to a legitimate degree of belief about whether the Q_i assigned under the information resident in $\mathcal{M}_k$ are really correct. The α_i parameters help provide this probability where, again, there is no uncertainty about the α_i. The α_i parameters either work correctly within Equation (47.8) as a component of the Dirichlet distribution, or they don't.

The λ_j and the α_i are "walled–off," as it were, living in separate universes where no communication can or need take place between them. They are part of a deductive mathematical framework. The idea that an IP maintains degrees of belief about mathematical components and thus attaches probabilities to these parameters is nonsense. Thus, any effort to find an uninformative prior probability for a model by manipulations on the λ_j is doomed to failure.

47.3 What Upset Jeffreys?

Where lies the origin of "Jeffreys's prior?" What was Jeffreys objecting to so strenuously, and why?

There are several places scattered throughout Jeffreys's book

Theory of Probability

where he lays out his thoughts on this issue. He objected to what he considered an uncritical acceptance of the Bayes–Laplace flat prior for models. My first mention of this was in Chapter Thirteen of Volume I where Jeffreys's famous tirade about "animals with feathers" was introduced. Then in Chapter Fifteen, the impact of varying the α_i parameters in the Dirichlet distribution was studied intensively. Most recently, we revisited these issues in Chapter Forty Four.

Jeffreys shared the near universal contempt for Laplace's oft–quoted probability for the Sun to rise tomorrow given its obvious success some number of times in the past. Moreover, granting precedence to Karl Pearson for first mentioning these things, he found it hard for a scientist to stomach Laplace's *Rule of Succession* for the probability of repeating some physical process after having once observed the process in question take place.

And then there were his objections, spelled out in scathing and contemptuous detail, contained within his "animals with feathers" example. He was concerned with what was observed in sampling from a finite population, and the resulting probabilities when taking the Bayes–Laplace flat prior as an assumption.

He found that even when treating the case of sampling from a finite population without replacement and using Laplace's prescription for ignorance about the total number in the population, he still ended up with the *Rule of Succession*. This conclusion upset him so much that he decided that the culprit in this story had to be the adoption of Laplace's uniform prior.

The only thing for it was to perform some radical surgery. He did just that by considering various reallocations of the prior probability. The most well–known surgery was to allocate more probability to the end points where a population was homogeneous in one type or the other. Thus, as a consequence, we see in the "Jeffreys's prior" for the binary case, a heightened prior probability made to assignments q that happen to be close to 0 or 1.

Chronologically, in presenting these first examples, Jeffreys chose to make a direct modification to the uniform prior probability. Initially, there was no appeal to any "parameter invariance" argument involving the Fisher information matrix.

This rationale appeared later and is conceptually distinct (as well as wrong) from his simple displeasure with a uniform prior probability. Jeffreys disagreed with both Bayes's and Laplace's rationale, but he did not break conceptually with the notion that it is at the level of $P(\mathcal{M}_k)$ where one expresses that disagreement.

The simplest and clearest example of what Jeffreys was objecting to, as well as his counter proposal to the Bayes–Laplace flat prior, is contained in his "test of whether a suggested value of chance is correct." Here he sets up a basic binary problem which we might as well take to be our pedagogical coin flip scenario.

In Jeffreys's opinion, there are certain situations where the IP should place a lump of prior probability on a specific well–disposed hypothesis. Any remaining prior probability should be distributed over the alternative hypotheses.

For example, suppose that the IP is favorably inclined to think that, more often than not, coins really are fair. In consequence, spreading the prior probability evenly over all conceivable values of q, that is, evenly over all q from 0 to 1 for the probability of HEADS, as Laplace advises us to do, is unjustified. Instead, one might consider placing a lump of prior probability, say 1/2, on a specific value of q, here that would also be $Q_i = 1/2$ for the favored hypothesis of a fair coin, and distribute the remaining 1/2 of the prior probability over all values of q.

Thus, there are two hypotheses (and Jeffreys used the extant terminology due to Fisher of a null and an alternative hypothesis) each with prior probability of 1/2. In our notation, we have one model $\mathcal{M}_A$ asserting that the correct value of q is 1/2, and a second model $\mathcal{M}_B$ asserting that the correct value of q could be anywhere in the interval from 0 to 1 inclusive. The prior probability of the first model is $P(\mathcal{M}_A) = 1/2$ with the prior probability of the second model also $P(\mathcal{M}_B) = 1/2$.

Jeffreys derives the ratio favoring the null hypothesis of a fair coin over the alternative hypothesis of a biased coin with his unnecessarily confusing notation,

$$K = \frac{P(q \mid \theta H)}{P(q' \mid \theta H)} = \frac{(x+y+1)!}{x!\, y!} p^x (1-p)^y \qquad (47.13)$$

I devote a few exercises to translating his obscure notation to a more familiar one. In our notation, the above result in Equation (47.13) is,

$$\frac{P(\mathcal{M}_A \mid \mathcal{D})}{P(\mathcal{M}_B \mid \mathcal{D})} = \frac{(N+1)!}{N_1!\, N_2!} Q_1^{N_1} Q_2^{N_2} \qquad (47.14)$$

where $Q_1 = Q_2 = 1/2$ under the null hypothesis.

He then provides some numerical examples of this formula in the vein of: If 5 HEADS and no TAILS have been observed in 5 coin flips, then the ratio favoring the null hypothesis of a fair coin is less than 0.20. If 5 HEADS and 5 TAILS have been observed in 10 coin flips, then the ratio favoring the null hypothesis of a fair coin is greater than 2.7. These outcomes do not conflict with our intuition.

Another revealing example where Jeffreys directly modifies the prior probability is when he derives the probability in a finite sample. All I want do at this point is to show how his result corresponds to my result for calculating the probability of the data. Here, the IP would be interested in the probability of some number of HEADS and TAILS in N tosses of the coin. But is nothing known about the coin, or the manner in which it will be tossed?

Jeffreys provides this answer, as presented below in his notation, as,

$$P(r \mid nH) = \frac{(r - 1/2)! \, (n - r - 1/2)!}{\pi \, r! \, (n-r)!} \qquad (47.15)$$

As I have said before in my books, I hope to provide the reader with at least the contribution of establishing where some common ground lies amidst all of the varied and idiosyncratic notation. In that vein, here is the same answer provided by Jeffreys in Equation (47.15) in my notation,

$$P(N_1, N_2) = \frac{(N_1 - 1/2)! \, (N_2 - 1/2)!}{\Gamma(\tfrac{1}{2}) \, \Gamma(\tfrac{1}{2}) \, N_1! \, N_2!} \qquad (47.16)$$

The inferential problem that Jeffreys was grappling with was the probability of N_1 HEADS and N_2 TAILS in N tosses of a coin, or, in other words, the probability of the data. Because the formal manipulation rules demand $P(\mathcal{M}_k)$ in the derivation, in other words, demand a prior probability for the models, Jeffreys had to find his own personal resolution for this issue. Obviously, his resolution could not agree with how Bayes and Laplace resolved the issue. Exercise 47.8.8 provides all of the, as it turns out, quite interesting details.

In what was eventually labeled as "Jeffreys's prior," he placed more probability on the end points near 0 and 1 when considering all the possible assignments from 0 to 1. Bayes and Laplace had invoked a "flat prior" placing equal probability over the entire spectrum from 0 to 1. So it appears that, contrary to the first example when he wanted to place a lump of prior probability on one specific suggested value of a chance like 1/2 for a fair coin, he now wants to emphasize models which lend some prominence to coins which will show either all HEADS or all TAILS.

Mathematically, in order to arrive at the result in Equation (47.15) above, he chose a *beta distribution* for $P(\mathcal{M}_k)$ with the α_i parameters set not equal to $\alpha_i = 1$ as would be appropriate for a uniform distribution, but rather set to $\alpha_i = 1/2$. Let me stress that, once again, conceptually I have no argument with the formalism he adopts. He correctly focuses on $P(\mathcal{M}_k)$, while at the same time, exhibits his disagreement with Bayes and Laplace by choosing different parameters for the *beta distribution*, the conjugate distribution for Q_1 and Q_2.

47.4 Parametric Invariance

Rather than starting with Jeffreys's displeasure over the consequences of adopting a uniform prior probability for the model space as I have chosen to do above, most commentators settle on the later rationale of *parametric invariance*. This notion seems to be relied upon almost universally as the justification for Jeffreys's priors. First, though, ask yourself this question: *What really must remain invariant in inferencing?*

The numerical assignment to the joint statements made under the information provided by some model must remain invariant. If some model should make the assignments Q_i, changing the parameters and constraint functions appropriately should make no difference to the assignments. Thus, it is not the parameters which are invariant, but the assignment under a given model, together with the implied information under that model, which must remain unchanged.

The MEP represents a beautiful example of this principle. The MEP might assign a definite set of Q_i under the information resident in some model $\mathcal{M}_k$ through the Lagrange multipliers λ_j and the set of constraint functions $F(X = x_i)$. But the parameters λ_j or the dual parameters $\langle F_j \rangle$ can change along with the constraint functions to provide the same assignment. The parameters are not invariant, but the assignment under the specified model must remain invariant.

We have already seen early examples of this attractive feature when the MEP was introduced in Volume II. It is not necessary to bring up anything more complicated than our ubiquitous coin toss to illustrate the point. Refer back to Tables 17.1 and 18.1.

In Table 17.1, the arbitrary constraint function for HEADS was $F(X = x_1) = 1$ and for TAILS was $F(X = x_2) = 2$ with information under some model $\mathcal{M}_k$ that $\langle F \rangle = 1.25$. This information resulted in an assignment of $Q_1 = 0.75$ and $Q_2 = 0.25$ with the dual parameter forced to take the value of $\lambda = -1.09861$.

But this assignment of $Q_1 = 0.75$ and $Q_2 = 0.25$ remains invariant although the parameters have been changed to $\lambda = 0.137$, and $\langle F \rangle = 1.00$ with the change to the constraint function of $F(X = x_1) = -3$ and $F(X = x_2) = 5$. Moreover, there is no sense that equal intervals for the λ parameter should result in equal intervals for the Q_i. For example, in Table 17.1, when λ changes by $+1$ from $\lambda = -3$ to $\lambda = -2$, Q_1 changes from $Q_1 = 0.9526$ to $Q_1 = 0.8808$. However, when the change from $\lambda = 0$ to $\lambda = 1$ is made, Q_1 changes from $Q_1 = 0.5000$ to $Q_1 = 0.2689$.

For Jeffreys, everything about a prior probability for a model was bound up inextricably with the idea of a probability distribution over the parameters. So you see contorted explanations as to why expressions like $P(\mu, \sigma)$ for the parameters of the Gaussian distribution have to take certain forms based on arguments related to equal lengths of these parameters over say $-\infty$ to $+\infty$ for μ or 0 to $+\infty$ for σ.

Putting it rather absurdly, but rather bluntly, an IP is allowed to assign a prior probability to a model of say $P(\mathcal{M}_A) = 0.70$ where $\mathcal{M}_A$ assigns a Gaussian distribution with $\mu = 10,000$ and $\sigma = 0.001$, a prior probability to a model of $P(\mathcal{M}_B) = 0.20$ where $\mathcal{M}_B$ assigns a Gaussian distribution with $\mu = -36.678$ and $\sigma = 10,000$, and the remaining prior probability evenly distributed over all models where μ ranges from $\mu = -\infty$ to $-700,000$ and $\sigma = 10^{-9}$.

It is clear that the information inserted into these probability distributions by the model parameters μ and σ has nothing to do with the prior probability of these models. In addition, the allowable range of the parameters, and consideration of any intervals over the real line, have nothing to do with the prior probability of these models.

47.5 Geometrical Rationale for Jeffreys's Prior

Jeffreys himself never really seemed to provide any kind of extended geometrical rationale for the prior probability of models. It seems that it was more the concept of "parametric invariance" accompanied by the mathematics of the transformation of variables within probability distributions that motivated him.

One rationale that I am aware of for Jeffreys's prior from the perspective of Information Geometry comes from a short commentary by Amari [2]. A second related, but more complicated treatment, was given by Kass and Vos [25]. The central geometrical idea is that Jeffrey's prior probability for parameters can be characterized mathematically as a uniform distribution over a hypersphere.

It is just possible with a lot of squinting and hand–waving to sketch out a correspondence between what we have been doing all along by invoking the Dirichlet distribution with the α_i parameters equal to $1/2$ as Jeffreys's prior probability for models, and the Information Geometry approach to a uniform distribution over a hypersphere. I devote the next Chapter to some initial musings along this line. There, I make some conjectures concerning my proposed analogies between Amari's brief geometrical comments about Jeffreys's prior and the Dirichlet distribution.

47.6 Important Concepts

Here are the important expressions and conceptual distinctions to keep uppermost in your mind when thinking about "Bayesian priors."

There are statements, and then there are degrees of belief about how true those statements are. $(X = x_i)$, $(X = x_i \,|\, \mathcal{M}_k)$, and $(\mathcal{M}_k)$ are *statements*. $P(X = x_i)$, $P(X = x_i \,|\, \mathcal{M}_k)$, and $P(\mathcal{M}_k)$ are *degrees of belief* as held by the IP about how true those statements are. $P(X = x_i)$ is a *marginal* probability, $P(X = x_i \,|\, \mathcal{M}_k)$ is a *conditional* probability, and $P(\mathcal{M}_k)$ is a *prior* probability.

The relationship among these three probabilities is spelled out by the formal manipulation rules that led to Equation (47.6). The number of models $\mathcal{M}$ is a number which tends to infinity, and when we let it approach infinity, the sum in Equation (47.6) gets transformed into an integral. The integration takes place over the entire region where the sum of the q_i equals 1. So it is the q_i, the numerical assignments that are made to the probabilities of the statements, that is integrated over and every single conceivable q_i is involved in this sum. The origin of each q_i involved in the integration, for example, by tracing its genesis back to the MEP formula, is irrelevant at this point.

There are no probability expressions like $P(\lambda_j)$ or $P(\alpha_i)$ for the set of *parameters* λ_j or α_i. Nor are they *estimated*. In opposition to the conventional wisdom held by Jeffreys, Fisher, Jaynes, and to my knowledge everyone else who has ever written on this topic, I banish the phrases **probability estimates** and **parameter estimates** from our vocabulary. They can keep company in the rubbish bin with other phrases like **measure theory**, **random variables**, and **probability as frequency**.

These parameters are found as part of a mathematical procedure. In addition, keep uppermost the separate and distinct nature of the λ_j or α_i parameters. The λ_j insert information from *one* model in order to tell us a legitimate degree of belief in the truth of a statement conditioned on assuming the truth of another statement, namely, $P(X = x_i \mid \mathcal{M}_k)$. The α_i insert information to tell us a legitimate degree of belief in the truth of the model, namely, $P(\mathcal{M}_k)$.

What could the *posterior probability of a parameter*, $P(\lambda \mid \mathcal{D})$, possibly mean? Since λ is, by its very definition, helping to insert information into a probability distribution independent of, and prior to, any data, and conceptually nothing would change about this role of λ even after data were known, the expression $P(\lambda \mid \mathcal{D})$, as Pauli would have it, is not even wrong!

Parameters and models do not exist in the real world. They are convenient fictions. An IP cannot measure, observe, or categorize them as it can the $(X = x_i)$. A model is a necessary mathematical expediency introduced for the sole purpose of getting on with the computation. They exist solely in a mathematical world to facilitate the need to assign numerical values from an informational perspective. And then to expunge this embarrassment, the models are dispensed with by the process of marginalization in a final *coup de grâce*.

Because numerical computation cannot take place with abstract probabilities, we must resort to models. However, the foundations of the theory do not permit a degree of belief about something like parameters that exist in a purely deductive world of mathematics.

An IP *is allowed* to attach probabilities to statements that eventually turn out to be true or false when observed. In a mathematical world, things are either true or not true not by their observation; but rather by whether they function correctly mathematically.

Parameters *are not allowed* to exist in a state of superposition with varying truth values as are statements about our real world. An IP cannot measure, observe, or categorize a parameter that exists mathematically, but it can do so for some statement about the real world. A statement made by a model may also have a probability attached to it because such a statement about numerical assignments is not part of a deductive framework as λ_j and the α_i are.

47.7 Connections to the Literature

We begin by referring to a passage early in Jeffreys's book [24, pg. 28] in order to ensure that his concepts and notation match up with ours. They are, in fact, equivalent, but more interestingly, there can be little doubt about what the phrase *prior probability* means to Jeffreys, at least in these early stages.

> This is the *principle of inverse probability*,
>
> $$P(q_r \,|\, pH) \propto P(q_r \,|\, H)\, P(p \,|\, q_r H) \qquad (4)$$
>
> first given by Bayes in 1763. It is the chief rule involved in the process of learning from experience. It may also be stated, by means of the product rule, as follows:
>
> $$P(q_r \,|\, pH) \propto P(pq_r \,|\, H) \qquad (5)$$
>
> This is the form used by Laplace, by way of statement that the posterior probabilities of causes are proportional to the probabilities *a priori* of obtaining the data by way of those causes. In the form (4), if p is a prescription of a set of observations and the q_r a set of hypotheses, the factor $P(q_r \,|\, H)$ may be called the *prior probability*, $P(q_r \,|\, pH)$ the *posterior probability*, and $P(p \,|\, q_r H)$ the likelihood, a convenient term introduced by Professor R. A. Fisher ... [Emphasis in the original.]

The equivalancies in concept and notation are listed below as,

$$P(q_r \,|\, H) \equiv P(\mathcal{M}_k) \qquad \text{Prior probability}$$

$$P(p \,|\, q_r H) \equiv P(\mathcal{D} \,|\, \mathcal{M}_k) \qquad \text{Likelihood}$$

$$P(q_r \,|\, pH) \equiv P(\mathcal{M}_k \,|\, \mathcal{D}) \qquad \text{Posterior probability}$$

$$P(q_r \,|\, pH) \propto P(q_r \,|\, H)\, P(p \,|\, q_r H) \equiv P(\mathcal{M}_k \,|\, \mathcal{D}) = \frac{P(\mathcal{D} \,|\, \mathcal{M}_k)\, P(\mathcal{M}_k)}{\sum_{k=1}^{\mathcal{M}} P(\mathcal{D} \,|\, \mathcal{M}_k)\, P(\mathcal{M}_k)}$$

Jeffreys [24, pg. 29] goes on to remark that,

> The use of this principle is easily seen in general terms. If there is originally no ground to believe in one set of alternatives rather than another, the prior

probabilities are equal. The most probable, when evidence is available, will then be the one that was most likely to lead to that evidence. We shall be most ready to accept the hypothesis that requires the fact that the observations have occurred to be the least remarkable coincidence.

These are sentiments which, following in the footsteps of Laplace, we agree with wholeheartedly. But the subsequent evolution of Jeffreys's thought led to a rejection of this initial sympathetic viewpoint and to its replacement with the *Jeffreys's prior*.

Because Jeffreys's entire argument leading to his prior hinges on the notion of *sampling from a population*, we review his introductory thoughts and notation on this matter [24, pg. 59].

> Suppose that we have a population, composed of members of two types ϕ and $\sim\phi$, in known numbers. A sample of given number is drawn in such a way that any set of that number in the population is equally likely to be taken. What, on these data, is the probability that the numbers of the two types will have a given pair of values?
>
> Let r and s be the numbers of type ϕ and $\sim\phi$ in the population, l and m those in the sample. The number of possible samples, subject to the conditions, is the number of ways of choosing $l+m$ things from $r+s$ which we denote by $^{r+s}C_{l+m}$. The number of them that will have precisely l things of type ϕ and m of type $\sim\phi$ is $^{r}C_{l}\,^{s}C_{m}$. Now on data H any two particular samples are exclusive alternatives and are equally probable; and some sample of total number $l+m$ must occur. Hence the probability that any particular sample will occur is $1/\,^{r+s}C_{l+m}$; and the probability that the actual numbers will be l and m is obtained, by the addition rule, by multiplying this by the total number of samples with these numbers. Hence
>
> $$P(l, m \,|\, H) = {}^{r}C_{l}\,{}^{s}C_{m} \,/\, {}^{r+s}C_{l+m}$$

As a quick illustration of what Jeffreys is talking about here, suppose we define a population of ten balls, with $r = 7$ red balls and $s = 3$ black balls as the two types. What is the probability of drawing a sample, without replacement, of $l = 2$ red balls and $m = 2$ black balls? The probability of any sample is then,

$$\frac{1}{{}^{r+s}C_{l+m}} = \frac{1}{{}^{10}C_{4}}$$

$$^{10}C_{4} = \frac{10!}{(10-4)!\,4!}$$

$$\frac{1}{{}^{10}C_{4}} = \frac{1}{210}$$

The number of ways of getting two red balls and two black balls is,

$$^{7}C_{2}\,{}^{3}C_{2} = \frac{7!}{(7-2)!\,2!} \times \frac{3!}{(3-2)!\,2!} = 21 \times 3$$

Thus, the probability of obtaining two red balls and two black balls in four draws from the population of seven red balls and three black balls is,

$$P(l = 2, m = 2 \,|\, H) = \frac{63}{210}$$

I devote Exercises 47.8.1 through 47.8.4 to a rather extensive numerical example that tries to explicate these introductory notions offered by Jeffreys on sampling from a population without replacement.

As confirmation on these numbers, ask yourself what are the other possibilities for four draws from the population? It's impossible to have drawn four black balls and no red balls, and we have already calculated that there are 63 ways for two red balls and two black balls. There are only three cases left. (1) One red and three black in 7 possible ways (obvious), (2) three red and one black in 105 ways, and (3) all four red in 35 ways. These total to the required 210 equally possible cases of drawing four balls from the ten in the population, $0 + 7 + 63 + 105 + 35 = 210$.

Here is another way to confirm these numbers, a way that I prefer over Jeffreys's combinatorial type argument because it emphasizes the probabilities that Jeffreys is assuming at each draw from the population. The easiest case is the probability for drawing a red ball on all four draws. From Jeffreys's formula we know that this probability is,

$$P(l = 4, m = 0 \,|\, H) = \frac{35}{210}$$

Given the stipulation in Jeffreys's description of the problem and sampling without replacement, the probability of a red ball at the first draw is 7/10, at the second draw 6/9, at the third draw 5/8, and, at the fourth and final draw 4/7. Thus the probability of drawing all four red balls from the total population of seven red balls and three black balls is,

$$\left(\frac{7}{10} \times \frac{6}{9} \times \frac{5}{8} \times \frac{4}{7}\right) = \frac{840}{5040} = \frac{35}{210} = \frac{1}{6}$$

For the case of drawing one red ball and three black balls, the combinatorial formula yields,

$$P(l = 1, m = 3 \,|\, H) = \frac{7}{210}$$

There are four ways of getting one red ball and three black balls in four draws, namely getting the one red ball at the first, second, third, or fourth draw, and the three black balls at the other three remaining draws. The probability for each of these four possibilities is 42/5040. For example, the probability of obtaining the red ball at the third draw, and the black balls at the first, second, and fourth draw is,

$$\left(\frac{3}{10} \times \frac{2}{9} \times \frac{7}{8} \times \frac{1}{7}\right) = \frac{42}{5040}$$

Adding up the probability for all four possibilities confirms the probability obtained from the combinatorial formula,

$$P(l=1, m=3 \mid H) = \left(4 \times \frac{42}{5040}\right) = \frac{168}{5040} = \frac{7}{210}$$

Let's skip ahead to where Jeffreys presents his rationale for determining the probability of seeing any future number of the two types after having sampled from a finite population. He ends up repeating Laplace's use of a uniform prior probability, but curiously employs a rather long, tortured argument based mainly on combinatorial reasoning to arrive at exactly the same formula. Essentially, he has re–derived Laplace's *Rule of Succession* in all of its detail. As just mentioned, he accomplished this by relying upon the uniform prior probability over model space.

My derivation of the very same formula, based on the easier reasoning of Jaynes and Laplace, was first presented as Equation (12.3), Volume I, and repeated in Exercises 20.9.1, 22.6.8, and 22.6.9 in Volume II. The posterior predictive probability formula was most recently used in the Chapter Forty Four Exercises. Further derivations are explored in the upcoming Exercises 47.8.5 and 47.8.6.

I will illustrate the parallels in the two formulas by finding the same answers as Jeffreys in the quote below [24, pp. 127–128], but arrived at independently through my formula.

> Hence the probability given the sample that the next $l' + m'$ will consist of just l' of the first type, in any order, is
>
> $$P(l', m' \mid l, m, N, H) = \frac{(l' + m')!}{l'! \, m'!} \frac{(l+1)\ldots(l+l')(m+1)\ldots(m+m')}{(l+m+2)\ldots(l+m+l'+m'+1)}$$
>
> This leads to some further interesting results. Suppose that $m = 0$, so that the sample is all of one type. Then the probability given the sample that the next will be of the type is $(l+1)/(l+2)$, which will be large if the sample is large. The probability that the next l' will all be of the type ($m' = 0$) is $(l+1)/(l+l'+1)$. Thus given that all members yet examined are of the type, there is a probability $\frac{1}{2}$ that the next $l+1$ will be of the type ...

After matching up Jeffreys's notation with mine, everything thereafter flows without discord. The state space is $n = 2$. The future frequency counts for the two types are,

$$l' \equiv M_1$$

$$m' \equiv M_2$$

$$l' + m' \equiv M$$

The data from the sampling is,

$$l \equiv N_1$$

$$m \equiv N_2$$

$$l + m \equiv N$$

The posterior predictive probability formula for $n = 2$ is,

$$P(M_1, M_2 \mid N_1, N_2) = \frac{M!\,(N+1)!}{N_1!\,N_2!\,(M+N+1)!} \times \frac{(M_1+N_1)!\,(M_2+N_2)!}{M_1!\,M_2!}$$

Thus, for Jeffreys's case where $m = 0$, "the sample is all of one type", we have, $N_1 = N$, and $N_2 = 0$, and "the next will be of the type," we have $M_1 = 1$, $M_2 = 0$.

$$P(M_1 = 1, M_2 = 0 \mid N_1 = N, N_2 = 0) = \frac{M!\,(N+1)!}{N_1!\,N_2!\,(M+N+1)!} \times \frac{(M_1+N_1)!\,(M_2+N_2)!}{M_1!\,M_2!}$$

$$= \frac{1!\,(N+1)!}{N!\,0!\,(N+2)!} \times \frac{(N+1)!\,0!}{1!\,0!}$$

$$= \frac{N+1}{N+2}$$

$$= \frac{N_1+1}{N_1+2}$$

$$\equiv \frac{(l+1)}{(l+2)}$$

The same formula is used for Jeffreys's second case where he wants to determine the probability that the next M will all be of the same type as the sample.

$$P(M_1 = M, M_2 = 0 \mid N_1 = N, N_2 = 0) = \frac{M!\,(N+1)!}{N_1!\,N_2!\,(M+N)!} \times \frac{(M_1+N_1)!\,(M_2+N_2)!}{M_1!\,M_2!}$$

$$= \frac{M!\,(N+1)!}{N!\,0!\,(M+N+1)!} \times \frac{(M+N)!\,0!}{M!\,0!}$$

$$= \frac{N+1}{M+N+1}$$

$$= \frac{N_1+1}{N_1+M_1+1}$$

$$\equiv \frac{(l+1)}{(l+l'+1)}$$

As Jeffreys claims, it is true that the probability for the next $M_1 = N + 1$ is $1/2$.

$$P(M_1 = N+1, M_2 = 0 \mid N_1 = N, N_2 = 0) = \frac{M!\,(N+1)!}{N_1!\,N_2!\,(M+N)!} \times \frac{(M_1+N_1)!\,(M_2+N_2)!}{M_1!\,M_2!}$$

$$= \frac{N_1 + 1}{N_1 + M_1 + 1}$$

$$= \frac{N_1 + 1}{N_1 + 1 + N_1 + 1}$$

$$= 1/2$$

Retaining for the time being the stilted language of "types," if the population happens to consist of 10,001 things of two types, and a sample of 5000 reveals that all of them were of the first type, then the probability is $1/2$ that all the remaining 5001 things are of the first type. Laplace would agree with this answer if we told him that we knew absolutely nothing about the two types, **or if we knew nothing about how the population was being sampled**.

In the first few pages of Chapter 6 in his book [22, pp. 149–155], Jaynes treats this problem of sampling without replacement in a manner very similar to Jeffreys to also arrive at Laplace's *Rule of Succession*. Curiously, this Chapter is entitled "Elementary parameter estimation" but for the life of me I can't figure out what the contents have to do with any "parameter estimation." Far more cogent and instructive, in my opinion, is Jaynes's analysis of this problem through the formal manipulation rules. Refer to Exercise 47.8.2 for an introductory illustration of Jaynes's hypergeometric distribution giving the same answers as Jeffreys.

After digesting the consequences of his "animals with feathers" example, Jeffreys [24, pg. 129] is ready to chuck Bayes's and Laplace's recommendation for a uniform prior overboard. It seems that Jeffreys's ire is directed mainly at Laplace.

> Now I say that for that reason the uniform assessment must be abandoned for ranges including the extreme values ... An adequate theory of scientific investigation must leave it open for any hypothesis whatever that can be clearly stated to be accepted on a moderate amount of evidence. It must not rule out a clearly stated hypothesis, such as that a class is homogeneous, until there is definite evidence against it. ... *Any clearly stated law has a positive prior probability, and therefore an appreciable posterior probability until there is definite evidence against it.* This is the fundamental statement of the simplicity postulate. The remarkable thing, indeed, is that this was not seen by Laplace, who in other contexts is referred to as the chief advocate of extreme causality. Had he applied his analysis of sampling to the estimation of the composition of an entire finite population, it seems beyond question that he would have seen that it could never lead to an appreciable probability for a single general law, and is therefore unsatisfactory. [Emphasis in the original.]

Jeffreys makes some quite remarkable statements here. I would argue that Laplace's requirement for a uniform distribution over all causes is doing exactly what Jeffreys demands. This is to attribute the same positive probability to any clearly stated law. For me, "clearly stated laws" mean the same thing as Laplace's *causes* for the *events* in the form of the $q_1, \cdots, q_n$ that have been defined within our framework for general inferencing as any legitimate numerical assignment to probabilities for all statements in the state space.

Jeffreys wants all or none of the same type to be accepted "on a moderate amount of evidence." But why should this situation be treated any differently than any other disposition of the types? Why should 1/2 of one type and 1/2 of the other type be accorded special treatment on a moderate amount of evidence? Why should 1/10 of one type and 9/10 of the other type be accorded any special treatment on a moderate amount of evidence? And so on.

Every disposition of the two types is accorded the proper status on **any** amount of evidence when Laplace's suggestion for the prior probability as well as the formal manipulation rules of probability are followed. I don't know what more one could ask of a rational procedure.

In most presentations that I have read, there is an immediate leap to discussing the determinant of Fisher's information matrix in searching for the origin of the "Jeffreys's prior." Yes, it is true that this is where Jeffreys eventually ended up, and where we will end up as well because of its relationship to Information Geometry.

Nevertheless, it pays dividends to search through his **Theory of Probability** looking for examples where his reasoning is somewhat more transparent. I have presented in section 47.3 a couple of examples where Jeffreys dealt with the coin tossing scenario. At least for me, inferences involving coin tossing serve as an anchor point where I can better pinpoint where any conceptual overlap, as well as departures, take place between my development so far and Jeffreys's. This is especially so with regard to the prior probability for models.

Section 47.3 entered into a discussion prompted by an example occurring rather late in Jeffreys's book. This example provided insight into his objections to the flat prior, as well as being an example that was easier to comprehend in the well understood context of the coin tossing scenario. [24, pg. 256].

5.1. Test of whether a suggested value of a chance is correct.

An answer in finite terms can be obtained in the case where the parameter in question is a chance, and we wish to know whether the data support or contradict a value suggested for it. Suppose that the suggested value is p, that the value on q', which is so far unknown, is p', and that our data consist of a sample of x members of one type and y of the other. Then on q', p' may have any value from 0 to 1. Thus

$$P(q \mid H) = 1/2, \quad P(q' \mid H) = 1/2, \quad P(dp' \mid q', H) = dp', \quad (1)$$

$$\text{whence} \quad P(q', dp' \mid H) = \frac{1}{2} dp'. \quad (2)$$

Also, if θ denotes the observational evidence,

$$P(\theta \mid qH) = {}^{x+y}C_x \, p^x (1-p)^y \qquad (3)$$

$$P(\theta \mid q', p', H) = {}^{x+y}C_x \, p'^x (1-p')^y \qquad (4)$$

whence $\quad P(q \mid \theta H) \propto p^x (1-p)^y \qquad (5)$

$$P(q', dp' \mid \theta H) \propto p'^x (1-p')^y \, dp' \qquad (6)$$

and by integration

$$P(q' \mid \theta H) \propto \int_0^1 p'^x (1-p')^y \, dp' = \frac{x! \, y!}{(x+y+1)!} \qquad (7)$$

Hence $\quad K = \dfrac{P(q \mid \theta H)}{P(q' \mid \theta H)} = \dfrac{(x+y+1)!}{x! \, y!} p^x (1-p)^y \qquad (8)$

We will base several exercises on this quote in order to ascertain whether we can shoehorn Jeffreys's explanation into our corpus of probability theory as developed so far. I do this so that you might share my difficulty in precisely nailing down Jeffreys's notation in order to better understand what he is getting at.

Here is another quote that offers up revealing insights about Jeffreys attitude towards the prior probability for models if we are willing to take the trouble to make a precise translation. Despite what happens later, Jeffreys does sometimes modify the prior probability for the models directly.

Once again, his notation obscures the immediate appreciation of this fact. As with the above quote, he is still dealing with a state space of $n = 2$ (the coin tossing scenario), but confuses us further by actually finding the probability for some number of *future* events conditioned on no data [24, pg. 188].

These rules [for finding prior probabilities of models] do not cover the sampling of a finite population. The possible number of one type are then all integers and differentiation is impossible. The difficulty does not appear insuperable. Suppose that the population is of number n and contains r members with the property. Treat this as a sample of n derived from chance. Then

$$P(d\alpha \mid nH) = \frac{d\alpha}{\pi \sqrt{\{\alpha(1-\alpha)\}}},$$

$$P(r \mid nH) = \frac{n!}{r!(n-r)!} \alpha^r (1-\alpha)^{n-r},$$

$$P(r \, d\alpha \mid nH) = \frac{n!}{\pi \, r! \, (n-r)!} \alpha^{r-1/2} (1-\alpha)^{n-r-1/2} \, d\alpha,$$

$$P(r \mid nH) = \frac{(r-1/2)! \, (n-r-1/2)!}{\pi \, r! \, (n-r)!} \qquad (50)$$

This is finite for both $r = 0$ and $r = n$.

Pay close attention here to the fact that Jeffreys's α and $(1-\alpha)$ correspond to my q_1 and q_2, and not to the α_i parameters that appear in the *beta distribution*. That these α_i parameters are set to $1/2$ is evident from the first equation above. Also, $r \equiv N_1$ and $n \equiv N$.

Most important is that this derivation is exactly the same as the one for finding $P(N_1, N_2)$, the number of future HEADS and TAILS in N tosses not conditioned on any previous observations of the coin toss, when an alternative to the flat prior is adopted. That alternative, when the $\alpha_i = 1/2$ in the "conjugate" choice for $P(\mathcal{M}_k)$ (the *beta distribution*), is labeled as a "Jeffreys's prior."

At the very beginning of Jeffreys's Chapter III, entitled **Estimation Problems**, we have a clear statement that he will definitely set up probability distributions over parameters [24, pg. 117]. On the other hand, what I want to find is the posterior probability for the models given the observations, $P(\mathcal{M}_k \,|\, \mathcal{D})$.

> A problem of estimation is one where we are given the form of a law, in which certain parameters can be treated as unknown, no special consideration needing to be given to any particular values, and we want the probability distribution of these parameters, given the observations.

This choice by Jeffreys to find probability distributions for the parameters in a model is where our paths diverge. Instead, I make the choice to find probability distributions over models, and, as a consequence, deny the conceptual rationale for ever considering the need to set up a probability distribution over parameters. Furthermore, as a semantic by-product of this choice, I also banish the phrase "estimation problems" as non-sensical.

And, as he contemplates various measures to serve as probability distributions over the parameters, he finds it necessary to assess the range of the parameter. Thus, it becomes important to know whether the parameter has a finite range, ranges infinitely from $-\infty$ to $+\infty$, or infinitely from 0 to $+\infty$.

And how should we find a distribution over the parameters if a different choice of parameters is acceptable? This leads into the confused notion of "parametric invariance" and the need to find distributions over parameters that are not adversely impacted by an arbitrary choice of parameters. It is truly frightening for any rational being to observe what happens when we happen to make some initial false turn, and then stubbornly try to follow this path all the way to its end [24, pg. 117].

> Our first problem is to find a way of saying that the magnitude of a parameter is unknown, when none of the possible values need special attention. Two rules appear to cover the commonest cases. If the parameter may have any value in a finite range, or from $-\infty$ to $+\infty$, its prior probability should be taken as uniformly distributed. If it arises in such a way that it may conceivably have any value from 0 to ∞, the prior probability of its logarithm should be taken as uniformly distributed. There are cases of estimation where a law can be equally well expressed in terms of several different sets of parameters, and it is desirable to have a rule that will lead to the same results whichever set we choose.

I plead guilty to taking the Gordian knot solution. I simply deny at the outset the efficacy of starting down the path of setting up a probability distribution over the parameters. So for me none of these issues are cause for concern.

If you inform me correctly that my λ parameter can range over values from $-\infty$ to $+\infty$, that has no bearing on my setting up a prior probability for the models. If you inform me correctly that a different set of parameters might be chosen to arrive at the same numerical assignments to probabilities, that also has no bearing on my setting up the same prior probability for the models.

If a model happens to assign $Q_1 = 0.75$ and $Q_2 = 0.25$, then the MEP will arrive mathematically at $\lambda = -1.0986$ for my specified set of constraint functions. I might decide to assign a prior probability of 0.90 to this model. The λ value of -1.0986 had nothing whatsoever to do with this prior probability, and, *a fortiori*, setting up a probability distribution over any of the λ_j values, even if that made any sense, would have nothing to do with setting up this prior probability for a model.

Finally, in another too frequent example of the misunderstanding surrounding Laplace's *Rule of Succession*, we have the following quote from Bernardo & Smith [6, pg. 322]. The essence of the inferential problem remains the same as Jeffreys's animal with feathers scenario.

> **Example 5.16 (*Induction*).** Consider a large, finite dichotomous population, all of whose elements individually may or may not have a specified property. A random sample is taken without replacement from the population, the sample being large in absolute size, but still relatively small compared with the population size. *All* the elements sampled turn out to have the specified property. Many commentators have argued that, in view of the large absolute size of the sample, one should be led to believe quite strongly that all elements of the *population* have the property, irrespective of the fact that the population size is greater still, **an argument related to Laplace's rule of succession**. [Emphasis in the original, but I have chosen my emphasis as indicated by the phrase in **bold** at the end].

But this is the exact opposite of the lessons learned from applying the *Rule of Succession*! Through Laplace's operational definition of what it means to be totally uninformed about any and all causes, there remains a substantial probability that at least one of the remaining unsampled elements does NOT have the property in question, even though every single element of the large sample size has possessed that property.

Compare their hypergeometric formula with Jaynes's shown in Exercise 47.8.2. Are they the same? Close scrutiny will uncover some discrepancies.

47.8 Solved Exercises for Chapter Forty Seven

Exercise 47.8.1: Translate Jeffreys's notation in order to present a very detailed example of his combinatorial formula designed for calculating probabilities when sampling without replacement from finite populations consisting of just two "types."

Solution to Exercise 47.8.1

As we have just recently been investigating in **Connections to the Literature**, Jeffreys gave us the following combinatorial formula [24, pg. 59],

$$P(l, m \,|\, H) = {}^rC_l \, {}^sC_m \,/\, {}^{r+s}C_{l+m}$$

for the probability of obtaining particular sample counts when sampling without replacement from a population consisting of a total of $r+s$ things. The propositions l and m refer to the actual sample numbers of type 1 and type 2, respectively.

The proposition to the right of the conditioned upon symbol, H, is the notation Jeffreys commonly used to indicate the detailed background description for the particular problem scenario under consideration. Here we are dealing with the problem of sampling from a finite population without replacement.

We practiced translating Jeffreys's idiosyncratic combinatorial notation like ${}^rC_l$ in section 47.7. The notation r refers to the total number of type 1 in the population, while s refers to the total number of type 2 in the population. Be careful to note that Jeffreys uses $N = r + s$ for the total number of both types in the population, and $n = l + m$ for the total of the sample taken. Today, we tend to use a combinatorial notation like,

$$\binom{x}{y} \equiv \frac{x!}{y!\,(x-y)!}$$

instead of Jeffreys's ${}^xC_y$

Back in section 47.7, we examined the case of sampling from a population of $N = 10$. The sample size was $n = 4$. Keep in mind that this notation clashes with our standard notation for the meaning of N and n, but this is what Jeffreys used, and at least for the duration of these exercises, we will adhere to his notation.

For an even easier example, let's pick some small numbers where it is possible to check the correctness of the combinatorial formula. Consider the classical sampling situation of two differently colored balls in an urn. Suppose that there are a total of $N = 6$ balls in the urn consisting $r = 4$ black balls and $s = 2$ red balls. (To keep you on your toes, I have switched the colors for r and s.) Draw $n = 3$ balls from the urn and ask for the probability of obtaining $l = 2$ black balls and $m = 1$ red ball where $l + m = n$.

Jeffreys's derivation of the probability for the sample numbers is expressed strictly as a combinatorial argument.

$$^rC_l = \frac{r!}{l!\,(r-l)!}$$

$$^sC_m = \frac{s!}{m!\,(s-m)!}$$

$$P(l, m \mid H) = \frac{^rC_l \;^sC_m}{^{r+s}C_{l+m}}$$

$$P(l=2, m=1 \mid H) = \frac{\frac{4!}{2!\,2!} \times \frac{2!}{1!\,1!}}{\frac{6!}{3!\,3!}}$$

$$= \frac{3}{5}$$

Jeffreys's answer for the probability of sampling two black balls and one red ball from the total of six balls in the urn is 3/5. In like manner, the probability for sampling one black and two red balls is 1/5, the probability for sampling three black and no red balls is also 1/5. The probability for sampling no black balls and three red balls must, of course, equal 0. The probability of the certain proposition, sampling zero through three black balls in three draws from the urn, is 1 as it must be. Whenever feasible, verify that the probability calculations make sense.

He says that the total number of possible samples is,

$$^{r+s}C_{l+m} = {^NC_n} = {^6C_3} = \frac{6!}{3!\,3!} = 20$$

with an equal probability of obtaining every possible sample. Figure 47.1 at the top of the next page illustrates all 20 possible samples when broken down into all four possibilities for three black balls, all four possibilities for one black ball, and all twelve possibilities for two black balls. The subscript identifies the particular numbered black or red ball.

Thus, Jeffreys must have meant, in imagining the actual physical process by which balls are extracted from the urn, that all three balls would be drawn at the same time, as opposed to drawing the first ball, putting it aside, then drawing the second ball, and so on.

Under the latter drawing procedure of extracting one ball at a time, there would be six possible ways for each one of the twenty ways shown in Figure 47.1. For example, $b_1 b_2 b_3$ represents drawing black ball #1, black ball #2, and black ball #3 in one grab. But black ball #3 might have been drawn in the first draw when extracting just one ball at a time, followed by black ball #2 on the second draw, and by black ball #1 on the third and final draw as one of the six possibilities.

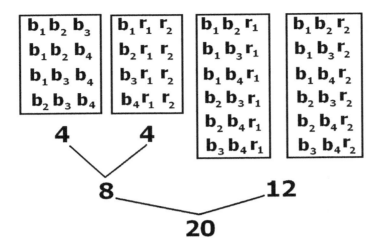

Figure 47.1: *A breakdown of all possibilities in sampling three balls from a total of four black balls and two red balls when drawing all three balls "in one grab."*

After taking into account the changing nature of the state space when sampling without replacement, we see that, for example, in the case of drawing two red balls and one black ball, that the black ball could have been drawn on the first draw with probability 4/6, and then the two red balls with probability $2/5 \times 1/4$. The overall probability is then 8/120.

But there are two other possibilities. The black ball could have been drawn on the second draw, or the third draw. But the probabilities for all three possibilities are the same at 8/120 because the numerator is always $1 \times 2 \times 4$ and the denominator is always $6 \times 5 \times 4$. Thus, the probability of drawing one black ball and two red balls is $3 \times (8/120) = 24/120 = 1/5$ the same answer as before.

Exercise 47.8.2: Work out the same example with Jaynes's version of the formula.

Solution to Exercise 47.8.2

Jaynes [22, pg. 150] presents his Equation (6.1), a hypergeometric distribution, as,

$$P(D \mid NRI) = \binom{N}{n}^{-1} \binom{R}{r} \binom{N-R}{n-r}$$

The data D are the sample of 2 black balls and 1 red ball. The probability for the data are conditioned on knowing the population size, and the number of black and red balls in the population. $N = 6$ is the population size, $n = 3$ is the sample size, $R = 4$ is the number of black balls in the population, and $r = 2$ is the number of black balls drawn in the sample.

$$P(r = 2 \text{ black}, n - r = 1 \text{ red} \,|\, N = 6, R = 4, I) = \binom{6}{3}^{-1} \binom{4}{2}\binom{2}{1}$$

$$= \frac{3}{5}$$

Jaynes had the balls in his urn colored red and white corresponding to my black and red. My apologies for any confusion that arises when you try to follow this example because of the color changes.

Exercise 47.8.3: Building on this first exercise, how did Jeffreys then fold in the notion of a uniform prior probability?

Solution to Exercise 47.8.3

Jeffreys starts with the following formula [24, pg. 125] in preparation for deriving the probability for seeing the next example of type 1 given the known sample numbers of type 1 and type 2.

$$P(l, m \,|\, NrH) = {}^{r}C_{l} \, {}^{N-r}C_{m} \,/\, {}^{N}C_{l+m}$$

Now this probability expression for a sample is conditioned on knowing N, the total number of balls in the urn, as well as r, the total number of black balls in the urn. It is the same formula we examined in the first exercise since $N = r + s$, and where we selected a specific N and r.

At the next step, he invokes a prior probability, but it is the prior probability for the number of black balls given the total number of balls,

$$P(r \,|\, NH) = 1/(N + 1)$$

We immediately recognize Laplace's answer for the probability of $0, 1, \ldots, N-1, N$ occurrences of HEADS in N future tosses of a coin.

Jeffreys does acknowledge that he has introduced this expression as a prior probability for the number of black balls to concur with Laplace's definition of total ignorance about how many black balls might be in the urn. Notice however that Laplace arrived at his answer after first adopting a uniform prior probability for all conceivable numerical assignments q_i to the black and red balls.

For Laplace (and for me) this is not a prior probability, but rather a *marginal* probability constructed from conditional and prior probabilities. We have usually written this marginal probability for the observed data as,

$$P(N_1, N_2, \ldots, N_n) = \int \cdot \int_{\sum q_i = 1} P(N_1, N_2, \ldots, N_n \,|\, q_1, q_2, \ldots, q_n) \, P(q_1, q_2, \ldots, q_n) \, dq_i$$

But Jeffreys relied upon the notion in his combinatorial argument that each of the twenty possibilities was *equally probable* and then counted up the number of favorable cases for say two black balls and one red ball (12) over all cases (20).

In sampling from a finite population, Jeffreys wants to find the probability of the total number of type 1 in the population (total number of black balls in the urn), r, given the results of the sample, l and m. In addition, this probability will also be conditioned on the known number of both types in the population, (the total number of the population), N, and all of the background details in the statement of the problem, H. This probability he wrote as $P(r \mid l, m, N, H)$.

Jeffreys did not show this explicitly, but an application of Bayes's Theorem (he prefers to call it the principle of inverse probability), results in,

$$P(r \mid l, m) = \frac{P(l, m, r)}{P(l, m)}$$

$$= \frac{P(l, m \mid r) P(r)}{P(l, m)}$$

$$= \frac{P(l, m \mid r) P(r)}{\sum_{r=0}^{N} P(l, m \mid r) P(r)}$$

In the above, we have omitted always placing N and H to the right of the conditioned upon symbol to unclutter the expressions.

In the numerator of Bayes's Theorem, we have already examined the first term, the direct probability for seeing the sample numbers conditioned on knowledge of r and N, that is, the total number of type 1 elements and the total number in the population,

$$P(l, m \mid r, N, H)$$

The second term $P(r)$ in the numerator is most interesting. The quote below shows how Jeffreys prefaced his interpretation of the Bayes–Laplace uniform prior probability [24, pg. 125].

> We have no information initially to say that one value of r, given N, is more likely than another. Hence we must take all their prior probabilities equal, and
>
> $$P(r \mid NH) = 1/(N+1)$$

Substituting this value of the prior probability of r into the result provided by

SOLVED EXERCISES FOR CHAPTER 47

Bayes's Theorem, we find that it cancels, leaving us with,

$$P(r\,|\,l,m) = \frac{P(l,m\,|\,r)\,P(r)}{\sum_{r=0}^{N} P(l,m\,|\,r)\,P(r)}$$

$$= \frac{P(l,m\,|\,r)\,\frac{1}{N+1}}{\sum_{r=0}^{N} P(l,m\,|\,r)\,\frac{1}{N+1}}$$

$$= \frac{P(l,m\,|\,r)}{\sum_{r=0}^{N} P(l,m\,|\,r)}$$

Since Jeffreys uses his own idiosyncratic combinatorial notation, which we have been practicing in previous exercises, this gets turned into,

$$P(l,m\,|\,r) = {}^{r}C_{l}\,{}^{N-r}C_{m}$$

$$P(r\,|\,l,m) = {}^{r}C_{l}\,{}^{N-r}C_{m}\Big/\sum_{r=0}^{N} {}^{r}C_{l}\,{}^{N-r}C_{m}$$

Jeffreys then goes on to show that the sum in the denominator is equal to,

$$\sum_{r=0}^{N} {}^{r}C_{l}\,{}^{N-r}C_{m} = {}^{N+1}C_{l+m+1}$$

$$= {}^{N+1}C_{n+1}$$

Jeffreys then presents a rather obscure and idiosyncratic argument all based on combinatorial formulas [24, pp. 125–127] to reach the following conclusion. The probability that the *next* sample is of the first type given knowledge of the sample numbers l and m,

$$P(\text{next sample is type 1}) = \frac{l+1}{l+m+2}$$

The first surprising realization was that this formula, through arrived at through Jeffreys's unique argument, was the same as Laplace's *Rule of Succession*,

$$P(M_1 = 1, M_2 = 0\,|\,N_1 = l, N_2 = m) = \frac{N_1 + 1}{N + 2}$$

A numerical example of the *Rule of Succession* for the coin toss scenario given in Volume II asked for the probability that the *next* toss of the coin would be HEADS given that 62 HEADS and 38 TAILS had already been observed. Applying the *Rule of Succession* yields,

$$P(M_1 = 1, M_2 = 0\,|\,N_1 = 62, N_2 = 38) = \frac{N_1 + 1}{N + 2}$$

$$P(\text{HEADS on 101st toss}\,|\,\text{data}) = \frac{63}{102}$$

The second shocker was that Jeffreys's formula,

$$P(\text{next sample is type 1}) = \frac{l+1}{l+m+2}$$

did not depend on $r + s = N$, the total number of balls actually in the urn.

Exercise 47.8.4: Use the same numerical example as before to illustrate this last formula.

Solution to Exercise 47.8.4

There are $N = 6$ balls in the urn. A sample of three balls drawn from the urn resulted in two black balls ($l = 2$) and one red ball ($m = 1$). What is the probability, $P(r \mid l, m, N, H)$, that $r = 0$ through $r = 6$ black balls are in the urn conditioned on knowing the sample and the total population?

We can dispense with three probabilities immediately.

$$P(r = 0 \mid l, m, N, H) = P(r = 1 \mid l, m, N, H) = P(r = 6 \mid l, m, N, H) = 0$$

The sample already told us that there were at least two black balls in the urn so it is impossible that there were only none or one black balls. It is also impossible that all six balls in the urn were black since the sample told us that at least one was red. Thus, the formula will be used to find the probability for $r = 2$ through $r = 5$ black balls in the urn.

The common denominator for all of these probabilities will be,

$$^{N+1}C_{n+1} = \frac{(N+1)!}{(n+1)!\,(N+1-(n+1))!} = \frac{7!}{4!\,3!} = 35$$

We will show one numerator in detail. Pick the $r = 5$ numerator so that,

$$^{r}C_{l}\,^{N-r}C_{m} = \frac{5!}{3!\,2!} \times \frac{1!}{1!\,0!} = 10$$

Thus,

$$P(r = 5 \mid l, m, N, H) = \frac{10}{35}$$

The remaining three probabilities are,

$$P(r = 2 \mid l, m, N, H) = \frac{4}{35}$$

$$P(r = 3 \mid l, m, N, H) = \frac{9}{35}$$

$$P(r = 4 \mid l, m, N, H) = \frac{12}{35}$$

SOLVED EXERCISES FOR CHAPTER 47

As Jaynes was fond of saying, any application of Bayes's Theorem ought to accord well with our intuition, after we have mulled things over a bit. So it is least likely that we have managed to pull out the only two black balls in the urn in our sample. It is most likely that there are four black balls and two red balls in the urn, and we pulled out two black balls and one red ball in our sample of three, that is, in the same proportion to the population. This outcome is slightly more probable than five black balls and only one red ball in the urn, where we would be lucky enough to pull out the only red ball in our sample draw.

Can my posterior predictive probability formula reproduce these same answers? It had better if I want to validate the claim that my derivation of the posterior predictive probability is exactly the same as Jeffreys's formula despite the surface dissimilarities.

Consider the case where the posterior probability of there actually being $r = 4$ black balls and $s = 2$ red balls given the known sample of $l = 2$ black balls and $m = 1$ red ball was found via Jeffreys's approach as,

$$P(r = 4 \,|\, l, m, N, H) = \frac{12}{35}$$

I will show a *Mathematica* program shortly that would immediately compute that my formula produces the same answer, but first we will do it the long way.

If the population is $N_{\text{pop}} = 6$, and the sample size is $N_{\text{sam}} = 3$, then the number of remaining balls is $M = 3$. If there are actually $r = 4$ black balls in the population, and two of them have been drawn already, that leaves $M_1 = 2$ and $M_2 = 1$ together with $N_1 = 2$ and $N_2 = 1$ as the values to enter into the formula.

$$P(M_1 = 2, M_2 = 1 \,|\, N_1 = 2, N_1 = 1) = C \times \frac{\prod_{i=1}^{n} \Gamma(M_i + N_i + \alpha_i)}{\prod_{i=1}^{n} \Gamma(N_i + \alpha_i)}$$

$$\text{where } C = W(M) \times \frac{\Gamma(N + \mathcal{A})}{\Gamma(M + N + \mathcal{A})}$$

$$C = \frac{3!}{2!\,1!} \times \frac{\Gamma(3 + 2)}{\Gamma(3 + 3 + 2)}$$

$$\frac{\prod_{i=1}^{n} \Gamma(M_i + N_i + \alpha_i)}{\prod_{i=1}^{n} \Gamma(N_i + \alpha_i)} = \frac{\Gamma(2 + 2 + 1)\,\Gamma(1 + 1 + 1)}{\Gamma(2 + 1)\,\Gamma(1 + 1)}$$

$$P(M_1 = 2, M_2 = 1 \,|\, N_1 = 2, N_1 = 1) = 3 \times \frac{4!}{7!} \times \frac{4!\,2!}{2!\,1!}$$

$$= \frac{12}{35}$$

Exercise 47.8.5: Derive the general formula for the posterior predictive probability of the future frequency counts for any α_i given the data.

Solution to Exercise 47.8.5

This is the formula that we depended on to solve the previous exercise and that will form the basis of the upcoming *Mathematica* program. In the previous exercise, $n = 2$, and since we were following Laplace's prescription for a uniform distribution over model space, $\alpha_1 = \alpha_2 = 1$.

From the formal manipulation rules of probability theory, in other words, Bayes's Theorem, we have that,

$$P(M_1, M_2, \cdots, M_n \mid N_1, N_2, \cdots, N_n) = \frac{P(M_1, M_2, \cdots, M_n, N_1, N_2, \cdots, N_n)}{P(N_1, N_2, \cdots, N_n)}$$

The denominator, $P(N_1, N_2, \cdots, N_n)$, that is, the probability for the data, will be derived as a sub–program in the next exercise. So all that's left is to derive a similar formula for the joint probability $P(M_1, M_2, \ldots, M_n, N_1, N_2, \ldots, N_n)$ in the numerator.

Taking the same tack in the derivation as seen before,

$$P(M_1, M_2, \cdots, M_n, N_1, N_2, \cdots, N_n) =$$

$$W(M) \times W(N) \times \frac{\Gamma(\mathcal{A})}{\prod_{i=1}^{n} \Gamma(\alpha_i)} \times \frac{\prod_{i=1}^{n} \Gamma(M_i + N_i + \alpha_i)}{\Gamma(M + N + \mathcal{A})}$$

When we divide this joint probability in the numerator by the probability for the data in the denominator, the second and third terms in the numerator conveniently cancel out. Add the final term from the denominator to arrive at,

$$P(M_1, M_2, \cdots, M_n \mid N_1, N_2, \cdots, N_n) =$$

$$W(M) \times \frac{\prod_{i=1}^{n} \Gamma(M_i + N_i + \alpha_i)}{\Gamma(M + N + \mathcal{A})} \times \frac{\Gamma(N + \mathcal{A})}{\prod_{i=1}^{n} \Gamma(N_i + \alpha_i)}$$

This formula could be rearranged to put all the expressions where the constants for a particular problem, M, N, and $\mathcal{A}$, appear together in a single constant term C,

$$P(M_1, M_2, \cdots, M_n \mid N_1, N_2, \cdots, N_n) = C \times \frac{\prod_{i=1}^{n} \Gamma(M_i + N_i + \alpha_i)}{\prod_{i=1}^{n} \Gamma(N_i + \alpha_i)}$$

$$\text{where } C = W(M) \times \frac{\Gamma(N + \mathcal{A})}{\Gamma(M + N + \mathcal{A})}$$

SOLVED EXERCISES FOR CHAPTER 47

Exercise 47.8.6: Provide a derivation for the probability of the data in order to complete the proof in the last exercise.

Solution to Exercise 47.8.6

Jaynes presents a rigorous proof, but based as it is on the Laplace transform and its inversion, it is rather difficult. Here is an easier proof from a probabilistic standpoint if one is willing to accept the Dirichlet distribution at face value with its parameters of α_i as the prior probability for the models.

To begin, we show the *marginal* probability for the data as first, a sum over the *conditional* probability of the data given a model times the *prior* probability of the models, and second, as the integration over the model's assignments of the $q_1, q_2, \ldots, q_n$ with the explicit recognition of the Dirichlet distribution as the prior probability density function over model space.

$$P(N_1, N_2, \ldots, N_n) = \sum_{k=1}^{\mathcal{M}} P(N_1, N_2, \ldots, N_n \mid \mathcal{M}_k)\, P(\mathcal{M}_k)$$

$$\equiv \int \cdots \int_{\sum q_i = 1} W(N)\, q_1^{N_1} q_2^{N_2} \cdots q_n^{N_n}\, C_D\, q_1^{\alpha_1 - 1} q_2^{\alpha_2 - 1} \cdots q_n^{\alpha_n - 1}\, \mathrm{d}q_i$$

The derivation follows the by now familiar steps we have taken advantage of many times before. My version given below was motivated by studying Jaynes's analysis.

$$P(N_1, N_2, \cdots, N_n) = \int \cdots \int_{\sum q_i = 1} W(N)\, q_1^{N_1} q_2^{N_2} \cdots q_n^{N_n}\, C_D\, q_1^{\alpha_1 - 1} q_2^{\alpha_2 - 1} \cdots q_n^{\alpha_n - 1}\, \mathrm{d}q_i$$

$$= W(N) \times C_D \times \int \cdots \int_{\sum q_i = 1} q_1^{N_1} q_2^{N_2} \cdots q_n^{N_n} q_1^{\alpha_1 - 1} q_2^{\alpha_2 - 1} \cdots q_n^{\alpha_n - 1}\, \mathrm{d}q_i$$

$$= W(N) \times C_D \times \int \cdots \int_{\sum q_i = 1} q_1^{N_1 + \alpha_1 - 1} q_2^{N_2 + \alpha_2 - 1} \cdots q_n^{N_n + \alpha_n - 1}\, \mathrm{d}q_i$$

$$= W(N) \times C_D \times \frac{\prod_{i=1}^{n} \Gamma(N_i + \alpha_i)}{\Gamma(\sum_{i=1}^{n} N_i + \alpha_i)}$$

$$= W(N) \times \frac{\Gamma(\mathcal{A})}{\prod_{i=1}^{n} \Gamma(\alpha_i)} \times \frac{\prod_{i=1}^{n} \Gamma(N_i + \alpha_i)}{\Gamma(\sum_{i=1}^{n} N_i + \alpha_i)}$$

$$= W(N) \times \frac{\Gamma(\mathcal{A})}{\prod_{i=1}^{n} \Gamma(\alpha_i)} \times \frac{\prod_{i=1}^{n} \Gamma(N_i + \alpha_i)}{\Gamma(N + \mathcal{A})}$$

The first two terms in $P(N_1, N_2, \cdots, N_n)$ conveniently cancel out the second and third terms in $P(M_1, M_2, \ldots, M_n, N_1, N_2, \ldots, N_n)$ of the previous exercise. Division by the third term above provides the final term in posterior predictive probability equation,

$$\frac{\Gamma(N + \mathcal{A})}{\prod_{i=1}^{n} \Gamma(N_i + \alpha_i)}$$

Exercise 47.8.7: Recall the implications for an IP who is totally ignorant about the physical causes of an event. The IP has also not yet observed any data.

Solution to Exercise 47.8.7

Before observing any events, an IP who doesn't know anything about what might be causing these events must possess the *same* degree of belief about the truth of any conceivable event. For example, suppose that the state space of the coin tossing scenario has been expanded to include a possible observation that the coin lands on its edge. What is the IP's degree of belief that the coin will land on its EDGE on each of the first four tosses of the coin? It believes the truth of this statement just as much as it believes that the coin will show two HEADS and two TAILS.

As another example where the probability of an event is the same as the above event, what is the probability for seeing two HEADS, one TAILS, and one EDGE in the first four tosses of the coin? Applying the formula derived in the last exercise to find the probability for any set of data, we find that the probability when the uniform probability for models is employed, in other words, when all $\alpha_i = 1$, is equal to,

$$P(N_1 = 2, N_2 = 1, N_3 = 1) = W(N) \times \frac{\Gamma(\mathcal{A})}{\prod_{i=1}^{n} \Gamma(\alpha_i)} \times \frac{\prod_{i=1}^{n} \Gamma(N_i + \alpha_i)}{\Gamma(N + \mathcal{A})}$$

$$= \frac{N!}{N_1!\, N_2!\, N_3!} \times \frac{\Gamma(3)}{0!\, 0!\, 0!} \times \frac{\prod_{i=1}^{3} \Gamma(N_i + 1)}{\Gamma(N + 3)}$$

$$= \frac{N!}{N_1!\, N_2!\, N_3!} \times 2! \times \frac{N_1!\, N_2!\, N_3!}{\Gamma(N + 3)}$$

$$= \frac{4! \times 2!}{\Gamma(4 + 3)}$$

$$= \frac{1}{15}$$

Of course, we know that, in general, when we adopt the uniform prior probability for models, the probability for the data is going to be,

$$P(N_1, N_2, \cdots, N_n) = \frac{N!\,(n-1)!}{(N+n-1)!}$$

The derivation of the *Rule of Succession*, following Laplace, demanded that the uniform distribution be employed as the prior probability over model space. This is how the IP implements the idea that all possible numerical assignments to the probabilities in the state space must be judged on an equal basis. It is also the way the IP implements the idea that all causes for the coin to shows HEADS, TAILS, or EDGE must be considered on an equal basis.

SOLVED EXERCISES FOR CHAPTER 47

As always, when employing the uniform prior, the probability for the data is independent of any particular frequency counts that might be called for. Thus, whether the IP inquires about the probability of 1) four EDGEs, 2) two HEADS and two TAILS, 3) two HEADS, one TAILS, and one EDGE, or for that matter, any one of the fifteen possible contingency tables, the probability will be always be 1/15.

$$P(N_1, N_2, N_3) = \frac{N!\,(n-1)!}{(N+n-1)!}$$

$$= \frac{4!\,2!}{(4+3-1)!}$$

$$= \frac{1}{15}$$

Inverting the above formula tells us the total number of possible contingency tables; thus, there are a total of fifteen events that might occur in the first four tosses of the coin. Table 47.1 presents a detailed list of all fifteen contingency tables.

Table 47.1: *An exhaustive listing of all fifteen possible contingency tables in an $n = 3$ coin flip scenario involving $N = 4$ flips.*

Frequencies	Frequencies	Frequencies
H T E	H T E	H T E
4 0 0	2 1 1 ⋆	0 1 3
3 1 0	1 3 0	0 4 0
3 0 1	1 2 1	0 3 1
2 2 0	1 1 2	0 2 2
2 0 2	1 0 3	0 0 4

The starred contingency table with two HEADS, one TAILS, and one EDGE was the particular frequency count of the fifteen picked out for the numerical example. But, as we now know, all fifteen contingency tables, that is, any data, have the same probability.

The justification for these claims that stem from an IP's ignorance always begins with the formal manipulation rules of probability theory. In short, the marginal probability for the first N occurrences with N_1 occurrences of type 1, N_2 occurrences of type 2, ..., and N_n occurrences of type n are found by the **Sum Rule** and the **Product Rule**,

$$P(N_1, N_2, \ldots, N_n) = \sum_{k=1}^{\mathcal{M}} P(N_1, N_2, \ldots, N_n \mid \mathcal{M}_k)\, P(\mathcal{M}_k)$$

$$\equiv \int \cdot \int_{\sum q_i = 1} W(N)\, q_1^{N_1} q_2^{N_2} \ldots q_n^{N_n}\, C_{\mathrm{D}}\, q_1^{\alpha_1 - 1} q_2^{\alpha_2 - 1} \cdots q_n^{\alpha_n - 1}\, dq_i$$

Exercise 47.8.8: Discuss Jeffreys's expression for the probability of the data with his prior as presented in Equation (47.15). Do this in the context of my expression for the probability of the data as derived in Exercise 47.8.6.

Solution to Exercise 47.8.8

I quoted Jeffreys on the topic for this Exercise in the **What Upset Jeffreys?** section, and again in the **Connections to the Literature** section. The topic centered on a notion seemingly involving an expression for the probability of the data where Jeffreys was definitely not going to employ Laplace's prior, but according to him could not apply his "new rules" either. I like this particular quote from Jeffreys because it fascinates me for almost diametrically opposed reasons.

Jeffreys had just laid out his counter argument, in his section **3.10**, labeled as the **Invariance Theory**, as a rationale for amending Laplace's uniform prior. This is the argument you see today culminating in the expression involving the square root of the determinant of the Fisher information matrix, $\| g_{ik} \|^{1/2}$.

Surprisingly, he mentions that these rules do not cover sampling from a finite population. This is strange!

I have taken pains to reproduce some of the effort that Jeffreys went to in trying to locate the source of his displeasure with Laplace's uniform distribution. The genesis, as our many numerical examples have emphasized, lay in those problems that Jeffreys has just excluded from his new rules for finding prior probabilities, namely, sampling without replacement from a finite population. Not to worry, says Jeffreys, let me show you how to deal with this obstacle.

He then proceeds to demonstrate in the clearest manner yet, after everything he has said about prior probabilities, the few mathematical steps one must take. When I say the "clearest manner yet," I hasten to add that this clarity occurs only after some serious translation of notation, followed by a careful check that his terms match up with my derivation of the posterior predictive probability equation.

First we have [24, pg. 188],

$$P(d\alpha \,|\, nH) = \frac{d\alpha}{\pi \sqrt{\{\alpha(1-\alpha)\}}}$$

which is, in fact, the assignment for the prior probability for a q and $(1-q)$ based on the *beta distribution* with parameters $\alpha_1 = 1/2$ and $\alpha_2 = 1/2$. This expression confirms that this was indeed his choice for the parameters of the prior probability, and what Jeffreys's had in mind as a "Jeffreys's prior" to replace Laplace's uniform prior with its choice of $\alpha_1 = 1$ and $\alpha_2 = 1$.

I have always written the prior probability for this $n = 2$ case as a special case of the Dirichlet distribution,

$$P(\mathcal{M}_k) \equiv \text{pdf}\,(dq) = C_{\text{Beta}} \times q_1^{\alpha_1 - 1} q_2^{\alpha_2 - 1}\, dq$$

SOLVED EXERCISES FOR CHAPTER 47

Confusion arises because of the notation where Jeffreys's α is equivalent to my q. Additionally, $\alpha_i = 1/2$ which lead to $q_1^{-(1/2)} q_2^{-(1/2)}$ are implicitly mirrored in Jeffreys's expression of $\frac{1}{\sqrt{\{\alpha(1-\alpha)\}}}$. The value of C_{Beta} in pdf (dq) is,

$$C_{\text{Beta}} = \frac{\Gamma(1/2 + 1/2)}{\Gamma(1/2)\,\Gamma(1/2)} = \frac{1}{\sqrt{\pi}\,\sqrt{\pi}} = \frac{1}{\pi}$$

so that we have in consonance with Jeffreys,

$$P(\mathcal{M}_k) = \text{pdf}\,(dq) = \frac{dq}{\pi\,\sqrt{q\,(1-q)}}$$

Next up is Jeffreys's expression for the conditional probability of r, conditioned on assuming the truth of α,

$$P(r\,|\,n, \alpha H) = \frac{n!}{r!\,(n-r)!}\,\alpha^r (1-\alpha)^{n-r}$$

For me, this is the conditional probability of obtaining frequency counts N_1 and N_2 with $N_1 + N_2 = N$ given the truth of some model $\mathcal{M}_k$.

$$P(N_1, N_2\,|\,\mathcal{M}_k) = \frac{N!}{N_1!\,N_2!}\,q_1^{N_1}\,q_2^{N_2}$$

where the equivalencies are $n \equiv N$, $r \equiv N_1$, $n - r \equiv N_2$, $\alpha \equiv q_1$, and $(1 - \alpha) \equiv q_2$.

With the conditional probability and prior probability in place, Jeffreys can now establish through a backwards application of the **Product Rule**,

$$P(r\,d\alpha\,|\,nH) = \frac{n!}{\pi\,r!\,(n-r)!}\,\alpha^{r-1/2}\,(1-\alpha)^{n-r-1/2}\,d\alpha$$

In my way of doing things, the probability for the data before the integration takes place becomes the product of the conditional probability and prior probability,

$$P(N_1, N_2, \mathcal{M}_k) = P(N_1, N_2\,|\,\mathcal{M}_k)\,P(\mathcal{M}_k)$$

$$= \frac{N!}{N_1!\,N_2!}\,q_1^{N_1}\,q_2^{N_2} \times C_{\text{Beta}} \times q_1^{\alpha_1 - 1}\,q_2^{\alpha_2 - 1}\,dq$$

$$= \frac{N!}{N_1!\,N_2!}\,q_1^{N_1}\,q_2^{N_2} \times \frac{1}{\pi} \times q_1^{-(1/2)}\,q_2^{-(1/2)}\,dq$$

$$= W(N) \times \frac{1}{\pi} \times q_1^{N_1 - 1/2}\,q_2^{N_2 - 1/2}\,dq$$

At the final step, Jeffreys is ready to integrate over α to arrive at the probability of the data r,

$$P(r\,|\,nH) = \frac{(r - 1/2)!\,(n - r - 1/2)!}{\pi\,r!\,(n-r)!}$$

The equivalent integration over q for me also results in the probability of the data,

$$P(N_1, N_2) = \int_0^1 W(N) \times C_{\text{Beta}} \times q^{N_1+\alpha-1} (1-q)^{N_1+\beta-1} dq$$

I have switched over to specifying the parameters in the *beta distribution* as $\alpha_1 \equiv \alpha$ and $\alpha_2 \equiv \beta$ in order to bring it more in line with what you see conventionally.

The major conceptual change, as it is implemented analytically, is to choose $\alpha = \beta = 1/2$ in order to implement Jeffreys's prior instead of the usual $\alpha = \beta = 1$ in the prior probability $P(\mathcal{M}_k)$ to implement Laplace's prior. So, in summary, first bring out from under the integral the constants not dependent on q,

$$P(N_1, N_2) = W(N) \times C_{\text{Beta}} \times \int_0^1 q^{N_1+\alpha-1} (1-q)^{N_1+\beta-1} dq$$

Now, collecting all of the steps together in one place, and with $\alpha = \beta = 1/2$, the derivation can proceed as,

$$\int_0^1 q^{N_1+\alpha-1} (1-q)^{N_1+\beta-1} dq = \frac{\Gamma(N_1+1/2)\,\Gamma(N_2+1/2)}{\Gamma(N_1+N_2+1)}$$

$$C_{\text{Beta}} = \frac{\Gamma(1/2+1/2)}{\Gamma(1/2)\,\Gamma(1/2)}$$

$$W(N) = \frac{N!}{N_1!\,N_2!}$$

$$P(N_1, N_2) = \frac{N!}{N_1!\,N_2!} \times \frac{\Gamma(1/2+1/2)}{\Gamma(1/2)\,\Gamma(1/2)} \times \frac{\Gamma(N_1+1/2)\,\Gamma(N_2+1/2)}{\Gamma(N_1+N_2+1)}$$

$$= \frac{N!}{N_1!\,N_2!} \times \frac{1}{\sqrt{\pi}\sqrt{\pi}} \times \frac{(N_1-1/2)!\,(N_2-1/2)!}{N!}$$

$$= \frac{1}{N_1!\,N_2!} \times \frac{1}{\pi} \times (N_1-1/2)!\,(N_2-1/2)!$$

$$= \frac{(N_1-1/2)!\,(N_2-1/2)!}{\pi\,N_1!\,N_2!}$$

This result is to be compared to Jeffreys's expression in Equation (47.15),

$$P(r \mid nH) = \frac{(r-1/2)!\,(n-r-1/2)!}{\pi\,r!\,(n-r)!}$$

Hence the common refrain in these books: My objections are not to the mathematics *per se*, but rather to the concepts and supporting rationale for the argument as developed by the mathematics.

SOLVED EXERCISES FOR CHAPTER 47

Exercise 47.8.9: Write the *Mathematica* code to compute the probability for any set of future frequency counts for any α_i when conditioned on the data.

Solution to Exercise 47.8.9

We will do a direct translation into *Mathematica* of the formula for the posterior predictive probability,

$$P(M_1, M_2, \cdots, M_n \mid N_1, N_2, \cdots, N_n) =$$

$$W(M) \times \frac{\prod_{i=1}^{n} \Gamma(M_i + N_i + \alpha_i)}{\Gamma(M + N + \mathcal{A})} \times \frac{\Gamma(N + \mathcal{A})}{\prod_{i=1}^{n} \Gamma(N_i + \alpha_i)}$$

```
posteriorPredictive[future_List, data_List, alpha_List] :=
    Module[{largeN, M, A, term1, term2, term3},
        M = Total[future];
        largeN = Total[data];
        A = Total[alpha];
        term1 = Apply[Multinomial, future];
        term2 = Apply[Times, Gamma[future + data + alpha]] /
                        Gamma[M + largeN + A];
        term3 = Gamma[largeN + A] /
                        Apply[Times, Gamma[data + alpha]];
        N[term1 term2 term3]]
```

If you prefer to see the result in terms of fractions, then don't use **N[ ]** in the last line.

Exercise 47.8.10: Use the *Mathematica* program to verify the answer from the derivation presented in section 47.7 for the probability that the next $M + 1$ sampled will all be of the same type as the first N sampled.

Solution to Exercise 47.8.10

Suppose that the first sample consisted of 5000 things of the first type and none of the second type, so that $N_1 = 5000$ and $N_2 = 0$. What is the probability that the *next* 5001 will all be of the first type? Evaluating,

```
posteriorPredictive[{5001, 0}, {5000, 0}, {1, 1}]
```

returns the answer **0.5**.

The next twelve exercises through Exercise 47.8.22 are numerical examples of the impact of the prior probability as illustrated for the simple kangaroo scenario.

Exercise 47.8.11: What was the impact of adopting Jeffreys's prior for the simple kangaroo scenario in Volume I?

Solution to Exercise 47.8.11

In Chapter Fifteen of Volume I, the $n = 4$ dimensional state space of the simplest kangaroo scenario served as a convenient numerical example for illustrating the impact of the prior probability over model space. One of the regions examined during the course of systematically changing the α_i parameters in the Dirichlet distribution for the prior probability over model space was the region from $\alpha_i = 1$ to $\alpha_i \to 0$. We noted that one special way point on this journey was setting all of the α_i equal to $1/2$. This particular value of the parameters for the prior probability over the models was called the "Jeffreys's prior."

The discussion began with how a "flat," or uniform, prior probability density function over the models was implemented by setting $\alpha_i = 1$. The consequence on the marginal probability for seeing, say, the first $M = 16$ kangaroos, was an equal degree of belief in any pattern of behavioral traits whatsoever. Thus, the probability for seeing all 16 kangaroos as right handed Foster's drinking kangaroos was the same as seeing an even breakdown of four kangaroos allocated to each of the four traits.

There was a total of 969 possible contingency tables, or possible future data, and each and every one of these possible future data, from all 16 kangaroos exhibiting the same behavioral trait to four kangaroos exhibiting each of the four traits, possessed the same probability. This was the inescapable result of adopting the Bayes–Laplace prescription of **total ignorance** about what was "causing" the kangaroos to exhibit any pattern of traits.

However, any departure from the Bayes–Laplace uniform distribution over model space meant that the IP was, in fact, not totally ignorant about what was behind the kangaroos's beer and hand preferences. If the Jeffreys's prior of $\alpha_i = 1/2$ was adopted as an alternative to "total ignorance," the ensuing consequence was that contingency tables with all 16 kangaroos exhibiting one of the four traits were favored over other patterns.

As a typical numerical example, the marginal probability for seeing the first 16 kangaroos as left handed Corona drinking kangaroos was some 25 times more probable than seeing them evenly distributed four to each trait. Under the uniform distribution of "total ignorance," the degree of belief would have been the same for these two possible future data. Jeffreys desired a prior probability over models that resulted in a larger probability for observing a "homogeneous population." His prior, as just mentioned, accomplished that goal.

SOLVED EXERCISES FOR CHAPTER 47

Exercise 47.8.12: Using the *Mathematica* program, calculate the effect of a higher probability for a homogeneous population when using Jeffreys's prior versus Laplace's prior.

Solution to Exercise 47.8.12

Say that 128 kangaroos have been sampled from a finite population of kangaroos, and that these data indicated that all 128 kangaroos were of the same type, that is, left handed Foster's drinkers. Using Jeffreys's prior, the probability that the next 128 kangaroos will also all be left handed Foster's drinkers turns out be 0.3551 after evaluating,

```
posteriorPredictive[{0, 0, 128, 0}, {0, 0, 128, 0},
                   {1/2, 1/2, 1/2, 1/2}]
```

Using Laplace's prior, the probability that the next 128 kangaroos will all be left handed Foster's drinkers is substantially lower at 0.1279.

```
posteriorPredictive[{0, 0, 128, 0}, {0, 0, 128, 0},
                   {1, 1, 1, 1}]
```

Jeffreys's prior is accomplishing what he intended for it to do, namely, returning a higher probability for a future sample after taking into account a limited sample size. Greater probability has been allocated to the hypothesis that the population is indeed homogeneous through modification of the prior probability for models.

Exercise 47.8.13: What modification to the prior probability for models produces even "better results" than Jeffreys's prior?

Solution to Exercise 47.8.13

If the α_i parameters of the Dirichlet distribution encapsulating the information about the probability of the models are allowed to approach 0, then an even higher probability for a homogeneous population can be achieved.

Setting all four $\alpha_i = 1/10$ in the *Mathematica* program results in a probability of 0.8120 that the next 128 kangaroos will all be left handed Foster's drinkers.

```
posteriorPredictive[{0, 0, 128, 0}, {0, 0, 128, 0},
                   {0.1, 0.1, 0.1, 0.1}]
```

Letting the $\alpha_i \to 0$ was J.B.S. Haldane's take on how to "fix" the Laplace uniform prior over model space. Apparently, Jeffreys was not willing to go this far in amending Laplace's uniform prior.

Exercise 47.8.14: Calculate the probability that the very first sixteen kangaroos are all of the same type versus the probability that they are evenly divided among all four types under a Jeffreys's prior.

Solution to Exercise 47.8.14

Suppose that we want to calculate the probability that the first sixteen kangaroos are all left handed Corona drinkers. Then, we need to first calculate,

$$P(M_1 = 0, M_2 = 0, M_3 = 0, M_4 = 16 \mid \mathcal{D})$$

under the Jeffreys's prior, followed by a comparison with the even division of traits,

$$P(M_1 = 4, M_2 = 4, M_3 = 4, M_4 = 4 \mid \mathcal{D})$$

However, we are presented with the simplification that no data $\mathcal{D}$ have yet been observed. This means that $N_1 = N_2 = N_3 = N_4 = 0$ and $N = 0$.

This predictive probability for the very first M kangaroos can be calculated with the formula first presented as Equation (15.3) in Volume I,

$$P(M_1, M_2, M_3, M_4) = \frac{M!\,\Gamma(\mathcal{A})}{\Gamma(M + \mathcal{A})} \times \prod_{i=1}^{4} \frac{\Gamma(M_i + \alpha_i)}{M_i!\,\Gamma(\alpha_i)}$$

It's easy to see that this formula is a consequence of the general posterior predictive probability as just derived,

$$P(M_1, M_2, \cdots, M_n \mid N_1, N_2, \cdots, N_n) =$$

$$W(M) \times \frac{\prod_{i=1}^{n} \Gamma(M_i + N_i + \alpha_i)}{\Gamma(M + N + \mathcal{A})} \times \frac{\Gamma(N + \mathcal{A})}{\prod_{i=1}^{n} \Gamma(N_i + \alpha_i)}$$

after substituting in the fact that all $N_i = 0$.

$$P(M_1, M_2, M_3, M_4) = W(M) \times \frac{\prod_{i=1}^{4} \Gamma(M_i + \alpha_i)}{\Gamma(M + \mathcal{A})} \times \frac{\Gamma(\mathcal{A})}{\prod_{i=1}^{4} \Gamma(\alpha_i)}$$

$$= \frac{M!}{M_1!\,M_2!\,M_3!\,M_4!} \times \frac{\prod_{i=1}^{4} \Gamma(M_i + \alpha_i)}{\Gamma(M + \mathcal{A})} \times \frac{\Gamma(\mathcal{A})}{\prod_{i=1}^{4} \Gamma(\alpha_i)}$$

$$= \frac{M!\,\Gamma(\mathcal{A})}{\Gamma(M + \mathcal{A})} \times \prod_{i=1}^{4} \left[\frac{\Gamma(M_i + \alpha_i)}{M_i!\,\Gamma(\alpha_i)} \right]$$

Future frequency counts really should be disambiguated as M and M_i which, in fact, is how they correctly appear in Equation (15.3). N and N_i should be reserved for data that have actually already occurred. In the statement of this problem, all the $N_i = 0$, and, of course, $N = 0$.

SOLVED EXERCISES FOR CHAPTER 47

After substituting the following numerical values for the Jeffreys's prior,

$$M = 16$$

$$\mathcal{A} = 2$$

$$\alpha_i = 1/2$$

we can commence the computation,

$$P(M_1, M_2, M_3, M_4) = \frac{M!\,\Gamma(\mathcal{A})}{\Gamma(M+\mathcal{A})} \times \prod_{i=1}^{4} \left[\frac{\Gamma(M_i + \alpha_i)}{M_i!\,\Gamma(\alpha_i)} \right]$$

$$\frac{M!\,\Gamma(\mathcal{A})}{\Gamma(M+\mathcal{A})} = \frac{16!\,\Gamma(2)}{\Gamma(18)}$$

$$= \frac{1}{17}$$

$$\prod_{i=1}^{4} \left[\frac{\Gamma(M_i + \alpha_i)}{M_i!\,\Gamma(\alpha_i)} \right] = \frac{\Gamma(0+1/2)}{0!\,\Gamma(1/2)} \times \cdots \times \frac{\Gamma(16+1/2)}{16!\,\Gamma(1/2)}$$

$$= \frac{\Gamma(16.5)}{16!\,\Gamma(1/2)}$$

$$P(M_1 = 0, M_2 = 0, M_3 = 0, M_4 = 16) = \frac{1}{17} \times \frac{\Gamma(16.5)}{16!\,\Gamma(1/2)}$$

$$= 0.008232$$

In a similar manner, the calculation of the probability for the second set of future frequency counts, $P(M_1 = 4, M_2 = 4, M_3 = 4, M_4 = 4)$, results in,

$$\prod_{i=1}^{4} \left[\frac{\Gamma(M_i + \alpha_i)}{M_i!\,\Gamma(\alpha_i)} \right] = \frac{\Gamma(4+1/2)}{4!\,\Gamma(1/2)} \times \cdots \times \frac{\Gamma(4+1/2)}{4!\,\Gamma(1/2)}$$

$$= \prod_{i=1}^{4} \frac{\Gamma(4.5)}{4!\,\Gamma(1/2)}$$

$$P(M_1 = 4, M_2 = 4, M_3 = 4, M_4 = 4) = \frac{1}{17} \times \prod_{i=1}^{4} \frac{\Gamma(4.5)}{4!\,\Gamma(1/2)}$$

$$= 0.0003288$$

Under the Jeffreys's prior, the marginal probability for one of the "homogeneous population" contingency tables where all 16 kangaroos exhibit the same trait of being left handed Corona drinkers is about 25 times larger than another possible contingency table with a breakdown of equal frequency counts for all four traits. You can see that this accomplishes Jeffreys's objective of raising the predictive probability for a certain class of contingency tables prior to any data whatsoever. Under the uniform prior, both of the contingency tables in this example, as well as the other 967 contingency tables, would have had the same probability of $1/969$.

Exercise 47.8.15: Using the formula for the probability of the data, what is the probability for data where all sixteen kangaroos are of one type and sixteen kangaroos are evenly distributed over the four types under Laplace's advice that nothing is known about what causes the kangaroos to fall into one type or another?

Solution to Exercise 47.8.15

As a validity check on the formula developed in Exercise 47.8.7, we must recover the known answer that the probability for both of these cases is equal to $1/969$. For the first case where, say, all sixteen kangaroos fall into the second type of right handed Corona drinkers, we have $N = 16$, $N_1 = 0$, $N_2 = 16$, $N_3 = 0$, and $N_4 = 0$. Also, $n = 4$ and all $\alpha_i = 1$, so that the sum $\mathcal{A} = 4$,

$$P(N_1 = 0, N_2 = 16, N_3 = 0, N_4 = 0) =$$

$$W(N) \times \frac{\Gamma(\mathcal{A})}{\prod_{i=1}^{n} \Gamma(\alpha_i)} \times \frac{\prod_{i=1}^{n} \Gamma(N_i + \alpha_i)}{\Gamma(N + \mathcal{A})}$$

$$= \frac{N!}{N_1!\, N_2!\, N_3!\, N_4!} \times \frac{\Gamma(\mathcal{A})}{\prod_{i=1}^{4} \Gamma(\alpha_i)} \times \frac{\prod_{i=1}^{4} \Gamma(N_i + \alpha_i)}{\Gamma(N + \mathcal{A})}$$

$$= \frac{16!}{0!\, 16!\, 0!\, 0!} \times \frac{\Gamma(4)}{1} \times \frac{\Gamma(1) \times \Gamma(17) \times \Gamma(1) \times \Gamma(1)}{\Gamma(20)}$$

$$= \Gamma(4) \times \frac{\Gamma(17)}{\Gamma(20)}$$

$$= \frac{3!\, 16!}{19!}$$

$$= 1/969$$

For the second case where all sixteen kangaroos are evenly distributed over all four types, $N = 16$, $N_1 = 4$, $N_2 = 4$, $N_3 = 4$, and $N_4 = 4$. The other components remain the same at $n = 4$ and all $\alpha_i = 1$, so that the sum $\mathcal{A} = 4$,

$$P(N_1 = 4, N_2 = 4, N_3 = 4, N_4 = 4) =$$

$$\frac{N!}{N_1!\,N_2!\,N_3!\,N_4!} \times \frac{\Gamma(\mathcal{A})}{\prod_{i=1}^{4} \Gamma(\alpha_i)} \times \frac{\prod_{i=1}^{4} \Gamma(N_i + \alpha_i)}{\Gamma(N + \mathcal{A})}$$

$$= \frac{16!}{4!\,4!\,4!\,4!} \times \frac{\Gamma(4)}{1} \times \frac{\Gamma(5) \times \Gamma(5) \times \Gamma(5) \times \Gamma(5)}{\Gamma(20)}$$

$$= 16! \times \frac{\Gamma(4)}{\Gamma(20)}$$

$$= \frac{16!\,3!}{19!}$$

$$= 1/969$$

Exercise 47.8.16: Using this formula for the probability of the data, what is the probability for data where all sixteen kangaroos are of one type and sixteen kangaroos are evenly distributed over the four types under Jeffreys's advice that homogeneous populations should be favored?

Solution to Exercise 47.8.16

As a validity check on the above formula, we must recover the known answer that the probability for the first case is equal to 0.008232 and the probability for the second case is equal to 0.0003288. For the first case where all sixteen kangaroos fall into the second type of right handed Corona drinkers, we have $N = 16$, $N_1 = 0$, $N_2 = 16$, $N_3 = 0$, and $N_4 = 0$. Also, $n = 4$, but now all $\alpha_i = 1/2$, so that the sum $\mathcal{A} = 2$,

$$P(N_1 = 0, N_2 = 16, N_3 = 0, N_4 = 0)$$

$$= \frac{N!}{N_1!\,N_2!\,N_3!\,N_4!} \times \frac{\Gamma(\mathcal{A})}{\prod_{i=1}^{n} \Gamma(\alpha_i)} \times \frac{\prod_{i=1}^{n} \Gamma(N_i + \alpha_i)}{\Gamma(N + \mathcal{A})}$$

$$= \frac{16!}{0!\,16!\,0!\,0!} \times \frac{\Gamma(2)}{4 \times \Gamma(1/2)} \times \frac{\Gamma(1/2) \times \Gamma(16.5) \times \Gamma(1/2) \times \Gamma(1/2)}{\Gamma(18)}$$

$$= 0.008232$$

For the second case where all sixteen kangaroos are evenly distributed over all four types, $N = 16$, $N_1 = 4$, $N_2 = 4$, $N_3 = 4$, and $N_4 = 4$. The other components remain the same at $n = 4$ and all $\alpha_i = 1/2$, so that the sum $\mathcal{A} = 2$,

$$P(N_1 = 4, N_2 = 4, N_3 = 4, N_4 = 4)$$

$$= \frac{N!}{N_1!\,N_2!\,N_3!\,N_4!} \times \frac{\Gamma(\mathcal{A})}{\prod_{i=1}^{n}\Gamma(\alpha_i)} \times \frac{\prod_{i=1}^{n}\Gamma(N_i + \alpha_i)}{\Gamma(N + \mathcal{A})}$$

$$= \frac{16!}{4!\,4!\,4!\,4!} \times \frac{\Gamma(2)}{4 \times \Gamma(1/2)} \times \frac{\Gamma(4.5) \times \Gamma(4.5) \times \Gamma(4.5) \times \Gamma(4.5)}{\Gamma(18)}$$

$$= 0.0003288$$

Exercise 47.8.17: What is obvious from these past few exercises?

Solution to Exercise 47.8.17

It is hard to escape the fact that the probability for the future frequency counts when there are no data, $P(M_1, M_2, M_3, M_4)$ is exactly the same as the probability for the data $P(N_1, N_2, N_3, N_4)$. The formulas are also exactly the same even though they were presented differently.

Exercise 47.8.18: What modification to the prior probability for models goes in the opposite direction of Jeffreys's prior?

Solution to Exercise 47.8.18

At the other end of the spectrum from Jeffreys and Haldane, there were individuals who felt that the probability of 1/969 given to an even distribution of types by Laplace's prior tended to ignore the overwhelmingly larger multiplicity factor for this situation as compared to the case where all were of the same type. They also can have their wishes satisfied within the freedom permitted by the general framework by pushing the prior probability for models in the opposite direction than that favored by Jeffreys and Haldane.

For example, let all four $\alpha_i = 100$. Then, the probability for seeing any kind of relatively even distribution of kangaroos is enormously higher than the probability that they are all of the same type. The probability that the first sixteen kangaroos are evenly distributed is 0.0138, from an evaluation by,

```
posteriorPredictive[{4, 4, 4, 4},
                    {0, 0, 0, 0}, {100, 100, 100, 100}]
```

versus the extremely depressed probability of 5.43×10^{-10} that the first sixteen kangaroos are all of the same type,

```
posteriorPredictive[{0, 0, 16, 0},
                    {0, 0, 0, 0}, {100, 100, 100, 100}]
```

Even when there is a substantial amount of data to support the even distribution, the probability changes ever so slightly to 0.0140. There is hardly a change from the above probability calculated on no data whatsoever.

```
posteriorPredictive[{4, 4, 4, 4},
                    {25, 25, 25, 25}, {100, 100, 100, 100}]
```

This is because the α_i parameters are serving as *virtual observations*. If an IP, at the very outset, assumes a relatively higher probability for one model over all the others, then less data are required to achieve higher probabilities. Here, the models that lend a great deal of respect to the multiplicity factor have their stature raised, and this is all done *prior to any data*. Which prior probability over model space do you think is doing the right thing by trying to fulfill the notion that the IP is completely uninformed?

Exercise 47.8.19: Supposing that some sampling of the kangaroos already supports a homogeneous population, what is the probability that the *next* kangaroo is the same type?

Solution to Exercise 47.8.19

Let's say that a sample of $N = 16$ kangaroos has resulted in all of them being right handed Foster's drinkers. What is the probability that the next kangaroo to be sampled is also a right handed Foster's drinker?

Applying Laplace's *Rule of Succession* to answer this question, we have,

$$
\begin{aligned}
P(M_1 = 1, M_2 = 0, M_3 = 0, M_4 = 0 \mid N_1 = 16, N_2 = 0, N_3 = 0, N_4 = 0) &= \frac{N_1 + 1}{N + n} \\
&= \frac{17}{20} \\
&= 0.85
\end{aligned}
$$

It is a certainty that the next kangaroo belongs to one of the four types. With the determination that the probability that the next kangaroo belongs to the first type is 0.85, the probability that the next kangaroo belongs to one of the final three types is 0.05 for each remaining type. This result is, of course, also found by the same application of the *Rule of Succession*. The probability that the next kangaroo belongs to one of the four types is, as hoped, equal to 1.

As we have stressed on numerous occasions, the formula for Laplace's *Rule of Succession* was derived under the assumption of a uniform prior probability for the models. Jeffreys wasn't too upset with a probability of 0.85 for the *next* kangaroo to be of the same type as the sample. But he was not happy when the hypothesis of a homogeneous population that covered all of the remaining kangaroos received a very low probability. He pointed the finger of blame at the adoption of the uniform prior probability.

Exercise 47.8.20: Supposing that some sampling of the kangaroos already supports a homogeneous population, what is the probability that the *next* number of kangaroos, equal in number to the original sample size, is the same type?

Solution to Exercise 47.8.20

For a change of pace, let's say that a sample of $N = 16$ kangaroos has resulted in all of them being left handed Foster's drinkers. What is the probability that the next $M = 16$ kangaroos to be sampled are also left handed Foster's drinker?

Applying Laplace's generalized *Rule of Succession* to answer this question, and letting $\mathcal{D}$ refer to the already sampled kangaroos, we have,

$$
\begin{aligned}
P(M_1 = 0, M_2 = 0, M_3 = 16, M_4 = 0 \mid \mathcal{D}) &= \frac{M!\,(N+n-1)!}{N_1!\cdots N_4!\,(M+N+n-1)!} \times \frac{\prod_{i=1}^{4}(M_i+N_i)!}{\prod_{i=1}^{4} M_i!} \\
&= \frac{16!\,(16+4-1)!}{0!\,0!\,16!\,0!\,(16+16+4-1)!} \times \frac{0!\,0!\,32!\,0!}{0!\,0!\,16!\,0!} \\
&= \frac{19!}{35!} \times \frac{32!}{16!} \\
&= \frac{19 \times 18 \times 17}{35 \times 34 \times 33} \\
&= 0.15
\end{aligned}
$$

Once again, Jeffreys felt that this probability was too low given that the population might be *all* left handed Foster's drinking kangaroos.

Exercise 47.8.21: Supposing that some sampling of the kangaroos already supports a homogeneous population, what is the probability that the rest of the kangaroos in the population are the same type?

Solution to Exercise 47.8.21

Here we have to specify the population as a known number, say, 16,000 kangaroos in all. Since we have already sampled $N = 16$ kangaroos, there are 15,984 left. The probability that all 15,984 are left handed Foster's drinkers given the data from the sample, is calculated as,

$$P(M_1 = 0, M_2 = 0, M_3 = 15984, M_4 = 0 \mid \mathcal{D}) = \frac{M!\,(N+n-1)!}{N_1!\cdots N_4!\,(M+N+n-1)!} \times \frac{\prod_{i=1}^{4}(M_i+N_i)!}{\prod_{i=1}^{4} M_i!}$$

$$= \frac{15984!\,(16+4-1)!}{0!\,0!\,16!\,0!\,(15984+16+4-1)!} \times \frac{0!\,0!\,16000!\,0!}{0!\,0!\,15984!\,0!}$$

$$= \frac{15984!\,19!}{16!\,16003!} \times \frac{16000!}{15984!}$$

$$= \frac{19 \times 18 \times 17}{16003 \times 16002 \times 16001}$$

$$= 1.42 \times 10^{-9}$$

If Jeffreys was upset with the last probability, he was really livid over this extremely small probability that the entire population was, in fact, homogeneous given the results from the original sample.

Who was to blame for this fiasco? Why, it must be Bayes and Laplace for that insidious uniform prior probability over the models. There was an urgent need, he said, for corrective surgery. There was obviously an insufficient prior probability on all of the homogeneous populations as illustrated in the above numerical example. This defect was due to the fact that they did not inherit their just due from the uniform prior.

Personally, though, I don't feel any special *angst* over this small probability given the fact that there are still 15,984 kangaroos left to be categorized, and I haven't been provided with any supporting rationale as to why the population should be homogeneous. Remember that I am "totally ignorant" about what causes kangaroos to have particular beer and hand preferences. Were there any special circumstances in how those first 16 kangaroos were sampled? By definition, we don't know the answer to that question.

Exercise 47.8.22: Provide a numerical example using the kangaroos where the sample constitutes a rather large fraction of the whole population.

Solution to Exercise 47.8.22

I have not yet given an example where the sample constituted a large fraction of the whole population, so let me rectify that oversight. In the previous exercises, I have shown Laplace's *Rule of Succession* in action to verify Jeffreys's complaints with the uniform prior over the models.

I chose the simple kangaroo scenario instead of Jeffreys's "animals with feathers," so with that change in mind, but retaining the details contained in the previous exercises, what is the probability that the final 10 kangaroos in the population are the same type given that 15,990 of the total population of 16,000 kangaroos have been sampled and were of the same type?

$$P(M_1 = 0, M_2 = 0, M_3 = 10, M_4 = 0 \,|\, N_1 = 0, N_2 = 0, N_3 = 15990, N_4 = 0)$$

$$= \frac{M!\,(N+n-1)!}{N_1!\cdots N_4!\,(M+N+n-1)!} \times \frac{\prod_{i=1}^{4}(M_i+N_i)!}{\prod_{i=1}^{4} M_i!}$$

$$= \frac{10!\,(15990+4-1)!}{0!\,0!\,15990!\,0!\,(10+15990+4-1)!} \times \frac{0!\,0!\,16000!\,0!}{0!\,0!\,10!\,0!}$$

$$= \frac{15993 \times 15992 \times 15991}{16003 \times 16002 \times 16001}$$

$$= 0.9981$$

Even here, I am not absolutely certain that all of the final ten kangaroos to be sampled will be left handed Foster's drinkers. Were those last ten kangaroos the last ten because of some difficulty in locating and querying them? If so, then maybe there is a small chance that they will exhibit a different trait.

Exercise 47.8.23: In preparation for future exercises concerning Jeffreys's example concerning "a suggested value of a chance," begin translating Jeffreys's unique notation into what has become our standard notation.

Solution to Exercise 47.8.23

Bound hand in hand with the low level task of translating notation, there is the far more important task of translating *analogous concepts*. Let's begin with Jeffreys's discussion of the coin tossing scenario, and the priors he suggests for this situation as quoted above in **Connections to the Literature**. This is an easy to understand *entré* into Jeffreys's objections to the usual flat prior probability for model space.

Jeffreys writes $P(q\,|\,H)$ for the prior probability of a model. We have always written this as $P(\mathcal{M}_k)$. Jeffreys seems to generally include an H to stand for any proposition laying out all of the background assumptions and details for whatever problem is of current focus.

I choose not to include any explicit statement analogous to H in my probability notation because doing so would seem to demand that one must also ultimately write down an expression $P(H)$ which is something I think is best avoided. Like most everyone else, I simply issue a blanket *caveat* that the background information H must be understood as whatever the author provides as the detailed explanation of the problem scenario.

Jeffreys's primary goal is to avoid allocating probability evenly across all models. He accomplishes this objective by placing half of all the prior probability available to the models on one specific assigned value for the probability of the coin through his prior probability expression $P(q\,|\,H) = 1/2$. That one assigned numerical value under the model is the value p, and Jeffreys calls this "the suggested value of a chance." This suggested value of a chance p might be the definitely assigned probability value of, say, 1/2 or 3/4, or anything between 0 and 1. For that matter, p might even be 0 or 1.

The remaining half of the prior probability over model space is allocated to a flat prior where all possible assigned numerical values from 0 to 1 are made under this second model, $P(q' \mid H) = 1/2$.

Under this second model, we know that as a Dirichlet distribution (the *beta distribution* for the coin tossing scenario) the pdf for the model has the flat value of 1 for every assigned value between 0 and 1 inclusive. Thus, $P(dp' \mid q', H) = dp'$. Then, applying the **Product Rule** in order to find the joint probability of this model and a numerical assignment,

$$P(dp' \mid q', H)\, P(q' \mid H) = P(q', dp' \mid H) = 1/2\, dp'$$

Jeffreys has set up this expression because he eventually wants to use the **Sum Rule** and integrate over dp'.

Exercise 47.8.24: Now translate Jeffreys's notation for the observational evidence.

Solution to Exercise 47.8.24

Jeffreys's formulas (3) and (4) give the probability of the data conditioned on each of his two models. Jeffreys denotes what he calls the observational evidence, that is, the data, as θ.

The probability for the data conditioned on the first model is then,

$$P(\theta \mid qH) = {}^{x+y}C_x\, p^x (1-p)^y$$

which we would write as,

$$P(\mathcal{D} \mid \mathcal{M}_A) = \frac{N!}{N_1!\, N_2!}\, Q_1^{N_1}\, Q_2^{N_2}$$

Thus, $x \equiv N_1$, $y \equiv N_2$, and $x + y \equiv N_1 + N_2 = N$. Because the prior probability $P(\mathcal{M}_A)$ is a delta function, q_{1A} and q_{2A} are assigned the definite numerical values Q_1 and Q_2, where Jeffreys's $p \equiv Q_1$, the "suggested value of a chance."

The probability for the data conditioned on the second model would be,

$$P(\theta \mid q', p', H) = {}^{x+y}C_x\, p'^x (1-p')^y$$

which we would write as,

$$P(\mathcal{D} \mid \mathcal{M}_B) = \frac{N!}{N_1!\, N_2!}\, q_{1B}^{N_1}\, q_{2B}^{N_2}$$

Exercise 47.8.25: Translate Jeffreys's notation into our notation for the posterior probability of the two models.

Solution to Exercise 47.8.25

Jeffreys's formulas (5) and (6) give the updated probability for the two models after conditioning on any known data through Bayes's Theorem. Jeffreys disregards the constant probability of the data in the denominator, the multiplicity factor, and the explicit prior probability for the first model so that the posterior probability of the first model is proportional to the product of the probability of the data given the model,

$$P(q \mid \theta H) \propto p^x (1-p)^y$$

We would prefer to express this as,

$$P(\mathcal{M}_A \mid \mathcal{D}) \propto P(\mathcal{D} \mid \mathcal{M}_A) \, P(\mathcal{M}_A)$$

$$\propto \frac{N!}{N_1! \, N_2!} Q_1^{N_1} Q_2^{N_2} \times 1/2$$

The posterior probability of the second model is first expressed by Jeffreys as the joint probability in his Equation (6),

$$P(q', dp' \mid \theta H) \propto p'^x (1-p')^y \, dp'$$

which we would write as,

$$P(\mathcal{M}_B, q \mid \mathcal{D}) = \frac{N!}{N_1! \, N_2!} q_1^{N_1} q_2^{N_2} \times C_{\text{Beta}} \, q_1^{\alpha-1} q_2^{\beta-1} \times 1/2$$

$$= \frac{N!}{N_1! \, N_2!} C_{\text{Beta}} \, q_1^{N_1+\alpha-1} q_2^{N_2+\beta-1} \times 1/2$$

Jeffreys is setting up the expression in his Equation (6),

$$P(q', dp' \mid \theta H) \propto p'^x (1-p')^y \, dp'$$

for an integration over dp' so that the actual posterior probability for the second model can be found in his Equation (7) as,

$$P(q' \mid \theta H) \propto \int_0^1 p'^x (1-p')^y \, dp' = \frac{x! \, y!}{(x+y+1)!}$$

In my version, the two parameters for the *beta function* are presented in the more usual α and β form, rather than as α_1 and α_2. We would then write this posterior probability with $\alpha = \beta = 1$ as,

$$P(\mathcal{M}_B \mid \mathcal{D}) = \int_0^1 \frac{N!}{N_1! \, N_2!} C_{\text{Beta}} \, q_1^{N_1+\alpha-1} q_2^{N_2+\beta-1} \, dq$$

$$= \frac{N!}{N_1! \, N_2!} C_{\text{Beta}} \int_0^1 q_1^{N_1+\alpha-1} q_2^{N_2+\beta-1} \, dq$$

$$= \frac{N!}{N_1! \, N_2!} \frac{\Gamma(\alpha+\beta)}{\Gamma(\alpha)\Gamma(\beta)} \frac{\Gamma(N_1+\alpha)\Gamma(N_2+\beta)}{\Gamma(N_1+\alpha+N_2+\beta)}$$

$$= \frac{1}{N+1}$$

Exercise 47.8.26: Review the solution to the *beta function*.

Solution to Exercise 47.8.26

We have had many occasions to take advantage of the known analytical solution to the Dirichlet integral. Jeffreys relies upon it as well in his Equation (7),

$$P(q' \mid \theta H) \propto \int_0^1 p'^x (1-p')^y \, dp' = \frac{x! \, y!}{(x+y+1)!}$$

For the case here, the *beta function* will be applied because there are only two statements in the state space,

$$\int_0^1 q^{\alpha-1}(1-q)^{\beta-1} \, dq = \frac{\Gamma(\alpha)\Gamma(\beta)}{\Gamma(\alpha+\beta)}$$

$$\int_0^1 q^x (1-q)^y \, dq = \frac{\Gamma(x+1)\Gamma(y+1)}{\Gamma(x+1+y+1)}$$

$$= \frac{x! \, y!}{(x+y+1)!}$$

Exercise 47.8.27: Translate Jeffreys's notation into our notation for the ratio of the posterior probability of the two models.

Solution to Exercise 47.8.27

With the background development in Equations (1) through (7), it is a relatively easy matter for Jeffreys to state the final result needed for his numerical examples in his Equation (8) reproduced below,

$$K = \frac{P(q \mid \theta H)}{P(q' \mid \theta H)} = \frac{(x+y+1)!}{x! \, y!} p^x (1-p)^y$$

For us, Jeffreys's K appearing on the left hand side is the familiar ratio of the posterior probability for two models,

$$\frac{P(\mathcal{M}_A|\mathcal{D})}{P(\mathcal{M}_B|\mathcal{D})} = \frac{(N+1)!}{N_1!\,N_2!}\,Q_1^{N_1}Q_2^{N_2}$$

Exercise 47.8.28: Show the easier derivation of Jeffreys's Equation (8) when carried out in our notation.

Solution to Exercise 47.8.28

We begin with the standard characterization of the posterior probability for two models after applying Bayes's Theorem and canceling out $P(\mathcal{D})$,

$$\frac{P(\mathcal{M}_A|\mathcal{D})}{P(\mathcal{M}_B|\mathcal{D})} = \frac{P(\mathcal{D}|\mathcal{M}_A)}{P(\mathcal{D}|\mathcal{M}_B)} \times \frac{P(\mathcal{M}_A)}{P(\mathcal{M}_B)}$$

$$= \frac{P(\mathcal{D}|\mathcal{M}_A)}{P(\mathcal{D}|\mathcal{M}_B)} \times \frac{1/2}{1/2}$$

$$P(\mathcal{D}|\mathcal{M}_A) = \frac{N!}{N_1!\,N_2!}\,Q_1^{N_1}Q_2^{N_2}$$

$$P(\mathcal{D}|\mathcal{M}_B) = \frac{1}{N+1}$$

$$\frac{P(\mathcal{D}|\mathcal{M}_A)}{P(\mathcal{D}|\mathcal{M}_B)} = \frac{\frac{N!}{N_1!\,N_2!}\,Q_1^{N_1}Q_2^{N_2}}{\frac{1}{N+1}}$$

$$\frac{P(\mathcal{M}_A|\mathcal{D})}{P(\mathcal{M}_B|\mathcal{D})} = \frac{(N+1)!}{N_1!\,N_2!}\,Q_1^{N_1}Q_2^{N_2}$$

At this juncture, I would like to highlight what I consider a somewhat strange aspect to Jeffreys's argument, made more obvious when translated into my notation as shown above. The model $\mathcal{M}_B$, Jeffreys's "alternative hypothesis" to the "null hypothesis" reflected in model $\mathcal{M}_A$ of the suggested value of chance written as p, covers all possible assignments to p from 0 through 1.

Therefore, the definitive assignment to p under $\mathcal{M}_A$ is not compared via K to some other definitive assignment under model $\mathcal{M}_B$, but rather to all possible legitimate assignments. Thus, in some sense, you could look at it as not really a comparison between competing models based on the data, but simply the ordinary updating of the prior probability of model $\mathcal{M}_A$ to a posterior probability,

$$P(\mathcal{M}_A \mid \mathcal{D}) = \frac{P(\mathcal{D} \mid \mathcal{M}_A)\, P(\mathcal{M}_A)}{P(\mathcal{D})}$$

$$= \frac{P(\mathcal{D} \mid \mathcal{M}_A) \times 1/2}{[P(\mathcal{D} \mid \mathcal{M}_A) \times 1/2] + [P(\mathcal{D} \mid \mathcal{M}_B) \times 1/2]}$$

$$= \frac{P(\mathcal{D} \mid \mathcal{M}_A)}{P(\mathcal{D} \mid \mathcal{M}_A) + P(\mathcal{D} \mid \mathcal{M}_B)}$$

$$= \frac{P(\mathcal{D} \mid \mathcal{M}_A)}{P(\mathcal{D} \mid \mathcal{M}_A) + [1/(N+1)]}$$

So, in this formulation, the IP updates the prior probability of model $\mathcal{M}_A$ from a beginning value of 1/2 to some higher or lower value depending on what the data had to say. As a probability, of course, the posterior probability $P(\mathcal{M}_A \mid \mathcal{D})$ can never leave the region from 0 to 1. After examining some numerical examples from the standpoint of Jeffreys's characterization as a *ratio of posterior probabilities for two models*, we will look at the different interpretation afforded by simply viewing the calculation as updating the prior probability for the single model $\mathcal{M}_A$.

Exercise 47.8.29: Replicate some of the numerical examples that Jeffreys used to illustrate his ideas concerning his K ratio, and what has since been labeled as a "Bayes factor."

Solution to Exercise 47.8.29

Jeffreys chose the value of $Q_1 = 1/2$ (p for Jeffreys) as "the suggested value of a chance" for the coin toss scenario. He initially examined just a small number of experimental observations (Jeffreys's observational evidence θ) for some numerical examples of his Equation (8).

He examined $N = 5$ data points where the number of successes (x), in this case, say HEADS, ranged from $N_1 = 0$ through $N_1 = 5$ as the number of TAILS (y) remained fixed at 0. For example, if there were five HEADS observed in five tosses of a coin, then the value of $K = 3/16$ expresses the relative favoring of the model where $Q_1 = 1/2$ is the "suggested value of a chance."

$$\frac{P(\mathcal{M}_A \mid \mathcal{D})}{P(\mathcal{M}_B \mid \mathcal{D})} = \frac{(N+1)!}{N_1!\, N_2!} Q_1^{N_1} Q_2^{N_2}$$

$$= \frac{6!}{5!\, 0!} (1/2)^5 (1/2)^0$$

$$= 3/16$$

This result makes intuitive sense since no TAILS in five tosses would certainly not favor model $\mathcal{M}_A$ where Q_1 and Q_2 have been assigned a value of 1/2.

Jeffreys was curious about when this ratio favoring an even chance crossed below the threshold of 0.1. With the experimental observations of $N_1 = 7$ HEADS and $N_2 = 0$ TAILS in $N = 7$ tosses of the coin, this ratio becomes,

$$\frac{P(\mathcal{M}_A \mid \mathcal{D})}{P(\mathcal{M}_B \mid \mathcal{D})} = \frac{(N+1)!}{N_1! \, N_2!} Q_1^{N_1} Q_2^{N_2}$$

$$= \frac{8!}{7! \, 0!} (1/2)^7 (1/2)^0$$

$$= 1/16$$

On the other hand, if the data were $N_1 = 5$ HEADS and $N_2 = 5$ TAILS in $N = 10$ tosses of the coin, these experimental observations would seem to favor model $\mathcal{M}_A$. They do, in fact, favor the suggested value of chance equal to 1/2 in a ratio of about 2.7 to 1.

$$\frac{P(\mathcal{M}_A \mid \mathcal{D})}{P(\mathcal{M}_B \mid \mathcal{D})} = \frac{(N+1)!}{N_1! \, N_2!} Q_1^{N_1} Q_2^{N_2}$$

$$= \frac{11!}{5! \, 5!} (1/2)^5 (1/2)^5$$

$$= 693/256$$

Exercise 47.8.30: With *Mathematica*'s assistance, continue to rely on Equation (8) for larger and larger number of experimental observations.

Solution to Exercise 47.8.30

Jeffreys developed an approximation to Equation (8) for large N_1 and N_2 using the Stirling approximation. But *Mathematica* will continue to compute Equation (8) correctly even when the number of experimental observations becomes very large.

For example, Jeffreys wanted to find out where the ratio favoring the suggested value of chance equal to 1/2 crossed a threshold greater than 10. Jeffreys say that this occurs at the point of 80 successes and 80 failures.

To confirm this, have *Mathematica* evaluate, in an available abbreviated syntax similar to what you might enter into a hand held calculator,

$$161! \, / \, (80! \, 80!) \, 0.5^{80} \, 0.5^{80}$$

which returns 10.1398.

Jeffreys comments that it is easier to find strong evidence against a specific value like 1/2 (for example, seven HEADS and no TAILS) than to find equally strong evidence in favor of it (for example, 80 HEADS and 80 TAILS). My intuitive justification for this result differs however than the one Jeffreys provided. My sense is that the broad spectrum of numerical assignments allowed by model $\mathcal{M}_B$ presents so many more opportunities for the experimental observations to support it when compared to the one very specific value under model $\mathcal{M}_A$.

Here is a more severe test of *Mathematica*'s computational abilities. Jeffreys mentions "a remarkable series of experiments" on whether the probability of a FIVE or a SIX for dice is really 1/3. We are still in the realm of the binary scenario with a FIVE or SIX a success, and a ONE, TWO, THREE, or FOUR a failure.

The experimental observations consisted of 106,602 successes (a FIVE or SIX) and 209,070 failures (not a FIVE or a SIX) in 315,672 attempts. The "suggested value of a chance" for a fair die captured in model $\mathcal{M}_A$ is $Q_1 = 1/3$ and $Q_2 = 2/3$. Let *Mathematica* evaluate,

`N[ScientificForm[315673! / (106602! 209070!) (1/3)`^{106602}` (2/3)`^{209070}`]]`

A value of $K = 6.46 \times 10^{-4}$ is returned by *Mathematica* as compared to Jeffreys's reported value, based on his approximation, of $K = 6.27 \times 10^{-4}$. So even though the frequency ratio of a FIVE or SIX based on the data is 0.337699, very close to 1/3, the ratio favoring the model that assigns a numerical value of $Q_1 = 1/3$ is very small when compared to the "alternative hypothesis" that the assignment lies somewhere else within the entire spectrum of 0 to 1.

It just seems not very helpful to me to say at the end of such a computation that the "null hypothesis" of an exact probability of 1/3 for a FIVE or a SIX is not supported by these data. To the contrary, these data overwhelmingly support an extremely large posterior probability for models very close to $Q_1 = 1/3$. The posterior predictive probability of a FIVE or SIX when averaged over *all models*,

$$P(X_{315673} = x_1 \mid \mathcal{D}) = \frac{106602 + 1}{315672 + 6} = 0.3377$$

can be compared to the probability based on one model,

$$P(X = x_1 \mid \mathcal{M}_k) = 1/3$$

Exercise 47.8.31: Examine Jeffreys's coin examples from the perspective of simply updating the prior probability of model $\mathcal{M}_A$.

Solution to Exercise 47.8.31

Back in Exercise 47.8.28, we advanced the notion that instead of calculating a "Bayes factor," that is, Jeffreys's K ratio of two posterior probabilities, the scenario might be better viewed as simply updating the prior probability of model $\mathcal{M}_A$. Model $\mathcal{M}_A$ was Jeffreys's q that assigned a definite numerical value to Q_1, or as Jeffreys put it, the hypothesis of "a suggested value of chance." If, for the sake of a numerical example, the suggested value of a chance is $p \equiv Q_1 = 1/2$, then Jeffreys found ratios, supporting or not as the case may be, for this value of p after some data had been observed.

Continuing on from the results of Exercise 47.8.28, the IP may straightforwardly compute the posterior probability for model $\mathcal{M}_A$,

$$P(\mathcal{M}_A \mid \mathcal{D}) = \frac{1}{1 + \frac{P(\mathcal{D} \, \mathcal{M}_B)}{P(\mathcal{D} \, \mathcal{M}_A)}}$$

The IP can easily observe how the data are shifting the degree of belief that $1/2$ is, in fact, the correct assignment for the suggested value of a chance. Table 47.2 illustrates the calculation of the posterior probability for model $\mathcal{M}_A$ using,

$$P(\mathcal{M}_A \mid \mathcal{D}) = \frac{1}{1 + \frac{1/(N+1)}{W(N) \, Q_1^{N_1} \, Q_2^{N_2}}}$$

Table 47.2: *The posterior probability for model $\mathcal{M}_A$ that specifies the assignment $Q_1 = Q_2 = 1/2$. These are Jeffreys's answers when the only other model is a model $\mathcal{M}_B$ that implements Laplace's position.*

Case	N_1	N_2	N	$P(\mathcal{M}_A \mid \mathcal{D})$
1	1	0	1	0.500000
2	5	0	5	0.157895
3	1	1	2	0.600000
4	2	2	4	0.652174
5	5	5	10	0.730242
6	80	80	160	0.910231

Case 1 shows that, for Jeffreys, HEADS on the first toss doesn't change the prior probability. Case 2 shows that the posterior probability retreats from its initial value of $1/2$ down to about 0.16 if five HEADS and no TAILS are seen in the first five coin tosses. The next four cases show the ever increasing posterior probability for model $\mathcal{M}_A$ as the data lend more and more support for the "suggested value of a chance."

Contrary to the impression you get from Jeffreys, Laplace's *Rule of Succession* seems to provide exactly what intuition would suggest. For any even number of N tosses, if the data support the supposition of $Q_1 = 1/2$, the posterior predictive probability that the very next toss will be HEADS always remains at $1/2$. Whether the IP has tossed the coin twice with one HEADS and one TAILS, or 100 times with 50 HEADS and 50 TAILS, the probability that the coin will come up HEADS on the next toss is always $1/2$,

$$P(\text{HEADS next toss} \mid \mathcal{D}) \equiv P(M_1 = 1, M_2 = 0 \mid N_1, N_2) = \frac{N_1 + 1}{N + n} = \frac{1}{2}$$

Chapter 48

Harmony of the Spheres?

48.1 Introduction

This final Chapter finishes up my thoughts on the Jeffreys's prior and its role in Bayesian data analysis. As I tried to point out in the previous Chapter, it appears to me that, in his initial deliberations on the matter, Jeffreys was content to merely modify Bayes's and Laplace's uniform distribution for the prior probability over model space. He was motivated by the results one obtained in sampling from finite populations if one were to adopt the uniform distribution. His trenchant commentary, "This does not correspond to my state of belief or anybody else's." reveals his motivation for finding a replacement for the Bayes–Laplace rationale.

We examined in great detail how, in the question of "testing whether a suggested value of a chance is correct," Jeffreys carried out this surgery by placing a lump of prior probability on one specific model. He then distributed the remaining prior probability, as Bayes and Laplace had done, over all of the conceivable "suggested values of a chance."

However, sometime later, and we know that by 1946 he was advocating this newer rationale, Jeffreys was invoking a geometric notion of squared distance in curvilinear coordinates as a justification for replacing the uniform distribution. This was what he labeled as the **Invariance Theory**. Let's therefore investigate whether, in leveraging some of the ideas from Information Geometry as presented in this Volume, we are able to dispel some of the confusion that continues to swirl around "Jeffreys's prior" to this day.

In using Information Geometry, one would also like to actually see the summit rather than blindly groping toward some unknown goal. For me, this goal is to achieve a deeper insight into Jeffreys's prior. On first exposure, I thought that Amari might have provided a pathway that allows one to leverage his α–connections to a better comprehension of Jeffreys's prior. Alas, as my fumbling arguments in this final Chapter attest to, that hope has been dashed.

Jeffreys's revised deliberations concerning prior probabilities and "invariance theory" began with a fundamental concept in Information Processing, namely, the relative entropy between probability distributions. From an intuitive standpoint, it is perfectly acceptable to think about relative entropy as some sort of "similarity or discrepancy measure" between any two probability distributions. From the purview of Information Geometry, we have discovered that a pretty strict analogy exists between "entropy" or "missing information" and distances between points in a flat Riemannian manifold.

Furthermore, and easy to visualize in a geometric way, I think it important to emphasize that **relative entropy** between *two* probability distributions p and q is a triangular relationship among *three* points p, q, and u, where u is that probability distribution that contains no information whatsoever, or alternatively, contains the maximum amount of missing information. Relative entropy is always grounded in the Shannon–Jaynes definition of information entropy through the Pythagorean theorem as it is viewed through the lens of Information Geometry.

Moreover, the relative entropy between two nearby distributions has been shown to have a close connection with the element of squared distance, or in other words, with the metric tensor, or Fisher information matrix. We investigated this in some detail back in Chapter Thirty Eight. But Jeffreys attached himself to the idea that it was the *prior probability over parameters* that should be considered as proportional to the square root of the determinant of the Fisher information matrix. Is it possible to connect all these dots in some coherent fashion?

My tentative answer is that the convincing coherent argument has not yet been put forward. I will show that trying to follow Amari's geometric rationale for Jeffreys's prior still leaves one in a conceptual muddle.

48.2 Lengths and Volumes

Jeffreys introduced his **Invariance Theory** [24, pg. 179] with a definition of an expression labeled as J which turned out to be very close to what was later defined as relative entropy. Kullback [28, pg. 6] clears this up by telling us that Jeffreys's J in his notation would look like this,

$$J(1,2) = \int (f_1(x) - f_2(x)) \log \frac{f_1(x)}{f_2(x)} \, d\lambda(x) \tag{48.1}$$

while for us it looks like this,

$$J = 1/2 \, [\, KL(p,q) + KL(q,p) \,] \tag{48.2}$$

that is, an average of the relative entropy between distributions p and q and its dual formulation.

We have established the relationship between the operational definition of the relative entropy as,

$$KL(p,q) = \sum_{i=1}^{n} p_i \ln\left(\frac{p_i}{q_i}\right) \qquad (48.3)$$

and the squared distance between these two distributions,

$$ds^2 = \sum_{r=1}^{m}\sum_{c=1}^{m} g_{rc}(p)\, d\theta^r\, d\theta^c \qquad (48.4)$$

with an approximation by an element of arc length if q is close to p,

$$ds = \sqrt{\sum_{r=1}^{m}\sum_{c=1}^{m} g_{rc}(p)\, d\theta^r\, d\theta^c} \approx \sqrt{2KL(p,q)} \qquad (48.5)$$

So we already have had exposure to an expression which involves *lengths* with the square root of the metric tensor.

Shortly after any abstract definition of the *determinant* of a matrix is given in Linear Algebra, it is usually quickly followed by the author trying to dispel the abstraction with the idea that the determinant is actually a measure of some hypervolume spanned by vectors. Our matrix of concern is the Fisher information matrix, or the metric tensor. Thus, the determinant of the metric tensor is somehow tied up with the definition of a generalized volume in the Riemannian manifold.

48.3 Determinant of the Metric Tensor

An important notion from Differential Geometry needed to understand the rationale for Jeffreys's prior is the determinant of a matrix, or more specifically, the square root of the determinant of the metric tensor. We will employ this notation, $\sqrt{|g|}$, for the square root of determinant of a metric tensor.

For example, for a manifold with two coordinates, the 2×2 metric tensor is generically,

$$g_{rc}(p) = \begin{pmatrix} g_{11} & g_{12} \\ g_{21} & g_{22} \end{pmatrix} \qquad (48.6)$$

The square root of the determinant of this matrix is, by definition,

$$\sqrt{|g|} = \sqrt{g_{11}g_{22} - g_{12}g_{21}} = \sqrt{g_{11}g_{22} - g_{12}^2} \qquad (48.7)$$

This can be confirmed through *Mathematica* with,

```
matrixG = Array[g##&, {2, 2}];
Sqrt[Det[matrixG]]
```

For another example that involves the square root of the determinant of a 3×3 diagonal matrix looking like this,

$$g_{rc}(p) = \begin{pmatrix} 1/g_{11} & 0 & 0 \\ 0 & 1/g_{22} & 0 \\ 0 & 0 & 1/g_{33} \end{pmatrix} \tag{48.8}$$

the answer is,

$$\sqrt{|g|} = \frac{1}{\sqrt{g_{11}}} \times \frac{1}{\sqrt{g_{22}}} \times \frac{1}{\sqrt{g_{33}}} \tag{48.9}$$

and verified with *Mathematica* through,

```
vecG = 1 / Array[g#, #&, 3];
PowerExpand[Sqrt[Det[DiagonalMatrix[vecG]]]]
```

A key concept involved in the transformation of variables and the justification for Jeffreys's prior is this equality,

$$\sqrt{|g|}\, dx^1\, dx^2 \cdots dx^n = \sqrt{|\bar{g}|}\, d\bar{x}^1\, d\bar{x}^2 \cdots d\bar{x}^n \tag{48.10}$$

where the overbar indicates the representation in an alternative coordinate system. This expression simplifies when the unbarred coordinate system is the Cartesian coordinate system.

In three dimensions, the unbarred Cartesian coordinates $x^1, x^2, \cdots, x^n$ become x, y, z with the metric tensor for the Cartesian coordinate system becoming the identity matrix, all leading to,

$$\sqrt{|g|}\, dx^1\, dx^2 \cdots dx^n = \sqrt{|\bar{g}|}\, d\bar{x}^1\, d\bar{x}^2 \cdots d\bar{x}^n$$

$$\sqrt{|g|}\, dx^1\, dx^2 \cdots dx^n = \sqrt{\det\begin{pmatrix} 1 & 0 & 0 \\ 0 & 1 & 0 \\ 0 & 0 & 1 \end{pmatrix}}\, dx\, dy\, dz$$

$$dx\, dy\, dz = \sqrt{|\bar{g}|}\, d\bar{x}^1\, d\bar{x}^2\, d\bar{x}^3 \tag{48.11}$$

When volumes need to be computed by integrating with respect to $dx^1\, dx^2 \cdots dx^n$ in the Cartesian coordinate system, the equivalent volumes are found by integrating the expression on the right hand side. Finding the volume of a hypersphere serves as the normalizing factor in Jeffreys's prior probability for models.

An example of the above finds the volume element $dV \equiv dx\, dy\, dz$ for the sphere in three dimensional space as,

$$dx\, dy\, dz = \sqrt{|\bar{g}|}\, d\bar{x}^1\, d\bar{x}^2\, d\bar{x}^3$$

$$dV = r^2 \sin\theta\, dr\, d\theta\, d\phi \tag{48.12}$$

If the radius of the sphere is fixed and does not change, this expression becomes the small element of a surface area dA,

$$dA = k \sin\theta \, d\theta \, d\phi \tag{48.13}$$

To find the total area of the manifold, (the surface area of the sphere with the fixed radius of r), integrate over the allowable range of 0 through π for the angle θ and over the allowable range of 0 through 2π for the angle ϕ,

$$\text{Area} = \int_0^{2\pi} \int_0^{\pi} k \sin\theta \, d\theta \, d\phi \tag{48.14}$$

The details of using the Cartesian coordinate system and the spherical coordinate system are presented in Exercises 48.9.1 through 48.9.4.

48.4 Jeffreys's Prior for the Gaussian

Quite understandably, one of the very first attempts within the Bayesian statistical community to apply these invariance ideas of Jeffreys via the square root of the determinant of the Fisher information matrix was to the Gaussian, or Normal, distribution. You will get to experience varying levels of obfuscation in almost every single Bayesian textbook you happen to pick up when the author gets around to discussing "noninformative priors." And, furthermore, it is then obligatory to show how the Fisher information matrix is formed for the Normal distribution.

To reemphasize Jeffreys's explanation that he was in the business of finding a "prior probability for parameters" you will typically see notation like $p(\mu, \sigma)$ for the prior probability for the two parameters of the Gaussian distribution, and a corresponding posterior probability showing these two parameters conditioned on the observed data as $p(\mu, \sigma \mid \mathcal{D})$. These two concepts enveloping a prior and posterior probability for parameters are nonsensical, or as Pauli would have it, "not even wrong!"

I will repeat myself here, and not for the last time, that there can be no "degree of belief" attached to a parameter which is the outcome of a mathematical procedure. Our model parameters, the Lagrange multipliers λ_j, or the dual parameters, the constraint function averages, $\langle F_j \rangle$, are derived via the optimization technique of the method of undetermined multipliers. There is no "uncertainty" about these parameters, we have either found them exactly and correctly through a proper application of the optimization technique, or they do not exist.

When the MEP formula finds that the numerical assignment to the probability for HEADS is $1/4$ under the information resident in the model $\mathcal{M}_k$,

$$P(\text{HEADS} \mid \mathcal{M}_k) = Q_1 = \frac{\exp[\lambda F(X = x_1)]}{\exp[\lambda F(X = x_1)] + \exp[\lambda F(X = x_2)]}$$

the value of the Lagrange parameter is $\lambda = -1.098216$. There is no distribution of probabilities for λ. There is never a "prior probability" expression that looks like $P(\lambda)$ appearing in the formal manipulation rules. If there never can be a prior probability for a parameter, *a fortiori*, there can never be a posterior probability.

Contemplate for a second what an absurdity the posterior probability for a parameter would be anyway. The distribution of probabilities for the parameters in a model would depend on the data when the very *raison d'être* for the existence of the λ_j in the first place is to insert information under a particular model. The capability of assigning $P(\text{HEADS} \,|\, \mathcal{M}_k) = 1/4$ through λ would somehow have to change after knowing what the data are. It doesn't make any sense whatsoever!

What does permit a legitimate "degree of belief" and the associated notation of a probability is the idea that the information in the model $\mathcal{M}_k$ leading to the specific assignment of $Q_1 = 1/4$ and $Q_2 = 3/4$ may be wrong. Therefore, even though there is no probability distribution over the λ_j, there can and should be a probability distribution over the $\mathcal{M}_k$. Thus we are led inexorably to the fully acceptable and non–mysterious notion *and notation* of a $P(\mathcal{M}_k) \equiv P(q_1, q_2)$.

This $P(\mathcal{M}_k)$ is, in fact, what a *prior probability* should mean, with the perfectly agreeable assertion that $P(q_1 = 1/4, q_2 = 3/4)$ could equal say 0.65, with the remaining prior probability spread out over any number of other models. Aligning ourselves with the reasoning of Laplace, it seems quite non–controversial to further assert that if we want to operationally implement the notion of an "uninformative prior," then a probability density function,

$$P(\mathcal{M}_k) \equiv \text{pdf}\,(q) = C_{\text{Beta}}\, q^{\alpha-1}\,(1-q)^{\beta-1} \qquad (48.15)$$

with $\alpha = \beta = 1$ would be the right way to proceed. Any conceivable assignment of Q_1 and Q_2 through the fixed value of some λ is thought to be granted an equal status with any other conceivable assignment through the fixed value of a different λ via the above non–informative prior probability.

To give you a good example of the rampant confusion that hovers inexorably over "the prior probability," I quote from two Bayesian experts in the upcoming **Connections to the Literature**. You could pick up almost any book on Bayesian statistics and find similarly incoherent discussions focusing on noninformative prior probabilities, but I prefer to focus on the "acknowledged experts" in the field as the more enjoyable target. They cannot even put together a couple of coherent sentences that exhibit a minimal understanding of the distinction between a distribution of probabilities over the statements constituting the state space, and the distribution of probabilities over the model space.

Having unburdened myself on this issue, I will now proceed to go ahead and reconstruct how this faulty idea of a prior probability for the parameters of the Normal distribution is implemented through a Jeffreys's prior. Since no divergence of opinion ensues, I will use the definition of a metric tensor as we have learned it from Information Geometry as opposed to what you might typically see in a Bayesian textbook that emphasizes the Fisher information matrix.

With this in mind, we have points living in a flat Riemannian manifold. At each different point, a different metric tensor is constructed at each different tangent space. There are only going to be two coordinates for this manifold. Thus, each metric tensor will be relatively simple consisting of a symmetric 2×2 matrix $g_{rc}(p)$. The metric tensor is what allows lengths to be calculated for close by points in the tangent space.

First reviewing Amari's definition of the metric tensor as it has evolved within Information Geometry, (albeit heavily influenced by a desired analogy with the Fisher information matrix), we have,

$$g_{ij}(\xi) = \int \partial_i l(x;\xi)\, \partial_j l(x;\xi)\, p(x;\xi)\, dx \qquad (48.16)$$

Translating this into our preferred MEP notation, we have,

$$g_{rc}(p) = E_p \left[\frac{\partial \ln P(X = x_i \mid \mathcal{M}_k)}{\partial \lambda_r} \times \frac{\partial \ln P(X = x_i \mid \mathcal{M}_k)}{\partial \lambda_c} \right] \qquad (48.17)$$

After the calculations have been performed in Exercise 48.9.7, the metric tensor is seen to be,

$$g_{rc}(p) = \begin{pmatrix} \frac{1}{\sigma^2} & 0 \\ 0 & \frac{2}{\sigma^2} \end{pmatrix} \qquad (48.18)$$

Finding the determinant of this square matrix followed by the square root yields a Jeffreys's noninformative "prior probability" for the two parameters of the Normal distribution,

$$p(\mu, \sigma) = \sqrt{|g|} = \| g_{rc}(p) \|^{1/2} = \sqrt{\frac{1}{\sigma^2} \times \frac{2}{\sigma^2}} = \frac{\sqrt{2}}{\sigma^2} \propto \sigma^{-2} \qquad (48.19)$$

But even Jeffreys did not like this outcome and recommended instead a "prior probability" of $p(\mu, \sigma) \propto \sigma^{-1}$.

Here is the main conceptual sticking point. A "prior probability" over parameter space is supposedly found, as was just done for the Normal distribution, by calculating the Fisher information matrix based on the joint distribution of the statements in the state space conditioned on a model, $P(X = x_i \mid \mathcal{M}_k)$. Finding elements of volume *in this space* would seem to have nothing to do with any prior probability. Furthering the confusion instilled by this approach using the Fisher information matrix, you will see "prior probabilities" found by differentiating the log of the *likelihood* function throughout the literature.

A "volume element" in a Riemannian space was indeed found by calculating this particular $\sqrt{|g|}$. But this metric tensor was calculated in the space $\mathcal{S}^n$ where all of the points p, q, r, ..., are *already* well-defined probability distributions! Moreover, these are the well–defined conditional distributions $P(X = x_i \mid \mathcal{M}_k)$, and not any kind of prior probabilities. Why is another definition based on this geometric notion of the volume of the space the points live in required for already defined probability distributions?

But if the "prior" is correctly understood as the prior probability over model space $P(\mathcal{M}_k) \equiv P(q_1, q_2, \cdots, q_n)$, and not incorrectly thought of as $P(X = x_i \,|\, \mathcal{M}_k)$, then we can see in developing the marginal probability $P(X = x_1)$, where again the point is made if we take the $n = 2$ state space,

$$P(X = x_1) = \sum_{k=1}^{\mathcal{M}} P(X = x_1 \,|\, \mathcal{M}_k) \, P(\mathcal{M}_k)$$

$$\equiv \int P(X = x_1 \,|\, q) \, P(q) \, \mathrm{d}q$$

$$= \int_0^1 \frac{\exp\left[\lambda F(X = x_1)\right]}{Z(\lambda)} \, C_{\text{Beta}} \, q^{\alpha-1} (1-q)^{\beta-1} \, \mathrm{d}q$$

$$= \int_0^1 q \, C_{\text{Beta}} \, q^{\alpha-1} (1-q)^{\beta-1} \, \mathrm{d}q \tag{48.20}$$

Nothing in the prior probability is modifying the λ parameter! Certainly, q is changing and therefore the parameter is changing, but not because of the prior probability. It is changing solely due to the demands of the integration that q cover the entire interval from 0 to 1.

48.5 Surface Areas and Volumes for Spheres

Our ultimate goal, and the one that we have been rambling on about in a discursive manner so far, is to find some relationship that connects all of the following: the prior probability for models, Jeffreys's Rule, the determinant of the metric tensor, Amari's α representations, and the classical geometry for spherical objects. It is no wonder that confusion sets in when trying blend all of these disparate concepts into some coherent whole.

For a start, look up the formulas from classical geometry to determine the surface area and volumes for circles and spheres with fixed radius r. The "volume" of a circle, in other words, its area, is,

$$\text{Volume} = \pi r^2 \tag{48.21}$$

and the volume of a sphere is,

$$\text{Volume} = \frac{4\pi r^3}{3} \tag{48.22}$$

The "surface area" for a circle, in other words, its circumference is,

$$\text{Surface area} = 2\pi r \tag{48.23}$$

and the surface area of a sphere is,

$$\text{Surface area} = 4\pi r^2 \tag{48.24}$$

Even here, we have to be careful with our language because mathematicians call the surface area of, say, a three dimensional sphere like the globe of the earth, *the 2-sphere*, and what we haphazardly called the sphere is in fact called the *n-ball*, here the 3-ball.

Fortunately, there are relatively simple formulas for the volume and surface area that allow us to generalize for any n and any r for n-balls. These are for the surface area,

$$\text{Surface area} = \frac{2\pi^{n/2} r^{n-1}}{\Gamma(n/2)} \tag{48.25}$$

and for the volume,

$$\text{Volume} = \frac{\pi^{n/2} r^n}{\Gamma(n/2+1)} \tag{48.26}$$

Let's confirm that these general formulas work for the specific cases where we know the answer. For a circle, let $n = 2$. The circumference, the surface area for a circle, is then,

$$\frac{2\pi^{n/2} r^{n-1}}{\Gamma(n/2)} = \frac{2\pi r}{\Gamma(1)}$$

$$\text{Circumference} = 2\pi r$$

For a sphere, let $n = 3$ and the surface area over the sphere is,

$$\frac{2\pi^{n/2} r^{n-1}}{\Gamma(n/2)} = \frac{2\pi^{3/2} r^2}{\Gamma(3/2)}$$

$$\text{Surface area} = 4\pi r^2$$

Now, apply the volume formula to determine the volume of a circle and a sphere,

$$\text{Volume} = \frac{\pi^{n/2} r^n}{\Gamma(n/2+1)}$$

For the $n = 2$ circle, the volume, the area in case of a circle, is,

$$\frac{\pi^{n/2} r^n}{\Gamma(n/2+1)} = \frac{\pi r^2}{\Gamma(2)}$$

$$\text{Volume} = \pi r^2$$

For the $n = 3$ sphere, the volume is,

$$\frac{\pi^{n/2} r^3}{\Gamma(n/2+1)} = \frac{\pi^{3/2} r^3}{\Gamma(5/2)}$$

$$\text{Volume} = \frac{4}{3}\pi r^3$$

48.6 From Spheres to the Dirichlet Distribution

Now, try to find some similarity in these formulas about surface areas and volumes of hyperspheres with a comment by Amari on Jeffreys's prior. He talks about a "Volume" V serving as the normalizing factor for a probability density function over parameters defining a point in a Riemannian manifold. He says that this V is the surface area of an n–dimensional sphere of radius 2 divided by 2^{n+1} and gives the formula of,

$$V = \frac{\pi^{(n+1)/2}}{\Gamma(\frac{n+1}{2})}$$

For a numerical example, refer to Exercise 48.9.4 where the surface area for our $n = 3$ sphere with radius $r = 2$ was found to equal 16π. For Amari's V we find that,

$$V = \frac{\pi^{3/2}}{\Gamma(\frac{3}{2})} = 2\pi$$

which is indeed the surface area of an 3–dimensional sphere of radius 2, 16π, divided by $2^{n+1} = 8$.

We ask the question of how to transition from Amari's formula for surface area of spheres to a similar formula for the partition function of a known probability density function. Amari had told us that V was the normalizing factor,

$$V = \int_\Xi \sqrt{\det G(\xi)}\, d\xi$$

for a prior probability, and furthermore, that this normalizing factor is not just for any prior probability, but specifically for Jeffreys's prior. The partition function that serves as the normalizing factor for the Dirichlet distribution is,

$$Z = \int \cdot \int_{\sum q_i = 1} q_1^{\alpha_1 - 1} q_2^{\alpha_2 - 1} \cdots q_n^{\alpha_n - 1}\, dq_i = 1/C_D = \frac{\prod_{i=1}^n \Gamma(\alpha_i)}{\Gamma(\sum_{i=1}^n \alpha_i)}$$

If all $\alpha_i = 1/2$, then,

$$Z(\alpha_i) = \frac{\prod_{i=1}^n \Gamma(\alpha_i)}{\Gamma(\sum_{i=1}^n \alpha_i)} = \frac{\pi^{n/2}}{\Gamma(n/2)}$$

Amari's discrete state space was defined by him as $\mathcal{X} = \{0, 1, 2, \cdots, n\}$. This is the reason for the extra 1 appearing in his formula. Thus, as in the above numerical example when Amari specifies $n = 2$, it is comparable to a state space of dimension $n = 3$, and we have the same formulas.

I have always used the Dirichlet distribution as the prior probability over model space, with the notation of $P(\mathcal{M}_k) \equiv P(q_1, q_2, \cdots, q_n)$. So it appears, as I have been arguing for, that the prior probability over model space should be constructed as a Dirichlet distribution. If an IP wants to implement Jeffreys's rule for the

Dirichlet, then, with $\alpha_i = 1/2$,

$$P(\mathcal{M}_k) = \text{pdf}\,(q_1, q_2, \cdots, q_n)$$

$$= \frac{\Gamma(\sum_{i=1}^n \alpha_i)}{\prod_{i=1}^n \Gamma(\alpha_i)}\, q_1^{-1/2} q_2^{-1/2} \cdots q_n^{-1/2}$$

$$V = \frac{\pi^{n/2}}{\Gamma(n/2)}$$

$$Q(\xi) = \frac{1}{V}\sqrt{\det G(\xi)}$$

But this is a different space than $\mathcal{S}^n$. It is, instead, the space where the points represent the probability for models. Like all Riemannian spaces, the points must have a coordinate system.

Our coordinate system is expressed through the two coordinates α and β with α and $\beta > 0$. One point in model space might have the following coordinates $\alpha = 1/2, \beta = 1/2$, while another point in model space might have the coordinates $\alpha = 1, \beta = 1$. Arguing for the dominant conceptual idea of finding the volume of a hypersphere where one point is singled out with coordinates $\alpha = 1/2, \beta = 1/2$, while ignoring the coordinates for other points is confusing!

As a primary example of where clarity should have been emphasized, but was not, originates with Jeffreys himself. Deeply buried in the middle of his section on **Invariance Theory** [24, pg. 184], and after giving his example for the Normal distribution followed by more obscure and difficult mathematics, out of the blue he comes up with two formulas for a "comparison of two chances," or "for a set of chances" α_r. The first is,

$$P(d\alpha \mid H) = \frac{1}{\pi}\,\frac{d\alpha}{\sqrt{\{\alpha(1-\alpha)\}}}$$

while the second is,

$$P(d\alpha_1 \ldots d\alpha_{m-1} \mid H) \propto \frac{d\alpha_1 \ldots d\alpha_{m-1}}{\sqrt{(\prod_1^m \alpha_r)}}$$

These two formulas are quite obviously examples of the Dirichlet distribution with parameters $\alpha_i = 1/2$. Take care to dispel any possible confusion by distinguishing Jeffreys's α_r as my q_i with my α_i implicit in Jeffreys's first formula as his first term, $\Gamma(\alpha = 1/2) \times \Gamma(\beta = 1/2) = \sqrt{\pi} \times \sqrt{\pi} = \pi$.

But he never makes clear, well at least not to me, whether his rule was supposed to apply (incorrectly) at forming the Fisher information matrix at the level of the $P(X = x_i \mid \mathcal{M}_k)$ as he did for the Normal distribution, or (correctly) at the level of $P(\mathcal{M}_k) \equiv P(q_1, q_2, \ldots, q_n)$ as he does in this latter example.

48.7 Amari's $\alpha = 0$ Representation

Another piece of the puzzle in trying to leverage Information Geometry to decipher Jeffreys's prior is Amari's α representations. We have touched throughout this Volume on the $\alpha = 1$ and $\alpha = -1$ representations. Before the short paragraph on Jeffreys's prior, Amari had spent some time talking about an $\alpha = 0$ representation. He said that Jeffreys's prior was an illustration of an $\alpha = 0$ representation.

Even earlier, while introducing and defining the Fisher information matrix for a point ξ in the Riemannian manifold [2, pg. 28, Equation (2.6)], and presented previously as Equation (48.16),

$$g_{ij}(\xi) = \int \partial_i l(x; \xi) \, \partial_j l(x; \xi) \, p(x; \xi) \, \mathrm{d}x$$

is expressed here in my alternative MEP notation as used throughout this Volume, and previously presented as Equation (48.17),

$$g_{rc}(p) = E_p \left[\frac{\partial \ln P(X = x_i \mid \mathcal{M}_k)}{\partial \lambda_r} \times \frac{\partial \ln P(X = x_i \mid \mathcal{M}_k)}{\partial \lambda_c} \right]$$

$$= \sum_{i=1}^{n} \left[\frac{\partial \ln P(X = x_i \mid \mathcal{M}_k)}{\partial \lambda_r} \times \frac{\partial \ln P(X = x_i \mid \mathcal{M}_k)}{\partial \lambda_c} \right] \times Q_i$$

Amari claimed that there existed another important representation for the Fisher information matrix written as [2, pg. 28, Equation (2.10)],

$$g_{ij}(\xi) = 4 \int \partial_i \sqrt{p(x; \xi)} \, \partial_j \sqrt{p(x; \xi)} \, \mathrm{d}x$$

Now anyone with even the slightest exposure to quantum mechanics experiences a slight rise in blood pressure when expressions like this appear containing *probability amplitudes*. But I have a hard enough time trying to understand all of this in a straightforward probability setting, so that kind of speculation will be left for others more competent than I. Exercise 48.9.8 engages in a hopefully not too repulsive idiosyncratic derivation of this alternative representation.

48.8 Connections to the Literature

In section 48.4, I mentioned that I would quote from a well–regarded book on Bayesian theory in order to illustrate the seemingly pervasive misunderstanding about so basic an issue as to what the prior probability refers to. In a section devoted to technical discussions about noninformative prior probabilities, Bernardo and Smith present us with this, [6, pp. 365–366].

Jaynes (1968) introduced a more general formulation of the problem. He allowed for the existence of a certain amount of initial "objective" information and then tried to determine a prior which reflected this initial information, but nothing else ... Jaynes considered the entropy of a distribution to be the appropriate measure of uncertainty subject to any "objective" information one might have. If no such information exists and ϕ can only take a finite number of values, Jaynes' *maximum entropy* solution reduces to the Bayes–Laplace postulate.

OK, let's see if we can begin to parse out the confusion here. As we have shown throughout Volume II, and also many times here in Volume III, Jaynes's MEP was used to assign probabilities, not to any **PRIOR probability**, such as $P(\mathcal{M}_k)$, nor to any **MARGINAL probability** such as $P(N_1, N_2)$, but rather to a **CONDITIONAL probability** such as $P(X = x_i \mid \mathcal{M}_k)$. It is still used today by the practitioners of the MEP to assign numerical values to probabilities of joint statements as defined by the state space when conditioned on the information resident in the model $\mathcal{M}_k$.

Furthering the confusion, the ϕ that Bernardo and Smith are referring to in the above quote is a *parameter*, and not any statement in the state space. It appears that they think the MEP should be applied to find probability distributions for parameters. The resulting maximum entropy solution, so they claim, reduces to the *Bayes–Laplace postulate*. Previously, they had told us what they meant by the Bayes-Laplace postulate [6, pg. 357],

... suppose we require that $p(\theta)$ be chosen such that

$$p(s) = \int_0^1 \binom{n}{s} \theta^s (1-\theta)^{n-s} p(\theta) \, d\theta$$

does not depend on s. Any easy calculation shows that this is satisfied if $p(\theta)$ is taken to be uniform on $(0,1)$—the so–called *Bayes (or Bayes–Laplace) Postulate*.

Of course, in our notation, the above probability given by Bernardo and Smith is simply the marginal probability for the data,

$$P(\mathcal{D}) = \sum_{k=1}^{\mathcal{M}} P(\mathcal{D} \mid \mathcal{M}_k) P(\mathcal{M}_k)$$

$$P(N_1, N_2) = \int_0^1 W(N) \, q^{N_1} (1-q)^{N_2} P(q) \, dq$$

Now, it is clear that the θ appearing in their equation cannot be a parameter. It is an assigned probability that finds its justification from the model statement $\mathcal{M}_k = \{q, (1-q)\}$, while $P(\mathcal{M}_k) \equiv p(\theta)$ is the **prior probability for that model**. *If* the uniform prior probability of Bayes and Laplace were to be adopted, *then* $P(\mathcal{D}) = 1/(N+1)$ (the Bayes–Laplace postulate).

The Bayes–Laplace postulate takes place at the level of $P(\mathcal{M}_k)$. The resulting marginal probability of $P(\mathcal{D})$ is **NOT** Jaynes's MEP solution at the "lower" level of finding numerical assignments to any **CONDITIONAL probability** in the state space, $P(X = x_i \,|\, \mathcal{M}_k)$. Jaynes's maximum entropy solution, contrary to Bernardo and Smith's conclusion, does **NOT** reduce to the Bayes–Laplace postulate.

If the state space consists of a finite number of n statements, then *one particular* model, namely the one with model parameters of $\lambda_j = 0$, results in the assignment of $Q_i = 1/n$ as the probability for every statement in the state space. This assignment has nothing to do with the Bayes–Laplace postulate. The Bayes–Laplace postulate, however, does say something about the prior probability $P(\mathcal{M}_k)$ of the model that assigned the $1/n$.

This prior probability $P(\mathcal{M}_k)$ might have a value of 1, 0, or for that matter, any other legitimate probability. The Bayes–Laplace postulate does lead the IP to assign a prior probability of $1/\mathcal{M}$ for a finite number of models in the model space. Compare this probability to the conceptually different probability of $1/n$. The prior probability naturally goes to a probability density function over all conceivable values that $q_1, q_2, \cdots, q_n$ might take on. If the IP is totally uninformed about which model is making the "correct" assignments $q_1, q_2, \cdots q_n$, it makes this probability density function, a "flat," or uniform distribution pdf $(q_1, q_2, \cdots, q_n)$.

The Bayes–Laplace postulate says exactly this. It provides a probability density function, the Dirichlet distribution,

$$\text{pdf}\,(q_1, q_2, \cdots, q_n) = C_D\, q_1^{\alpha_1-1} q_2^{\alpha_2-1} \cdots q_n^{\alpha_n-1}$$

that reduces to a uniform distribution when all the α_i parameters equal 1.

The Bayes–Laplace postulate says that when an IP is "uninformed," all of these assignments when adopting a binary state space,

$$Q_i = \{(1,0), (0.99, 0.01), (0.98, 0.02), \cdots, (1/2, 1/2), \cdots, (0.01, 0.99), (0,1)\}$$

as well as everything in between have the same prior probability. This concept is quite different from the very misleading implication in the quote above that one unique *maximum entropy* solution resulted from the application of the Bayes–Laplace postulate. Every one of these assignments in the set above resulted from an MEP solution based on information inserted from some model, and **NOT** from any application of the Bayes–Laplace postulate.

This information has a clearly defined meaning within the MEP formalism as the defined values for the constraint function averages, $\langle F_j \rangle$. Specifying this information in the set of $\langle F_j \rangle$ is synonymous with specifying the model parameters in the set of λ_j. *Parameters* do NOT have any probability distribution, prior, posterior, or otherwise associated with them because there is no uncertainty that they reflect the correct value for the information specified as known and given in $\mathcal{M}_k$.

The confusion originates in not keeping the distinction between the statements existing in the state space and the statements existing in the model space always front and center. If this basic confusion does intrude, then it migrates over into confusion about the probabilities for these different statements.

HARMONY OF THE SPHERES?

Having tried to indict Bernardo and Smith of confusion over the distinction between a probability assignment to a statement conditioned on the information in a model, and the probability assigned to the model itself prior to any data, in the end I must mitigate their sentence because they were misled by an accomplice to their crime. And the identity of that miscreant turns out to be none other than Jaynes himself!

The ϕ that Bernardo and Smith refer to is a parameter, but did Jaynes ever use the MEP to find a prior probability for parameters? You might naturally assume, as I did, that he used it only to find a distribution of probabilities over the statements in the state space conditioned on *knowing* a model.

In the above quote, they reference a 1968 article by Jaynes entitled, believe it or not, **Prior Probabilities** [16]. In this article, Jaynes essentially tackles the question of how to convert information into *a prior distribution over parameters*! Compounding the error, Jaynes then illustrates how the MEP can be utilized to find such a prior.

If that 1968 article were the only one you ever read, then you might very well draw exactly the same incorrect conclusion about prior probabilities as Bernardo and Smith did. However, they should have known better because they are experts in this field, and they list as well in their references almost all of Jaynes's more relevant applications of the MEP as probability assignments to joint statements in the state space.

In this final Chapter, I have attempted to shed some light on Jeffreys's prior as it might be dissected from the standpoint of Information Geometry. It was one short, almost throw away, paragraph in Amari and Nagaoka's book **Methods of Information Geometry** that inspired me to investigate Jeffreys's prior from the perspective of Information Geometry, independently of the more naive, straightforward analysis that I gave in the last Chapter.

After much work on my part, (as documented in this Chapter), and supposing some merit in studying Information Geometry to augment a somewhat conventional probabilistic approach, unfortunately I have to admit that I did not achieve any new and valuable insight into the whole miasma surrounding Jeffreys's prior.

Let's now see exactly what Amari had to say about Jeffreys's prior [2, pg. 44].

> Here, we discuss Jeffreys' prior distribution to illustrate an application of the observation made above. Let $S = \{p_\xi \,|\, \xi = [\xi^1, \cdots, \xi^n] \in \Xi\}$ be a statistical model, and let $G(\xi)$ denote the Fisher information matrix at point ξ. Now suppose that the volume $V \stackrel{\text{def}}{=} \int_\Xi \sqrt{\det G(\xi)}\, d\xi$ of S with respect to the Fisher metric is finite (where the integral is implicitly n–fold.) Then $Q(\xi) \stackrel{\text{def}}{=} \frac{1}{V}\sqrt{\det G(\xi)}$ defines a probability density function on Ξ. Since this is invariant over the choice of a coordinate system $[\xi^i]$, we may consider it as a probability distribution on the model S. This distribution is called **Jeffreys' prior** within the field of Bayesian statistics, ... Now consider the case when $S = \mathcal{P}(\mathcal{X})$ and $\mathcal{X} = \{0, 1, \cdots, n\}$. Then we can see that V equals to the sur-

face area of the n–dimensional sphere of radius 2 divided by 2^{n+1}, and letting Γ be the gamma function we have

$$V = \frac{\pi^{(n+1)/2}}{\Gamma(\frac{n+1}{2})}$$

In addition, $Q(\xi)\,\mathrm{d}\xi$ is then the uniform distribution on the sphere.

The discussions that arise in Differential Geometry or Tensor Calculus about the transformations between coordinate systems are germane to Information Geometry. Most particularly, we are interested in how these notions got applied by Jeffreys in the development of his prior probability for models.

Calculating the Jacobian matrix for Cartesian and spherical coordinates, the Jacobian as its determinant, and then the square root of the determinant of the metric tensor all help when Jeffreys serves up his **Invariance Theory** without much prior warning (no pun intended). If we have accustomed ourselves to the plausibility of the steps leading up to the equality between barred and unbarred coordinate systems expressed in,

$$\sqrt{|g|}\,dx^1\,dx^2\cdots dx^n = \sqrt{|\bar{g}|}\,d\bar{x}^1\,d\bar{x}^2\cdots d\bar{x}^n$$

then perhaps Jeffreys's [24, pp. 180–181] explanation goes down a little easier at the mathematical level, but still not at the conceptual level because of his insistence on advancing the notion of a prior probability over parameters.

But in the transformation of a multiple integral

$$d\alpha_1\,d\alpha_2\ldots d\alpha_m = \left\|\frac{\partial \alpha_i}{\partial \alpha'_j}\right\|\,d\alpha'_1\ldots d\alpha'_m$$

$$= \left(\frac{\|g'_{jl}\|}{\|g_{ik}\|}\right)^{1/2} d\alpha'_1\ldots d\alpha'_m$$

Hence $\|g_{ik}\|^{1/2}\,d\alpha_1\ldots d\alpha_m = \|g'_{jl}\|^{1/2}\,d\alpha'_1\ldots d\alpha'_m$

> This expression is therefore invariant for all non–singular transformations of the parameters. ...
>
> In consequence of this result, if we took the prior probability density for the parameters to be proportional to $\|g_{ik}\|^{1/2}$, it could be stated for any law that is differentiable with respect to all parameters in it, and would have the property that the total probability in any region of the α_i would be equal to the total probability in the corresponding region of the α'_j; in other words, it satisfies the rule that equivalent propositions have the same probability.

A good introduction to basic concepts about coordinate systems, and spherical coordinate systems in particular, the Jacobian matrix, and the metric tensor can be found in Chapters 2, 3, and 5 of David Kay's **Tensor Calculus** [26].

48.9 Solved Exercises for Chapter Forty Eight

Exercise 48.9.1: Extend Exercise 34.9.1 on the polar coordinate system by explaining the spherical coordinate system.

Solution to Exercise 48.9.1

Given the groundwork laid down in examining polar coordinates, let's see if we can generalize from the circle to the sphere. We are jumping up from the two dimensional space of the circle to the three dimensional space of the sphere.

The Cartesian coordinate system is expressed in x, y, and z coordinates. The spherical coordinate system retains the radius r from the circle, but now thought of as the radius of the sphere. Two angles θ and ϕ are demanded for a sphere. Thus, one coordinate system, the Cartesian coordinate system, is (x, y, z) and a second coordinate system, the spherical coordinate system, is (r, θ, ϕ).

It's best to draw upon our familiarity with the globe of the earth and make the analogy that the angle ϕ represents the longitude, while the angle θ represents the latitude. We visualize ϕ rotating 360° from the center of the earth to mark the circle for the equator and indicating where the longitude intersects the equator, while θ rotates 180° from the North Pole to the South Pole to indicate the latitude. If θ rotated all the way around, the points on the surface of the sphere would be counted twice.

Although it quickly becomes tedious, right triangles can be drawn inside the sphere to set up trigonometric relationships just as we did for the circle. This leads to the transformation between spherical coordinates and Cartesian coordinates,

$$x = r \sin\theta \cos\phi \qquad r = \sqrt{x^2 + y^2 + z^2}$$

$$y = r \sin\theta \sin\phi \qquad \theta = \cos^{-1}(z/r)$$

$$z = r \cos\theta \qquad \phi = \tan^{-1}(y/x)$$

The right triangle easiest to visualize is the one that is constructed to find the z coordinate. As θ is tracing out a circle covering different latitudes at some fixed longitude, the hypotenuse of the triangle extends from the center of the globe to the surface of the globe and this is r. The adjacent side of the right triangle is the z–axis running through the center of the earth from the North Pole to the South Pole. Thus,

$$\cos\theta = \frac{\text{adjacent}}{\text{hypotenuse}}$$

$$= \frac{z}{r}$$

$$z = r\cos\theta$$

But the latitude is measured as the angle $\theta' = 90° - \theta$. For example, at a latitude of $\theta' = 30°$ for some fixed longitude on the globe, the Cartesian z coordinate, taking the radius of the earth at roughly 4000 miles, would be about $z = 4000 \cos 60° = 2000$ miles. At this latitude, the z coordinate would be about halfway from the North Pole to the center of the Earth.

Assuming we want the z coordinate to run from 0 to 4000 miles from the center of the Earth to the North Pole, and from 0 to -4000 miles from the center of the Earth to the South Pole, then at a latitude of $\theta' = -30°$, the z coordinate is located at $4000 \cos 120° = -2000$ miles. The latitude θ would range from $90°$ at the North Pole through $0°$ at the Equator and from $0°$ through $-90°$ at the South Pole.

Exercise 48.9.2: Drawing upon the example of the polar coordinates, what is the Jacobian matrix for the spherical coordinates?

Solution to Exercise 48.9.2

The same pattern applies as was employed for the circle and polar coordinates when moving from two coordinates to three coordinates. Symbolically, showing the partial differentiations of each Cartesian coordinate with respect to each polar coordinate,

$$\mathbf{J} = \begin{pmatrix} \frac{\partial x}{\partial r} & \frac{\partial x}{\partial \theta} & \frac{\partial x}{\partial \phi} \\ \frac{\partial y}{\partial r} & \frac{\partial y}{\partial \theta} & \frac{\partial y}{\partial \phi} \\ \frac{\partial z}{\partial r} & \frac{\partial z}{\partial \theta} & \frac{\partial z}{\partial \phi} \end{pmatrix}$$

After all the partial differentiations have been carried out, the Jacobian matrix $\mathbf{J}$ looks like this,

$$\mathbf{J} = \begin{pmatrix} \cos \phi \sin \theta & r \cos \theta \cos \phi & -r \sin \theta \sin \phi \\ \sin \theta \sin \phi & r \cos \theta \sin \phi & r \sin \theta \cos \phi \\ \cos \theta & -r \sin \theta & 0 \end{pmatrix}$$

Mathematica will calculate this Jacobian matrix by evaluating,

```
D[{r Sin[θ] Cos[φ], r Sin[θ] Sin[φ], r Cos[θ]}, {{r, θ, φ}}]
```

Exercise 48.9.3: What is the metric tensor for the spherical coordinates?

Solution to Exercise 48.9.3

It turns out that the Jacobian matrix $\mathbf{J}$ is, in some sense, the square root of the metric tensor. So, the metric tensor $\mathbf{g}$ is formed from the matrix multiplication of the transpose of the Jacobian matrix with the Jacobian matrix $\mathbf{g} = \mathbf{J}^T \mathbf{J}$. We will ask *Mathematica* to evaluate,

```
j = D[{r Sin[θ] Cos[φ], r Sin[θ] Sin[φ], r Cos[θ]}, {{r,θ, φ}}];
jT = Transpose[j];
g = Simplify[Dot[jT, j]];
MatrixForm[g]
```

which returns the metric tensor for the spherical coordinate system as,

$$\mathbf{g} = \begin{pmatrix} 1 & 0 & 0 \\ 0 & r^2 & 0 \\ 0 & 0 & r^2(\sin\theta)^2 \end{pmatrix}$$

The important feature revealed here in the metric tensor is that the spherical coordinates are orthogonal since all the off–diagonal coefficients are 0. Also, the determinant of the Jacobian matrix $\mid J\mid = r^2 \sin\theta$ is the same as the square root of the determinant of **g** verified by,

PowerExpand[Simplify[Sqrt[Det[g]]]]

returning $\sqrt{\mid g\mid} = \sqrt{1 \times r^2 \times r^2 \sin(\theta)^2} = \sqrt{r^4 \sin(\theta)^2} = r^2 \sin\theta$.

Exercise 48.9.4: What is the small element of surface area of a sphere?

Solution to Exercise 48.9.4

Answering this question becomes important when the discussion once again turns to the underlying rationale for Jeffreys's prior. The essential idea is that in assigning a prior probability for a model $P(\mathcal{M}_k)$, the argument for Jeffreys's prior says that this probability is proportional to the surface area of a sphere for a state space where $n = 3$, and more generally, to the surface area of a hypersphere for general n.

As we discovered in the last Exercise, the determinant of the Jacobian matrix **J** was found to be,

$$\mid J\mid = \sqrt{\mid \overline{g}\mid} = r^2 \sin\theta$$

where we are labeling the spherical coordinates $\{r, \theta, \phi\}$ as the barred coordinates and the Cartesian coordinates $\{x, y, z\}$ as the unbarred coordinates. The volume element $dV \equiv dx\,dy\,dz$ is then equal to,

$$dx\,dy\,dz = \sqrt{\mid \overline{g}\mid}\,d\overline{x}^1\,d\overline{x}^2\,d\overline{x}^3$$

$$dV = r^2 \sin\theta\,dr\,d\theta\,d\phi$$

Integrate over these volume elements to find the total volume of the manifold,

$$V = \int \cdot \int_{\overline{R}} \sqrt{\mid \overline{g}\mid}\,d\overline{x}^1\,d\overline{x}^2\cdots d\overline{x}^n$$

For some fixed r, and say that $r = 2$, this becomes the small element of surface area dA,

$$dA = 4 \sin\theta \, d\theta \, d\phi$$

To find the total area of the manifold, (the surface area of the 3–dimensional sphere with the fixed radius of $r = 2$), integrate,

$$\text{Area} = \int_0^{2\pi} \int_0^{\pi} 4 \sin\theta \, d\theta \, d\phi$$
$$= 16\pi$$

remembering that the angle θ covers $180°$ and angle ϕ covers $360°$.

This answer can be double–checked by the formula for the surface area with $n = 3$,

$$A(n) = \frac{2\pi^{(n/2)} r^{(n-1)}}{\Gamma(n/2)}$$

$$A(3) = \frac{2\pi^{(3/2)} r^2}{\Gamma(3/2)}$$

$$= \frac{8\pi^{(3/2)}}{\Gamma(3/2)}$$

$$\Gamma(3/2) = \frac{\sqrt{\pi}}{2}$$

$$A(3) = \frac{8\pi^{(3/2)}}{\frac{\sqrt{\pi}}{2}}$$

$$= 2 \times 8\pi^{(3/2)} \times \pi^{(-1/2)}$$

$$= 16\pi$$

But after having seen this result, it reminds us of the similarity between Amari's Volume definition in his explanation of Jeffreys's prior,

$$V = \frac{\pi^{(n+1)/2}}{\Gamma(\frac{n+1}{2})}$$

and the factor C_D for the Dirichlet distribution,

$$C_D = \frac{\Gamma(\sum_{i=1}^n \alpha_i)}{\prod_{i=1}^n \Gamma(\alpha_i)}$$

The Volume, $1/C_D$, for $n = 3$ and a Jeffreys's prior with $\alpha_i = 1/2$ becomes,

$$\frac{1}{C_D} = \frac{\prod_{i=1}^{3} \Gamma(\alpha_i)}{\Gamma(\sum_{i=1}^{3} \alpha_i)}$$

$$= \frac{\prod_{i=1}^{3} \Gamma(1/2)}{\Gamma(\sum_{i=1}^{3} 1/2)}$$

$$= \frac{\pi^{3/2}}{\frac{\sqrt{\pi}}{2}}$$

$$= 2\pi$$

$$= \text{Surface area} / 2^{n+1} \text{ for Amari's } n = 2, \mathcal{X} = \{0, 1, 2\}$$

Thus, a Jeffreys's prior probability for a model space with three statements in the state space would be,

$$\text{pdf}(q_1, q_2, q_3) = \frac{1}{2\pi} q_1^{-1/2} q_2^{-1/2} q_3^{-1/2}$$

Exercise 48.9.5: Use *Mathematica* to confirm that the formula for the surface area is the partial derivative of the formula for the volume with respect to the radius.

Solution to Exercise 48.9.5

Directly substitute the general formula for the volume of the n–ball into the `D[ ]` operator, and take the derivative with respect to the radius r,

```
D[(Power[π, n/2] Power[r, n]) / Gamma[(n/2) + 1], r]
                                                   // FullSimplify
```

which does return the formula for the surface area, the "$n - 1$ sphere."

Exercise 48.9.6: Find the surface area and volume for the hyperspheres of four, five, and six dimensions with radius 2.

Solution to Exercise 48.9.6

In section 48.5, the surface area and volume for the $n = 2$ circle and the $n = 3$ sphere were calculated. The surface area and volumes of higher dimensional hyperspheres can be calculated with the same formula. So, the surface areas and volumes for hyperspheres of dimension $n = 4, 5$, and 6 of radius $r = 2$ are shown at the top of the next page in Table 48.1.

Table 48.1: *The surface areas and volumes of n–balls with radius equal to 2.*

n	$\frac{2\pi^{(n/2)} r^{n-1}}{\Gamma(n/2)}$	Surface Area	$\frac{\pi^{(n/2)} r^n}{\Gamma(n/2+1)}$	Volume
4	$(2\pi^{4/2} \times 8)/1$	$16\pi^2 \, m^3$	$(\pi^{4/2} \times 16)/2$	$8\pi^2 \, m^4$
5	$(2\pi^{5/2} \times 16)/(3\sqrt{\pi}/4)$	$128\pi^2/3 \, m^4$	$(\pi^{5/2} \times 32)/(15\sqrt{\pi}/8)$	$256\pi^2/15 \, m^5$
6	$(2\pi^{6/2} \times 32)/2$	$32\pi^3 \, m^5$	$(\pi^{6/2} \times 64)/6$	$32\pi^3/3 \, m^6$

Here is a more detailed solution for the volume of the 5–ball.

$$n = 5$$

$$r = 2$$

$$\text{Volume} = \frac{\pi^{5/2} r^5}{\Gamma(\frac{7}{2})}$$

$$\Gamma(7/2) = \frac{15\sqrt{\pi}}{8}$$

$$= \frac{8 \times \pi^{5/2} \times 32}{15 \times \pi^{1/2}}$$

$$= \frac{8 \times \pi^{(5/2-1/2)} \times 32}{15}$$

$$\text{Volume} = \frac{256 \, \pi^2}{15}$$

An arbitrary unit of measurement, a meter, is attached to these surface areas and volumes. So don't be misled by a larger number in front of the units for a surface area as compared to a volume because the surface area will always be one power less in meters when compared to the volume. For example, for the 4–ball, the surface area is $16\pi^2$ cubic meters, while the volume is $8\pi^2$ meters to the fourth power. One linear foot is 12 inches, but one square foot is 144 square inches.

Conceptually, the most important fact is that the surface area for the n–ball is the same as the normalizing factor $1/C_D$ in the Dirichlet distribution used as the operational implementation of the meaning of the prior probability over model space when the $\alpha_i = 1/2$.

For example, if the state space has dimension $n = 6$, ($n = 5$ for Amari), then,

$$\frac{1}{C_D} = \frac{\prod_{i=1}^{6} \Gamma(\alpha_i)}{\Gamma(\sum_{i=1}^{6} \alpha_i)}$$

$$= \frac{\prod_{i=1}^{6} \Gamma(1/2)}{\Gamma(\sum_{i=1}^{6} 1/2)}$$

$$= \frac{\sqrt{\pi}^6}{\Gamma(3)}$$

$$= \frac{\pi^3}{2}$$

which is equal to the surface area of the 6–ball of radius 2 divided by 2^6,

$$\frac{32\pi^3}{2^6} = \frac{\pi^3}{2}$$

Matching up across the board, may we come to the important conclusion that Amari's probability density function on Ξ defined as,

$$Q(\xi) = \frac{1}{V}\sqrt{\det G(\xi)}$$

must be the same as the Dirichlet probability density function representing the prior probability for the assignments q_i? In other words, the same as Jeffreys's prior viewed as the modification to the Bayes–Laplace uniform prior,

$$Q(\xi) \equiv P(\mathcal{M}_k) \equiv \text{pdf}\,(q_1, q_2, \cdots, q_6) = \frac{2}{\pi^3} \times q_1^{-1/2} q_2^{-1/2} \cdots q_6^{-1/2}$$

Exercise 48.9.7: Use *Mathematica* to calculate all of the elements in the Fisher information matrix for the Normal distribution.

Solution to Exercise 48.9.7

To begin, review the *Mathematica* characterization of the Gaussian, or the Normal distribution, with its two parameters μ and σ,

PDF[NormalDistribution[μ, σ], x]

which returns the following familiar expression,

$$\text{pdf}\,(x\,|\,\mathcal{M}_k) = \frac{e^{-\frac{(x-\mu)^2}{2\sigma^2}}}{\sqrt{2\pi}\,\sigma}$$

We will eventually require the partial derivative of the log of this expression for the Normal distribution,

$$\ln\left[\frac{e^{-\frac{(x-\mu)^2}{2\sigma^2}}}{\sqrt{2\pi}\,\sigma}\right] = -\frac{(x-\mu)^2}{2\sigma^2} - \ln(\sqrt{2\pi}\,\sigma)$$

$$= \frac{1}{2}\left(-\frac{(x-\mu)^2}{\sigma^2} - \ln 2\pi - 2\ln\sigma\right)$$

Mathematica confirms this with,

PowerExpand[Log[PDF[NormalDistribution[μ, σ], x]]]
// FullSimplify

We will find that constant invocation of the **PowerExpand** and **FullSimplify** functions are necessary in order to force *Mathematica* to return the expressions we would prefer to see.

Next up, is the derivative of the log of the probability distribution with respect to each of its parameters. For the $r = 1, c = 1$ element in the metric tensor, this will be the derivative of the probability expression for the Normal distribution with respect to μ,

D[PowerExpand[Log[PDF[NormalDistribution[μ, σ], x]]], μ]

returning $(x - \mu)/\sigma^2$.

But the element at the $g_{11}(p)$ position is, by definition,

$$E_p\left[\frac{\partial \ln \text{Normal}}{\partial \mu} \times \frac{\partial \ln \text{Normal}}{\partial \mu}\right]$$

so we will want to find the expectation of,

$$E_p\left[\frac{(x-\mu)^2}{\sigma^4}\right]$$

with respect to the Normal distribution.

Now, we are ready to ask *Mathematica* to perform this integration with respect to the Normal distribution,

PowerExpand[Integrate[PDF[NormalDistribution[μ, σ], x]
            Power[D[PowerExpand[Log[
            PDF[NormalDistribution[μ, σ], x]]], μ], 2],
            {x, $-\infty$, ∞}, Assumptions $\to$ Re[σ^2] > 0]]

After *Mathematica* has done all of the hard work (it took nearly 24 seconds on my computer), we are ready to fill in the first of the four elements in the (2 × 2) metric tensor matrix, with,

$$E_p\left[\frac{(x-\mu)^2}{\sigma^4}\right] = g_{11}(p) = \frac{1}{\sigma^2}$$

Calculation of the remaining three elements of the metric tensor for the Normal distribution follows the same pattern. Since the metric tensor is symmetric, $g_{12}(p) = g_{21}(p)$. This common value turns out to equal 0 as we now show,

```
PowerExpand[Integrate[PDF[NormalDistribution[μ, σ], x]
        D[PowerExpand[Log[
            PDF[NormalDistribution[μ, σ], x]]], μ],
        D[PowerExpand[Log[
            PDF[NormalDistribution[μ, σ], x]]], σ],
        {x, -∞, ∞}, Assumptions → Re[σ²] > 0]]
```

The only element remaining in the metric tensor left for *Mathematica* to calculate is $g_{22}(p) = 2/\sigma^2$.

```
PowerExpand[Integrate[PDF[NormalDistribution[μ, σ], x]
        Power[D[PowerExpand[Log[
            PDF[NormalDistribution[μ, σ], x]]], σ], 2],
        {x, -∞, ∞}, Assumptions → Re[σ²] > 0]]
```

Filling in these values, the metric tensor is seen to be, as announced in section 48.4,

$$g_{rc}(p) = \begin{pmatrix} \frac{1}{\sigma^2} & 0 \\ 0 & \frac{2}{\sigma^2} \end{pmatrix}$$

For the final step, the square root of the determinant of the above matrix is,

$$\sqrt{|g|} = \sqrt{\frac{1}{\sigma^2} \times \frac{2}{\sigma^2}} = \frac{\sqrt{2}}{\sigma^2}$$

as found by *Mathematica* through,

```
PowerExpand[Sqrt[Det[{{1/σ², 0}, {0, 2/σ²}}]]]
```

Thus, Jeffreys's rule has resulted in a "noninformative prior probability for the two parameters" of the Normal distribution of $p(\mu, \sigma) \propto \sigma^{-2}$. Jeffreys didn't like this outcome so he changed it to $p(\mu, \sigma) \propto \sigma^{-1}$. I don't like it either, but for different reasons.

Exercise 48.9.8: Derive an important variation of the Fisher information matrix mentioned by Amari in the 0–representation.

Solution to Exercise 48.9.8

We will begin the proof in my MEP notation, but then transition after a couple of steps to a shorter condensed notation in Amari's style. Start off with the definition of the Fisher information matrix that relies upon the log transform of the assigned probabilities under the information provided by a given model,

$$g_{rc}(p) = E_p \left[\frac{\partial \ln P(X = x_i \mid \mathcal{M}_k)}{\partial \lambda_r} \times \frac{\partial \ln P(X = x_i \mid \mathcal{M}_k)}{\partial \lambda_c} \right]$$

The first transformation involves the relationship between the partial derivative of the log of a function and the function itself in the generic expression,

$$\frac{\partial \ln f(x)}{\partial x} = \frac{1}{f(x)} \frac{\partial f(x)}{\partial x}$$

as can be verified through *Mathematica* with,

D[Log[f[x]], x] == (1 / f[x]) (D[f[x], x])

returning **True**. With this result in hand, the next step is,

$$g_{rc}(p) = E_p \left[\frac{1}{P(X = x_i \mid \mathcal{M}_k)} \frac{\partial P(X = x_i \mid \mathcal{M}_k)}{\partial \lambda_r} \times \frac{1}{P(X = x_i \mid \mathcal{M}_k)} \frac{\partial P(X = x_i \mid \mathcal{M}_k)}{\partial \lambda_c} \right]$$

Here, for the sake of clarity, adopt Amari's notational style for the final steps,

$$g_{rc}(p) = \int \frac{\partial_r P}{P} \times \frac{\partial_c P}{P} \times P \, \mathrm{d}P$$

$$= \int \frac{\partial_r P}{\sqrt{P}} \times \frac{\partial_c P}{\sqrt{P}} \, \mathrm{d}P$$

$$= 4 \int \frac{\partial_r P}{2\sqrt{P}} \times \frac{\partial_c P}{2\sqrt{P}} \, \mathrm{d}P$$

$$= 4 \int \partial_r \sqrt{P} \times \partial_c \sqrt{P} \, \mathrm{d}P$$

The last step can be verified through *Mathematica* with,

D[Sqrt[f[x]], x] == (D[f[x], x]) / (2 Sqrt[f[x]])

returning **True**. Match up this expression in the final step with Amari's Equation (2.10) [2, pg. 28],

$$g_{ij}(\xi) = 4 \int \partial_i \sqrt{p(x;\xi)} \, \partial_j \sqrt{p(x;\xi)} \, \mathrm{d}x$$

It seems then that the x and $\mathrm{d}x$ in Amari's $p(x;\xi)$ must match up with pdf $(q_i \mid \alpha_i)$ and $\mathrm{d}q_i$ in the Dirichlet distribution.

Exercise 48.9.9: Follow the analogy to Amari's probability formula over hyperspheres to its logical conclusion.

Solution to Exercise 48.9.9

Make the tentative assumption that what Amari calls a probability density function on parameter space is, in fact, the Dirichlet probability density function for model space. Amari uses this notation to define a probability density function over the parameter space Ξ,

$$Q(\xi) \stackrel{\text{def}}{=} \frac{\sqrt{\det G(\xi)}}{V}$$

where V is defined as,

$$V \stackrel{\text{def}}{=} \int_{\Xi} \sqrt{\det G(\xi)}\, d\xi$$

In our interpretation take V as,

$$V \stackrel{\text{def}}{=} \int \cdot \int_{\sum q_i = 1} q_1^{\alpha_1 - 1} q_2^{\alpha_2 - 1} \ldots q_n^{\alpha_n - 1}\, dq_i$$

with the result,

$$V = \frac{\prod_{i=1}^{n} \Gamma(\alpha_i)}{\sum_{i=1}^{n} \Gamma(\alpha_i)}$$

For Jeffreys's prior, take $\alpha_i = 1/2$ to find that,

$$V = \frac{\prod_{i=1}^{n} \Gamma(1/2)}{\sum_{i=1}^{n} \Gamma(1/2)}$$

$$= \frac{\prod_{i=1}^{n} \sqrt{\pi}}{\Gamma(\frac{n}{2})}$$

$$= \frac{\pi^{n/2}}{\Gamma(\frac{n}{2})}$$

The n for us appearing in the formula above has to be bumped up by 1 to correspond to Amari's n so that we have equivalent formulas,

$$V = \frac{\pi^{(n+1)/2}}{\Gamma(\frac{n+1}{2})}$$

Guided by the analogy, we would want the Fisher information matrix $G(\xi) \equiv g_{rc}(p)$ to look something like this,

$$g_{rc}(p) = \begin{pmatrix} \frac{1}{q_1} & 0 & \cdots & 0 \\ 0 & \frac{1}{q_2} & \cdots & 0 \\ \vdots & & \cdots & \vdots \\ 0 & 0 & \cdots & \frac{1}{q_n} \end{pmatrix}$$

so that $\det G(\xi)$ looks like,

$$\mid g_{rc}(p) \mid = \prod_{i=1}^{n} \frac{1}{q_i}$$

and $\sqrt{\det G(\xi)}$ looks like,

$$\sqrt{\mid g_{rc}(p) \mid} = \prod_{i=1}^{n} q_i^{-1/2} = q_1^{\alpha_1 - 1} q_2^{\alpha_2 - 1} \ldots q_n^{\alpha_n - 1}$$

with $\alpha_i = 1/2$.

Amari's **0–representation** is essentially a coordinate transformation taking a *coordinate* $4q_i$ to another representation as $2\sqrt{q_i}$. The Jacobian matrix $\boldsymbol{J}$ is the matrix of the partial derivatives of one set of coordinates $\bar{x}^i$ with respect to the transformed set x^i. Generically the Jacobian looks like this,

$$\boldsymbol{J} = \begin{pmatrix} \frac{\partial \bar{x}^1}{\partial x^1} & \frac{\partial \bar{x}^1}{\partial x^2} & \cdots & \frac{\partial \bar{x}^1}{\partial x^m} \\ \\ \frac{\partial \bar{x}^2}{\partial x^1} & \frac{\partial \bar{x}^2}{\partial x^2} & \cdots & \frac{\partial \bar{x}^2}{\partial x^m} \\ \\ \vdots & \vdots & \cdots & \vdots \\ \\ \frac{\partial \bar{x}^m}{\partial x^1} & \frac{\partial \bar{x}^m}{\partial x^2} & \cdots & \frac{\partial \bar{x}^m}{\partial x^m} \end{pmatrix}$$

For this situation, the barred coordinates $\bar{x}^i$ are $4\,q_i$, while the unbarred coordinates x^i are the $2\sqrt{q_i}$. The partial derivatives to insert into the J_{11} location of the matrix as found by *Mathematica* are,

$$\texttt{D[4 q}_1 \texttt{ / (2 Sqrt[q}_1\texttt{]), q}_1\texttt{]}$$

equal to $1/\sqrt{q_1}$. In like manner, J_{12} is found as,

$$\texttt{D[4 q}_1 \texttt{ / (2 Sqrt[q}_1\texttt{]), q}_2\texttt{]}$$

equal to 0. The rest of the first row to J_{1m} is then going to contain 0s as well. It's pretty clear to see the pattern developing here. J_{22} will contain,

$$\texttt{D[4 q}_2 \texttt{ / (2 Sqrt[q}_2\texttt{]), q}_2\texttt{]}$$

equal to $1/\sqrt{q_2}$ and J_{21} through J_{3m} will contain 0s. For our $n = 3$ and Amari's $n = 2$, the Jacobian matrix will look like this,

$$\boldsymbol{J} = \begin{pmatrix} \frac{1}{\sqrt{q_1}} & 0 & 0 \\ 0 & \frac{1}{\sqrt{q_2}} & 0 \\ 0 & 0 & \frac{1}{\sqrt{q_3}} \end{pmatrix}$$

We saw an example in Exercise 48.9.3 where the determinant of the Jacobian matrix was equal to the square root of the determinant of the metric tensor,

$$\mid \boldsymbol{J} \mid = \sqrt{\mid g_{rc}(p) \mid}$$

It's easy to see that this determinant is,

$$\mid \boldsymbol{J} \mid = \frac{1}{\sqrt{q_1}} \times \frac{1}{\sqrt{q_2}} \times \frac{1}{\sqrt{q_3}}$$

which is in the desired format to match up with Amari's numerator in the probability density function for the coordinates ξ, our q_i and $P(q_i)$,

$$Q(\xi) = \frac{1}{V}\sqrt{\det G(\xi)}$$

and the relevant terms in the Dirichlet distribution,

$$q_1^{\alpha_1-1} q_2^{\alpha_2-1} \cdots q_n^{\alpha_n-1}$$

with $\alpha_i = 1/2$.

The Jacobian matrix is the "square root" of the metric tensor through the relationship,

$$\mathbf{g} = \mathbf{J^T J}$$

In matrix algebra, such a matrix square root is calculated by what is called the *Cholesksy decomposition*. This is implemented in *Mathematica* as,

CholeskyDecomposition[m]

where here the one argument is a square matrix m.

Under Amari's **0–representation**, the metric tensor is constructed from the inner product of tangent vectors $X^{(0)}(x)$ and $Y^{(0)}(x)$ which are functions of,

$$\frac{1}{\sqrt{q_i}}$$

This is an alternative approach from Information Geometry corresponding to the more straight forward invocation of the Dirichlet distribution for a prior probability over model space.

Exercise 48.9.10: Show that as a prior probability over model space, the *beta distribution* can be constructed via the MEP.

Solution to Exercise 48.9.10

Recall that we are trying to find an assignment to $P(\mathcal{M}_k) \equiv P(q_1, q_2) \equiv P(q)$. The statement $\mathcal{M}_k$ asserts that the correct probability for the two joint statements are q_1 and q_2. The *beta probability density function* capturing the degree of belief in any model $\mathcal{M}_k$ is,

$$\text{pdf}(\mathcal{M}_k) = C_{\text{Beta}}\, q^{\alpha-1}(1-q)^{\beta-1}$$

If we re–express,

$$q^{\alpha-1}(1-q)^{\beta-1}$$

as,

$$\exp\left[(\alpha-1)\ln q + (\beta-1)\ln(1-q)\right]$$

we see that we have the numerator of a probability density function pdf (x) in the standard MEP form of,

$$\text{Numerator} = e^{\lambda_1 F_1(x) + \lambda_2 F_2(x)}$$

where the parameters are $\lambda_1 \equiv \alpha - 1$, $\lambda_2 \equiv \beta - 1$, and the constraint functions are $F_1(x) \equiv \ln q$ and $F_2(x) \equiv \ln(1-q)$. The partition function $Z(\alpha, \beta)$ is then found by integrating the numerator over all possible q,

$$Z(\alpha, \beta) = \int_0^1 \exp\left[(\alpha - 1) \ln q + (\beta - 1) \ln(1-q)\right] dq$$

Mathematica finds the partition function as the expected,

$$Z(\alpha, \beta) = \frac{\Gamma(\alpha)\Gamma(\beta)}{\Gamma(\alpha + \beta)}$$

through the evaluation of,

```
Integrate[Exp[(α - 1) Log[q] + (β - 1) Log[1 - q]], {q, 0, 1},
    Assumptions → α > 0 && β > 0] // FullSimplify
```

Just as some function of the joint statement $F_j(X = x_i)$ makes an appearance in the numerator of the MEP formula when assigning $P(X = x_i \mid \mathcal{M}_k)$, so functions of q, namely $\ln(q)$ and $\ln(1-q)$, appear in the analogous positions in the numerator of the MEP formula for $P(\mathcal{M}_k)$.

Also take note that this assignment to the probability of a model under the rubric of the MEP adopts only *two* parameters over the real numbers between 0 and 1, and only n parameters for n statements about the q_i. There is no injunction against literally an infinite number of parameters over the real line between 0 and 1. Thus, the Dirichlet density function, just like the Normal, Gamma, and all the rest of the probability density functions over the real line are simplifications of what could have been possible.

According to our well–used MEP definition of the constraint function average, $\langle F_1 \rangle$ should equal,

$$\frac{\partial \ln Z}{\partial \alpha} = \int_0^1 \ln q \, C_{\text{Beta}} \, q^{\alpha - 1} (1-q)^{\beta - 1} \, dq$$

Putting *Mathematica* through its paces, we find that,

```
D[Log[Gamma[α] Gamma[β] / Gamma[α + β]], α] // Simplify
```

is the same as,

```
NIntegrate[Log[q] PDF[BetaDistribution[1/2, 1/2], q], {q, 0, 1}]
```

yielding a value of $\langle F_1 \rangle = -\ln 4 = -1.38629$. The second dual parameter $\langle F_2 \rangle$ has the same value.

Exercise 48.9.11: The final exercise.

Solution to Exercise 48.9.11

As a final exercise that focuses on Information Geometry, illustrate a fundamental relationship between the metric tensor as it has come down to us through IG when compared to the just completed MEP characterization of the prior probability. We will first verify Amari's two representations for the metric tensor in model space. To keep things as simple as possible we will rely on the *beta distribution* for $P(\mathcal{M}_k)$. In other words, the dimension of the state space is $n = 2$.

First, we have the definition of the metric tensor as the Fisher information matrix,

$$g_{\alpha\beta}[P(\mathcal{M}_k)] = \int_0^1 \frac{\partial \ln p(q)}{\partial \alpha} \times \frac{\partial \ln p(q)}{\partial \beta} p(q) \, dq$$

Secondly, we have the Amari's alternate representation as,

$$g_{\alpha\beta}[P(\mathcal{M}_k)] = 4\int_0^1 \partial_\alpha \sqrt{p(q)} \times \partial_\beta \sqrt{p(q)} \, dq$$

Mathematica evaluates the first element in the metric tensor, g_{11}, following the Fisher information prescription,

```
Integrate[PDF[BetaDistribution[α, β], q]
    D[Log[PDF[BetaDistribution[α, β], q]], α]
    D[Log[PDF[BetaDistribution[α, β], q]], α], {q, 0, 1},
    Assumptions → Re[α] > 0 && Re[β] > 0] // FullSimplify
```

to yield this result,

```
PolyGamma[1, α] - PolyGamma[1, α + β]
```

Mathematica evaluates the second element in the metric tensor, g_{12}, following Amari's alternative representation,

```
4 PowerExpand[Integrate[
    D[Sqrt[PDF[BetaDistribution[α, β], q]], α]
    D[Sqrt[PDF[BetaDistribution[α, β], q]], β], {q, 0, 1},
    Assumptions → Re[α] > 0 && Re[β] > 0]] // FullSimplify
```

to yield this result,

$$\text{- PolyGamma[1, } \alpha + \beta\text{]}$$

All four elements of the metric tensor *are calculated by either method* as,

$$g_{\alpha\beta}[P(\mathcal{M}_k)] = \begin{pmatrix} \psi^{(1)}(\beta) - \psi^{(1)}(\alpha+\beta) & -\psi^{(1)}(\alpha+\beta) \\ -\psi^{(1)}(\alpha+\beta) & \psi^{(1)}(\beta) - \psi^{(1)}(\alpha+\beta) \end{pmatrix}$$

By far an easier method confirming these findings for the metric tensor derive from the MEP formalism. Back in the opening Chapter, we listed this formula for the metric tensor,

$$\frac{\partial^2 \ln Z}{\partial \lambda_r \, \partial \lambda_c} = g_{rc}(p)$$

Applying this definition to the prior probability for models, we have,

$$\frac{\partial^2 \ln Z}{\partial \alpha \, \partial \beta} = g_{\alpha\beta}[P(\mathcal{M}_k)]$$

Differentiations are always easier than integrations, so ask *Mathematica* for the fourth element in the metric tensor g_{22},

```
D[Log[Gamma[α] Gamma[β] / Gamma[α + β]], β, β] // Simplify
```

to yield this result,

```
PolyGamma[1, α] - PolyGamma[1, α + β]
```

The full metric tensor using this method inspired by the MEP is the same as found above by using the integration required by the Fisher information matrix, or as required by Amari's 0–representation.

Specifically, the metric tensor at the point in Riemannian space that represents the model space with coordinates $\alpha = \beta = 1/2$ is,

$$g_{\alpha\beta}[P(\mathcal{M}_k)] \equiv \frac{\partial^2 \ln Z}{\partial \alpha \, \partial \beta} = \begin{pmatrix} +3.28987 & -1.64493 \\ -1.64493 & +3.28987 \end{pmatrix}$$

What is the IP's degree of belief in the q_i specified under the information from a particular model? In other words, what is the degree of belief in any one of the assignments $\{(q = 1, 1 - q = 0), (q = 0.9, 1 - q = 0.1), \cdots, (q = 0, 1 - q = 1)\}$ for the case of $n = 2$?

For Laplace, with $\alpha = \beta = 1$, the pdf $(\mathcal{M}_k)$ has the constant value of 1 at every value of q, the "flat" curve extending from $q = 0$ to $q = 1$. For Jeffreys, with $\alpha = \beta = 1/2$, the pdf $(\mathcal{M}_k)$ has the U–shaped curve over all values of q. It is easy to understand why the IP's degree of belief about the truth that q lies in the interval from 0.4 to 0.6 is 0.2 when the prior probability is "flat."

$$\int_{0.4}^{0.6} \text{pdf}\,(q, [1-q])\,|\,\alpha = 1, \beta = 1)\,dq = \int_{0.4}^{0.6} \frac{\Gamma(\alpha+\beta)}{\Gamma(\alpha)\,\Gamma(\beta)}\, q^0 (1-q)^0 \, dq = 0.2$$

as compared to the IP's lesser degree of belief in the truth of that same interval under Jeffreys's prior,

$$\int_{0.4}^{0.6} \text{pdf}\,(q, [1-q])\,|\,\alpha = 1/2, \beta = 1/2)\,dq = \int_{0.4}^{0.6} \frac{\Gamma(\alpha+\beta)}{\Gamma(\alpha)\,\Gamma(\beta)}\, q^{-1/2} (1-q)^{-1/2} \, dq = 0.128$$

Appendix A

Working with Vectors and Matrices in *Mathematica*

This appendix discusses *Mathematica* syntax for the various vector and matrix operations commonly called upon in Information Geometry. For example, these operations are needed in the Chapter Thirty Four exercises illustrating how the metric tensor **G** is involved in distance calculations. We will show how *Mathematica* finds the solutions to Exercises 34.9.4 through 34.9.6.

Either the **Array[ ]** or the **Table[ ]** commands can be used to create the lists *Mathematica* uses to represent both vectors and matrices. I prefer to use the **Array[ ]** command because it has a slightly simpler syntax. I also tend to mentally pigeon–hole the **Table[ ]** command to perform iterations. **Array[ ]** will also generalize to produce tensors when we get to the point of needing them.

I think it better to actually start out with symbolic manipulations rather than numeric computations because it's the general structure of vectors and matrices that we are interested in. For example, symbolic entries can show us how vector and matrix multiplications are equivalent to corresponding summation formulas.

At the most primitive level, we could simply create a list {**a, b, c**} to represent, say, three components of a vector,

$$\text{vector = List[a, b, c]}$$

However, we will choose this method to create a vector **v** with three elements,

$$\text{vectorV = Array[v, 3]}$$

This also produces the requisite list {**v[1], v[2], v[3]**} to represent **v**.

There are some distinctions to be made here. Any element in the vector **v**, say, **v[2]**, is an *indexed object*. **v[2]** looks very much like how we defined a function by **head[arg]** with a *head* **v** and *arg* 2. Suppose a function $f(x) = x^2 + 4$ is defined as looking like,

$$v[x\_] := Plus[Power[x, 2], 4]$$

If the expression **v[2]** is evaluated it returns **8**. But **v[2]** exists here as an element of a vector and is an indexed object. We can set it to any value we want with, say, **v[2] = 49**.

If we ask for the second element in the list we write,

$$Part[vectorV, 2] \text{ or } vectorV[[2]]$$

we get **v[2]**. If we ask for **v[2]** we get **49**.

Typically, we will represent a vector as consisting of a list of indexed objects like **{v[1], v[2], v[3]}**. At other times, however, we might like to match the visual forms that we are more accustomed to. For example, we might prefer to represent the vector of Lagrange multipliers with subscripts as $\{\lambda_1, \lambda_2, \lambda_3\}$.

There is a slightly more complicated way to create this visual form. For example, create a vector **w**, with three subscripted elements, by,

$$vectorWsub = Array[Function[x, Subscript[w, x]], 3]$$

to produce the list $\{w_1, w_2, w_3\}$ in place of **{w[1], w[2], w[3]}**. To see the vector displayed in the standard typography, wrap **MatrixForm[ ]** around the vector as in **MatrixForm[vectorWsub]** to produce,

$$\mathbf{w} = \begin{pmatrix} w_1 \\ w_2 \\ w_3 \end{pmatrix}$$

Even though this vector is displayed as a column vector, **w** can be either a row or column vector. There is no need for *Mathematica* to distinguish between row and column vectors when doing computations, but it *is* helpful for us to keep things straight.

As always, there is a syntactic short–cut available,

$$vectorVsub = Array[v_\#\&, 3]$$

with **MatrixForm[vectorVsub]** for

$$\mathbf{v} = \begin{pmatrix} v_1 \\ v_2 \\ v_3 \end{pmatrix}$$

VECTORS AND MATRICES

After these tedious preliminary remarks, we are now ready for some more serious business. To multiply two vectors, take the dot product,

$$\texttt{Dot[vectorV, vectorW]}$$

which returns `v[1]w[1] + v[2]w[2] + v[3]w[3]`. This result is easily seen to be the same as produced by,

$$\texttt{Sum[v[i] w[i], \{i, 1, 3\}]}$$

The sum indicates that a *row* vector **v** has multiplied a *column* vector **w** to produce a scalar. This operation was able to take place because the vectors were conformable. **v** is a (1×3) row vector and **w** is a (3×1) column vector resulting in a (1×1) scalar. DO NOT wrap **MatrixForm[ ]** around a vector before multiplication.

Mathematica also possesses the generalization of the **Dot[ ]** command called **Inner[ ]**. The four arguments to **Inner[ ]** look like this,

$$\texttt{Inner[f, list1, list2, g]}$$

By retaining the two functions **f** and **g** in their symbolic form,

$$\texttt{Inner[f, \{a, b, c\}, \{x, y, z\}, g]}$$

we can observe the general pattern that **Inner[ ]** produces,

$$\texttt{g[f[a, x], f[b, y], f[c, z]]}$$

Recreate the specialized **Dot[vectorV, vectorW]** function with **Inner[ ]**,

$$\texttt{Inner[Times, vectorV, vectorW, Plus]}$$

by following the above symbolic template `g[f[a, x], f[b, y], f[c, z]]` and substituting $g \equiv \texttt{Plus[ ]}$, $f \equiv \texttt{Times[ ]}$, $a \equiv \texttt{v[1]}$, $x \equiv \texttt{w[1]}$, and so on, to arrive at,

$$\texttt{Plus[Times[v[1], w[1]], Times[v[2], w[2]], Times[v[3], w[3]]]}$$

to finally evaluate to the required `v[1]w[1] + v[2]w[2] + v[3]w[3]`. Recall that *Mathematica* will keep on evaluating expressions until it can no longer do so.

Move up to the next level to see how *Mathematica* represents matrices. To no great surprise, *Mathematica* represents matrices as a nested list of lists. The counterpart for matrices to the primitive creation of a vector as a list is to directly create a list of lists,

```
matrix = List[List[a, b, c], List[d, e, f], List[x, y, z]]
```

which produces,

```
matrix = {{a, b, c}, {d, e, f}, {x, y, z}}
```

An easier way is to once again use the **Array[ ]** command. To create a matrix **G**, it is necessary to specify two arguments for the number of rows and the number of columns. Since we will be dealing mainly with symmetric matrices, we will set the number of rows and columns equal. The syntax changes to either,

```
matrixG = Array[g, {3, 3}]
```

for a 3 row by 3 column matrix shown as lists of indexed objects,

```
matrixG = {{g[1, 1], g[1, 2], g[1, 3]},
           {g[2, 1], g[2, 2], g[2, 3]},
           {g[3, 1], g[3, 2], g[3, 3]}}
```

Or, if you prefer subscripted entries in the matrix,

```
matrixGsub = Array[g##&, {3, 3}]
```

In either case, *Mathematica* must represent both matrices as a nested list. For the standard typography, **MatrixForm[matrixGsub]** will show the list of lists laid out in standard fashion as,

$$\mathbf{G} = \begin{pmatrix} g_{1,1} & g_{1,2} & g_{1,3} \\ g_{2,1} & g_{2,2} & g_{2,3} \\ g_{3,1} & g_{3,2} & g_{3,3} \end{pmatrix}$$

The **Dot[ ]** command is used once again to find any required vector–matrix multiplication, say, $\mathbf{v} \cdot \mathbf{G}$. Even though *Mathematica* may have shown us vector **v** as if it were a column vector when we wrapped it in **MatrixForm[ ]**, it really is a row vector. Once again, as far as the *Mathematica* computation is concerned, no special care needs to be taken,

```
vectorVG = Dot[vectorV, matrixG]
```

outputs a (1×3) row vector $\mathbf{v} \cdot \mathbf{G}$ as the list with three elements,

```
{g[1,1]v[1] + g[2,1]v[2] + g[3,1]v[3],
 g[1,2]v[1] + g[2,2]v[2] + g[3,2]v[3],
 g[1,3]v[1] + g[2,3]v[2] + g[3,3]v[3]}
```

VECTORS AND MATRICES

For the problem of finding the distance between two points, we will need the vector–matrix–vector multiplication $\mathbf{v} \cdot \mathbf{G} \cdot \mathbf{w}^\mathbf{T}$. Thus, the (1×3) row vector $\mathbf{v} \cdot \mathbf{G}$ will multiply the (3×1) column vector $\mathbf{w}^\mathbf{T}$ to result in a (1×1) scalar. Therefore, we use the **Dot[ ]** command twice, as in,

```
scalarVGW = Expand[Dot[Dot[vectorV, matrixG], vectorW]]
```

The **Expand[ ]** function was needed to force *Mathematica* to multiply out all of the terms in the way we would prefer to see them. The final result, a scalar, is a sum of nine terms,

```
g[1, 1]v[1]w[1] + g[2, 1]v[2]w[1] + g[3, 1]v[3]w[1] +
g[1, 2]v[1]w[2] + g[2, 2]v[2]w[2] + g[3, 2]v[3]w[2] +
g[1, 3]v[1]w[3] + g[2, 3]v[2]w[3] + g[3, 3]v[3]w[3]
```

The multiplication of each element in the column vector $\mathbf{w}^\mathbf{T}$ by the corresponding element in the row vector $\mathbf{v} \cdot \mathbf{G}$ can be clearly seen.

We would like to check that the alternative summation format for the squared distance between two points is the same as **scalarVGW**. Translate the sum formula for the vector–matrix–vector multiplication,

$$\mathbf{v} \cdot \mathbf{G} \cdot \mathbf{w}^\mathbf{T} = \sum_{k=1}^{3} \sum_{j=1}^{3} g_{jk} v^j w^k$$

into *Mathematica* as,

```
Sum[g[j, k] v[j] w[k], {k, 1, 3}, {j, 1, 3}]
```

Ask *Mathematica* if the evaluation of this expression is the same as the symbolic sum computed above with **Dot[ ]** through the application of the **Equal[ ]** function. The symbolic forms on both sides of **Equal[ ]** must be *exactly* the same.

```
Equal[scalarVGW, Sum[g[j, k] v[j] w[k], {k, 1, 3}, {j, 1, 3}]]
```

The answer returned is **True**. *Mathematica*'s ability to perform these kind of vector–matrix–vector multiplications was used in the solution to Exercise 34.9.6.

Mathematica has an important built–in function called **Outer[ ]**. This function will be used to compute the metric tensor as IG interprets it as the expectation over all possible combinations of basis functions. The argument template for **Outer[ ]** looks like this,

Outer[*f, list1, list2, ... , listi* **]**

The vital role of **Outer[ ]** is to form all possible combinations of the elements in the lists as arguments to an expression with the head **f**. So, for example,

$$\text{Outer}[f, \{a, b\}, \{c, d\}]$$

creates,

$$\{\{f[a, c], f[a, d]\}, \{f[b, c], f[b, d]\}\}$$

Notice that the matrix–like structure of lists within a list is formed.

Another simple example of forming all possible combinations of elements in the lists submitted as arguments to **Outer[ ]** acting as arguments to f is,

$$\text{Outer}[\text{Plus}, \{1, 2\}, \{3, 4\}, \{5, 6\}]$$

This evaluation produces the result,

$$\{\{\{9, 10\}, \{10, 11\}\}, \{\{10, 11\}, \{11, 12\}\}\}$$

where the first entry of 9 results from **Plus[1, 3, 5]**, the second entry of 10 results from **Plus[1, 3, 6]**, and so on. Notice here the increasingly complicated structure of lists within lists, essentially a matrix with elements that are themselves column vectors.

It is interesting to speculate on whether it is legitimate to do things the other way around and multiply a *row* vector **v** by a *column* vector **w**. The same heuristic applies of multiplying columns by rows, but now the first row is simply w_1, while the first column is simply v_1 resulting in $w_1 \times v_1$ as the first element in a 3×3 matrix. This is still a conformable operation since we have a (3×1) column vector multiplying a (1×3) row vector yielding the resulting (3×3) matrix.

Mathematica performs this kind of vector–vector multiplication, called an *outer product*, with,

$$\text{Outer}[\text{Times}, \text{vectorW}, \text{vectorV}]$$

Wrapping **MatrixForm[ ]** around this results in the 3×3 matrix,

$$\begin{pmatrix} w_1 v_1 & w_1 v_2 & w_1 v_3 \\ w_2 v_1 & w_2 v_2 & w_2 v_3 \\ w_3 v_1 & w_3 v_2 & w_3 v_3 \end{pmatrix}$$

The next example will show how to form the metric tensor for the Cartesian coordinate system. Let the lists as arguments to **Outer[ ]** be the orthonormal basis vectors of the Cartesian system. Thus, the *first* list, *list1*, is the list of lists,

$$\{\{1, 0, 0\}, \{0, 1, 0\}, \{0, 0, 1\}\}$$

VECTORS AND MATRICES

and the *second* list, *list2*, is exactly the same. f must be the head **Dot**. Since **Outer[ ]** creates every possible combination of the elements of these lists as the arguments to **Dot[ ]**, we will get the dot product of {1, 0, 0} with {1, 0, 0} as the first component of the metric tensor matrix, and so on, which is exactly what we want.

However, in order to tell *Mathematica* to treat the basis vector {1, 0, 0} as the lowest level element, and not the 1s or 0s, it is necessary to amend the argument template with,

$$\text{Outer[} f, \text{ } list1, \text{ } list2, \text{ } \ldots, \text{ } listi, \text{ } n \text{]}$$

with $n = 1$. This tells *Mathematica* to go up one level so to speak and treat the basis vector {1, 0, 0} as the lowest level element. Thus the full expression is,

$$\text{Outer[Dot, \{\{1, 0, 0\}, \{0, 1, 0\}, \{0, 0, 1\}\},}$$
$$\text{\{\{1, 0, 0\}, \{0, 1, 0\}, \{0, 0, 1\}\}, 1]}$$

which evaluates to the metric tensor,

$$g = \begin{pmatrix} 1 & 0 & 0 \\ 0 & 1 & 0 \\ 0 & 0 & 1 \end{pmatrix}$$

I would now like to demonstrate the utility of **Outer[ ]** with the following derivation of the relationship,

$$\langle \mathbf{v}, \mathbf{w} \rangle = \mathbf{v} \cdot \mathbf{G} \cdot \mathbf{w}^\mathbf{T}$$

as first explored in Exercise 34.9.6.

Set up a basis vector with **basisF = Array[F, 3]** so that any vector in the given vector space can be constructed from the coordinates v^j and the basis vectors through $\sum_{j=1}^{3} v^j f_j$. Thus, **Dot[vectorV, basisF]** which produces,

$$\text{F[1]v[1] + F[2]v[2] + F[3]v[3]}$$

and **Dot[vectorW, basisF]** which produces,

$$\text{F[1]w[1] + F[2]w[2] + F[3]w[3]}$$

specifies two vectors in the vector space.

The outer product of these two vectors,

$$\text{Outer[Times, Dot[vectorV, basisF], Dot[vectorW, basisF]]}$$

results in the *scalar* inner product $\langle \sum_{j=1}^{3} v^j \mathbf{f}_j, \sum_{k=1}^{3} w^k \mathbf{f}_k \rangle$,

```
F[1]F[1]v[1]w[1] + F[1]F[2]v[2]w[1] + F[1]F[3]v[3]w[1] +
F[1]F[2]v[1]w[2] + F[2]F[2]v[2]w[2] + F[1]F[3]v[3]w[2] +
F[1]F[3]v[1]w[3] + F[2]F[3]v[2]w[3] + F[3]F[3]v[3]w[3]
```

However, first forming the metric tensor as in,

```
Outer[Times, basisF, basisF]
```

and then the two dot products,

```
Dot[Dot[vectorV, Outer[Times, basisF, basisF]], vectorW]
```

to implement $\mathbf{v} \cdot \mathbf{G} \cdot \mathbf{w}^{\mathbf{T}}$ evaluates to exactly the same symbolic expression. Thus,

$$\langle \mathbf{v}, \mathbf{w} \rangle = \sum_{j=1}^{3} \sum_{k=1}^{3} v^j \langle \mathbf{e}_j, \mathbf{e}_k \rangle w^k = \mathbf{v} \cdot \mathbf{G} \cdot \mathbf{w}^{\mathbf{T}}$$

Appendix B

Using *Mathematica* for Differentiation in IG

Mathematica has a straightforward syntax for differentiation. The simplest case looks like this, **FullForm[D[f[x], x]]**, with *Mathematica* returning the rather opaque symbolic expression **Derivative[1][f][x]** for $f'(x)$. The head of the *Mathematica* symbolic expression for finding the derivative is designated by **D** with its two arguments **D[*arg1, arg2*]** clearly evident.

More transparently, suppose $f(x) = x^3 + 3x^2 + 10$. Then, the derivative is,

$$\frac{\mathrm{d}f(x)}{\mathrm{d}x} = f'(x) = 3x^2 + 6x$$

as found by **D[Plus[Power[x, 3], 3 Power[x, 2], 10], x]**

When dealing with the MEP formalism, the function $f(\mathbf{x})$ is often the log of the partition function,

$$f(\mathbf{x}) \equiv \ln Z(\lambda_1, \lambda_2, \cdots, \lambda_m)$$

where the vector of arguments $\mathbf{x}$ consists of the m Lagrange multipliers. These are the $m < n - 1$ parameters of some model in the sub–space of the full n–dimensional manifold. The traditional mathematical symbolism for a *partial* differentiation with respect to some argument x_j is $\frac{\partial f(\mathbf{x})}{\partial x_j}$. But *Mathematica* still uses **D[f[x], x]** for partial differentiation as well.

As a first example, recall where the MEP formalism does introduce a partial differentiation. The fundamental relationship between the partial derivative of the log of the partition function with respect to some parameter of the model is given by,

$$\frac{\partial \ln Z}{\partial \lambda_r} = \langle F_r \rangle$$

In the notation from Information Geometry, this has an alternative symbology of,

$$\left\{\frac{\partial \ln Z}{\partial \lambda_r} = \langle F_r \rangle\right\} \equiv \{\partial_r \psi = \eta_r\}$$

Thus, the replacements **f** → **Log[z]** and **x** → **λ_r** would be made to construct the partial differentiation **D[Log[z], λ_r]**.

We will now deconstruct **Log[z]**. First, we have to make the symbols for the Lagrange multipliers, and for ease in following the numerical example, let $n = 4$ and $m = n - 1 = 3$ of the simplest kangaroo scenario. Use,

Map[Subscript[λ, #]&, Range[3]]

to produce the list $\{\lambda_1, \lambda_2, \lambda_3\}$.

Next, we perform the vector–matrix multiplication of the Lagrange multipliers with the matrix of the constraint functions **cm** from the $n = 4$ dimensional state space of the simple kangaroo scenario,

Dot[Map[Subscript[λ, #]&, Range[3]], cm]

where **cm** is the matrix consisting of the list of lists,

cm = {{1, 1, 0, 0}, {1, 0, 1, 0}, {1, 0, 0, 0}}

The result of this (1×3) row vector of Lagrange multipliers times the (3×4) matrix of constraint functions is a (1×4) row vector in the list,

$$\{\lambda_1 + \lambda_2 + \lambda_3, \lambda_1, \lambda_2, 0\}$$

Finally, after exponentiating and summing,

zsymbolic = Total[Exp[Dot[Map[Subscript[λ, #]&, Range[3]], cm]]]

we have the partition function Z,

$$Z = e^{\lambda_1+\lambda_2+\lambda_3} + e^{\lambda_1} + e^{\lambda_2} + 1$$

We are now in a position to actually implement the partial differentiation of $\ln Z$ with respect to one of its parameters which was our original objective. Suppose that we seek the solution to $\partial_2 \psi = \eta_2 = 0.75$ from some particular model. Evaluating,

D[Log[z], λ_2]

returns,

$$\frac{\partial \ln Z}{\partial \lambda_2} = \frac{e^{\lambda_2} + e^{\lambda_1+\lambda_2+\lambda_3}}{e^{\lambda_1+\lambda_2+\lambda_3} + e^{\lambda_1} + e^{\lambda_2} + 1}$$

DIFFERENTIATION

This, of course, is the correct symbolic answer. But insert the known Lagrange multipliers for a particular model where,

$$\lambda_1 = -1.386294$$

$$\lambda_2 = -1.386294$$

$$\lambda_3 = +4.025352$$

with,

`ReplaceAll[D[Log[z], `λ_2`],`
`{Rule[`λ_1`, -1.386294], Rule[`λ_2`, -1.386294], Rule[`λ_3`, 4.025352]}]`

The answer returned by evaluating this code is 0.75 which we know is, in fact, $\langle F_2 \rangle \equiv \eta_2$ under this model. The syntactic short-cut for the above is,

`D[Log[z], `λ_2`] /. {`λ_1` → -1.386294, `λ_2` → -1.386294, `λ_3` → 4.025352}`

Having accomplished this much, it is quite easy to have *Mathematica* address the differentiation within the MEP formalism where we would like to take the second partial differentiation of the log of the partition function with respect to two of the parameters,

$$\frac{\partial \ln Z}{\partial \lambda_r} \times \frac{\partial \ln Z}{\partial \lambda_c} \equiv \frac{\partial^2 \ln Z}{\partial \lambda_r \, \partial \lambda_c} \equiv \frac{\partial \langle F_c \rangle}{\partial \lambda_r}$$

To differentiate the multivariate function $f(\mathbf{x})$ with respect to *two* of its arguments, say, x_r and x_c, *Mathematica* provides the template,

`D[f, `x_r`, `x_c`]`

For a change of pace from the example in Chapter Thirty Three, suppose that we want to calculate the component $g_{13}(p)$ in the first row and third column of the metric tensor. With $r = 1$ and $c = 3$,

`D[Log[z], `λ_1`, `λ_3`]`

evaluates to the already complicated symbolic answer of,

$$\frac{(e^{\lambda_1+\lambda_2+\lambda_3})(e^{\lambda_2+1})}{(e^{\lambda_1+\lambda_2+\lambda_3} + e^{\lambda_1} + e^{\lambda_2} + 1)^2}$$

Fortunately,

`D[Log[z], `λ_1`, `λ_3`] /.`
    `{`λ_1` → -1.386294, `λ_2` → -1.386294, `λ_3` → 4.025352}`

evaluates to what we know is the correct answer of $g_{13}(p) = 0.175$. Refer back to Table 28.2, pg. 394, in Volume II for confirmation of this result.

Mathematica makes it easy to use `D[ ]` to compute the Jacobian matrix. The syntax template is,

$$D[\{f_1, f_2, \ldots\}, \{\{x_1, x_2, \ldots\}\}]$$

A first example was the calculation of the Jacobian matrix for polar coordinates in Exercise 34.9.2. There,

$$f_1 \equiv r \cos \theta$$

$$f_2 \equiv r \sin \theta$$

$$x_1 \equiv r$$

$$x_2 \equiv \theta$$

A second example was the calculation of the Jacobian matrix for the case of spherical coordinates as carried out in Exercise 48.9.2. There,

$$f_1 \equiv r \sin \theta \cos \phi$$

$$f_2 \equiv r \sin \theta \sin \phi$$

$$f_3 \equiv r \cos \theta$$

$$x_1 \equiv r$$

$$x_2 \equiv \theta$$

$$x_3 \equiv \phi$$

The Jacobian matrix for the spherical coordinate system is evaluated with,

$$D[\{r\ \text{Sin}[\theta]\ \text{Cos}[\phi], r\ \text{Sin}[\theta]\ \text{Sin}[\phi], r\ \text{Cos}[\theta]\}, \{\{r, \theta, \phi\}\}]$$

The concept of the *total differential* has already made an appearance several times in this tutorial on Information Geometry. For example, the total differential appeared in Chapter 37 when we were discussing Amari's proof that the partial differentiation of the negative entropy with respect to some constraint function average was the corresponding Lagrange multiplier.

The essential idea behind the total differential is that it is an approximation to an exact finite difference. Thus, if a finite difference Δy is defined as the difference in the functional value after an increment of Δx has been added to the initial argument x,

$$\Delta y = f(x + \Delta x) - f(x)$$

DIFFERENTIATION

then its approximation is the differential,

$$dy = f'(x)\, dx$$

Mathematica has the operator `Dt[ ]` available to find the total differential.

In this case, *Mathematica*'s symbolic capabilities come to the fore. *Mathematica* evaluates symbolic expressions that match what we would expect from looking at the mathematical formula for a total differential. Starting with the simplest case, evaluating `FullForm[Dt[f[x]]]` returns,

`Times[Dt[x], Derivative[1][f][x]]`

`Derivative[1][f][x]` is the way *Mathematica* expresses $f'(x)$ and `Dt[x]` is the way it expresses dx. Even though it seems a bit excessive in this simple case to revert to the `FullForm[ ]` symbolic expressions, the pattern carries over very nicely to functions with more than one argument.

Moving on to a function $f(x, y)$ with two arguments, a standard mathematical formula for the total differential will look like,

$$df = \frac{\partial f}{\partial x}\, dx + \frac{\partial f}{\partial y}\, dy$$

We first introduced the general formula for the total differential in Exercise 27.7.4 in Volume II [8, pp. 341–343] in connection with explaining Erwin Schrödinger's statistical thermodynamics. It was presented as,

$$df = \sum_{i=1}^{n} \frac{\partial f}{\partial x_i}\, dx_i$$

Again, it is very easy to just literally substitute `f[x, y]` as the argument to `Dt[ ]`. `FullForm[Dt[f[x, y]]]` then returns,

`Plus[Times[Dt[y], Derivative[0, 1][f][x, y]],`
`     Times[Dt[x], Derivative[1, 0][f][x, y]]]`

Induction, then, would lead us to expect that the total differential for a function of three arguments $f(x, y, z)$, as *Mathematica* sees it, should be,

`Plus[Times[Dt[z], Derivative[0, 0, 1][f][x, y, z]],`
`     Times[Dt[y], Derivative[0, 1, 0][f][x, y, z]],`
`     Times[Dt[x], Derivative[1, 0, 0][f][x, y, z]]]`

Notice the idiosyncratic way that *Mathematica* reverses the order of the arguments x, y, and z. It is only a short step at this point to the *chain rule*,

$$\frac{df(t)}{dt} = \frac{\partial f}{\partial x_1}\frac{dx_1}{dt} + \frac{\partial f}{\partial x_2}\frac{dx_2}{dt} + \cdots + \frac{\partial f}{\partial x_n}\frac{dx_n}{dt}$$

The **chain rule** was used in Chapter Thirty Five in the derivation of curve length.

Appendix C

Mathematica and the Metric Tensor

Mathematica was used in Exercise 36.9.9 to compute the metric tensor for Jaynes's model of the Wolf's dice data scenario. There, only the *Mathematica* code and the correct answers that it produced were presented. We promised to delve more into the inner workings of this code for the metric tensor as part of our introductory tutorial on how *Mathematica* serves as an indispensable partner for computation.

The easier beginning was the formation of the tangent vectors,

$$F_j(X = x_i) - \langle F_j \rangle$$

through,

```
tangentvec = MapThread[#1 - #2 &, {cm, cfa}]
```

where **cm** and **cfa** were, respectively, the matrix of constraint functions and the vector of constraint function averages. These two entities appear earlier in the MEP algorithm program, prior to the current focus on the metric tensor computation, and, therefore, are already available to this expression.

Deconstructing this short piece of code, we look first at,

```
#1 - #2 &
```

This is the short–hand form of the ubiquitous **Function[ ]**. Filling in the generic arguments to **Function[ ]**, we have the template,

```
Function[{x, y, ... }, body ]
```

The first argument is a list of formal parameters $\{x, y, \ldots\}$, while *body* is the second argument and stands for whatever expression **Function[ ]** is supposed to compute.

For the situation here, we require only two parameters x and y, while the *body* is simply $x - y$. With the parameter list and body now specified we have,

$$\texttt{Function[\{x, y\}, x - y]}$$

The **Function[ ]** does not evaluate the *body* until the function is applied to its arguments. So nothing happens until we write an expression like,

$$\texttt{Function[\{x, y\}, x - y][a, b]}$$

which evaluates to $a - b$.

Thus, this **Function[{x, y}, x - y]** appears as the first argument to the *Mathematica* built–in function **MapThread[ ]**. This very useful function is quite similar to **Map[ ]**. The arguments to **MapThread[**f, $\{list1,\ list2,\ ...\}$**]** are first, as just mentioned, some function f, followed by any number of lists. The individual elements of each list given as an argument to **MapThread[ ]** are the arguments required by the function f.

For our current case, the function f requires the subtraction $x - y$ so f has two arguments. x will be the constraint function value at $F_j(x_i)$, and y will be the constant constraint function average $\langle F_j \rangle$ for all x_i in $F_j(x_i)$. Start filling in the argument template for **MapThread[ ]** with the function f,

$$\texttt{MapThread[Function[\{x, y\}, x - y], \{}list1,\ list2\texttt{\}]}$$

Now fill in both lists. Let the first list containing x be **{a, c}**, and the second list containing y be **{b, d}** so that,

$$\texttt{MapThread[Function[\{x, y\}, x - y], \{\{a, c\}, \{b, d\}\}]}$$

evaluates to **{a - b, c - d}**. Thus, f takes as its first argument x all the elements in *list1* one by one, and takes as its second argument y all of the corresponding elements in *list2* one by one. A more explicit template is,

$$\texttt{MapThread[Function[\{x, y\}, x - y], \{\{}arg\ x\texttt{\}, \{}arg\ y\texttt{\}\}]}$$

The two lists occurring as the arguments to **MapThread[ ]** are seen to provide the arguments required by **Function[{x, y}, x - y]**. Previously, we specifically showed the arguments in the example **Function[{x, y}, x - y][a, b]**.

However, we want our *first* list to be the *matrix* of constraint functions, and the *second* list to be the *vector* of constraint function averages, so the template changes to this list structure,

$$\texttt{MapThread[Function[\{x, y\}, x - y], \{\{\{a, c\}, \{b, d\}\}, \{e, g\}\}]}$$

evaluating to the required **{{a - e, c - e}, {b - g, d - g}}**.

METRIC TENSOR

For Exercise 36.9.9 and Wolf's dice data, the first vector of constraint functions $F_1(X = x_i)$ together with the second vector of constraint functions $F_2(X = x_i)$ led to the construction of the matrix of constraint functions **cm**,

```
cm = {{-2.5, -1.5, -0.5, 0.5, 1.5, 2.5}, {1, 1, -2, -2, 1, 1}}
```

while the vector of the two constraint function averages, $\langle F_1 \rangle$ and $\langle F_2 \rangle$ led to,

```
cfa = {.0983, .1393}
```

Finally, evaluating,

```
tangentvec = MapThread[Function[{x, y}, x - y], {cm, cfa}]
```

or, if you prefer the short–hand version, evaluating,

```
tangentvec = MapThread[#1 - #2 &, {cm, cfa}]
```

produced the list of two tangent vectors as shown in that Exercise,

```
{{-2.5983, -1.5983, -0.5983, .4017, 1.4017, 2.4017},
 {.8607, .8607, -2.1393, -2.1393, .8607, .8607}}
```

Now that we see how the tangent vectors **tangentvec** were computed, we move on to the second line of code,

```
metrictensor = Outer[Total[(qi #1 #2)]&,
                    tangentvec, tangentvec, 1]
```

The above discussion of **Function[ ]** can be utilized immediately as it is the first argument to **Outer[** f, list1, list2, ..., level **]**. Once again, by filling out its template,

$$\text{Function}[\{x, y, \ldots\}, \text{body}]$$

we still want the parameter list as $\{x, y\}$, but the *body* is replaced by,

```
Total[Times[x, y, qi]]
```

leading to,

```
Function[{x,y}, Total[Times[x, y, qi]]]
```

Note that **x**, **y** and **qi** are themselves each lists of length n.

We repeat the fact that **Function[ ]** does not evaluate the *body* until the function is applied to its arguments. Nothing happens for this second function f until we write an expression like,

```
Function[{x, y}, Total[Times[x, y, qi]]][{1, 2}, {3, 4}]
```

which correctly evaluates to 5.

The numerical assignments to the probabilities of the joint statements `qi` are available because they were computed earlier in the MEP program. For the purposes of this example, suppose that `qi = {.6, .4}`. Then, by evaluating,

```
Function[{x, y}, Total[Times[x, y, qi]]][{1, 2}, {3, 4}]
```

Mathematica has computed,

$$(1 \times 3 \times 0.6) + (2 \times 4 \times 0.4) = 5$$

The two arguments x and y given to **Function[ ]** do not have to be supplied independently as appears to be the case in the above example where $x = \{1, 2\}$, and $y = \{3, 4\}$ had to be supplied. **Outer[** f, *list1*, *list2*, ..., *level* **]** automatically supplies these arguments with *list1* and *list2*. Each of these two lists will consist of $m = 2$ tangent vectors. Thus, for Wolf's dice data, both of these lists in the template for **Outer[ ]** are **tangentvec**.

Remember though that **tangentvec** collected all of the tangent vectors into a matrix like,

```
{{-2.5983, -1.5983, -0.5983, .4017, 1.4017, 2.4017},
 {.8607, .8607, -2.1393, -2.1393, .8607, .8607}}
```

in other words, a list of lists. Therefore, a more informative template in our current application for **Outer[ ]** is,

$$\text{Outer[} f, \overbrace{\{\{...\},\{...\}\}}^{list\ 1}, \overbrace{\{\{...\},\{...\}\}}^{list\ 2}, level \text{]}$$

The *level* argument is set to 1 in order to instruct **Outer[ ]** to use the appropriate lists to match up with the parameter list $\{x, y\}$. Both x and y should be lists consisting of all $n = 6$ components of the tangent vector and setting *level* equal to 1 accomplishes this. Otherwise, **Outer[ ]** would proceed all the way down to the bottom most level where each individual component of all the tangent vectors would be used. This is not what we want, so the level specification argument of 1 is inserted in **Outer[** f, *list1*, *list2*, ..., 1 **]**.

To wrap this up, consider how all of the above plays out when *Mathematica* finds the second component $g_{12}(p)$ of the metric tensor for Jaynes's model. In evaluating,

```
metrictensor = Outer[Total[(qi #1 #2)]&,
                          tangentvec, tangentvec, 1]
```

METRIC TENSOR

Outer[ ] will form all possible combinations from *list1* and *list2*. The expression **(qi #1 #2)** within parentheses can be reordered because specification of the slot values **#1 #2**, that is, the parameters **x** and **y**, will automatically match up with *list1* and *list2*.

For example, when computing $g_{12}(p)$,

$$\{-2.5983, -1.5983, -0.5983, 0.4017, 1.4017, 2.4017\}$$

will be taken from the first **tangentvec**, and,

$$\{0.8607, 0.8607, -2.1393, -2.1393, 0.8607, 0.8607\}$$

will be taken from the second **tangentvec** through the auspices of **Outer[ ]**. The **qi** have already been found previously by the MEP algorithm, so, finally, all three terms can be multiplied and summed. This will be carried out over all $n = 6$ statements in the state space.

The returned evaluation will find the component $g_{12}(p)$ in the metric tensor **G** at point p of Jaynes's model. It is conducting the same type of computation as was explicitly shown in Table 36.2.

I will now discuss some alternative *Mathematica* code to compute the metric tensor. This method depends upon the direct translation of Fisher's information matrix by computing the variances and covariances of the constraint functions. This alternative for calculating the metric tensor relies upon the second derivative of the log of the partition function.

Jaynes had derived for us the formula,

$$g_{rc}(p) = \frac{\partial^2 \ln Z}{\partial \lambda_r \, \partial \lambda_c}$$

Mathematica provides the evaluation of the second derivative as,

$$\text{D[Log[z]}, \lambda_r, \lambda_c\text{]}$$

Much of the detailed explanation given above can be carried over directly to this code. For example, the **Outer[ ]** function is used again, so all of the detail about its arguments are fresh in our minds. Once again, Jaynes's probability distribution for Wolf's dice data is used as the numerical example.

It is instructive to present five different versions of the *Mathematica* code. We start with the shortest version with the most short–cuts, and proceed up to the longest version with no short–cuts. These varying length versions possess a different appeal to different people, with most adopting some sort of hybrid that has a special aesthetic appeal for them. For example, I happen to prefer the second version because it explicitly spells out **Function[ ]**. The longest version does have the rather bizarre appeal of keeping to the nested **head[** *arg1, arg2, ...* **]** format.

1. The shortest version.

`Outer[D[Log[z], #1, #2]&, {`λ_1`, `λ_2`}, {`λ_1`, `λ_2`}]`
$$/. \{\lambda_1 \to .031724, \lambda_2 \to .071776\}$$

2. A hybrid version.

`Outer[Function[{x, y}, D[Log[z], x, y]], {`λ_1`, `λ_2`}, {`λ_1`, `λ_2`}]`
$$/. \{\lambda_1 \to .031724, \lambda_2 \to .071776\}$$

3. A long version.

`ReplaceAll[Outer[Function[{x, y}, D[Log[z], x, y]],`
`    {`λ_1`, `λ_2`}, {`λ_1`, `λ_2`}], {`$\lambda_1 \to .031724$`, `$\lambda_2 \to .071776$`}]`

4. A longer version.

`ReplaceAll[Outer[Function[{x, y}, D[Log[z], x, y]],`
`   {`λ_1`, `λ_2`}, {`λ_1`, `λ_2`}], {Rule[`λ_1`, .031724], Rule[`λ_2`, .071776]}]`

and then finally the longest version utilizing `List[ ]`

5. The longest version.

`ReplaceAll[Outer[Function[List[x, y], D[Log[z], x, y]],`
`       List[`λ_1`, `λ_2`], List[`λ_1`, `λ_2`]],`
`       List[Rule[`λ_1`, .031724], Rule[`λ_2`, .071776]]]`

All five versions necessarily output the same correct metric tensor $\mathbf{G} \equiv g_{rc}(q)$ for Jaynes's model of Wolf's dice data presented in section 36.5 as,

$$g_{rc}(q) = \begin{pmatrix} 3.0942 & 0.0778 \\ 0.0778 & 1.8413 \end{pmatrix}$$

The *Mathematica* code described here was used in section 36.5 and Exercises 36.9.9 and 36.9.11 to compute the components of the metric tensor for various models of Wolf's die. These computations relying on the second derivative of the log of the partition function verify that the metric tensor for the fair model is,

$$g_{rc}(p) = \begin{pmatrix} 2.9167 & 0 \\ 0 & 2 \end{pmatrix}$$

I would like to make a final comment on a peculiarity of *Mathematica* syntax that appeared in this Appendix. I have stressed the fundamental commonality of *Mathematica* syntax that relies upon the nested nature of the template for any expression,

`head1[arg11, arg12, head2[arg23, head3[arg34]],` $\cdots$ `]`

But it may have occurred to you that we have encountered expressions that violate this policy. Consider the extensive use of `Function[` *parameters, body* `]` where **Function[ ]** appeared strictly as a head that demanded its own arguments because it had to appear syntactically as,

Function[*parameters, body* **] [** *args* **]**

For a first example, we were exposed to **Function[{x, y}, x - y][a, b]** in forming the tangent vectors. Cutting straight to the chase, *Mathematica* allows expressions with heads that are not just arbitrary symbols like the head **f** in **f[x]** or built–in symbols like the head **Log** in **Log[z]**. As a matter of fact, any expression can be used as a head. The *Mathematica* documentation presents a valid example of **(a+b)[x]**, although I am hard pressed to understand what this would mean.

However, other important expressions with a head that is not just a symbol, involve differentiation. For example, **Derivative[2][f]** is the head, and **[x]** is the argument in the expression **Derivative[2][f][x]** for the second derivative of $f(x)$, sometimes written as $f''(x)$. This syntax came into play when *Mathematica* was called on to help derive Jaynes's Equation (11.73).

The result I gave in Exercise 36.9.15 for,

$$\frac{\partial^2 \ln f(x,y)}{\partial x \, \partial y} = \frac{1}{f(x,y)} \frac{\partial^2 f(x,y)}{\partial x \, \partial y} - \left[\frac{1}{f(x,y)} \frac{\partial f(x,y)}{\partial x} \times \frac{1}{f(x,y)} \frac{\partial f(x,y)}{\partial y} \right]$$

by evaluating,

D[Log[f[x, y]], x, y] // FullForm

was a translation back to traditional symbols from the actual, rather involved, *Mathematica* output,

Plus[Times[-1, Power[f[x, y], -2],
Derivative[0, 1][f][x, y], Derivative[1, 0][f][x, y]],
Times[Power[f[x, y], -1], Derivative[1, 1][f][x, y]]]

where, for example, **Derivative[1, 1][f][x, y]** matches up with,

$$\frac{\partial^2 f(x,y)}{\partial x \, \partial y}$$

There is a perverse fun to parsing these expressions. The last line of *Mathematica* code is the first term in the traditional expression,

$$\frac{1}{f(x,y)} \frac{\partial^2 f(x,y)}{\partial x \, \partial y}$$

Times[*arg1, arg2* **]** has two arguments. **Power[f[x, y], -1]** is seen to be the first argument. The second argument is then **Derivative[1, 1][f][x, y]**. Conforming to Wolfram's enhanced definition of a "pure anonymous function" within *Mathematica*, **Derivative[1, 1][f]** must be the head, and **[x, y]** the argument. Aware that one must be in a constant mind–set to adopt nested structures, all of this is merely the second argument to **Times[ ]**.

Bibliography

[1] Amari, Shun–ichi. *Differential Geometrical Theory of Statistics*. Published as Chapter 2 in *Differential Geometry in Statistical Inference*, Institute of Mathematical Statistics, Lecture Notes–Monograph Series, Volume 10, ed. by S. S. Gupta, Hayward, CA, 1987.

[2] Amari, Shun–ichi and Nagaoka, Hiroshi. *Methods of Information Geometry*. Originally published in Japanese by Iwanami Shoten Publishers, Tokyo, 1993. Translated by D. Harada and published by Oxford University Press, 2000.

[3] Amari, Shun–ichi. Information Geometry of the EM and em Algorithms for Neural Networks. *Neural Networks*, Vol. 8, No. 9, pp. 1379–1408, 1995.

[4] Amari, Shun–ichi. Information Geometry on Hierarchy of Probability Distributions. *IEEE Transactions on Information Theory*, Vol. 47, No. 5, July 2001.

[5] Amari, Shun–ichi, Ikeda, S., and Shimokawa, H. Information Geometry of α–Projection in Mean Field Approximation, in *Advanced Mean Field Methods – Theory and Practice*, ed. by Copper, M. and Saad, D., pp. 241-257, MIT Press, Cambridge, MA, 2001.

[6] Bernardo, J. M. and Smith, A. F. M. *Bayesian Theory*. John Wiley & Sons, Ltd., Chichester, England, 1994.

[7] Blower, David J. *Information Processing: Boolean Algebra, Classical Logic, Cellular Automata, and Probability Manipulations. Volume I*. CreateSpace, Amazon.com, May 2011.

[8] Blower, David J. *Information Processing: The Maximum Entropy Principle. Volume II*. CreateSpace, Amazon.com, June 2013.

[9] Calin, O. and Udrişte, C. *Geometric Modeling in Probability and Statistics*. Springer International Publishing, Switzerland, 2014.

[10] Cover, T. M. and Thomas, J. A. *Elements of Information Theory*. Second Edition. John Wiley & Sons, Hoboken, NJ, 2006.

[11] Frankel, Theodore. *The Geometry of Physics. An Introduction*. Revised edition, Cambridge University Press, Cambridge, UK, 2001.

[12] Jaynes, Edwin T. *E.T. Jaynes: Papers on Probability, Statistics and Statistical Physics*. Edited by R. D. Rosenkrantz, Kluwer Academic Publishers, Dordrecht, The Netherlands, 1983.

[13] Jaynes, Edwin T. Information Theory and Statistical Mechanics I, *Physical Review*, 106, pp. 171–190, 1957, Chapter 2 in *Papers*.

[14] Jaynes, Edwin T. Information Theory and Statistical Mechanics II, *Physical Review*, 108, pp. 620–630, 1957, Chapter 3 in *Papers*.

[15] Jaynes, Edwin T. Information Theory and Statistical Mechanics, *1962 Brandeis Summer Institute in Theoretical Physics.*, ed. by K. Ford, Benjamin Cummings Publishing, 1962, Chapter 4 in *Papers*.

[16] Jaynes, Edwin T. Prior Probabilities, *IEEE Trans. on Systems, Science and Cybernetics.*, SSC–4, 227–241, 1968, Chapter 7 in *Papers*.

[17] Jaynes, Edwin T. Where Do We Stand on Maximum Entropy?, *The Maximum Entropy Formalism*, ed. by R. D. Levine and M. Tribus, MIT Press, Cambridge, MA, 1978, Chapter 10 in *Papers*.

[18] Jaynes, Edwin T. Concentrations of Distributions at Entropy Maxima, *19th NBER–NSF Seminar on Bayesian Statistics.*, Montreal, October 1979, Chapter 11 in *Papers*.

[19] Jaynes, Edwin T. On the Rationale of Maximum–Entropy Methods. *Proc. of the IEEE*, Vol. 70, No. 9, September 1982.

[20] Jaynes, Edwin T. Where Do We Go From Here? *Maximum Entropy and Bayesian Methods in Inverse Problems*, ed. by C. Ray Smith and W. T. Grandy, pp. 21–58, D. Reidel Publishing, Dordrecht, The Netherlands, 1985.

[21] Jaynes, Edwin T. Monkeys, Kangaroos, and N. *Maximum Entropy and Bayesian Methods in Applied Statistics*, ed. by J. H. Justice, pp. 27–58, Cambridge University Press, 1986.

[22] Jaynes, Edwin T. *Probability Theory: The Logic of Science.* ed. by G. Larry Bretthorst, Cambridge University Press, New York, NY, 2003.

[23] Jaynes, Edwin T. Some Random Observations. *Synthese*, 63, 115–138, D. Reidel Publishing, 1985.

[24] Jeffreys, Harold. *Theory of Probability.* Third Edition, Oxford University Press, 1961.

[25] Kass, R. E. and Vos, P.W. *Geometrical Foundations of Asymptotic Inference.* John Wiley & Sons, New York, NY, 1997.

[26] Kay, David C. *Tensor Calculus*, Schaum's Outline Series, McGraw–Hill Publishing, New York, 2011.

[27] Kreyszig, Erwin. *Differential Geometry.* This Dover edition, first published in 1991, is an unabridged republication of the 1963 printing of the work first published by the University of Toronto Press, Toronto, Canada, in 1959 as No. 11 in their series, *Mathematical Expositions*, Dover, Mineola, NY, 1991.

[28] Kullback, S. *Information Theory and Statistics.* This Dover edition, first published in 1997, is an unabridged republication of the Dover 1968 edition which was an unabridged republication of the work originally published in 1959 by John Wiley & Sons, New York, with a new preface and corrections and additions by the author, Dover Publications, Mineola, NY, 1997.

[29] MacKay, D. J. C. *Information Theory, Inference, and Learning Algorithms.*, Cambridge University Press, Cambridge, UK, 2003.

[30] Murray, M. K. and Rice, J. W. *Differential Geometry and Statistics.* Monographs on Statistics and Applied Probability 48. Chapman and Hall / CRC, Boca Raton, FL, 1993.

[31] Neal, Radford M. *Bayesian Learning for Neural Networks.*, Lecture Notes in Statistics, 118, Springer–Verlag, New York, NY, 1996.

[32] Nocedal, J. and Wright, S. J. *Numerical Optimization.*, Springer–Verlag, New York, 1999.

BIBLIOGRAPHY

[33] Rao, C. R. *Advanced Statistical Methods in Biometric Research.* John Wiley & Sons, New York, 1952.

[34] Rumelhart, D.E., McClelland, J. L. *Parallel Distributed Processing: Volume 1,* The MIT Press, Cambridge, MA, 1986.

[35] Schrödinger, Erwin. *Statistical Thermodynamics.* Dover Publications, Mineola, NY, 1989. A reprint of a work first published in 1946 by the Cambridge University Press, Cambridge, UK, entitled, "A Course of Seminar Lectures Delivered on January – March 1944, At the School of Theoretical Physics, Dublin Institute for Advanced Study."

[36] Shannon, Claude E. and Weaver, Warren. *The Mathematical Theory of Communication.* University of Illinois Press, Urbana, IL, 1949.

[37] Stein, D. L. *Lectures in the Sciences of Complexity.*, ed. by D. L. Stein, Santa Fe Institute Studies in the Sciences of Complexity, Addison–Wesley, 1989.

[38] Stoker, J. J. *Differential Geometry.* Wiley Classics Library. Wiley–Interscience, 1989.

[39] Tipler, Frank J. *The Physics of Immortality: Modern Cosmology, God and the Resurrection of the Dead,* Doubleday, New York, NY, 1994.

[40] Wolfram, Stephen. *A New Kind of Science.* Wolfram Media, Inc., Champaign, IL, 2002.

639